1928 Paul A. M. Dirac proposes a relativistic quantum theory.

1929 Edwin Hubble reports evidence for the expansion of the universe.

1931 Carl Anderson discovers the positron (antielectron).

1931 Wolfgang Pauli suggests existence of neutral particle (neutrino) emitted in beta decay.

1932 James Chadwick discovers the neutron.

1932 John Cockcroft and Ernest Walton produce the first nuclear reaction using a high-voltage accelerator.

1935 Hideki Yukawa proposes existence of medium-mass particles (mesons).

1938 Otto Hahn and Fritz Strassmann discover nuclear fission.

1938 Hans Bethe proposes thermonuclear fusion reactions as the source of energy in stars.

1940 Edwin McMillan, Glenn Seaborg, and colleagues produce first synthetic transuranic elements.

1942 Enrico Fermi and colleagues build first nuclear fission reactor.

1945 Detonation of first fission bomb in New Mexico desert.

1946 George Gamow proposes big-bang cosmology.

1948 John Bardeen, Walter Brattain, and William Shockley demonstrate first transistor.

1952 Detonation of first thermonuclear fusion bomb at Eniwetok atoll.

1956 Frederick Reines and Clyde Cowan demonstrate experimental evidence for existence of neutrino.

1958 Rudolf L. Mössbauer demonstrates recoilless emission of gamma rays.

1960 Theodore Maiman constructs first ruby laser; Ali Javan constructs first helium-neon laser.

1964 Allan R. Sandage discovers first quasar.

1964 Murray Gell-Mann and George Zweig independently introduce three-quark model of elementary particles.

1965 Arno Penzias and Robert Wilson discover cosmic microwave background radiation.

1967 Jocelyn Bell and Anthony Hewish discover first pulsar.

1967 Steven Weinberg and Abdus Salam independently propose a unified theory linking the weak and electromagnetic interactions.

1974 Burton Richter and Samuel Ting and co-workers independently discover first evidence of fourth quark.

1977 Leon Lederman and colleagues discover new particle believed to show evidence for fifth quark.

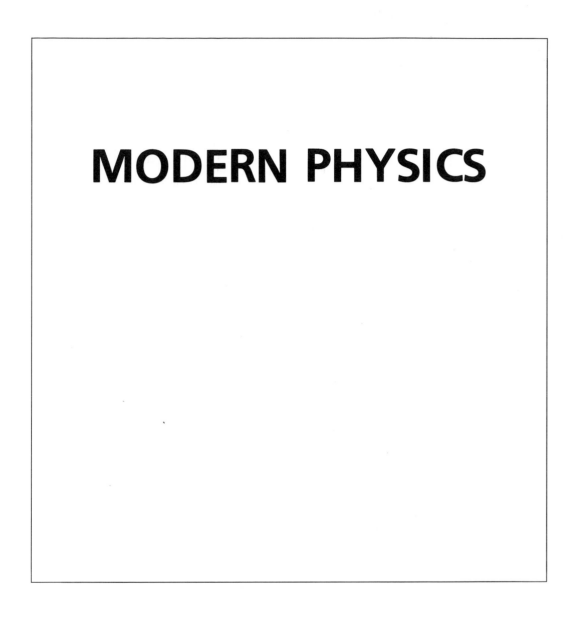

MODERN PHYSICS

Library of Congress Cataloging in Publication Data:

Krane, Kenneth S.
　Modern physics.

　Includes indexes.
　1. Physics.　I. Title.
QC21.2.K7　1983　　　539　　　82-11096
ISBN 0-471-07963-4

Printed in the United States of America

10

PREFACE

This textbook was written for an introductory course in modern physics, including relativity, quantum physics, and applications. Such a course would normally follow immediately the standard introductory course in calculus-based classical physics. Special efforts have been made to keep the text at an introductory level, thereby easing the transition from classical to modern physics.

The major goal of this text, and of the course for which it is intended, is to instill in the student an appreciation of the concepts and methods of twentieth-century physics. The text has grown from a course at Oregon State University, taken generally during the sophomore year, which serves two functions for two different audiences. (1) Physics majors, who will later take a more rigorous course in quantum physics, find an introductory modern course helpful to broaden their perspective before undertaking the rigors of the traditional junior-year studies in classical mechanics and electromagnetism. (2) Nonmajors, who likely will take no further physics, are increasingly finding need for modern physics in their disciplines; an introductory classical course is hardly sufficient for chemists, computer scientists, nuclear and electrical engineers, and so forth.

Necessary prerequisities for undertaking the text include any standard comprehensive introductory calculus-based course covering mechanics, electromagnetism, thermal physics, and optics. Calculus is used extensively in this text; no previous training in differential equations is assumed. Brief mention is made of complex variables and partial derivatives, but no previous knowledge is assumed.

For a course length of one academic quarter, Chapters 1 to 11 (covering special relativity, quantum theory, atomic, nuclear, and particle physics) are recommended; sections marked with an asterisk can be omitted without loss of continuity. For a semester-long course, additional topics from Chapters 12 to 16 may be included. Statistical physics (Chapter 12) is a necessary introduction to the remaining chapters; although most of Chapter 13 (molecular structure) can be attempted without a background in statistics (only the last section on molecular spectroscopy uses Boltzmann statistics), Chapter 14 (solids) should not be undertaken without an understanding of Fermi-Dirac statistics. Chapters 15 and 16 use statistical physics extensively.

The ordering of topics within the text is standard, with possibly one exception—the Bohr atomic theory is grouped with Rutherford scattering under introductory atomic physics, rather than under introductory quantum physics. Although the Bohr theory perhaps belongs with the latter group by historical association, it is not essential to the traditional logical development of quantum theory (Planck-Einstein-Millikan-Compton-deBroglie-Davisson-Heisenberg-Schrödinger), and it therefore is delayed until Chapter 6, which treats introductory atomic physics. (Although deBroglie did use his "matter wave" theory to derive the Sommerfeld quantization condition for stable atomic orbits, it is not at all pivotal to deBroglie's work and does not occupy a prominent place in his writings.) Instructors who feel strongly otherwise may wish to cover Chapter 6 (Rutherford-Bohr) before Chapters 4 (deBroglie waves) and 5 (Schrödinger theory); since Chapter 6 refers to wave mechanics only in the discussion of the last section and in the questions and problems, this inversion can be done without serious loss of continuity.

In order to keep the length of the text within reasonable limits (and to preserve its introductory nature) many topics have been omitted. Although a complete course in modern physics must cover these topics, the essentials of modern physics can be appreciated without them, keeping in mind that the purpose of this text is *not* to train potential relativists, quantum mechanics, or particle physicists. Such topics as 4-vectors, spacetime diagrams, parity, symmetric and antisymmetric wave functions, total angular momentum, isotopic spin, hyperchange, and nuclear models have been eliminated; although these topics appear in other "introductory" modern physics texts, their inclusion can add to the already great potential for confusion and cannot be justified at the elementary level to which this text aspires. That is, if a student "understands" atomic structure based on l and s, is it necessary to introduce j and does its introduction contribute to enlightenment? I think not. With the student already reeling from the introduction of lepton number, baryon number, and strangeness, do we create enlightenment or confusion by introducing isospin and hyperchange? Can a student be expected to gain *any* insight into nuclear structure from a few descriptive paragraphs on the shell and collective models?

On the other hand, real insight into modern physics requires such concepts as degeneracy and the quark model, and these are treated extensively, even at the elementary level.

Throughout the text, the empirical basis of modern physics is repeatedly emphasized. This emphasis takes two forms. (1) The *experimental tests* of the theories of modern physics are presented and discussed. Examples include tests of special relativity and the experiments that support quantum theory. The usual experimental evidence for atomic shell structure (radii and ionization energies) is shown, but in addition the electrical conductivity and magnetic susceptibility are discussed. In other areas the experimental tests are also emphasized. (2) *Applications* of all basic phenomena are presented, including barrier penetration (wave mechanics), lasers (atomic physics), radioactive dating, transuranic and superheavy elements, neutron activation analysis (nuclear physics), liquid helium (statistics), molecular spectroscopy (molecular physics), semiconductor devices, ferromagnetism, and electrical conductivity (solids). Other unusual references to experiments can be found throughout the text as well as in the questions and problems.

A unique feature of the present text is the inclusion of introductory material on astrophysics and cosmology (Chapters 15 and 16). This material should not be seen as a separate study, but rather as the logical culmination of the previous 14 chapters, drawing not only on relativity and quantum theory, but also on topics from atomic, nuclear, particle, molecular, statistical, and solid-state physics.

Although many of our students learn *in spite of* the efforts of instructors and authors, others clearly can benefit from their efforts. A large fraction of our students learns best by example; each chapter therefore includes many worked examples that illustrate basic techniques. At the end of each chapter are many questions and problems. The questions can be used for self-study or for class discussion; they are designed to force the student to consider the material in the chapter in a nonmathematical way and perhaps from a slightly different viewpoint. The problems are written to cover a range of abilities, including simple "plug-in" problems, from which the student gains familiarity with the important formulas, mathematical manipulations and derivations, which promote the development of mathematical skills, and advanced topics, which present unusual applications or new material and often call for special insight. The problems are *not* keyed to specific sections of the text; such keys tend to encourage compartmentalization in the mind of the student, discourage the student from adopting a synthetic approach that (along with analytic ability) is a necessary and complementary part of the scientific method, and breed laziness in both students and instructors. An important part of the learning process is recognizing the approach to use for each type of problem; when reading "light of wavelength λ falls on a metal surface. . . ." the student should recognize a photoelectric effect problem, without being told to look at Section 3.3. The problems in each chapter are ordered roughly to correspond with the ordering of subjects in the chapter.

While it has been attempted to use SI units wherever possible (nanometers, rather than angstroms, for wavelength and atomic sizes and spacings; joules, rather than calories, for heat energy), several exceptions have been made. Electron volts are used for energy as a matter of convenience. Densities of ordinary materials are expressed in grams per cubic centimeter (g/cm³); occasionally kilograms per cubic meter (kg/m³) are used. Conventional units are used for astrophysical quantities.

It is, I think, unfortunate that novel and text share a common classification as "books," for the former is certainly a solitary enterprise representing the outlook and creative talents of one person, while the latter is a cooperative enterprise, requiring the active participation of many people. In this sense the effort more resembles that of a motion picture, and one would therefore like to roll the credits across the screen, acknowledging the work of those who have contributed to the finished product, beginning with Producer-Director Robert McConnin and including the efficient and creative editorial, design, artistic, and production staff at John Wiley and Sons. It has been a pleasant experience to work with them, and only the fear of slighting some by omission keeps me from trying to list them all by name. Special recognition is due typist Carole Vogel, who successfully rendered marginally legible handwritten pages into a finished typescript, correcting numerous errors in the process.

I have benefitted from the comments of many reviewers who read the manuscript at various stages, and I would like especially to thank the anonymous reviewers who contributed valuable advice and suggestions, and also my colleagues at Oregon State University, particularly Peter Fontana, John Gardner, and Carl Kocher, who helped to excise some of the more embarrassingly inaccurate statements and oversimplifications. Discussions with other colleagues at Oxford University and at the Los Alamos National Laboratory, where portions of this work were conceived and executed, were also important in shaping my thoughts, and I thank those individuals for contributing their time. Needless to say, the blame for any remaining inaccuracies or misstatements rests with the author.

For well over a decade I have benefitted from close professional associations with two physicists whose influence and teachings are, perhaps in subtle ways, reflected on nearly every page of this text, and I would like to thank them for their patient instruction and for their friendship and support—Rolf Steffen, who taught me the virtues of elegance, and Bill Steyert, who taught me the equally valuable lesson that apparently simple explanations are possible only through deep understanding. The balance between theory and experiment that I have tried to achieve was accomplished primarily through the philosophical yin-and-yang inspired by these two special colleagues.

Finally, I must mention a fact that is self-evident to those who have previously written textbooks but perhaps is not so apparent to those who approach this activity, as I did, with the naïveté of inexperience—preparing a text is not an activity that can be done peripherally to other academic and research activities without long-term sacrifices by one's family. With good cheer they have endured neglect and accepted the many demands on my time, and despite all they have been enthusiastic helpers in the production process. It is to them, and to my parents for their unflagging support and encouragement, that this work is dedicated.

Corvallis, Oregon
June 1982

Kenneth S. Krane

CONTENTS

5. The Schrödinger Equation 114

6. The Rutherford-Bohr Model of the Atom 147

7. The Hydrogen Atom in Wave Mechanics 178

8. Many-Electron Atoms 202

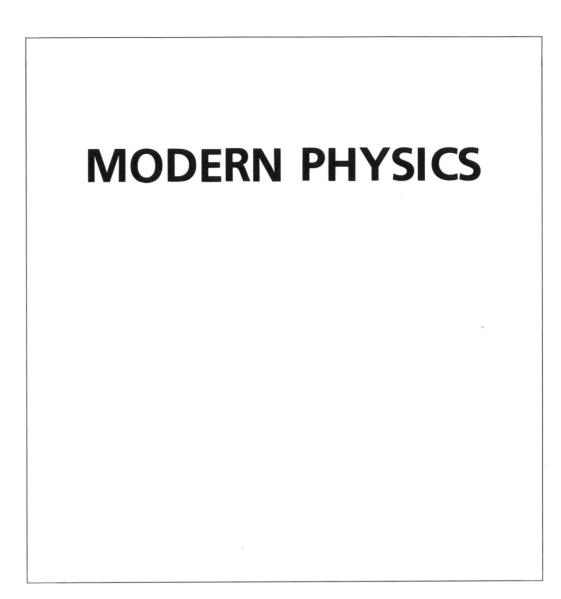

MODERN PHYSICS

At the end of the nineteenth century it seemed that most of what there was to know about physics had already been learned. Newton's dynamics had been carefully and repeatedly tested, and its success had provided a framework for a deep and consistent understanding of nature. Electricity and magnetism had been unified by Maxwell's theoretical work, and the electromagnetic waves predicted by Maxwell's equations had been discovered and investigated in the experiments conducted by Hertz. The laws of thermodynamics and kinetic theory had been particularly successful in providing a unified explanation of a wide variety of phenomena. More generally, the Industrial Revolution had introduced a measure of technological sophistication that would have profound influence on the lives and standard of living of people everywhere. After a period of economic and geographical expansion, the United States was beginning to assert its role as a world power. In Europe, strong monarchies had provided an environment in which industrialization could proceed at a rapid pace. However, beneath this apparent air of stability and optimism there were strong undercurrents, which, in a few years, would plunge the world into the brutal conflict of World War I; the rising tide of militarism, the forces of nationalism and revolution, and the gathering strength of Marxism would soon upset the established order. The fine arts were similarly in the middle of revolutionary change, as new ideas began to dominate the fields of painting, sculpture, and music. The understanding of even the very fundamental aspects of human behavior were subject to serious and critical modification by the Freudian psychologists. In the world of physics, too, there were undercurrents that would soon cause revolutionary changes in the apparently successful world view of the physicist. Several experiments gave results that were not explainable in terms of the successful theories of mechanics, electromagnetism, and thermodynamics. Although the properties of the electromagnetic waves of Maxwell and Hertz were well understood, experiments to study the properties of the medium that transmits those waves were not successful. Experiments to study the emission of electromagnetic waves by hot, glowing objects gave results that could not be explained by the classical theories of thermodynamics and electromagnetism. Experiments on the emission of electrons from surfaces illuminated with light also could not be understood using classical theories.

These few experiments may not seem significant, especially when viewed against the background of the many successful and well-understood experiments of the nineteenth century. However, these experiments were to have a profound and lasting effect, not only on the world of physics, but on all of science, on the political structure of our world, and on the way we view ourselves and our place in the universe. Within the short span of two decades, the results of these experiments were to lead to the special theory of relativity and to the quantum theory; soon after the revolutions inspired by these new theories came the development of atomic physics, nuclear physics, and solid-state physics, with the monumental impact that applications of research in these fields have had on our daily lives.

The designation *modern physics* usually refers to those developments that began with the relativity and quantum theories, and includes the applications of those modern theories to understanding the

Modern physics

properties of the atom, of the atomic nucleus and the particles of which it is composed, of collections of atoms in molecules and solids, and, on a cosmic scale, of the origin and evolution of the universe. Our discussion of modern physics in this text touches on each of these fields. We begin with the *relativity theory*, exploring its assumptions, implications, and experimental verification. After reviewing the experiments that signaled the inadequacy of classical concepts of particles and waves, we discuss the success of the *quantum theory*, or wave mechanics, as it is sometimes known, in resolving those failures. A complete discussion of wave mechanics requires mathematical proficiency beyond the level of this text, therefore we undertake only a superficial introduction to the techniques and applications of wave mechanics. The remainder of the text deals with applications of these principles, first to the study of the structure and properties of the atom, and next to the study of the structure and properties of the atomic nucleus and the elementary particles. We then show that many of the same principles can be applied to the study of groups of atoms, both small groups in molecules and large groups in solids. Finally, we turn from the microscopic to the cosmic and discuss the applications of modern physics to the understanding of some problems of astrophysics and cosmology.

As you undertake this study, keep in mind that the details of the story of modern physics have been written only during this century, and that many of the discoveries have been made during our lifetimes. This means that the story of modern physics is not yet complete and will continue to evolve. Many of the theories that are part of the discipline of modern physics are only approximations (although sometimes very good ones). Often we find that each time we look deeper or refine our techniques we learn something new and must revise our theories to account for the new discoveries. As a result, sometimes modern physics takes on the appearance and structure of a patchwork quilt with a different explanation for each of the effects that we study. Beneath it all, however, lies the fabric of wave mechanics, holding all these diverse fields together and forming the basis on which they are constructed.

Any field of science builds on the results of previous investigations, and modern physics is no exception to this principle. The previous work in our case is classical physics and so before we begin our study of modern physics we review some of the required principles of classical physics.

1.1 REVIEW OF CLASSICAL PHYSICS

Although we will find many areas in which the concepts of modern physics differ radically from those of classical physics, we will frequently find the need to refer back to concepts of classical physics. In order to identify the more important fundamentals of classical physics and to define the notation we will use, some of the concepts of classical physics we will be using are briefly reviewed.

Mechanics An object of mass m moving with velocity v has a *kinetic energy* defined by

$$K = \tfrac{1}{2}mv^2 \qquad (1.1)$$

and a *linear momentum* **p** defined by

$$\mathbf{p} = m\mathbf{v} \qquad (1.2)$$

When that object collides with another object, we analyze the collision by applying two fundamental conservation laws:

I. Conservation of Energy The total energy of an isolated system (on which no net external forces act) remains constant. This means (in this case) that the total energy of the two particles *before* the collision is equal to the total energy of the two particles *after* the collision.

Conservation laws

II. Conservation of Linear Momentum The total linear momentum of an isolated system remains constant; the total linear momentum of the two particles *before* the collision is equal to the total linear momentum of the two particles *after* the collision. Since linear momentum is a vector, application of this law usually gives us two equations, one for the x components and another for the y components.

These two conservation laws are of the most basic importance to understanding and analyzing a wide variety of problems in classical physics. Problems 1 through 5 at the end of this chapter review the use of these laws.

The importance of these conservation laws is both so great and so fundamental that, even though in Chapter 2 we will learn that the special theory of relativity modifies Equations (1.1) and (1.2), the laws of conservation of energy and linear momentum will still be valid.

Another application of the principle of conservation of energy occurs when a particle moves subject to an external force F. Corresponding to that external force there is often a potential energy V, defined such that (for one-dimensional motion)

$$F = -\frac{dV}{dx} \qquad (1.3)$$

The total energy E is just the sum of the potential and kinetic energies:

$$E = K + V \qquad (1.4)$$

As the particle moves, K and V may change, but E remains constant. (In Chapter 2, we find that the special theory of relativity gives us a new definition of total energy.)

When an object moving with linear momentum **p** is at a displacement **r** from the origin O, the *angular momentum* **l** about the point O is defined by

$$\mathbf{l} = \mathbf{r} \times \mathbf{p} \qquad (1.5)$$

There is a conservation law for angular momentum, just like that for linear momentum. In practice this has many important applications. For example, when a charged particle moves near, and is deflected by, another charged particle, the total angular momentum of the system (the two particles) remains constant if no net external torques act on the system. If the second particle is so much heavier than the first that its motion is unchanged by the influence of the first particle, the angular momentum of the first particle remains constant (because the second particle acquires no angular momentum). A similar situation occurs when a comet moves in the gravitational field of the sun. As it approaches the sun, r decreases, and so p must increase if l is to remain constant; the comet therefore accelerates as it approaches the sun.

Electricity and Magnetism   The electrostatic force (Coulomb force) between two charged particles q_1 and q_2 is

$$F = \frac{1}{4\pi\varepsilon_0}\frac{q_1q_2}{r^2} \tag{1.6}$$

In the SI system of units, which we are going to use, the constant $1/4\pi\varepsilon_0$ has the value

$$\frac{1}{4\pi\varepsilon_0} = 8.988 \times 10^9 \text{ N·m}^2/\text{C}^2$$

The corresponding potential energy is

$$V = \frac{1}{4\pi\varepsilon_0}\frac{q_1q_2}{r} \tag{1.7}$$

In all equations derived from Equation (1.6) or (1.7) as starting points, *the quantity* $1/4\pi\varepsilon_0$ *must appear.* In some texts and reference books, you may find electrostatic quantities in which this constant does not appear. In such cases, the centimeter-gram-second (cgs) system has probably been used, in which the constant $1/4\pi\varepsilon_0$ is *defined* to be 1. You should always be very careful in making comparisons of electrostatic quantities from different references and check that the units are identical.

An electric current i causes a magnetic field **B**. The case we are most concerned with is that of the circular loop of current of radius r; the magnetic field at the center of such a loop is

$$B = \frac{\mu_0 i}{2r} \tag{1.8}$$

In SI units, B is measured in teslas (one tesla (T) is one newton per ampere-meter). The constant μ_0 is

$$\mu_0 = 4\pi \times 10^{-7} \text{ N·s}^2/\text{C}^2$$

Be sure to remember that i is in the direction of the conventional (*positive*) current, opposite to the actual direction of travel of the negatively charged electrons that might produce the current. The direction of **B** is chosen according to the right-hand rule: if you hold the wire in the right hand with the thumb pointing in the direction of the current, the fingers point in the direction of the magnetic field.

It is often convenient to define the *magnetic moment* $\boldsymbol{\mu}$ of a current loop: **Magnetic moment**

$$|\boldsymbol{\mu}| = iA \tag{1.9}$$

where A is the geometrical area enclosed by the loop. The direction of $\boldsymbol{\mu}$ is perpendicular to the plane of the loop, according to the right-hand rule. When a current loop is placed in a uniform *external* magnetic field $\mathbf{B}_{\text{ext}}$, there is a torque $\boldsymbol{\tau}$ on the loop that tends to line up $\boldsymbol{\mu}$ with $\mathbf{B}_{\text{ext}}$:

$$\boldsymbol{\tau} = \boldsymbol{\mu} \times \mathbf{B}_{\text{ext}} \tag{1.10}$$

Another way to interpret this is to assign a potential energy to the magnetic moment $\boldsymbol{\mu}$ in the external field $\mathbf{B}_{\text{ext}}$:

$$V = -\boldsymbol{\mu} \cdot \mathbf{B}_{\text{ext}} \tag{1.11}$$

When the field $\mathbf{B}_{\text{ext}}$ is applied, $\boldsymbol{\mu}$ rotates so that its energy tends to a minimum value, which occurs when $\boldsymbol{\mu}$ and $\mathbf{B}_{\text{ext}}$ are parallel.

It is important for us to understand the properties of magnetic moments, because we will find that particles such as electrons or protons have magnetic moments. Although we don't imagine these particles to be tiny current loops, their magnetic moments do obey Equations (1.10) and (1.11).

A particularly important aspect of electromagnetism is *electromagnetic waves*. In Chapter 3 we discuss the properties of these waves in more detail; in particular the properties of interference and diffraction will be extremely fundamental to our discussions in this text. Electromagnetic waves travel in free space with speed c (the speed of light), which is related to the electromagnetic constants ε_0 and μ_0

$$c = (\varepsilon_0 \mu_0)^{-1/2} \tag{1.12}$$

The waves have a frequency ν and wavelength λ, which are related by

$$c = \lambda\nu \tag{1.13}$$

The wavelengths range from the very short (nuclear gamma rays) to the very long (radio waves). Figure 1.1 shows the electromagnetic spectrum with the conventional names assigned to the different ranges of wavelengths.

Kinetic Theory of Matter   The mean thermal kinetic energy of the molecules of an ideal gas at temperature T is

$$K = \tfrac{3}{2}kT \tag{1.14}$$

where k is the Boltzmann constant

$$k = 1.381 \times 10^{-23} \text{ J/K}$$

The SI unit of temperature is the kelvin (K), *not* "degree kelvin." Be careful not to confuse the symbol K for kelvin with the symbol for kinetic energy; also be careful about confusing the Boltzmann constant k with the wave number $k = 2\pi/\lambda$.

The quantity kT is often taken as a rough estimate of the kinetic energy per particle in a system of particles in thermal equilibrium at the

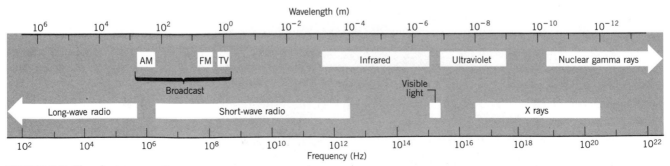

FIGURE 1.1 The electromagnetic spectrum.

temperature T. For example, at room temperature ($20°C = 293$ K), the mean thermal energy is about 4.0×10^{-21} J, while in the interior of a star where $T \sim 10^7$ K, the mean thermal energy is about 10^{-16} J.

A quantity of matter called a gram-molecular weight, or *mole*, is the amount of that substance with a mass in grams equal to the molecular weight. The molecular weight of hydrogen is about 2, since each molecule of hydrogen has two atoms, each of nearly unit mass. Therefore one mole of hydrogen is about 2 g of hydrogen. The molecular weight of iron is about 56; one mole of iron is 56 g of iron. One mole of *any* substance contains a number of molecules equal to Avogadro's number, N_A, where

$$N_A = 6.022 \times 10^{23} \text{ molecules/mole}$$

1.2 UNITS AND DIMENSIONS

Nearly all of the physical constants and variables we will be using have both *units* and *dimensions*. The dimensions of a constant or variable tell us something about the kind of constant or variable it is; a quantity that in one frame of reference has the dimension of length, for example, will have the dimension of length in *every* frame of reference, although its magnitude and the units in which we express it may vary. Although it is sometimes helpful to reduce quantities to their fundamental dimensions of mass m, length l, and time t, it is usually not necessary. What is *always* necessary in working problems is to be sure that your equations are dimensionally consistent; if, for example, you have an equation that contains the term $(v + m)$ where $v = $ velocity and $m = $ mass, it is a sure bet that you have made a mistake somewhere—two quantities can *never* be added unless they have the same dimensions. [However, if your equation contains the term $(av + m)$, where a is a constant, it may be dimensionally correct if a has the proper dimensions.]

Checking your results for dimensional consistency is a good habit to acquire. It is one of the simplest techniques to apply and can almost always be done quickly by inspection. If your results check for dimensional consistency, it doesn't guarantee that they are correct; the reverse, however, is true—lack of dimensional consistency *always* indicates an error.

It is sometimes possible for a quantity to have units, but no dimensions! Suppose your watch runs slow and loses 6.0 seconds each day. Its rate of loss is therefore $R = 6.0$ s/day. R is a dimensionless quantity—it has dimensions of t/t—but it has units, and its value changes as its units change; we could also express R as 0.10 min/day or 0.25 s/hour, or even in a unitless form as 6.9×10^{-5}, which gives the fractional loss of time in any interval. As another example, all conversion factors such as 25.4 mm/inch or 1000 g/kg are dimensionless.

The last several years have seen the adoption and widespread use of the International System of units, called "SI units." These units are in many cases identical with the old mks (meter-kilogram-second) system, but many units of convenience that were part of the old mks system have been dropped. Of course the inch and the foot and the pound and all the other units of the old "British" system have no place in the SI system, but in much of our work we must force ourselves to discard such convenient and familiar units as the *atmosphere* (atm) as a unit of pressure, the *gram per cubic centimeter* (g/cm³) as a unit of density, the *calorie* (cal) as a unit of heat, and so forth.

In modern physics we encounter similar problems in choosing units in which to express the quantities with which we will be working, but in our case there is an additional reason—for many of the topics we study, the SI units are simply too large to be useful. Typical energies associated with atomic or nuclear processes may be 10^{-19} to 10^{-12} J, and typical sizes of atomic and nuclear systems range from 10^{-10} to 10^{-15} m. Although there is no reason that we could not use these SI units for our study of modern physics, we will frequently give way to convenience and historical precedent and use a different set of units. Some of the constants and variables we will be using are discussed as follows.

Length The SI unit for length is of course the *meter* (m), but we will need lengths much smaller than the meter for atomic and nuclear systems. We will use the following units for lengths:

micrometer = μm = 10^{-6} m
nanometer = nm = 10^{-9} m
femtometer = fm = 10^{-15} m

Wavelengths are conveniently measured in nanometers—visible light has wavelengths in the range 400 to 700 nm. Atomic sizes are typically 0.1 nm, and nuclear sizes are about 1 to 10 fm. (The unit fm is sometimes known as the *fermi* in honor of Enrico Fermi, who was a pioneer experimental and theoretical nuclear physicist.) In your previous work you may have encountered the angstrom unit Å (10^{-10} m) as a measure of wavelength. Here is one particular case in which we try to use SI units; we will use nanometers rather than angstroms to measure wavelengths.

Energy The SI unit of energy is the *joule* (J), which is too large to be of convenience for atomic and nuclear physics. A much more convenient unit is the *electron-volt* (eV), defined as the energy gained by an electric charge equal to that of the electron in falling through a potential difference of one volt. (The volt is an SI unit, so we have not gone too far astray!) Since an electron has a charge of 1.602×10^{-19} C, and since 1 V = 1 J/C we have the equivalence

The electron-volt

$$1 \text{ eV} = 1.602 \times 10^{-19} \text{ J}$$

Some convenient multiples of the electron-volt are

keV = kilo electron-volt = 10^3 eV
MeV = mega electron-volt = 10^6 eV
GeV = giga electron-volt = 10^9 eV

(In some older works you may find reference to the BeV, for billion electron-volts; this is a source of confusion, for in the United States a billion is 10^9 while in Europe a billion is 10^{12}.)

Electric Charge The standard unit of electric charge is the *coulomb* (C), and the basic unit of charge, that of the electron, is $e = 1.602 \times 10^{-19}$ C. Very frequently we wish to find the potential energy of two basic charges separated by typical atomic or nuclear dimensions, and we wish to have the result expressed in electron-volts. Here is a convenient way of doing this. Let us find the potential energy of two electrons at a separation of $r = 1.00$ nm:

$$V = \frac{1}{4\pi\varepsilon_0}\frac{e^2}{r}$$

The quantity $e^2/4\pi\varepsilon_0$ can be expressed in a convenient form:

$$\frac{e^2}{4\pi\varepsilon_0} = \left(8.988 \times 10^9 \frac{\text{N}\cdot\text{m}^2}{\text{C}^2}\right)(1.602 \times 10^{-19}\ \text{C})^2$$

$$= 2.307 \times 10^{-28}\ \text{N}\cdot\text{m}^2$$

$$= 2.307 \times 10^{-28}\ \text{J}\cdot\text{m}\ \frac{1}{1.602 \times 10^{-19}\ \text{J/eV}} \cdot \frac{10^9\ \text{nm}}{\text{m}}$$

$$= 1.440\ \text{eV}\cdot\text{nm}$$

With this useful combination of constants it becomes very easy to calculate electrostatic potential energies. For two unit charges separated by 1.00 nm,

$$V = \frac{1}{4\pi\varepsilon_0}\frac{e^2}{r} = \frac{e^2}{4\pi\varepsilon_0}\frac{1}{r} = 1.440\ \text{eV}\cdot\text{nm}\ \frac{1}{1.00\ \text{nm}}$$

$$= 1.44\ \text{eV}$$

For calculations using typical nuclear sizes, the femtometer is a more convenient unit of distance, so

$$\frac{e^2}{4\pi\varepsilon_0} = 1.440\ \text{eV}\cdot\text{nm}\ \frac{1\ \text{m}}{10^9\ \text{nm}}\frac{10^{15}\ \text{fm}}{\text{m}}\frac{1\ \text{MeV}}{10^6\ \text{eV}}$$

$$= 1.440\ \text{MeV}\cdot\text{fm}$$

It is remarkable (and convenient to remember) that the quantity $e^2/4\pi\varepsilon_0$ has the same value of 1.440 whether we use typical atomic energies and sizes (eV·nm) or typical nuclear energies and sizes (MeV·fm).

Mass The *kilogram* (kg) is the basic SI unit of mass, but it is too large a unit to be particularly useful for work in atomic and nuclear physics. Another difficulty, as we discuss in Chapter 2, is that we are frequently interested in using Einstein's equation $E = mc^2$ to transform mass into energy and back again. Using c^2, a very large number, whenever we wish to do this conversion is inconvenient and can lead to mistakes. We avoid this problem by keeping the factor of c^2 with the mass units, and remembering that $m = E/c^2$. For example, you will often find in tables of the masses of the elementary particles that an electron has a mass of 0.511 MeV/c^2. Although this may look like energy units, it is really a measure of mass—the factor of c^2 does the conversion. What is important is that we can keep the factor of c^2 without bothering to put in its numerical value. You can check this result by converting MeV to joules and putting in the numerical value of c^2, and you should obtain 9.11×10^{-31} kg, the mass of the electron.

Another mass unit that we find convenient to use is the *atomic mass unit*, u. This is most convenient in calculating atomic and nuclear binding energies. The atomic mass unit is *defined* so that the mass of the most abundant isotope of carbon is exactly equal to 12 u. All other atomic masses are measured relative to this value. In Appendix B you

Atomic mass unit

will find a table of atomic masses. (In some older reference books you may find a different system of atomic mass units, defined so that the most common stable isotope of oxygen has a mass of exactly 16 u. This will make a small but important difference in calculated results. Be sure to check how the mass units are defined when using reference works!)

The Speed of Light   One of the fundamental constants of nature is the speed of light, c, which you will be using quite frequently throughout your study of modern physics. Its value is

$$c = 3.00 \times 10^8 \text{ m/s}$$

It will often be convenient for us to measure speeds by comparing them with the speed of light; in Chapter 2 you will find many examples of problems in which speeds are given as a fraction of c, for example, $v = 0.6c$. Fortunately, many of the equations of special relativity involve not v but v/c, and thus it is often not necessary to convert $0.6c$ to a numerical speed in meters per second.

Planck's Constant   Another of the fundamental constants of nature is Planck's constant, h, with a value of

$$h = 6.63 \times 10^{-34} \text{ J·s}$$

Planck's constant obviously has dimensions of energy × time, but with a little bit of manipulation, you can show that it also has dimensions of linear momentum × displacement and also angular momentum. Planck's constant will appear in many applications when we begin our study of quantum physics, and each of its different dimensions will represent an important application.

We have already mentioned our desire to use electron-volts rather than joules for energy, and so it is useful to express Planck's constant using eV:

$$h = 4.14 \times 10^{-15} \text{ eV·s}$$

We also encounter the product hc in many calculations. In our units of convenience, you should be able to derive its value:

$$hc = 1240 \text{ eV·nm}$$
$$= 1240 \text{ MeV·fm}$$

It is interesting to note that hc and $e^2/4\pi\varepsilon_0$ have the same dimensions, and we have in fact calculated both quantities in the same units, eV·nm. The ratio of these two quantities is a pure number, independent of the system of units we have chosen, and we will learn that this ratio is of fundamental importance in atomic physics. The dimensionless constant α, called the *fine structure constant*, is actually 2π times the ratio:

Fine structure constant

$$\alpha = 2\pi \frac{e^2/4\pi\varepsilon_0}{hc} \qquad (1.15)$$

$$= 2\pi \frac{1.440 \text{ eV·nm}}{1240 \text{ eV·nm}}$$

$$= 0.007297$$

This number is usually expressed as $\alpha = 1/137.0$.

There are two rules you must remember for using significant figures:

1. When adding or subtracting, the least significant digit of the number being added or subtracted determines the least significant digit of the sum or difference. *The number of significant figures does not matter.*
2. When multiplying or dividing, count the number of significant figures in the quantities being multiplied or divided. The number of significant figures in the product or quotient is determined by the factor with the smallest number of significant figures. *The location of the least significant digit does not matter.*

Here are some examples that illustrate the use of these rules.

Find the mass difference between a proton and a neutron. Express the result in u and in MeV/c².

EXAMPLE 1.1

The proton and neutron masses are (to seven significant figures):

Solution

$m_n = 1.008665$ u
$m_p = 1.007276$ u

The difference is

1.008665 u $- 1.007276$ u $= 0.001389$ u

In all cases, the last digit (indicated in boldface) is the least significant. In addition or subtraction, we do not count the number of significant figures. It is of *no importance* for the subtraction that each of the two masses has seven significant figures while the difference has only four.

The conversion factor from u to MeV/c² is

1 u $= 931.50$ MeV/c²

The mass difference is therefore

$$0.001389 \text{ u} \times \frac{931.50 \text{ MeV/c}^2}{1 \text{ u}} = 1.294 \text{ MeV/c}^2$$

The mass difference in u has four significant figures, the conversion factor has five significant figures, and the product can therefore have only four significant figures, according to the second rule given above.

A proton and an electron can combine to form an atom of hydrogen. Find the total mass of a proton and an electron.

EXAMPLE 1.2

The values of the masses are

Solution

$m_p = 1.007276$ u
$m_e = 5.4858 \times 10^{-4}$ u

The combined mass is

1.007276 u $+ 0.00054858$ u $= 1.007825$ u

Notice that the *position* of the least significant digit (the 6 in m_p) determines the *position* of the least significant digit in the sum.

The value of *hc* was given in the last section as 1240 eV·nm. Compute the value of *hc* to four significant figures and determine if the zero in the last digit is significant.

EXAMPLE 1.3

The values given previously for *h* and *c* contain only three significant figures and thus cannot be used for this calculation, for which we require four significant figures. To avoid round-off errors, we will use the values of the constants to five significant figures, and then round off the final result to four figures.

$$h = 6.6262 \times 10^{-34} \text{ J·s}$$
$$c = 2.9979 \times 10^{8} \text{ m/s}$$
$$1 \text{ eV} = 1.6022 \times 10^{-19} \text{ J}$$

Therefore,

$$hc = \frac{(6.6262 \times 10^{-34} \text{ J·s})(2.9979 \times 10^{8} \text{ m/s})(10^{9} \text{ nm/m})}{1.6022 \times 10^{-19} \text{ J/eV}}$$

$$= 1239.8 \text{ eV·nm}$$

Rounding off to four figures, the result is *hc* = 1240 eV·nm, and the zero *is* significant. (Of course, it is good practice not to leave zeros on the end of numbers such as this, where we don't know whether or not they are significant. It would be better to express this result as 1.240×10^{3} eV·nm.)

Notice that, if four significant figures are required for a calculation, the value $c = 3.00 \times 10^{8}$ m/s is not precise enough.

Proper attention to significant figures is a matter of habit, and the sooner you form this good habit the less trouble you will have in expressing the results of calculations. Simply because your calculator display shows eight digits does not make them all significant, and only the significant ones should be written down as the answer to a problem.

When you first began to study science, perhaps in your elementary or high school years, you may have learned about the "scientific method," which was supposed to be a sort of procedure by which scientific progress was achieved. The basic idea of the "scientific method" was that, on reflecting over some particular aspect of nature, the scientist would invent a *hypothesis* or *theory*, which would then be tested by *experiment* and if successful would be elevated to the status of *law*. This procedure is meant to emphasize the importance of doing experiments as a way of testing hypotheses and rejecting those that do not pass the tests. For example, the ancient Greeks had some rather definite ideas about the motion of objects, such as projectiles, in the Earth's gravity. Yet they tested none of these by experiment, so convinced were they that the power of reason *alone* could be used to discover the hidden and mysterious laws of nature and that once reason had been applied to understanding a problem, no experiments were necessary. If theory and experiment were to disagree, they would argue, then there must be something wrong with the experiment! This dominance of reason and faith were so pervasive that it was 2000 years before Galileo, using an inclined plane and a crude timer (equipment surely within the abilities of

1.4 THEORY, EXPERIMENT, LAW

the early Greeks to construct), discovered the laws of motion, which were later to be organized and analyzed by Newton.

Modern physics is an extreme example of the need for experimentation. None of the precepts of modern physics are obvious from reason alone, and it is only by doing often difficult and necessarily precise experiments that these unexpected and fascinating effects become known. In our study of modern physics we will therefore try to emphasize the experiments that have been done to study relativity and quantum physics. These experiments have been done to unprecedented levels of precision—of the order of one part in 10^6 or better—and it can certainly be concluded that modern physics has been tested far better in the twentieth century than classical physics was tested in all of the preceding centuries.

Nevertheless, there is a persistent and often perplexing problem associated with modern physics, one which stems directly from your previous acquaintance with the "scientific method." This concerns the use of the word "theory," as in "theory of relativity" or "quantum theory," or even "atomic theory," or "theory of evolution." There are two contrasting and conflicting definitions of the word "theory" in the dictionary:

1. A hypothesis or guess.
2. An organized body of facts or explanations.

The "scientific method" refers to the first kind of "theory," while when we speak of the "theory of relativity" we refer to the second kind. Yet there is often confusion between the two definitions, and therefore relativity and quantum physics are sometimes regarded by students as merely hypotheses, on which evidence is still being gathered, in the hope of someday submitting that evidence to some sort of international (or intergalactic) tribunal, which in turn might elevate the "theory" into a "law." Thus the "theory of relativity" might someday become the "law of relativity," like the "law of gravity." *Nothing could be further from the truth!*

The theory of relativity and the quantum theory, like the atomic theory or the theory of evolution, are truly "organized bodies of facts and explanations" and *not* "hypotheses." There is no question of these "theories" becoming "laws"—the "facts" (experiments, observations) of relativity and quantum physics, like those of atomism or evolution, are accepted by virtually all scientists today; whether one calls them theories or laws is merely a question of semantics and has nothing to do with their scientific merits. Like all scientific principles, they will continue to develop and change as new discoveries are made; that is the essence of scientific progress, and it must be remembered that the search for ultimate truths or eternal laws is *not* a goal of science.

Still, there are two remaining questions that you may find particularly bothersome as you study modern physics. First is the "how" of these theories. The experimental evidence that forms the basis of modern physics is almost always *indirect*—no one has ever "seen" a quantum or a pi meson or even a nucleus, and no one has ever traveled at nearly the speed of light and "seen" the effects of relativity; similarly, no one has ever "seen" single atoms joining to form compounds or "seen" one species evolving into another. Yet the experimental evidence for all of these effects, and others, is so compelling that no one who approaches them in the spirit of free and open inquiry can doubt them. Keep in mind that most of modern physics is supported by the indirect evidence gained from the *analysis* and *interpretation* of experimental results, rather than by direct observation of phenomena.

The second vexing question concerns the "why" of these theories. Why does Nature behave according to Einstein's relativity, rather than according to Galileo's? Why do particles sometimes behave as waves, and waves sometimes as particles? Why do atoms join to form compounds? Why do higher forms of life evolve from lower forms? Although scientists can provide extremely precise answers to the "how" of these theories, they cannot provide the answers to the "why," not because their powers of observation or experimental abilities are limited, but rather because the questions are outside the realm of experimentation. These questions are of extreme importance, and as potential practitioners of pure or applied science you should be aware of them and spend some time thinking about them. If answers to these questions are to be found at all, they will be found not in the field of science, but in the fields of philosophy or theology. As you begin to study the facts of modern science, you should keep these additional questions in mind and perhaps seek your instructor's opinions concerning them. Although such speculations are an exciting intellectual endeavor in their own right, they will not be discussed in this text.

If you feel the need to review some topics from classical physics, here are some introductory classical physics texts:

SUGGESTIONS FOR FURTHER READING

F. J. Bueche, *Introduction to Physics for Scientists and Engineers* (New York, McGraw-Hill, 1980).

D. Halliday and R. Resnick, *Fundamentals of Physics*, 2nd edition (New York, Wiley, 1981).

F. W. Sears, M. W. Zemansky, and H. D. Young, *University Physics*, 5th edition (Reading, Addison-Wesley, 1976).

P. A. Tipler, *Physics* (New York, Worth, 1976).

R. T. Weidner and R. L. Sells, *Elementary Physics* (Boston, Allyn and Bacon, 1975).

Some other modern physics books at about the same level as this text:

A. Beiser, *Concepts of Modern Physics*, 3rd edition (New York, McGraw-Hill, 1981).

P. A. Tipler, *Modern Physics* (New York, Worth, 1978).

R. T. Weidner and R. L. Sells, *Elementary Modern Physics*, 3rd edition (Boston, Allyn and Bacon, 1980).

Some more advanced modern physics texts:

R. Eisberg and R. Resnick, *Quantum Physics of Atoms, Molecules, Solids, Nuclei, and Particles* (New York, Wiley, 1974).

R. B. Leighton, *Principles of Modern Physics* (New York, McGraw-Hill, 1959).

F. K. Richtmeyer, E. H. Kennard, and J. N. Cooper, *Introduction to Modern Physics*, 6th edition (New York, McGraw-Hill, 1969).

H. Semat and J. R. Albright, *Introduction to Atomic and Nuclear Physics* (New York, Holt, Rinehart and Winston, 1972).

Some descriptive, historical, philosophical, and nonmathematical texts which give good background material and are great fun to read:

A. Baker, *Modern Physics and Anti-Physics* (Reading, Addison-Wesley, 1970).

F. Capra, *The Tao of Physics* (Berkeley, Shambhala Publications, 1975).

G. Gamow, *Thirty Years that Shook Physics* (New York, Doubleday, 1966).

R. March, *Physics for Poets* (New York, McGraw-Hill, 1978).

E. Segrè, *From X-Rays to Quarks: Modern Physicists and their Discoveries* (San Francisco, W. H. Freeman, 1980).

G. L. Trigg, *Landmark Experiments in Twentieth Century Physics* (New York, Crane, Russak, 1975).

F. A. Wolf, *Taking the Quantum Leap* (San Francisco, Harper & Row, 1981).

G. Zukav, *The Dancing Wu Li Masters, An Overview of the New Physics* (New York, Morrow, 1979). Also available in paperback.

Gamow, Segrè, and Trigg have contributed directly to the development of modern physics and their books are written from a perspective that only those who were part of that development can offer. The books by Capra and Zukav draw interesting parallels between modern physics (especially quantum theory and particle physics) and oriental philosophy.

PROBLEMS

1. An atom of mass m moving in the x direction with speed v collides elastically with an atom of mass $3m$ at rest. After the collision the first atom moves in the y direction. Find the direction of motion of the second atom and the speeds of both atoms (in terms of v) after the collision.

2. An atom of mass m moves in the positive x direction with speed v. It collides with and sticks to a mass $2m$ moving in the positive y direction with speed $2v/3$. Find the resultant speed and direction of motion of the combination, and find the kinetic energy lost in this inelastic collision.

3. An atom of beryllium ($m \cong 8.00$ u) splits into two atoms of helium ($m \cong 4.00$ u) with the release of 92.0 keV of energy. If the original beryllium atom is at rest, find the kinetic energy, speed, and momentum of the two helium atoms.

4. Suppose the beryllium atom of the previous problem were not at rest, but instead moved in the positive x direction and had a kinetic energy of 40.0 keV. One of the helium atoms is found to be moving in the positive x direction. Find the direction of motion of the second helium, and find the velocity of each of the two helium atoms. Solve this problem in two different ways:
   (a) By direct application of conservation of momentum and energy.
   (b) By applying the results of the previous problem to a frame of reference moving with the original beryllium atom and then switching to the reference frame in which the beryllium is moving.

5. The beryllium atom of Problem 3 now moves in the positive x direction and has kinetic energy 60.0 keV. One helium atom is found to move at an angle of 30° with respect to the x axis. Find the direction of motion of the second helium atom and find the velocity of each helium atom. Work this problem in two ways as you did the previous problem. [Hint for method (b): Consider one helium to be emitted with velocity components v_x and v_y in the beryllium rest frame. What is the relationship between v_x and v_y? How do v_x and v_y change when we move in the x direction at speed v?]

6. Express the following speeds as a fraction of the speed of light: (a) a typical automobile speed (100 km/h); (b) the speed of sound (330 m/s); (c) the escape velocity of a rocket from the Earth's surface (11 km/s); (d) the orbital speed of the Earth about the sun (Earth-sun distance = 1.5×10^8 km).

7. Show that Planck's constant h has dimensions of linear momentum $\times$ displacement.

8. Starting from Coulomb's law, show that $e^2/4\pi\varepsilon_0$ has dimensions of energy $\times$ distance.

9. (a) Starting from Newton's universal law of gravitation, show that Gm^2 has dimensions of energy $\times$ distance. (b) Evaluate Gm^2 in units of eV·nm using the proton mass. (c) Evaluate the ratio $Gm^2/(e^2/4\pi\varepsilon_0)$. Is this a pure number? What is its significance?

10. Use Avogadro's number to find the mass in kilograms equivalent to one atomic mass unit (u).

11. The orbital radius of an electron in an atom of hydrogen is $h^2\varepsilon_0/\pi m_e e^2$. (a) Show that this quantity has dimensions of length. (b) Compute its value to four significant figures. (Hint: Multiply numerator and denominator by common factors to allow use of some of the combinations of constants computed in this chapter.)

12. The *Bohr magneton* μ_B is equal to $eh/4\pi m_e$. (a) Show that μ_B has units of joules per tesla. (b) Show that μ_B has the same dimensions as a magnetic moment (current $\times$ area).

13. A fundamental unit frequently encountered is the Compton wavelength of the electron, $h/m_e c$. (a) Show that this quantity has the dimension of length. (b) Compute the value of $h/m_e c$ to four significant figures.

14. A helium nucleus consists of two protons and two neutrons, and has a mass of 4.001506 u. Find the mass difference between a helium nucleus and its constituents. Express your result in u and in MeV/c^2.

2

THE SPECIAL THEORY OF RELATIVITY

Einstein's special theory of relativity and Planck's quantum theory burst forth on the physics scene almost simultaneously during the first decade of the twentieth century. Both theories caused profound changes in the most fundamental way we view our universe.

In this chapter we study the special theory of relativity.* This theory has a completely undeserved reputation as being so exotic that few people can understand it. On the contrary, special relativity is simply a different system of kinematics and dynamics, based on a very different set of postulates than classical physics. The resulting formalism is not much more complicated than Newton's laws, but it does lead to several predictions which seem to go against our common sense. Even so, the special theory of relativity has been carefully and thoroughly tested by experiment and found to be correct in all its predictions.

We will first review classical Newtonian relativity and show why Einstein proposed to replace it. We will then discuss the mathematical aspects of special relativity, the predictions of the theory, and finally the experimental tests.

2.1 FAILURE OF CLASSICAL RELATIVITY

The Newtonian world view had provided a successful framework for understanding a large class of physical phenomena. This world view, which traces its origin to Galileo, asserts that space and time are absolute, that there is an absolute frame of reference, a grand universal Cartesian coordinate system, with absolute clocks attached, and that experiments performed in our frame of reference are only meaningful to the extent that they can be related to experiments performed in that absolute coordinate system. For example, it is an assertion, commonly known as Galileo's principle of inertia, that an object at rest tends to remain at rest unless acted on by an external force. Yet try this experiment in an accelerated frame of reference, such as a car coming to a sudden stop or a rapidly rotating merry-go-round, and you will find it not to hold. Newton's laws (including the principle of inertia) do not hold in accelerated reference frames, but they do hold in frames moving at constant velocity. Such frames of reference are known as *inertial frames*.

Inertial frames

Events viewed from different inertial frames may appear different to observers in these frames, but the observers will agree on the validity of Newton's laws, conservation of energy, and so forth. The comparison of these observations made in different inertial frames requires the *Galilean transformation*, which asserts that velocities add in the simplest possible way. Suppose an observer O in one inertial frame measures the velocity of an object to be **v**. An observer O' in a different inertial frame moving at *constant* velocity **u** relative to O would observe the same object to move with velocity $\mathbf{v}' = \mathbf{v} - \mathbf{u}$.

Galilean transformation

We will simplify the discussion of transformations somewhat by choosing the coordinate systems in our two frames of reference so that the relative motion **u** is always along the x direction. The Galilean

---

* What is "special" about special relativity is that it applies to the ordinary Euclidian (known as "flat") geometry with which we are familiar. The "curved" geometries of *general* relativity are discussed in Chapter 15.

transformation then becomes

$$v'_x = v_x - u \qquad (2.1a)$$

$$v'_y = v_y \qquad (2.1b)$$

$$v'_z = v_z \qquad (2.1c)$$

Only the x components of the velocities are affected. Integrating the first of these expressions gives

$$x' = x - ut \qquad (2.2)$$

and differentiating gives

$$\frac{dv'_x}{dt} = \frac{dv_x}{dt}$$

or

$$a'_x = a_x \qquad (2.3)$$

Equation (2.3) shows us immediately why Newton's laws hold in both frames. As long as u is constant $(du/dt = 0)$, the observers measure identical accelerations and will agree on the results of applying $\mathbf{F} = m\mathbf{a}$. Examples of the application of the Galilean transformation are given below:

Two cars are traveling at constant speed along a road in the same direction. Car A moves at 60 km/h and car B moves at 40 km/h, each measured relative to an observer on the ground. What is the speed of car A relative to car B?   **EXAMPLE 2.1**

Let O be the observer on the ground, who observes car A to move at $v = 60$ km/h. Assume O' to be moving with car B at $u = 40$ km/h. Then   ***Solution***

$$v' = v - u = 60 - 40 = 20 \text{ km/h}$$

A swimmer capable of swimming at a speed c in still water is swimming in a stream in which the current is u. Suppose the swimmer swims upstream a distance L and then returns downstream to the starting point. Find the time necessary to make the round trip, and compare it with the time to swim across the stream a distance L and return.   **EXAMPLE 2.2**

Let the frame of reference of O be the ground and the frame of reference of O' be the water, moving at speed u (Figure 2.1a). The swimmer always moves at speed c relative to the water, and thus $v' = -c$ for the upstream swim. (Remember u always defines the *positive x* direction.) According to Equation (2.1a), $v' = v - u$, so $v = v' + u = u - c$. (As expected, the speed relative to the ground is smaller than c; it is also *negative*, since the swimmer is swimming in the negative x direction, so $|v| = c - u$.) Therefore, $t_{\text{up}} = L/(c - u)$. For the downstream swim, $v' = c$, so $v = u + c$, $t_{\text{down}} = L/(c + u)$, and the total time is   ***Solution***

$$t = \frac{L}{c + u} + \frac{L}{c - u} = \frac{L(c - u) + L(c + u)}{c^2 - u^2} = \frac{2Lc}{c^2 - u^2} = \frac{2L}{c} \frac{1}{1 - u^2/c^2}$$

In order to swim directly across the stream, the swimmer's efforts must be directed somewhat upstream in order to counter the effect of the current (Figure 2.1b). That is, in the frame of reference of O we would like to have $v_x = 0$,

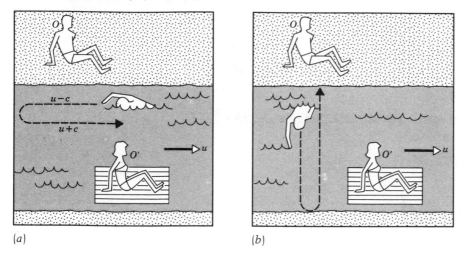

(a)   (b)

which requires $v'_x = -u$ according to Equation (2.1a). Since the speed relative to the water is always c, $\sqrt{v'^2_x + v'^2_y} = c$; thus $v'_y = \sqrt{c^2 - v'^2_x} = \sqrt{c^2 - u^2}$, and the round-trip time is

$$t = 2t_{\text{across}} = \frac{2L}{\sqrt{c^2 - u^2}} = \frac{2L}{c} \frac{1}{\sqrt{1 - u^2/c^2}}$$

Notice the difference *in form* between this result and the result for the upstream-downstream swim.

We can make a general characterization of a wave as the propagation of a periodic disturbance through a medium. The means by which the wave propagates depend on the forces exerted between the particles of the medium; no understanding of the propagation of waves is complete without an understanding of the dynamical behavior of the medium. It is therefore not surprising that, soon after Maxwell showed that electromagnetic waves were predicted on the basis of the equations of classical electromagnetism, efforts were made to study the properties of the medium responsible for the propagation of these waves. This medium was known as the *ether*; since it had not been observed in other experiments, it was postulated to be an invisible, massless medium pervading all space and having no function other than the propagation of electromagnetic waves. The concept of an ether is attractive for at least two reasons. First, it is difficult to imagine a wave propagating without a medium—try to picture water waves without the water!

Second, the basic notion of an ether ties in rather nicely with Newton's idea of absolute space—the ether is somehow associated with our Grand Universal Coordinate System. Thus as a side benefit of our search for the ether, detecting the motion of the Earth through the ether should reveal the motion of the Earth relative to "absolute space."

The most precise early experiment to search for evidence of the ether was performed in 1887 by the American physicist Albert A. Michelson and his associate E. W. Morley. Their experiment consisted essentially of a specially designed Michelson interferometer, a schematic diagram of which is shown in Figure 2.2. A monochromatic beam of

Albert A. Michelson (1852–1931, United States). He spent 50 years doing increasingly precise experiments with light, for which he became the first U. S. citizen to win the Nobel prize (1907).

Michelson-Morley experiment

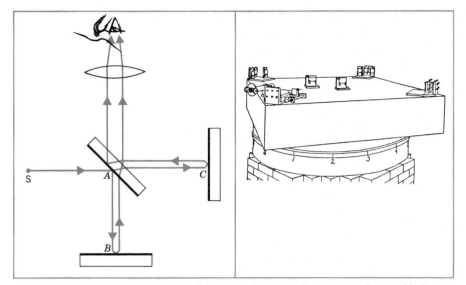

FIGURE 2.2 A schematic diagram of the Michelson interferometer. A beam of light from the source S is split in two at point A; one part is reflected by the mirror at B and the other is reflected at C. The beams are then recombined for observation of the interference. The right half of the figure shows a sketch of Michelson's apparatus. To improve sensitivity, mirrors were set up to make the beams travel each leg of the apparatus eight times, rather than just twice. To reduce vibrations from the surroundings, the interferometer was mounted on a stone slab about $1\frac{1}{2}$ m square floating in a pool of mercury.

light is split in two; the two beams are made to travel different paths and are then recombined. Due to the differing path lengths an interference pattern will be produced, as shown in Figure 2.3.

Let us imagine for the moment that the Earth happens to be moving through the ether along the direction AB in Figure 2.2. In the interference pattern, the dark bands appear where the two light beams interfere destructively and the bright bands, where the interference is constructive. Whether the interference is destructive or constructive depends on the *phase difference* between the two beams. There are two contributions to this phase difference. The first originates from the path difference $(AB - AC)$; one of the two beams simply has a longer distance to travel. The second contribution to the phase difference would still be present even if the two path lengths were exactly equal. We can understand this by referring to Example 2.2. A light beam "swimming" through the ether upstream and downstream takes a different time than one that travels cross-stream and back. If we could separate and measure this second contribution, then we could deduce the "speed" of the stream and hence the motion of the Earth through the ether. Unfortunately, no such separation is possible. However Michelson and Morley used a clever method to deduce this second component—they rotated the entire apparatus by 90°! The contribution to the phase difference due to the path difference of course doesn't change, but the contribution due to the ether motion changes sign, since now it is the beam along AC that is traveling along the stream while AB is now cross-stream. This change would be observed as a change in the pattern of light and dark fringes as the apparatus is rotated. Each change of light to dark or dark to light represents a phase change of 180° (one half cycle), which is equivalent to a

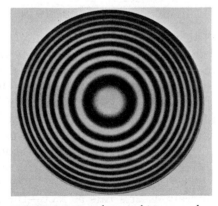

FIGURE 2.3 Interference fringes as observed with the Michelson interferometer of Figure 2.2. When the path length ACA changes by one-half wavelength relative to ABA, all light areas turn dark and all dark areas turn light.

time advance or delay of one half period. (This would amount to about 10^{-15} s for visible light.) From our relations for the difference between the upstream-downstream and cross-stream times, we could then deduce the speed of the Earth through the ether (see Problem 1 at the end of this chapter).

When Michelson and Morley performed their experiment, there was no observable change in the fringe pattern—they deduced a shift of less than 0.01 fringe, corresponding to a speed of the Earth through the ether of at most 5 km/s. As a last resort, Michelson and Morley reasoned that perhaps the orbital motion of the Earth just happened to cancel out the overall motion through the ether. If this were true, six months later the Earth would be moving in its orbit in the opposite direction, and the cancellation should not occur. When the experiment was repeated six months later, a null result was again obtained.

In summary, we have seen that there is a direct chain of reasoning that leads from Galileo's principle of inertia, through Newton's laws with their implicit assumptions about space and time, ending with the failure of the Michelson-Morley experiment to observe the motion of the Earth relative to the ether. Although several explanations were offered for the unobservability of the ether and the corresponding failure of the upstream-downstream and cross-stream velocities to add in the expected way, the most novel, revolutionary, and ultimately successful explanation requires a serious readjustment of our traditional concepts of space and time, and therefore alters some of the very foundations of physics.

2.2 EINSTEIN'S POSTULATES

Albert Einstein (1879–1955, Germany-United States). A gentle philosopher and pacifist, he was the intellectual guru to two generations of theoretical physicists and left his imprint on nearly every field of modern physics.

The resolution of the difficulties posed by the Michelson-Morley experiment forms the cornerstone of new concepts of space and time, known as the *special theory of relativity*. The theory is based on two postulates, put forth by Albert Einstein in 1905.

1. The principle of relativity: the laws of physics are identical in all inertial systems.
2. The constancy of the speed of light: the speed of light has the same value of c in all inertial systems.

The first postulate essentially asserts that there is *no* experiment by which we can measure velocity with respect to absolute space—all we can measure is the *relative* speed of two inertial systems. Whether or not absolute space exists then no longer becomes meaningful; there may in fact be a Grand Universal Reference System, but no experiment we can do will ever reveal its presence (or our relation to it) and so we might just as well discard the notion as an unnecessary complication.

The second postulate is both bold and deceptively simple. The Michelson-Morley experiment seems to indicate that the upstream-downstream and cross-stream speeds are identical. The second postulate merely asserts this as fact—the speed of light is the same for all observers, even when in relative motion. For example, suppose two rocket ships were approaching at a relative speed of $c/2$, when one of the ships fired a beam of light at the other. The second ship would not measure a speed of $c + (c/2)$ for the approaching light beam, as would be the case with Galilean relativity, Equation (2.1a), but simply c.

In the following section we explore some of the consequences of Einstein's postulates and discuss the mathematical transformation from which follows the constancy of the speed of light.

Consider two observers O and O'. O fires a beam of light at a mirror a distance L away and measures the time interval $2\,\Delta t$ for the beam to be reflected from the mirror and to return to O. (Of course, $L = c\,\Delta t$.) Observer O' is moving at constant velocity u as shown in Figure 2.4. As seen from the point of view of O, the beam is sent and received from the same point, while O' moves off in a perpendicular direction. Figure 2.5 shows the same experiment from the viewpoint of O', from whose perspective O is moving with velocity $-u$. The beam is sent from point A and received at point B a time $2\,\Delta t'$ later, according to O'. The distance AB is just $2u\,\Delta t'$. According to O, the light beam travels a distance $2L$ in a time $2\,\Delta t$. According to O', the light beam travels the path AMB, a distance of $2\sqrt{L^2 + (u\,\Delta t')^2}$ in a time $2\,\Delta t'$. According to *Galilean relativity*, $\Delta t = \Delta t'$, and O measures a speed c while O' measures a speed $\sqrt{c^2 + u^2}$. According to *Einstein's second postulate*, this is not possible—both O and O' must measure the same speed c. *Therefore* Δt *and* $\Delta t'$ *must be different.* We can find a relationship between Δt and $\Delta t'$ by setting the two speeds equal to c. According to O, $c = 2L/2\,\Delta t$, so $L = c\,\Delta t$. According to O', $c = 2\sqrt{L^2 + (u\,\Delta t')^2}/2\,\Delta t'$, so $c\,\Delta t' = \sqrt{L^2 + (u\,\Delta t')^2}$. Combining these, we find

$$c\,\Delta t' = \sqrt{(c\,\Delta t)^2 + (u\,\Delta t')^2}$$

and, solving for $\Delta t'$,

$$\Delta t' = \frac{\Delta t}{\sqrt{1 - u^2/c^2}} \qquad (2.4)$$

This relationship summarizes the effect known as *time dilation*. According to Equation (2.4), observer O' measures a longer time interval than O measures. This is a general result within special relativity, which we can state as follows. Consider an occurrence that has a duration Δt. An observer O fixed with respect to that occurrence (i.e., its beginning and end take place at the same point in space, according to O) measures the interval Δt, which is known as the *proper time*. An observer O' moving with velocity **u** with respect to O will measure a longer time interval $\Delta t'$ for the same occurrence. The interval $\Delta t'$ is always longer than Δt, no matter what the magnitude or direction of **u**.

It must be pointed out that this is a real effect that applies not only to clocks based on light beams but to time itself; all clocks will run more slowly according to an observer in relative motion, biological clocks included. Even the growth, aging, and decay of living systems are slowed by the time dilation effect.

The effects of time dilation can be observed in a variety of experiments. For example, we consider the birth and decay of the elementary particle known as the muon. (Muons can be produced in energetic collisions between other particles—more about this in Chapter 11.) In its own reference frame, the creation of the muon and its subsequent decay (into an electron and other particles called neutrinos) take place at the same point in space, and therefore the lifetime as measured in that

2.3 CONSEQUENCES OF EINSTEIN'S POSTULATES

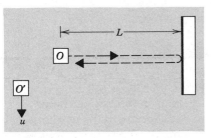

FIGURE 2.4 Observer O sends and receives a beam of light that is reflected by a mirror. Observer O' is in motion at speed u.

Time dilation

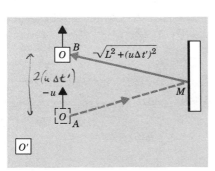

FIGURE 2.5 The experiment shown in Figure 2.4, as seen by observer O'. Observer O emits a light beam at point A and receives the reflected beam at B.

How fast must the muons be traveling to "survive" the trip to the surface of the Earth?

EXAMPLE 2.3

We assume that the muons are traveling at speeds close to c. To travel 100 km then takes a time of roughly $\Delta t' = 100$ km$/(3 \times 10^8$ m/s$) \cong 3 \times 10^{-4}$ s. Using Equation (2.4), we find

$$3 \times 10^{-4} = \frac{2 \times 10^{-6}}{\sqrt{1 - u^2/c^2}}$$

from which

$$u = 0.99998c$$

frame of reference is the proper time interval Δt. This can be measured in the laboratory to be about 2×10^{-6} s. Muons are also produced when the high-energy particles known as cosmic rays collide with the atoms of the upper atmosphere; these muons then rush toward the ground at speeds very close to the speed of light. If these muons lived for 2×10^{-6} s in our reference frame on the ground and were traveling close to 3×10^8 m/s, they could travel at most a distance of about 600 m, a small distance compared with the height of the atmosphere of more than 100 km. Therefore we should never see these muons at the surface of the Earth, yet we observe them in great quantities. The explanation for this lies in the time dilation effect. The muons do indeed live for only 2×10^{-6} s *in their own frame of reference*, but viewed from our frame of reference, rushing toward the muons at high speed, the time interval is much longer.

Let us now return to our original experiment with observers O and O'. Now we assume that O' moves parallel to the light beam, and Figure 2.6 illustrates the appearance of the experiment from the two frames of reference. Suppose we consider the experiment from the point of view of O'. In traveling toward the mirror for a time $\Delta t_1'$, the light beam is observed by O' to travel a distance of $L' - u \Delta t_1'$, since the mirror has moved toward the source by an amount $u \Delta t_1'$ in that time. Observer O' measures the speed of the light beam to be c, and deduces

$$c \, \Delta t_1' = L' - u \, \Delta t_1'$$

Similarly, the reflected light beam travels back to its source in a time $\Delta t_2'$; it must travel the distance $L' + u \, \Delta t_2'$, and therefore

$$c \, \Delta t_2' = L' + u \, \Delta t_2'$$

If we let $2 \, \Delta t'$ be the total time for the round trip (as measured by O'), then

$$2 \, \Delta t' = \Delta t_1' + \Delta t_2'$$

$$= \frac{L'}{c + u} + \frac{L'}{c - u} = L' \frac{2c}{c^2 - u^2}$$

We also know that O measures the same speed c for the light beam, which travels a distance $2L$ in a time $2 \, \Delta t$. Furthermore, we know $\Delta t' =$

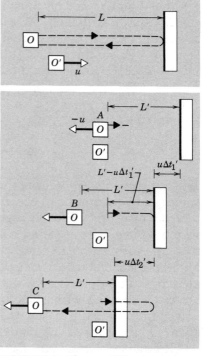

FIGURE 2.6 Observer O' now moves parallel to the light beam, as shown in the top of the figure from the reference frame of O. According to O', the beam is emitted when O is at A, reflected when O is at B, and received when O is at C. The bottom part of the figure shows the three instants of emission, reflection, and receipt of the light beam. The actual path of the light beam according to O' is shown in each of the three sketches.

$\Delta t / \sqrt{1 - u^2/c^2}$. Combining these results, we have:

$$\frac{2\,\Delta t}{\sqrt{1 - u^2/c^2}} = L' \frac{2c}{c^2 - u^2} = \left(\frac{L'}{c}\right) \frac{2}{1 - u^2/c^2}$$

$$L' = c\,\Delta t \sqrt{1 - u^2/c^2}$$

$$L' = L\sqrt{1 - u^2/c^2} \qquad (2.5)$$

Thus the length L' measured by O' is *shorter* than the length L measured by O. This result is known as *length contraction*.

Length contraction

Length contraction is a general result, and has nothing to do with the kind of length measurement we make. The length of an object measured in a frame of reference in which the object is at rest is known as its *proper length*. The length measured in all frames of reference moving at constant speed with respect to the object's rest frame will be shorter by an amount given by Equation (2.5). *Length contraction only occurs along the direction of motion*—all other lengths are unaffected. Figure 2.7 shows some idealized representations of length-contracted objects.

It must be emphasized that, like time dilation, this is a real effect which occurs for all observers in relative motion. To an observer in a rocket ship passing by the Earth, we might appear as shown in Figure 2.8, but we don't notice the effect because, of course, nothing changes for us in our frame of reference. Moreover, our view of the passing ship is as shown in Figure 2.9.

The views we have presented of moving objects are idealized—our eye would in fact not perceive the length contraction as we have discussed it, nor would a camera photograph it. To understand why this is so, remember that the retina of our eye, or the film of a camera, responds to a series of images that strike its surface *at the same instant*. Consider the moving cube as shown in Figure 2.10. Light from the bottom surface has a shorter distance to travel than does light from the front face, and so light that was emitted earlier from the front face reaches the film at the

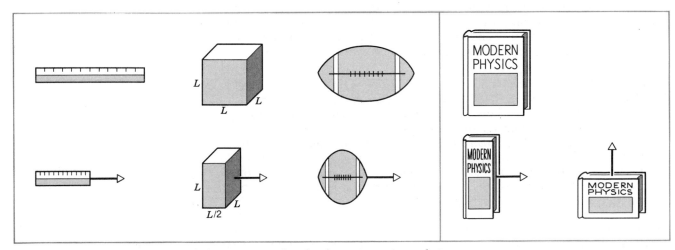

FIGURE 2.7 Some length-contracted objects. Notice that the shortening occurs only in the direction of motion.

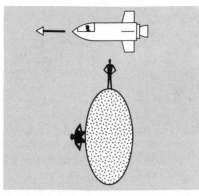

FIGURE 2.8 A passing rocket views the contracted Earth.

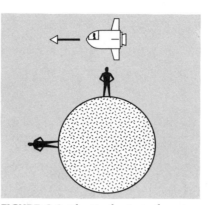

FIGURE 2.9 The Earth views the passing contracted rocket.

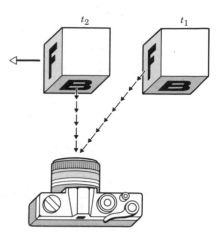

FIGURE 2.10 Photographing a moving cube. Light from the front face F emitted at time t_1 reaches the camera at the same time as light from the bottom face B emitted at the later time t_2. If the shutter is opened when the cube is directly above the camera, both B and F appear on the photograph.

same instant as light that was emitted at a later time from the bottom surface. Thus when the cube is directly overhead, we "see" the bottom surface and the front face simultaneously. The only way this is possible is if the cube appears to rotate slightly, and so the photograph would look like Figure 2.11.

Since clocks in relatively moving coordinate systems run at different rates, our concept of "absolute time" is no longer valid. Moreover, two events that are simultaneous in one frame of reference will not be simultaneous in relatively moving frames. The following example illustrates some of the difficulties that arise from this situation.

FIGURE 2.11 The view of the moving cube.

An interplanetary scout files the following report of an encounter with a passing space ship: "As the other ship approached, I maneuvered my own ship so it would pass right alongside. At a particular instant of time I noticed that both ends of our ships lined up exactly, as I show in the sketch. At that instant I fired my fore and aft laser beams, directing them over the bow and stern of the passing ship. As you know, firing simultaneous beams across bow and stern is the accepted intergalactic greeting of peace and friendship. Yet the passing ship did not respond in a friendly way and instead fired at my ship, inflicting serious damage." Analyze this incident from the perspective of the second ship.

EXAMPLE 2.4

Keep in mind that from the frame of reference of ship A, ship A has its proper length and all other objects in relative motion are shortened. Thus, even though the two ships *appear* to be of the same length from A's point of view, this is completely due to A's particular frame of reference—the *proper length* of ship A appears to be equal to the *shortened length* of ship B. (It is therefore obvious that the *proper length* of ship B must be greater than the *proper length* of ship A.) From B's frame of reference, of course, the reverse is true—ship B has its proper length and ship A is shortened, and an observer in ship B would give the following description: "As the other ship passed, a beam was fired across my bow. Later on, another beam was fired across my stern. Since the passing ship was much shorter than my own, our bows and sterns could not possibly line up simultaneously, and so the two beams could not have been fired simultaneously as a greeting. I therefore returned the fire."

Solution

EXAMPLE 2.5

An observer is standing on the station platform when a modern, high-speed train passes through at $u = 0.80c$. The observer, who measures the platform to be 60 m long, notices that the front and back of the train line up exactly with the ends of the platform. (a) How long does the ground-based observer measure for the train to pass a fixed point on the platform? (b) What is the proper length of the train? (c) What is the length of the platform, according to an observer on the train? (d) How long will it take for a point on the platform to pass the full length of the train, according to an observer on the train? (e) To an observer on the train, the ends of the train will *not* simultaneously line up with the ends of the platform. Find the time interval between the front end of the train lining up with one end of the platform and the back end of the train lining up with the other.

(a) To pass a given point, the train must travel a distance of its length as observed from the platform. Thus:

$$\Delta t = \frac{L}{0.8c} = \frac{60 \text{ m}}{2.4 \times 10^8 \text{ m/s}} = 2.5 \times 10^{-7} \text{ s}$$

(b) Since the ground-based observer is measuring a contracted length of the train (but the proper length of the platform) to be 60 m, the proper length of the train is, according to Equation (2.5):

$$L_t = \frac{L_t'}{\sqrt{1 - u^2/c^2}} = \frac{60}{\sqrt{1 - (0.8)^2}} = 100 \text{ m}$$

(c) The observer on the train observes the platform to have a contracted length L_p', related to its proper length L_p by

$$L_p' = L_p\sqrt{1 - u^2/c^2} = 60\sqrt{1 - (0.8)^2} = 36 \text{ m}$$

(d) Since the length of the train is 100 m,

$$\Delta t' = \frac{100 \text{ m}}{2.4 \times 10^8 \text{ m/s}} = 4.2 \times 10^{-7} \text{ s}$$

Note that we have called this $\Delta t'$ to indicate that it is *not* a proper time interval—the events of the point on the platform crossing first the front of the train and then the back of the train do *not* occur at the same point in space according to the observer on the train. Of course Δt from part (a) and $\Delta t'$ are related by the time dilation formula, as you should verify.

(e) The time interval between the front end of the train lining up with one end of the platform and the back end of the train lining up with the other is just the distance "traveled" by the station, $100 - 36 = 64$ m, divided by the relative speed

$$\Delta t = \frac{64 \text{ m}}{2.4 \times 10^8 \text{ m/s}} = 2.7 \times 10^{-7} \text{ s}$$

Thus two events that are simultaneous in one reference frame occur 2.7×10^{-7} s apart in another. In Section 2.6 we explore this phenomenon further.

Since two observers in relative motion differ in their measurements of time intervals, we might similarly ask if their measurements of frequency will also differ. In classical physics, you studied the Doppler effect for sound waves, in which the source and observer could move with speeds v_S and v_O with respect to the medium, and in which the frequency v' heard by the observer O is different from the frequency v emitted by the source S; the relationship is

$$\nu' = \nu \, \frac{v \pm v_O}{v \mp v_S} \qquad (2.6)$$

in which we choose the upper signs whenever S is moving toward O, or O toward S. (The speed v gives the speed of waves in the medium.) Because all speeds are measured with respect to a medium (still air, for example) source motion gives a different Doppler shift than does observer motion. For example, for sound in air $v = 340$ m/s. Suppose the source is emitting sound waves at 1000 Hz. If the source and observer are moving toward one another at 30 m/s, we can identify three situations from among many:

1. Source at rest in medium, observer moving at 30 m/s toward source:

$$\nu' = 1000 \left(\frac{340 + 30}{340} \right) = 1088 \text{ Hz}$$

2. Observer at rest, source moving toward observer at 30 m/s:

$$\nu' = 1000 \left(\frac{340}{340 - 30} \right) = 1097 \text{ Hz}$$

3. Source and observer each moving at 15 m/s relative to medium:

$$\nu' = 1000 \left(\frac{340 + 15}{340 - 15} \right) = 1092 \text{ Hz}$$

Notice that the values of ν' are different for these three cases—we can distinguish "absolute motion" with respect to the medium that carries the sound waves.

Einstein's first postulate asserts that this situation is not possible for light waves, since light waves have no medium (no "ether") and since no experiment can reveal absolute motion. We therefore require a different formula for the Doppler shift for light waves, a formula that does not distinguish between source motion and observer motion but rather includes only the relative motion.

Suppose our observer O has a source of radiation, emitting waves at a frequency (measured by O) equal to ν. O', moving at speed u relative to O, measures a larger frequency if moving toward O (more wave fronts are crossed per second) and a smaller frequency when moving away from O.

Let us examine the situation from the point of view of O', as the distance between O' and the source decreases (O' is moving toward O). If T' is the time between wave crests as measured by O' (Figure 2.12) and λ' is the wavelength seen by O', then according to O', the distance between wave crests is $(c - u)T'$, since a given wave crest moves a distance cT' before the source emits the next wave, while the source itself moves a distance of uT'. Thus

$$\lambda' = (c - u)T'$$

The time between crests T' as measured by O' is related to T, the time between crests as measured by O, according to the time dilation formula, Equation (2.4), $T' = T/\sqrt{1 - u^2/c^2}$; T is related to the frequency ν measured by O as $T = 1/\nu$. The wavelength λ' measured by O' is related

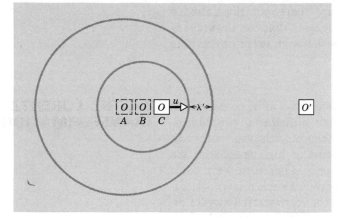

FIGURE 2.12 Observer O emits waves at frequency ν. When O is at point A, the first wave is emitted; at B the second wave is emitted. A third wave (not shown) is about to be emitted from point C. Observer O' measures the distance λ' between the wave crests.

to the frequency measured by O' according to $c = \lambda'\nu'$. Thus

$$\frac{c}{\nu'} = (c - u)\frac{T}{\sqrt{1 - u^2/c^2}} = \frac{1}{\nu}\frac{c - u}{\sqrt{1 - u^2/c^2}}$$

or

$$\nu' = \nu\frac{\sqrt{1 - u^2/c^2}}{1 - u/c} = \nu\sqrt{\frac{1 + u/c}{1 - u/c}} \qquad (2.7)$$

Equation (2.7) is the Doppler shift formula that is consistent with Einstein's postulates. Notice that (unlike the classical formula) it does *not* distinguish between source motion and observer motion and depends only on the relative speed u. (If the source is moving away from the observer, replace u by $-u$ in the Doppler shift formula.)

Relativistic Doppler shift

A distant galaxy is moving away from the Earth at such high speed that the blue hydrogen line at 434 nm is recorded at 600 nm, in the red range of the spectrum. What is the speed of the galaxy relative to the Earth?

EXAMPLE 2.6

Since $\lambda' > \lambda$, then $\nu' < \nu$, and Equation (2.7) indicates that the galaxy is moving away from the Earth. Thus we have

Solution

$$\nu' = \nu\sqrt{\frac{1 - u/c}{1 + u/c}}$$

or, using $\nu = c/\lambda$ and $\nu' = c/\lambda'$

$$\lambda' = \lambda\sqrt{\frac{1 + u/c}{1 - u/c}}$$

$$600 \text{ nm} = 434 \text{ nm}\sqrt{\frac{1 + u/c}{1 - u/c}}$$

or

$$\frac{u}{c} = 0.31$$

Thus the galaxy is moving away from Earth at a speed of 9.4×10^7 m/s.

Evidence obtained in this way indicates that nearly all the galaxies we observe are moving away from us. This suggests that the universe is expanding, and is usually taken to provide evidence in favor of the "Big Bang" theory of cosmology (see Chapter 16).

We have already seen how the Galilean transformation of coordinates, time, and velocities is not consistent with Einstein's postulates. Although the Galilean transformation is in accord with our "common sense," it is not in agreement with experiments at high speeds, as we will illustrate in the last section of this chapter. Therefore we need a new set of transformation equations that can predict such relativistic effects as length contraction, time dilation, and the relativistic Doppler effect. Also, since we know that the Galilean transformation (2.1) does work well at low speeds, our new transformation must give the same results as the Galilean transformation when the relative speed of O and O' is small.

2.4 THE LORENTZ TRANSFORMATION

The transformation that satisfies these requirements is known as the *Lorentz transformation* and, like the Galilean transformation, it relates the coordinates of an event (x, y, z, t) as observed from frame of reference O to the coordinates of the *same event* (x', y', z', t') as observed from a reference frame O' moving at constant velocity $\mathbf{u}$ with respect to O. As before, we assume that the relative motion is in the *positive x* (or x') direction (O' moves *away from O*).

The Lorentz transformation equations have the following form:

$$x' = \frac{x - ut}{\sqrt{1 - u^2/c^2}} \tag{2.8a}$$

$$y' = y \tag{2.8b}$$

$$z' = z \tag{2.8c}$$

Lorentz coordinate transformation

$$t' = \frac{t - (u/c^2)x}{\sqrt{1 - u^2/c^2}} \tag{2.8d}$$

(If O' moves *toward* O, replace u by $-u$.) Application of the Lorentz transformation requires the following: an "event" is observed by O with coordinates (x, y, z, t). Observer O' is moving at speed u with respect to O, and describes the *same event* with the coordinates (x', y', z', t'). These equations then allow us to compare the two descriptions. *We make no special assumptions about the relationship between* O *and the event*—for example, the object whose instantaneous coordinates are given by the event (x,y,z,t) need not be at rest relative to O.

Use the Lorentz transformation to derive the expression for length contraction, $L' = L\sqrt{1 - u^2/c^2}$.

EXAMPLE 2.7

The determination of the length of an object requires two observations—the coordinates of the ends of the object. Let us assume the object is at rest in coordinate system S. Observer O measures the coordinates of the ends of the object to be $x_1 = 0$ and $x_2 = L$. (Since the object is at rest with respect to O, these observations need not be made simultaneously—x_1 and x_2 will not change with time.)

Solution

Observer O' measures the coordinates x_1' (at t_1') and x_2' (at t_2') for the ends of the object, and thus $L' = x_2' - x_1'$. *In order for O' to measure correctly the length of the object, x_1' and x_2' must be measured simultaneously, since the object is in motion relative to O'.* That is, $t_2' = t_1'$. (Try to measure the length of a moving car by recording the position of the back end at one time and the front end a few minutes later!) Thus the following table of values is obtained, using Equations (2.8a) and (2.8d)

	Observer O	Observer O'
Event 1	$x_1 = 0$	$x_1' = (x_1 - ut_1)/\sqrt{1 - u^2/c^2}$
	at t_1	$t_1' = [t_1 - (u/c^2)x_1]/\sqrt{1 - u^2/c^2}$
Event 2	$x_2 = L$	$x_2' = (x_2 - ut_2)/\sqrt{1 - u^2/c^2}$
	at t_2	$t_2' = [t_2 - (u/c^2)x_2]/\sqrt{1 - u^2/c^2}$

$$L' = x_2' - x_1' = \frac{(x_2 - ut_2) - (x_1 - ut_1)}{\sqrt{1 - u^2/c^2}}$$

$$L' = \frac{L}{\sqrt{1 - u^2/c^2}} - \frac{u(t_2 - t_1)}{\sqrt{1 - u^2/c^2}}$$

where we have used $x_2 - x_1 = L$. Also, from the equations for t_1' and t_2',

$$t_2' - t_1' = 0 = \frac{[t_2 - (u/c^2)x_2] - [t_1 - (u/c^2)x_1]}{\sqrt{1 - u^2/c^2}}$$

$$0 = \frac{(t_2 - t_1)}{\sqrt{1 - u^2/c^2}} - \frac{u}{c^2}\frac{L}{\sqrt{1 - u^2/c^2}}$$

Substituting $(t_2 - t_1)$ from this expression into the above equation for L' and combining terms gives

$$L' = L\sqrt{1 - u^2/c^2}$$

A somewhat simpler derivation is outlined in Problem 2 at the end of the chapter; a similar exercise in deriving the time dilation equation may be found in Problem 3.

Use the Lorentz transformation to derive the expression for the relativistic Doppler effect.   **EXAMPLE 2.8**

Suppose a plane, monochromatic light wave has the mathematical form $E = E_0 \sin 2\pi(x/\lambda - \nu t)$ according to O and $E' = E_0' \sin 2\pi(x'/\lambda' - \nu't')$ according to O'. Then substituting the Lorentz expressions to transform from O' to O, we find   *Solution*

$$E' = E_0' \sin\left\{2\pi\left[\frac{(x - ut)/\sqrt{1 - u^2/c^2}}{\lambda'} - \nu'\frac{[t - (u/c^2)x]}{\sqrt{1 - u^2/c^2}}\right]\right\}$$

$$E' = E_0' \sin\left\{2\pi\left[\frac{x(1/\lambda' + \nu'u/c^2)}{\sqrt{1 - u^2/c^2}} - \frac{t(u/\lambda' + \nu')}{\sqrt{1 - u^2/c^2}}\right]\right\}$$

Comparing this transformed form to our expected form $E = E_0 \sin 2\pi(x/\lambda - \nu t)$ we require that the coefficient of x must be $1/\lambda$ and the coefficient of t must be ν. This gives us, for example,

$$\nu = \frac{u/\lambda' + \nu'}{\sqrt{1 - u^2/c^2}}$$

Since $1/\lambda' = \nu'/c$, it follows that

$$\nu = \frac{\nu'(1 + u/c)}{\sqrt{1 - u^2/c^2}}$$

or

$$\nu' = \frac{\nu\sqrt{1 - u^2/c^2}}{(1 + u/c)} = \nu\sqrt{\frac{1 - u/c}{1 + u/c}}$$

Thus, as expected, O' (moving away from O) measures a lower frequency ν'.

Suppose we have an object that is observed by O to move at a velocity $\mathbf{v} = (v_x, v_y, v_z)$. To find its velocity $\mathbf{v}' = (v'_x, v'_y, v'_z)$ as observed by O', we require the *Lorentz velocity transformation:*

Lorentz velocity transformation

$$v'_x = \frac{v_x - u}{1 - v_x u/c^2} \tag{2.9a}$$

$$v'_y = \frac{v_y\sqrt{1 - u^2/c^2}}{1 - v_x u/c^2} \tag{2.9b}$$

$$v'_z = \frac{v_z\sqrt{1 - u^2/c^2}}{1 - v_x u/c^2} \tag{2.9c}$$

These relationships follow directly from our original expressions for the Lorentz transformation. As an example, the expression for v'_y will be derived, and the derivations of v'_x and v'_z are left as exercises (Problem 10).

Derive the Lorentz velocity transformation for v_y.

EXAMPLE 2.9

The component v'_y is just dy'/dt'.

Solution

$$y' = y \longrightarrow dy' = dy$$

$$t' = \frac{t - (u/c^2)x}{\sqrt{1 - u^2/c^2}} \longrightarrow dt' = \frac{dt - (u/c^2)dx}{\sqrt{1 - u^2/c^2}}$$

So

$$v'_y = \frac{dy'}{dt'} = \frac{dy}{[dt - (u/c^2)dx]/\sqrt{1 - u^2/c^2}} = \sqrt{1 - u^2/c^2}\frac{dy}{dt - (u/c^2)dx}$$

$$= \sqrt{1 - u^2/c^2}\frac{dy/dt}{1 - (u/c^2)dx/dt} = \frac{v_y\sqrt{1 - u^2/c^2}}{1 - uv_x/c^2}$$

In using the Lorentz velocity transformation, it is important to keep careful account of the identities of the observers O and O' and the event being observed. We illustrate these concepts with the following examples.

Two rockets are approaching one another on a head-on collision course. Each is moving at velocity $0.5c$ relative to an independent observer midway between the two. With what velocity does one rocket observe the approach of the other?

EXAMPLE 2.10

Let O represent the independent observer, and let O' be one of the rockets. Then the "event" under observation is the approach of the second rocket, as in the following diagram.   *Solution*

Observer O sees rocket 2 moving with $v_x = -0.5c$. Observer O' (rocket 1) is moving relative to O with $u = 0.5c$. Then, using Equation (2.9a) for the transformation of v_x,

$$v'_x = \frac{v_x - u}{1 - v_x u/c^2} = \frac{(-0.5c) - (0.5c)}{1 - (-0.5c)(0.5c)/c^2} = -0.8c$$

Notice that this result is less than the relative velocity of $-0.5c - 0.5c = -c$ that the Galilean transformation predicts. Since special relativity demands that the value c be the limiting speed for all relative motion, the two rockets can never move with relative speed greater than c, and the form of the Lorentz velocity transformation guarantees this. For example, if instead of $0.5c$, the speeds were $0.999c$, then we would have

$$v'_x = \frac{-0.999c - 0.999c}{1 - (-0.999c)(0.999c)/c^2} = -0.9999995c$$

rather than $-1.998c$ as the Galilean transformation would give.

Observer O measures the value c for the speed of light. Prove that the Lorentz velocity transformation gives the same value as measured by O'.   **EXAMPLE 2.11**

If the system under observation is a light beam traveling in the x direction then O measures $v_x = c$, and we must find v'_x.   *Solution*

$$v'_x = \frac{v_x - u}{1 - v_x u/c^2} = \frac{c - u}{1 - cu/c^2} = c$$

Thus O' measures c, independent of the value of u. This feature of the Lorentz velocity transformation is, of course, in agreement with Einstein's postulate regarding the constancy of the speed of light.

Two rockets are leaving their space station along perpendicular paths, as measured by an observer on the space station. Rocket 1 moves at $0.60c$ and rocket 2 moves at $0.80c$, both measured relative to the space station. What is the velocity of rocket 2 as observed by rocket 1?   **EXAMPLE 2.12**

Observer O is the space station, observer O' is rocket 1 (moving at $u = 0.60c$) and the "event" is rocket 2, moving (according to O) in a direction perpendicular to rocket 1. We take this to be the y direction of the reference frame of O. Thus O observes rocket 2 to have velocity components $v_x = 0$, $v_y = 0.80c$:   *Solution*

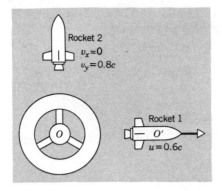

We can find v'_x and v'_y using the Lorentz velocity transformation:

$$v'_x = \frac{0 - 0.6c}{1 - 0(0.6c)/c^2} = -0.60c$$

$$v'_y = \frac{0.8c\sqrt{1 - (0.6c)^2/c^2}}{1 - 0(0.6c)/c^2} = 0.64c$$

Thus, according to O', the situation looks like this:

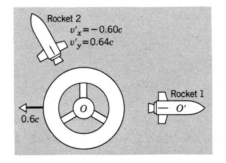

The speed of rocket 2 according to O' is $\sqrt{(0.60c)^2 + (0.64c)^2} = 0.88c$, less than c. According to the Galilean transformation, v'_y would be identical with v_y, and thus the speed would be $\sqrt{(0.6c)^2 + (0.8c)^2} = c$. Once again, the Lorentz transformation prevents relative speeds from reaching or exceeding the speed of light. Again, if the two speeds were $0.999c$, the transformation equations would give $v'_x = -0.999c$ and $v'_y = 0.0447c$, giving a relative speed of $0.999998c$. Notice how the Lorentz transformation causes the extreme shrinking of the y component from $0.999c$ to $0.0447c$. This results from the time dilation factor in the moving reference frame of O'.

2.5 RELATIVISTIC DYNAMICS

We have seen how Einstein's postulates have led to a new "relative" interpretation of such previously absolute concepts as length and time, and that our classical concepts of relative speeds no longer are true. It is reasonable then to ask how far this revolution is to go in changing our interpretation of physical concepts. We therefore discuss now the dynamical quantities of mass, energy, momentum, and force, in order to examine these from the point of view of special relativity. Are our familiar relationships, such as $p = mv$, $K = \frac{1}{2}mv^2$, and $F = ma$ (more properly, $F = dp/dt$), still valid, or must we have new concepts of dynamical quantities also? Moreover, what of the fundamental conservation laws of classical physics, such as conservation of energy and conservation of linear momentum? These concepts are so important to classical physics that we are very reluctant to discard them. These conservation laws (along with the conservation of angular momentum) can be shown to follow from our belief in the *homogeneity* and *isotropy* of the universe—if we correct for all local effects (variations in atmospheric or environmental conditions, for example), then experiments performed on one day should yield the same results as identical experiments performed on the following day, experiments performed in one laboratory should yield the same results as identical experiments performed in laboratories some distance away, and rotating our laboratory apparatus through some angle should likewise not change the results of our experiments. These notions of *invariance* with respect to *translations* in time and in space and with respect to *rotations* in space can be shown to be equivalent to our concepts of conservation of energy, linear momentum, and angular momentum, and discarding those concepts would imply that we live in a very strange sort of universe. Therefore, we make the assumption that our universe has the most agreeable sort of structure, and that these conservation laws are valid, but we keep in mind that special relativity may require a redefinition of the basic quantities. In fact, we can guess immediately that this will be necessary. Suppose we apply a constant force F to a mass m, giving it an acceleration $a = F/m$. If we apply that force for a long enough time, classical dynamics predicts that the particle will continue to gain speed until its speed exceeds the speed of light. However, we know that the Lorentz transformation gives meaningless results when $u \geq c$, and so we are going to need a new set of dynamical laws that prevents accelerating objects to speeds greater than that of light.

Let us begin by considering the following problem, which you have studied using Newtonian dynamics. Two identical masses approach each other, each moving at speed v. Following the collision, we have a mass of $2m$ at rest. This is the description according to observer O in the laboratory:

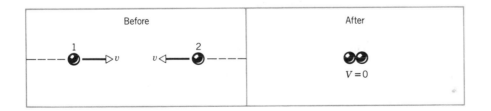

Let us now transform to a frame of reference moving with speed v toward the right. According to classical physics, mass 1 would appear to be at rest, and mass 2 would appear to be approaching at a speed of $2v$. However, the Lorentz transformation gives a different result. Let O' move to the right at $u = v$. Then according to O', the velocity of mass 1 is (all velocities are in the x direction, so we will drop the x subscript)

$$v'_1 = \frac{v_1 - u}{1 - v_1 u/c^2} = \frac{v - v}{1 - v^2/c^2} = 0$$

and the velocity of mass 2 is (with $v_2 = -v$ according to O)

$$v_2' = \frac{v_2 - u}{1 - v_2 u/c^2} = \frac{(-v) - (v)}{1 - (-v)v/c^2} = \frac{-2v}{1 + v^2/c^2}$$

The velocity of the combined mass $2m$ is

$$V' = \frac{V - u}{1 - Vu/c^2} = \frac{0 - v}{1 - 0(v)/c^2} = -v$$

Here then is the experiment as viewed by O':

According to O, the linear momentum before and after the collision are

$$p_{\text{initial}} = m_1 v_1 + m_2 v_2 = mv + m(-v) = 0$$

$$p_{\text{final}} = (2m)(V) = 0$$

According to O',

$$p_{\text{initial}}' = m_1 v_1' + m_2 v_2' = m(0) + m\left(\frac{-2v}{1 + v^2/c^2}\right) = \frac{-2mv}{1 + v^2/c^2}$$

$$p_{\text{final}}' = 2mV' = 2m(-v) = -2mv$$

and, since O' measures $p_{\text{initial}}' \neq p_{\text{final}}'$, linear momentum is *not* conserved according to O'.

According to our previous discussion, we would like to try to preserve conservation of linear momentum in *all* reference frames. We know we have handled the velocities correctly, and since linear momentum involves only mass and velocity, the fault must lie in our handling of the masses. In analogy with our discussion of length contraction and time dilation, let us assume that there is a *relativistic mass increase*, according to

$$m = \frac{m_0}{\sqrt{1 - u^2/c^2}} = \gamma m_0 \qquad (2.10) \qquad \textbf{Relativistic mass}$$

where m_0 is called the *rest mass* and is, like the proper length and proper time, measured in a frame of reference in which the object is at rest. In all other reference frames, the *relativistic mass* m will be larger than m_0. (Notice how this resolves our previous problem about accelerating objects beyond the speed of light—as the object approaches the speed of light, its mass becomes ever larger, and thus the force becomes less effective in producing an acceleration. As the mass becomes infinite, no acceleration can be produced by a finite force, and so we can never reach or exceed the speed of light.)

Let us examine how this definition of the relativistic mass preserves momentum conservation in the O and O' frames of reference. We denote the masses measured by O as m_1, m_2, and M (the combined mass), and those measured by O' as m_1', m_2', and M'. We assume the objects to have identical *rest masses* m_0. Then the masses according to O are

$$m_1 = \frac{m_0}{\sqrt{1 - v^2/c^2}} \quad \text{and} \quad m_2 = \frac{m_0}{\sqrt{1 - v^2/c^2}}$$

since $v_1 = v_2 = v$; also

$$M = m_1 + m_2 = \frac{2m_0}{\sqrt{1 - v^2/c^2}}$$

Since the combined mass is at rest in this reference frame, the mass M is just its rest mass, and we will denote this by M_0. According to O', m_1' is at rest, so $m_1' = m_0$. Since m_2' is moving at a speed of $v_2' = -2v/(1 + v^2/c^2)$,

$$m_2' = \frac{m_0}{\sqrt{1 - \frac{1}{c^2}\left(\frac{-2v}{1 + v^2/c^2}\right)^2}} = m_0 \left(\frac{1 + v^2/c^2}{1 - v^2/c^2}\right)$$

The combined mass M' is moving at $V' = -v$, so

$$M' = \frac{M_0}{\sqrt{1 - v^2/c^2}}$$

If we substitute our result $M_0 = 2m_0/\sqrt{1 - v^2/c^2}$, we obtain

$$M' = \frac{2m_0}{1 - v^2/c^2}$$

It is apparent that this new definition of mass will still conserve momentum according to O, since $p_{\text{initial}} = m_1 v_1 + m_2 v_2$ is still equal to zero, as is p_{final}. Let us therefore examine the initial and final momenta in the O' frame of reference:

$$p_{\text{initial}}' = m_1' v_1' + m_2' v_2'$$

$$= m_0(0) + m_0 \left(\frac{1 + v^2/c^2}{1 - v^2/c^2}\right)\left(-\frac{2v}{1 + v^2/c^2}\right)$$

$$= \frac{-2m_0 v}{1 - v^2/c^2}$$

and

$$p_{\text{final}}' = M'V' = \frac{2m_0}{1 - v^2/c^2}(-v) = \frac{-2m_0 v}{1 - v^2/c^2}$$

Hence $p_{\text{initial}}' = p_{\text{final}}'$, and our definition of the relativistic mass has enabled us to conserve linear momentum in both frames of reference. In fact, this definition of the relativistic mass gives momentum conservation in *all* frames of reference, not just the two particular frames we have considered in this example.

Instead of defining the relativistic mass as we did above, we could have simply redefined the relativistic momentum to be

$$p = \frac{m_0 v}{\sqrt{1 - v^2/c^2}} \quad (2.11) \quad \textbf{Relativistic momentum}$$

This definition is in fact the better one, for several reasons: it is easily extended to two or three dimensions and it avoids the confusion of using the relativistic mass in cases where it doesn't apply. Consider the following experiment. Two masses m_1 and m_2 separated by a distance r attract one another according to the law of gravitation, $F = Gm_1m_2/r^2$. The masses are connected by a spring scale, which records the force between them. Observer O' in a rocket ship travels away from the two masses, perpendicular to the line connecting m_1 and m_2. If we simply insert the relativistic expression for the mass into the classical expression for the force, we would conclude that O and O' would observe different readings on the same scale. This is clearly impossible! As we show as follows, it is *wrong* to treat all dynamical equations as we did the momentum equation—merely substituting the relativistic mass for the classical mass. In particular, it is *incorrect* to write the kinetic energy as $\frac{1}{2}mv^2$ using the relativistic mass.

The kinetic energy is defined in classical physics as the work done by an external force that changes the speed of an object. We keep the same definition in relativistic mechanics (restricting ourselves to one dimension). The change in kinetic energy $\Delta K = K_f - K_i$ is

$$\Delta K = W = \int F \, dx$$

If the object starts from rest, where $K_i = 0$, the final kinetic energy K is

$$K = \int F \, dx$$

We have not yet treated forces from the relativistic point of view, and so we are not yet certain how to proceed. Without proof or justification, we state that we will try to preserve Newton's second law in its most general form $(F = dp/dt)$ as the proper relativistic dynamical relationship (keeping in mind that we have already redefined p and therefore will surely need to redefine F). Thus we have

$$K = \int \frac{dp}{dt} \, dx = \int dp \, \frac{dx}{dt} = \int v \, dp$$

This last result can be modified if we use the standard technique of integrating by parts, with $d(pv) = v \, dp + p \, dv$

$$K = pv - \int_{v=0}^{v=v} p \, dv$$

$$= \frac{m_0 v}{\sqrt{1 - v^2/c^2}} \, v - \int_{v=0}^{v=v} \frac{m_0 v}{\sqrt{1 - v^2/c^2}} \, dv$$

Performing the integration, we have

$$K = \frac{m_0 v^2}{\sqrt{1 - v^2/c^2}} + m_0 c^2 \sqrt{1 - v^2/c^2} - m_0 c^2$$

which we can write in the following form

$$K = mc^2 - m_0 c^2 \qquad (2.12) \quad \textbf{Relativistic kinetic energy}$$

with the relativistic mass defined as in Equation (2.10). This equation gives us a fundamental result for the *relativistic kinetic energy:* The difference between the quantity mc^2 (which has units of energy) for a particle in motion at speed v, and the quantity m_0c^2 (which also has units of energy) for a particle at rest, is just the kinetic energy. It is in fact *exactly* what we have always meant by kinetic energy—the additional energy that a particle has due only to its motion. The quantity m_0c^2 is called the *rest energy* of the particle and is denoted by E_0. A particle in motion, then, has an energy E_0 plus an additional energy K, so its *relativistic total energy E* is just

$$E = E_0 + K = m_0c^2 + K = mc^2 \qquad (2.13)$$

Relativistic total energy

This is Einstein's famous result that the energy of an object is merely another measure of the mass of the object—energy and mass are equivalent, and a gain or loss of energy can be equally well regarded as a gain or loss of mass.

Suppose we enclose an ordinary 100-W light bulb and a power source for it in a sealed transparent container suspended from a very sensitive balance. Compute the change in mass if the bulb burns continuously for one year.

EXAMPLE 2.13

$$100 \text{ W} = 100 \text{ J/s}$$

Solution

Since one year $\cong \pi \times 10^7$ s (a useful factor to remember), in one year the energy radiated is about 3×10^9 J. Then

$$\Delta m_0 = \frac{\Delta E_0}{c^2} = \frac{3 \times 10^9}{9 \times 10^{16}} \cong 3 \times 10^{-8} \text{ kg}$$

At a distance equal to the radius of the Earth's orbit (1.5×10^{11} m), the sun's radiation has an intensity of about 1.4×10^3 W/m². Find the rate at which the sun is losing mass.

EXAMPLE 2.14

If we assume that the radiation from the sun is distributed uniformly over a sphere of radius 1.5×10^{11} m, then the total radiative power emitted by the sun is

Solution

$$4\pi(1.5 \times 10^{11} \text{ m})^2(1.4 \times 10^3 \text{ W/m}^2) = 4 \times 10^{26} \text{ W} = 4 \times 10^{26} \text{ J/s}$$

A rest energy loss ΔE_0 of 4×10^{26} J corresponds to a change in rest mass of

$$\Delta m_0 = \frac{\Delta E_0}{c^2} = \frac{4 \times 10^{26}}{9 \times 10^{16}} = 4 \times 10^9 \text{ kg}$$

Thus the sun loses rest mass at a rate of about 4 billion kilograms per second! If this rate were to remain constant, the sun (with a present mass of about 2×10^{30} kg) will shine "only" for another 10^{13} years.

Returning to the example of the collision we considered previously, show that total relativistic energy is conserved in both reference frames.

EXAMPLE 2.15

According to O, the energies are:

$$E_{m_1} = m_1 c^2 = \frac{m_0 c^2}{\sqrt{1 - v^2/c^2}}$$

$$E_{m_2} = m_2 c^2 = \frac{m_0 c^2}{\sqrt{1 - v^2/c^2}}$$

$$E_M = M_0 c^2 = \frac{2m_0 c^2}{\sqrt{1 - v^2/c^2}}$$

Thus

$$E_{m_1} + E_{m_2} = E_M$$

and energy is conserved.

According to O',

$$E'_{m_1} = m_0 c^2$$

$$E'_{m_2} = m_0 c^2 \left(\frac{1 + v^2/c^2}{1 - v^2/c^2}\right)$$

$$E'_M = \frac{2m_0 c^2}{1 - v^2/c^2}$$

$$E'_{m_1} + E'_{m_2} = m_0 c^2 + m_0 c^2 \left(\frac{1 + v^2/c^2}{1 - v^2/c^2}\right)$$

$$= m_0 c^2 \left(1 + \frac{1 + v^2/c^2}{1 - v^2/c^2}\right) = \frac{2m_0 c^2}{1 - v^2/c^2}$$

Thus

$$E'_{m_1} + E'_{m_2} = E'_M$$

and total relativistic energy is conserved in the O' frame also.

It is left as an exercise (Problem 17) to derive an important relationship between relativistic energy and momentum:

$$E^2 = p^2 c^2 + (m_0 c^2)^2 \tag{2.14}$$

This is one of the most useful expressions for relating relativistic energy and momentum, and is easily remembered as the Pythagorean theorem for a right triangle whose sides are pc and $m_0 c^2$ and whose hypotenuse is E.

Find the total energy, kinetic energy, and momentum of a proton $(m_0 c^2 = 938 \text{ MeV})$ moving at $v/c = 0.800$.

EXAMPLE 2.16

We find the total energy from Equation (2.13):

Solution

$$E = mc^2 = \frac{m_0 c^2}{\sqrt{1 - v^2/c^2}} = \frac{938 \text{ MeV}}{\sqrt{1 - (0.8)^2}} = 1563 \text{ MeV}$$

The kinetic energy follows directly:

$$K = E - m_0 c^2 = 1563 \text{ MeV} - 938 \text{ MeV} = 625 \text{ MeV}$$

(This is very different from the classical kinetic energy $\frac{1}{2}mv^2$, which gives 300 MeV.) The momentum can be found either from (2.11) or (2.14). We will use both expressions in order to show a way of simplifying the units. First we use Equation (2.11), which requires the rest mass m_0:

$$m_0 = \frac{938 \text{ MeV}}{c^2} = \frac{(938 \times 10^6 \text{ eV})(1.6 \times 10^{-19} \text{ J/eV})}{9 \times 10^{16} \text{ m}^2/\text{s}^2} = 1.67 \times 10^{-27} \text{ kg}$$

From (2.11)

$$p = \frac{(1.67 \times 10^{-27} \text{ kg})(0.8 \times 3 \times 10^8 \text{ m/s})}{\sqrt{1 - (0.8)^2}} = 6.68 \times 10^{-19} \text{ kg·m/s}$$

We could, as an alternative, use Equation (2.14) to solve for cp:

$$cp = \sqrt{E^2 - (m_0c^2)^2} = \sqrt{(1563 \text{ MeV})^2 - (938 \text{ MeV})^2} = 1250 \text{ MeV}$$

We could convert this into more familiar units by dividing by the numerical value of c, and of course we should obtain the same result we did using Equation (2.11) directly. However, in nearly all applications, we can use the product cp rather than p itself, and so there is no need to change units. Dividing on both sides by the *symbol* c (*not* by its numerical value) we find

$$p = 1250 \text{ MeV}/c$$

The units MeV/c or eV/c are perfectly acceptable for momentum.

We could also have found this result directly from Equation (2.11) with a bit of manipulation:

$$p = \frac{m_0v}{\sqrt{1 - v^2/c^2}} = \frac{1}{c} \frac{(m_0c^2)(v/c)}{\sqrt{1 - v^2/c^2}} = \frac{1}{c} \frac{(938 \text{ MeV})(0.8)}{\sqrt{1 - (0.8)^2}} - 1250 \text{ MeV}/c$$

These units may look unusual and artificial, but with a bit of practice you will learn how much of a convenience they can be.

Find the velocity and momentum of an electron with a kinetic energy of 10.0 MeV.

EXAMPLE 2.17

$$E = K + m_0c^2 = 10.0 \text{ MeV} + 0.511 \text{ MeV} = 10.5 \text{ MeV}$$

$$m = E/c^2 = 10.5 \text{ MeV}/c^2$$

Rewriting Equation (2.10)

$$\frac{v}{c} = \sqrt{1 - \left(\frac{m_0}{m}\right)^2} = \sqrt{1 - \left(\frac{0.511}{10.5}\right)^2} = 0.9988$$

$$cp = \sqrt{E^2 - (m_0c^2)^2} = \sqrt{(10.5)^2 - (0.511)^2} = 10.49 \text{ MeV} \cong 10.5 \text{ MeV}$$

$$p = 10.5 \text{ MeV}/c$$

Notice that, when $v \cong c$, the total relativistic energy E is very nearly equal to cp. This is known as the *extreme relativistic* approximation.

To summarize this section, we can preserve the following basic concepts of classical physics:

1. Conservation of energy.
2. Conservation of linear momentum.
3. Newton's second law, $F = dp/dt$.

if we introduce the following new relativistic concepts:

1. $p = m_0 v / \sqrt{1 - v^2/c^2}$
2. $m = m_0 / \sqrt{1 - v^2/c^2}$
3. $E = mc^2 = m_0 c^2 + K = (p^2 c^2 + m_0^2 c^4)^{1/2}$

Equations of relativistic dynamics

These relationships are the essential features of relativistic dynamics, which we will have occasion to use later in the text. As with all of our relativistic equations, kinematic as well as dynamic, when v is small compared with c, they must give the familiar results of classical physics. In particular, $K \cong \frac{1}{2} m_0 v^2$ when $v \ll c$. Problem 20 shows how to obtain this important result from Equation (2.13).

In this section, we consider two of the more challenging and perhaps irksome of the consequences of special relativity. The first concerns the notion of simultaneity and the synchronization of clocks. For most of us synchronizing our watches or clocks is not a difficult process; for example, we can set our watches by looking at the nearest clock. However, this method ignores the time that it takes the light from the face of the clock to travel to our eye so we can set our watches. If we are 1 m from the clock, our watch will be behind by about 3 ns (3×10^{-9} s). Although this short time lag may not make you late for your physics lecture, it can be a serious problem for the experimental physicist, for whom the measurement of time intervals smaller than 1 ns is routine. Let us therefore try to be more precise. We will build a device similar to that shown in Figure 2.13. Two clocks are located at $x = 0$ and $x = L$. A flashbulb is located precisely at $x = L/2$, and the clocks are set running when they receive the flash of light. Since the light takes precisely the same interval of time to reach the two clocks, the clocks start together precisely at a time $L/2c$ after the flash is emitted, and the clocks are exactly synchronized.

Now let us examine the same situation from the point of view of the moving observer O'. If you have been paying attention this far into this chapter, you are probably suspecting that O' does not agree that the clocks are synchronized. In the frame of reference of O, two events occur: the receipt of a light signal by clock 1 at $x_1 = 0, t_1 = L/2c$ and the receipt of a light signal by clock 2 at $x_2 = L, t_2 = L/2c$. Using the expressions for the Lorentz transformation, we find that O' observes clock 1 to receive its signal at

$$t_1' = \frac{t_1 - (u/c^2)x_1}{\sqrt{1 - u^2/c^2}} = \frac{L/2c}{\sqrt{1 - u^2/c^2}}$$

2.6 SIMULTANEITY AND THE TWIN PARADOX

Clock synchronization

FIGURE 2.13 A flash of light, emitted from a point midway between the two clocks, starts the two clocks simultaneously according to O. Observer O' sees clock 2 start ahead of clock 1.

while clock 2 receives its signal at

$$t_2' = \frac{t_2 - (u/c^2)x_2}{\sqrt{1 - u^2/c^2}} = \frac{L/2c - (u/c^2)L}{\sqrt{1 - u^2/c^2}}$$

Thus t_2' is smaller than t_1' and clock 2 appears to receive its signal earlier than clock 1, so that the clocks start at times that differ by

$$\Delta t' = t_1' - t_2' = \frac{uL/c^2}{\sqrt{1 - u^2/c^2}} \tag{2.15}$$

according to O'. It is important to keep in mind that this is *not* a time dilation effect—time dilation comes from the *first* term of the Lorentz transformation equation (2.8d) for t', while the lack of synchronization arises from the *second* term. O' observes *both* clocks to run slow, due to time dilation; O' *also* observes clock 2 to be a bit ahead of clock 1. The time $\Delta t'$ measured by O' between the start of the clocks gives, using Equation (2.4), $\Delta t = u\,L/c^2$ for the reading of clock 2 when O sees clock 1 read 0.

We therefore reach the following conclusion: two events that are simultaneous in one reference frame are not simultaneous in another reference frame moving with respect to the first, unless the two events occur at the same point in space. (In our previous example, if $L = 0$, so that both clocks are at the origin, then they are synchronized in all reference frames.) Clocks that appear to be synchronized in one frame of reference will not necessarily be synchronized in another frame of reference in relative motion.

We now turn briefly to what has become known as the twin paradox. Suppose there is a pair of twins on Earth. One, whom we shall call Casper, remains on Earth, while his twin sister Amelia sets off in a rocket ship on a trip to a distant planet. Casper, based on his understanding of special relativity, knows that his sister's clocks will run slow relative to his own and that therefore she should be younger than he when she returns; this is just what our discussion of time dilation would suggest. However, recalling that discussion, we know that for two observers in relative motion, *each* thinks *the other's* clocks are running slow. We could therefore study this problem from the point of view of Amelia, according to whom Casper and the Earth (accompanied by the solar system and galaxy) make a round-trip journey away from her and back again. Under such circumstances, she will think it is her brother's clocks (which are now in motion relative to her own) that are running slow, and will therefore expect her brother to be younger than she when they meet again. While it is possible to disagree over whose clocks are running slow relative to his or her own, which is merely a problem of frames of reference, when Amelia returns to Earth (or when the Earth returns to Amelia), all observers must agree as to which twin has aged less rapidly. This is the paradox—each twin expects the other to be younger.

The resolution of this paradox lies in considering the asymmetric role of the two twins. The laws of special relativity apply only to inertial frames, those moving relative to one another at constant velocity. We may supply Amelia's rockets with sufficient thrust so that they accelerate for a very short length of time, bringing the ship to a speed at which it can coast to the planet, and thus during her outward journey Amelia

Twin paradox

spends all but a negligible amount of time in a frame of reference moving at constant speed relative to Casper. However, in order to return to Earth, she must decelerate and reverse her motion. Although this also may be done in a very short time interval, Amelia's return journey occurs in a completely different inertial frame than her outward journey. It is Amelia's jump from one inertial frame to another that causes the asymmetry in the ages of the twins. Only Amelia has the necessity of jumping to a new inertial frame to return, and therefore *all observers will agree* that it is Amelia who is "really" in motion, and that it is her clocks that are "really" running slow; therefore she is indeed the younger twin on her return.

Let us make this discussion more quantitative with some numerical examples. We assume, as discussed above, that the acceleration and deceleration take negligible time intervals, so that all of Amelia's aging is done during the coasting. For simplicity, we will assume the distant planet is at rest relative to the Earth; this does not change the problem, but it avoids the need to introduce yet another frame of reference. Suppose the planet to be 12 light-years distant from Earth, and suppose Amelia travels at a speed of 0.6c. Then according to Casper it takes his sister 20 years (20 years × 0.6c = 12 light-years) to reach the planet and 20 years to return, and therefore she is gone for a total of 40 years. (However, Casper doesn't know his sister has reached the planet until the light signal carrying news of her arrival reaches Earth. Since light takes 12 years to make the journey, it is 32 years after her departure when Casper sees his sister's arrival at the planet. Eight years later she returns to Earth.) From the frame of reference of Amelia aboard the rocket, the distance to the planet is contracted by a factor of $\sqrt{1 - (0.6)^2} = 0.8$, and is therefore $0.8 \times 12 = 9.6$ light-years. At a speed of 0.6c, Amelia will measure 16 years for the trip to the planet, for a total round trip time of 32 years. Thus Casper ages 40 years while Amelia ages only 32 years and is indeed the younger on her return. We can confirm this analysis by having Casper send a light signal each year, on his birthday, to his sister. We know that the frequency of the signal as received by Amelia will be Doppler shifted. During the outward journey, she will receive signals at the rate of $(1/\text{year}) \times \sqrt{(1 - u/c)/(1 + u/c)} = (1/\text{yr}) \sqrt{0.4/1.6} = 0.5/\text{yr}$. During the return journey, the Doppler-shifted rate will be $(1/\text{yr}) \sqrt{(1 + u/c)/(1 - u/c)}$ or $2/\text{yr}$. Thus for the first 16 years, during Amelia's trip to the planet, she receives 8 signals, and during the return trip of 16 years, she receives 32 signals, for a total of 40. She receives 40 signals, indicating her brother has celebrated 40 birthdays during her 32-year journey.

It is left as an exercise (Problem 29) to consider the situation if it is Amelia who is sending the signals.

2.7 EXPERIMENTAL TESTS OF SPECIAL RELATIVITY

Nonexistence of the Ether   We have already discussed the Michelson-Morley experiment and its relationship to special relativity. Over the nearly 100 years since the original experiment, the basic experiment has been repeated many times with a number of variations and with continuously improved sensitivity. In all experiments, no evidence for an ether or for the variation of the speed of light with direction in space has yet been detected, in spite of the fact that the sensitivity of the experiments has been extended more than an order of magnitude beyond that of the original experiment.

In the following, we discuss some of the other experiments that have confirmed the predictions of special relativity.

Time Dilation   We have discussed the time dilation effect on the decay of muons produced by cosmic rays. As another example, the decay of rapidly moving elementary particles can be studied in the laboratory. One such particle, the pi meson, has a lifetime of about 26×10^{-9} s (26 ns). This is a very convenient time range for laboratory experiments—it is long enough so that the pi meson, produced in collisions between other particles, can be brought to rest before it decays, permitting a measurement of the *proper* lifetime; the lifetime is also short enough so that a pi meson moving at a speed approaching the speed of light will not travel further than the dimensions of a reasonable laboratory (10 to 20 m) in the time before it decays. A measurement of the proper lifetime (bringing the pi mesons to rest) gave the value of 26.0 ns. Pi mesons in flight at $v/c = 0.913$ were measured to have a lifetime of 63.7 ns in the laboratory frame of reference. This lifetime is longer than the proper lifetime because of time dilation in the reference frame of the moving pi mesons. The time dilation factor is $(1 - v^2/c^2)^{1/2} = 0.408$, so a measured lifetime of 63.7 ns is equivalent to a decay lifetime of $63.7 \times 0.408 = 26.0$ ns in the reference frame of the pi mesons. This is identical with the value measured for pi mesons at rest, and thus the time dilation effect is confirmed [see D. S. Ayres et al., *Physical Review D 3*, 1051 (1971)].

Relativistic Mass and Energy   Nearly every time the nuclear or particle physicist enters the laboratory, a direct or indirect test of the mass-energy relationships of special relativity is made. We briefly discuss some experiments done with electrons that were performed specifically to test the special relativity mass-energy relationships, and also discuss two other examples from nuclear and particle physics.

The earliest direct confirmation of special relativity came just a few years after Einstein's 1905 paper. The increase of mass with velocity, predicted by Equation (2.10), was tested by measuring the momentum and velocity of the high-energy electrons emitted in certain radioactive decay processes (nuclear beta decay—see Chapter 9). Figure 2.14 shows the results of many different investigations, which are in perfect agreement with the expected mass increase. Other more recent measurements, in which the kinetic energies of fast electrons were measured, are shown in Figure 2.15. Once again, the data at large velocities are in agreement with the equations of special relativity and in disagreement with the equations of classical, nonrelativistic mechanics.

As an additional example, we consider the atom of deuterium, or "heavy hydrogen," which consists of an atom of ordinary hydrogen with a single added neutron. The sum of the hydrogen and neutron rest masses is as follows:

$$m_H + m_n = (1.67356 \times 10^{-27} \text{ kg}) + (1.67496 \times 10^{-27} \text{ kg})$$
$$= 3.34852 \times 10^{-27} \text{ kg}$$

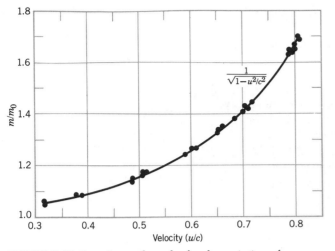

FIGURE 2.14 Experimental results for the variation of mass with velocity. The experimental data were obtained by measuring the momentum of electrons from the radii of curvature of their paths in a magnetic field. The agreement between the data and the prediction of special relativity is excellent. (*Source.* Resnick, *Basic Concepts in Relativity and Early Quantum Theory.*)

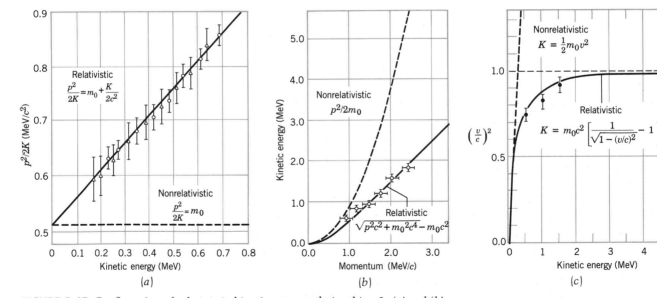

FIGURE 2.15 Confirmation of relativistic kinetic energy relationships. In (*a*) and (*b*) the momentum and energy of radioactive decay electrons were measured simultaneously. In these two independent experiments, the data were plotted in different ways, but the results are clearly in good agreement with the relativistic relationships and in poor agreement with the classical, nonrelativistic relationships. In (*c*) electrons were accelerated to a fixed energy through a large electric field (up to 4.5 million volts, as shown) and the velocities of the electrons were determined by measuring the flight time over 8.4 m. Notice that at small kinetic energies ($K \ll m_0 c^2$), the relativistic and nonrelativistic relationships become identical. [*Sources.* (*a*) K. N. Geller and R. Kollarits, *Am. J. Phys. 40,* 1125 (1972); (*b*) S. Parker, *Am. J. Phys. 40,* 241 (1972); (*c*) W. Bertozzi, *Am. J. Phys. 32,* 551 (1964).]

When the mass of the deuterium atom is measured directly, the result is:

$$m_D = 3.34455 \times 10^{-27} \text{ kg}$$

This is a case of the whole being less than the sum of its parts, by an amount $\Delta m = 0.00397 \times 10^{-27}$ kg (about four times the mass of an electron). This is equivalent to an energy $\Delta E = (\Delta m)c^2 = 2.23$ MeV, known as the *binding energy* of deuterium. In order to break up a deuterium atom into hydrogen plus a neutron, we would need to supply 2.23 MeV of energy, which would in the process of the breakup be converted into mass. The conversion of mass into energy, or more properly the conversion of mass energy into kinetic energy (as, for example, in a nuclear reactor or the interior of the sun) and the conversion of kinetic energy into mass energy (as when a high-energy physicist "manufactures" particles in the laboratory) are common occurrences in the physics laboratory, much as the conversion of chemical energy to heat energy.

As another example of the conversion of energy to mass, pi mesons with rest energies of about 140 MeV (about 274 times as massive as the electron) do not exist in nature under normal circumstances, but they are routinely produced at high-energy accelerators in collisions between ordinary particles such as protons, as shown in the following diagram.

In this process, about 140 MeV of kinetic energy of the protons is converted into mass energy of the pi meson.

Constancy of the Speed of Light
If the speed of light did indeed depend on the motion of the source or observer, we might represent this by $c' = c + ku$, where c is the speed of light in the rest frame of the source, c' is the speed measured in the moving frame of reference, and u is the relative speed of the two reference frames. Variable k is a number to be determined by experiment; according to special relativity, k is 0, while according to Galilean relativity, k is equal to 1.

One experiment of this type is to study the X rays emitted by a binary pulsar, a rapidly pulsating source of X rays in orbit about another star, which would eclipse the pulsar as it rotated in its orbit. If the speed of light (in this case, X rays) were to change as the pulsar moved first toward and later away from the Earth in its orbit, the beginning and end of the eclipse would not be equally spaced in time from the midpoint of the eclipse. No such effect is observed, and from observations of such systems one concludes that $k < 2 \times 10^{-9}$, in agreement with predictions of special relativity. Those observations were done at $u/c \cong 10^{-3}$ [see K. Brecher, *Physical Review Letters* 39, 1051 (1977)].

In another experiment of this type, the decay into gamma rays (another form of electromagnetic waves traveling at c) of the pi mesons is observed. When pi mesons (produced in laboratories with large accelerators) emit these gamma rays, they are traveling at speeds close to the speed of light, relative to the laboratory. Thus if Galilean relativity were valid, we should expect to find gamma rays emitted in the direction of motion of the decaying pi mesons traveling at a speed c' in the laboratory of nearly $2c$, rather than always with c as predicted by special relativity. The observed laboratory speed of these gamma rays in one experiment was $(2.9977 \pm 0.0004) \times 10^8$ m/s when the decaying pi mesons were moving at $u/c = 0.99975$. Thus k is zero within experimental uncertainty, and $c' = c$ as expected from special relativity. This experiment shows directly that an object moving at a speed of nearly c relative to the laboratory emits "light" which travels at a speed of c relative to both the object *and* the laboratory, giving direct evidence for Einstein's second postulate [see T. Alvager et al., *Physics Letters 12*, 260 (1964)].

Twin Paradox Although we cannot perform the experiment to test the twin paradox as we have described it, we can do an equivalent experiment. We take two clocks in our laboratory and synchronize them carefully. We then place one of the clocks in an airplane and fly it around the Earth. When we return the clock to the laboratory and compare the two clocks, we expect to find, if special relativity is correct, that the clock that has left the laboratory is the "younger" one—that is, it will tick away fewer seconds and appear to run behind its stationary twin. In this experiment, we use very precise clocks based on the atomic vibrations of cesium in order to measure the time differences between the clock readings, which amount to only about 10^{-7} s. This experiment is complicated by several factors, all of which can be computed rather precisely: the rotating Earth is *not* an inertial frame (there is a centripetal acceleration), clocks on the surface of the Earth are *already* moving because of the rotation of the Earth, and the *general* theory of relativity predicts that a change in the gravitational field strength, which our moving clock will experience as it changes altitude in its airplane flight, will also change the rate at which the clock runs. In this experiment, as in the others we have discussed, the results are entirely in agreement with the predictions of special relativity [see J. C. Hafele and R. E. Keating, *Science 177*, 166 (1972)].

SUGGESTIONS FOR FURTHER READING

Special relativity has perhaps been the subject of more popular, nonmathematical books than any other area of science. Here are a few you can read for fun and to broaden your background:

L. Barnett, *The Universe and Dr. Einstein* (New York, Time Inc., 1962).

G. Gamow, *Mr. Tompkins in Paperback* (Cambridge, Cambridge University Press, 1967). Especially recommended; the master of popular science writing takes us on a fanciful journey to a world where c is so small that special relativistic effects are commonplace.

L. Marder, *Time and the Space Traveler* (Philadelphia, University of Pennsylvania Press, 1971).

B. Russell, *The ABC of Relativity* (New York, New American Library, 1958).

J. T. Schwartz, *Relativity in Illustrations* (New York, New York University Press, 1962).

Other introductions to relativity, more complete but not particularly more difficult than the present level, are the following:

A. P. French, *Special Relativity* (New York, Norton, 1968).

R. Resnick, *Basic Concepts in Relativity and Early Quantum Theory* (New York, Wiley, 1972).

The best is saved for last:

E. F. Taylor and J. A. Wheeler, *Spacetime Physics* (San Francisco, Freeman, 1966). Elegant and witty; all of the relativity paradoxes you could ever want, carefully explained and diagrammed, with many worked examples.

QUESTIONS

1. Explain in your own words what is meant by the term "relativity." Are there different theories of relativity?

2. Does the Michelson-Morley experiment show that the ether does not exist or that it is merely unnecessary?

3. Suppose we made a pair of shears in which the cutting blades were many orders of magnitude longer than the handle. Let us in fact make it so long that, when we move the handles at angular velocity ω, a point on the tip of the blade has a tangential velocity $v = \omega r$ that is greater than c. Does this contradict special relativity? Justify your answer.

4. Light travels through water at a speed of about 2.25×10^8 m/s. Is it possible for a particle to travel through water at a speed v greater than 2.25×10^8 m/s?

5. Is it possible to have particles that travel at the speed of light? What does Equation (2.10) require of such particles?

6. Explain in your own words the terms *time dilation* and *length contraction*.

7. How large does the moon appear to a space traveler approaching it at $v = 0.99c$?

8. According to the time dilation effect, would the life expectancy of someone who lives at the equator be longer or shorter than someone who lives at the North Pole? By how much?

9. Criticize the following argument. "Here is a way to travel faster than light. Suppose a star is 10 light-years away. A radio signal sent from Earth would need 20 years to make the round trip to the star. If I were to travel to the star in my rocket at $v = 0.8c$, to me the distance to the star is contracted by $\sqrt{1 - (0.8)^2}$ to 6 light-years, and at that speed it would take me 6 light-years/$0.8c$ = 7.5 years to travel there. The round trip takes me only 15 years, and therefore I travel faster than light, which takes 20 years."

10. Is it possible to synchronize clocks that are in motion relative to each other? Try to design a method to do so. Which observers will believe the clocks to be synchronized?

11. In a rocket moving at high speed, would the travelers feel their own relativistic mass increase?

12. Which is more massive, an object at low temperature or the same object at high temperature? A spring at its natural length or the same spring under compression? A container of gas at low pressure or at high pressure? A charged capacitor or an uncharged one?

13. The data of Figure 2.14 were obtained by measuring the charge-to-mass ratio

e/m of moving electrons. (a) Why do we assume that the mass changes with velocity but the electric charge doesn't? (b) What properties of nature would be different if there were a relativistic transformation law for electric charge? (c) What experiments could be done to prove that electric charge does *not* change with velocity?

PROBLEMS

1. A shift of one fringe in the Michelson-Morley experiment corresponds to a change in the round-trip travel time along one arm of the interferometer by one period of vibration of light (about 2×10^{-15} s) when the apparatus is rotated by 90°. Using the results of Example 2.2, what velocity through the ether would be deduced from a shift of one fringe? (Take the length of the interferometer arm to be 11 m.)

2. A meter stick of length L_0 extends in the O' reference frame from x_1' to x_2'. An observer O (relative to whom O' moves at speed u) measures the length of the stick to be $L = x_2 - x_1$ by making a *simultaneous* measurement of x_1 and x_2. Use the Lorentz transformation equations to derive the length contraction relating L and L_0.

3. A light bulb in the frame of reference of O at point x blinks on and off at intervals $\Delta t = t_2 - t_1$. Observer O', moving relative to O at speed u, measures the interval to be $\Delta t' = t_2' - t_1'$. Use the Lorentz transformation expressions to derive the time dilation expression relating Δt and $\Delta t'$.

4. The distance from New York to Los Angeles is about 5000 km and should take about 50 h in a car driving at 100 km/h. (a) How much shorter than 5000 km is the distance to the car travelers? (b) How much less than 50 h do they age during the trip? (c) How much does the mass of the 1000-kg car increase during the trip?

5. How fast must an object move before its length is contracted to one-half its proper length?

6. An astronaut must journey to a distant planet, which is 200 light-years from Earth. What speed will be necessary if the astronaut wishes to age only 10 years during the round trip?

7. The proper lifetime of a certain particle is 100.0 ns. (a) How long does it live in the laboratory if it moves at $v = 0.960c$? (b) How far does it travel in the laboratory during that time? (c) How far does it travel in its own frame of reference?

8. High-energy particles are observed in laboratories by photographing the tracks they leave in certain detectors; the length of the track depends on the speed of the particle and its lifetime. A particle moving at $0.995c$ leaves a track 1.25 mm long. What is the proper lifetime of the particle?

9. A "cause" occurs at point 1 (x_1, t_1) and its "effect" occurs at point 2 (x_2, t_2). Use the Lorentz transformation to find $t_2' - t_1'$, and show that $t_2' - t_1' > 0$; that is, O' can never see the "effect" coming before its "cause."

10. Derive the Lorentz velocity transformations for v_x' and v_z'.

11. Observer O fires a light beam in the y direction $(v_y = c)$. Use the Lorentz velocity transformation to find v_x' and v_y' and show that O' also measures the value c for the speed of light. Assume O' moves relative to O with velocity u in the x direction.

12. In Example 2.5 the lining up of the ends of the train with the ends of the platform was simultaneous to O but not to O'. Show that the numerical results from that example are consistent with Equation (2.15).

13. Several spacecraft leave a space station at the same time. Relative to an observer on the station, A travels at $0.60c$ in the x direction, B at $0.50c$ in the y

direction, C at $0.50c$ in the negative x direction, and D at $0.50c$ at $45°$ between the y and negative x directions. Find the velocity components, directions, and speeds of B, C, and D as observed from A.

14. Supply the missing steps in the derivation of Equation (2.12).

15. For what range of velocities of a particle of rest mass m_0 can we use the classical expression for kinetic energy $\frac{1}{2}m_0v^2$ to within an accuracy of 1 percent?

16. For what range of velocities of a particle of rest mass m_0 can we use the extreme relativistic approximation $E \cong pc$ to within an accuracy of 1 percent?

17. Use Equations (2.11) and (2.13) to derive Equation (2.14).

18. Suppose an observer O measures a particle of rest mass m_0 moving in the x direction to have speed v, energy E, and momentum p. Observer O', moving at speed u in the x direction, measures v', E', and p' for the same object. (a) Use the Lorentz velocity transformation to find E' and p' in terms of m_0, u, and v. (b) Reduce $E'^2 - (p'c)^2$ to its simplest form and interpret the result.

19. Repeat Problem 18 for the mass moving in the y direction according to O. The velocity u of O' is still along the x direction.

20. Use the binomial expansion $(1 + x)^n = 1 + nx + [n(n - 1)/2!]x^2 + \cdots$ to show that Equation (2.12) for the relativistic kinetic energy reduces to the classical expression $\frac{1}{2}m_0v^2$ when $v \ll c$. This important result shows that our familiar expressions are correct at low speeds. By evaluating the first term in the expansion beyond $\frac{1}{2}m_0v^2$, find the speed necessary before the classical expression is off by 0.01 percent.

21. What is the change in mass when 1 g of copper is heated from 0 to $100°C$? The specific heat of copper is 0.40 J/g·K.

22. Find the kinetic energy of an electron moving at a speed of (a) $v = 1.00 \times 10^{-4}c$; (b) $v = 1.00 \times 10^{-2}c$; (c) $v = 0.300c$; (d) $v = 0.999c$.

23. The 3-km-long accelerator at Stanford University accelerates electrons to 25 GeV. (a) What is the speed of a 25-GeV electron? Express your result as the difference, in m/s, between the electron's speed and c. (b) How long would it take the electron to travel 3 km at that speed? (c) What is the length of the accelerator from the electron's frame of reference? (d) How long does it take, in the electron's frame of reference, to travel the 3 km?

24. Electrons are accelerated to high speeds by a two-stage machine. The first stage accelerates the electrons from rest to $v = 0.99c$. The second stage accelerates the electrons from $0.99c$ to $0.999c$. (a) How much energy does the first stage add to the electrons? (b) How much energy does the second stage add in increasing the velocity by only 0.9 percent?

25. An electron and a proton are each accelerated through a potential difference of 10.0 million volts. Find the momentum (in MeV/c) and the kinetic energy (in MeV) of each, and compare with the results of using the classical formulas.

26. Derive the relativistic expression $p^2/2K = m_0 + K/2c^2$, which is plotted in Figure 2.15a.

27. One of the strongest emission lines observed from distant galaxies comes from hydrogen and has a wavelength of 122 nm (in the ultraviolet region). (a) How fast must a galaxy be moving away from us in order for that line to be observed in the visible region at 366 nm? (b) What would be the wavelength of the line if that galaxy were moving toward us at the same speed?

28. A physics professor claims in court that the reason he went through the red light ($\lambda = 650$ nm) was that, due to his motion, the red color was Doppler shifted to green ($\lambda = 550$ nm). How fast was he going?

29. Suppose rocket traveler Amelia (Section 2.6) has a clock made on Earth.

Every year on her birthday she sends a light signal to brother Casper on Earth. (a) At what rate does Casper receive the signals during Amelia's outward journey? (b) At what rate does he receive the signals during her return journey? (c) How many of Amelia's birthday signals does Casper receive during the journey that he measures to last 40 years?

30. In a nuclear reactor, each atom of uranium releases about 200 MeV when it fissions. What is the change in mass when 1 kg of uranium is fissioned?

31. An electron and a positron (an antielectron) make a head-on collision, each moving at $v = 0.99999c$. In the collision the electrons disappear and are replaced by two muons ($m_0c^2 = 105.7$ MeV) which move off in opposite directions. What is the kinetic energy of each of the muons?

32. A pi meson has a rest energy of 135 MeV. It decays into two gamma rays, bursts of electromagnetic radiation that travel at the speed of light. A pi meson moving through the laboratory at $v = 0.98c$ decays into two gamma rays of equal energies, making equal angles θ with the direction of motion. Find the angle θ and the energies of the two gamma rays. (*Hint.* Gamma rays are electromagnetic radiation with $E = pc$.)

33. A helium nucleus (alpha particle) contains two protons and two neutrons and has a mass of 4.001506 u. (a) What is the binding energy of a helium nucleus? (b) What is the mass difference in kg between a helium nucleus and its constituents?

34. (a) How much energy is released when a neutron decays into a proton? (b) Part of that energy appears as the rest energy of an electron produced in the decay. Suppose the remainder of the energy goes to the electron's kinetic energy. What is the speed of the electron? (This process is an example of *nuclear beta decay*, discussed in Chapter 9.)

From the elegant theory of Albert Einstein to the precise experiments of Robert Millikan, the photoelectric effect is an example of the transition of a physical phenomenon from theory to experiment to application, in this case the television camera. The photo shows an example of the benefits to humanity that have resulted from applications of this basic physics.

3

THE PARTICLELIKE PROPERTIES OF ELECTROMAGNETIC RADIATION

The first consequence of wave mechanics we will encounter is the breakdown of the classical distinction between particles and waves. In this chapter we discuss the three early experiments that led to and verified the quantum theory. These experiments provided evidence that light, which we have treated as a wave phenomenon, has properties that we normally associate with particles. Instead of spreading its energy smoothly over a wave front, the energy is delivered in concentrated bundles like particles; a discrete bundle (*quantum*) of electromagnetic energy is known as a photon.

Before we begin to discuss the experimental evidence that supports the existence of the photon and the particlelike properties of light, we first review some of the properties of electromagnetic waves.

3.1 REVIEW OF ELECTROMAGNETIC WAVES

An electromagnetic field is characterized by its electric field strength **E** and magnetic field intensity **B**. For example, the radial electric field due to a point charge q at the origin is

$$\mathbf{E} = \frac{1}{4\pi\varepsilon_0}\frac{q}{r^2}\,\hat{\mathbf{r}} \tag{3.1}$$

where $\hat{\mathbf{r}}$ is a unit vector in the radial direction. The magnetic field at a distance r from a long, straight, current-carrying wire along the z-axis is

$$\mathbf{B} = \frac{\mu_0 i}{2\pi r}\,\hat{\boldsymbol{\theta}} \tag{3.2}$$

where $\hat{\boldsymbol{\theta}}$ is a unit vector in the θ direction in cylindrical coordinates.

If the charges are accelerated, or if the current varies with time, an electromagnetic wave is produced, in which **E** and **B** vary not only with **r** but also with t. The mathematical expression that describes such a wave may have one of many different forms, depending on the properties of the source of the wave and of the medium through which the wave travels. One special form is the *plane wave*, in which the wave fronts are planes. (A point source, on the other hand, produces spherical waves, in which the wave fronts are spheres.) A plane electromagnetic wave traveling in the z direction is described by the expressions

$$\mathbf{E} = \mathbf{E}_0 \sin{(kz - \omega t + \phi)}$$

$$\mathbf{B} = \mathbf{B}_0 \sin{(kz - \omega t + \phi)} \tag{3.3}$$

Plane electromagnetic wave

where the *wave number* k is found from the wavelength, $k = 2\pi/\lambda$, and the *circular frequency* ω is found from the frequency, $\omega = 2\pi\nu$. Since λ and ν are related by $c = \lambda\nu$, k and ω are also related by $c = \omega/k$.

The polarization of the wave is represented by the vector $\mathbf{E}_0$; the plane of polarization is determined by $\mathbf{E}_0$ and by the direction of propagation, the z axis in this case. Once we specify the direction of travel and the polarization $\mathbf{E}_0$, the direction of $\mathbf{B}_0$ is fixed by the requirements that **B** must be perpendicular to both **E** and the direction of travel, and that the vector product $\mathbf{E} \times \mathbf{B}$ point in the direction of travel. For example if $\mathbf{E}_0$ is in the x direction ($\mathbf{E}_0 = E_0\hat{\mathbf{i}}$, where $\hat{\mathbf{i}}$ is a unit vector in the x direction), then $\mathbf{B}_0$ must be in the y direction ($\mathbf{B}_0 = B_0\hat{\mathbf{j}}$). Moreover, the magnitude of $\mathbf{B}_0$ is determined by

$$B_0 = \frac{E_0}{c} \qquad (3.4)$$

where c is the speed of light.

An electromagnetic wave transmits energy from one place to another; the energy flux is specified by the *Poynting vector* $\mathbf{S}$:

$$\mathbf{S} = \frac{1}{\mu_0} \mathbf{E} \times \mathbf{B} \qquad (3.5)$$

For the plane wave, this reduces to

$$\mathbf{S} = \frac{1}{\mu_0} E_0 B_0 \sin^2 (kz - \omega t + \phi)\hat{\mathbf{k}} \qquad (3.6)$$

where $\hat{\mathbf{k}}$ is a unit vector in the z direction. The Poynting vector has dimensions of energy per unit time per unit area—for example, $J/s/m^2$ or W/m^2. Figure 3.1 shows the orientation of the vectors $\mathbf{E}$, $\mathbf{B}$, and $\mathbf{S}$ for this special case.

Let us imagine the following experiment. We place a detector of electromagnetic radiation (a radio receiver or a human eye) at some point on the z axis, and we determine the electromagnetic power that the above plane wave delivers to the receiver. The receiver is oriented so that its sensitive area A is perpendicular to the z axis, so that the maximum signal is received; we will correspondingly drop the vector notation of $\mathbf{S}$ and work only with its magnitude S. The power P is then just

$$P = SA$$
$$P = \frac{1}{\mu_0} E_0 B_0 A \sin^2 (kz - \omega t + \phi) \qquad (3.7)$$

which we can rewrite as

$$P = \frac{1}{\mu_0 c} E_0^2 A \sin^2 (kz - \omega t + \phi) \qquad (3.8)$$

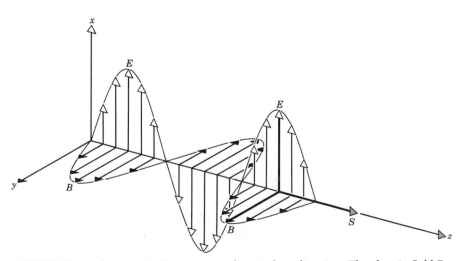

FIGURE 3.1 An electromagnetic wave traveling in the z direction. The electric field $\mathbf{E}$ lies in the xz plane and the magnetic field $\mathbf{B}$ lies in the yz plane.

There are two important features of this expression that you should recognize:

1. The intensity is proportional to E_0^2. This is a general property of waves: *the intensity is proportional to the square of the amplitude.* We will learn later that this same property also characterizes the waves that describe the behavior of material particles.

Intensity of electromagnetic waves

2. The intensity fluctuates with time, with the frequency $2\nu = 2(\omega/2\pi)$. Of course we don't usually observe this fluctuation; visible light, for example, has a frequency of about 10^{15} oscillations per second, and since our eye doesn't respond that quickly, we observe the time average of many (perhaps 10^{13}) cycles. If T is the observation time (perhaps 10^{-2} s in the case of the eye) then the average power is

$$P_{av} = \frac{1}{T} \int_0^T P \, dt \qquad (3.9)$$

and using Equation (3.8) we obtain

$$P_{av} = \frac{1}{2\mu_0 c} E_0^2 A \qquad (3.10)$$

since the average value of $\sin^2 \theta$ is $\frac{1}{2}$.

The property that makes waves a unique physical phenomenon is the *principle of superposition,* which, for example, allows two waves to meet at a point, to cause a combined disturbance at the point that might be greater or less than the disturbance produced by either wave alone, and finally to emerge from the point of "collision" with all of the properties of each wave totally unchanged by the collision. To appreciate the important distinction between material objects and waves, imagine trying that trick with two automobiles!

This important and special property of waves produces the phenomena of *interference* and *diffraction.* The simplest and best-known example of interference is *Young's double-slit experiment,* in which a monochromatic plane wave is incident on a barrier in which two narrow slits have been cut. (This experiment was first done with light waves, but in fact any wave will do as well, not only other electromagnetic waves, such as microwaves, but mechanical waves, such as water waves or sound waves. We will assume that the experiment is being done with light waves.)

Young's double-slit experiment

Figure 3.2 illustrates this experimental arrangement. The plane wave is *diffracted* by each of the slits, so that the light passing through each slit covers a much larger area on the screen than the geometric shadow of the slit. This causes the light from the two slits to overlap on the screen, causing the interference. For example, if we move away from the center of the screen just the right distance, we reach a point at which a wave crest passing through one slit arrives at exactly the same time as the previous wave crest which passed through the other slit. When this occurs, the intensity is a maximum, and a bright region appears on the screen. This is *constructive interference,* and will occur at the point on the screen that is exactly one wavelength further from one slit than from the other. That is, if X_1 and X_2 are the distances from the point on the screen to the two slits, then a condition for maximum constructive

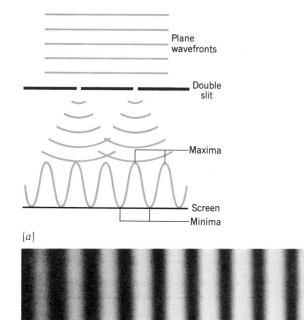

(a)

(b)

FIGURE 3.2 (a) Young's double-slit experiment. A plane wave front passes through both slits; the wave is diffracted at the slits, and interference occurs where the diffracted waves overlap on the screen. (b) The interference fringes observed on the screen. (*Source:* Cagnet, et al., *Atlas of Optical Phenomena.* Springer-Verlag, 1971.)

interference is just $|X_1 - X_2| = \lambda$. Constructive interference will occur when any wave crest from one slit arrives simultaneously with another from the other slit, whether it is the next, or the fourth, or the forty-seventh. The general condition for complete constructive interference is that the difference between X_1 and X_2 be an integral number of wavelengths:

$$|X_1 - X_2| = n\lambda \qquad n = 0, 1, 2, \ldots \qquad (3.11)$$ **Double-slit maxima**

It is also possible for the crest of the wave from one slit to arrive at a point on the screen simultaneously with the trough (valley) of the wave from the other slit. When this happens, the two waves will cancel, giving a dark region on the screen. This is known as *destructive interference*. (The existence of destructive interference with intensity minima immediately shows that we must add the vector amplitudes **E** of the waves from the two slits, and not their powers P, since P can never be negative.) Destructive interference occurs whenever the distances X_1 and X_2 are such that the phase of one wave differs from the other by one-half cycle, or by one and one-half cycles, two and one-half cycles, and so forth:

$$|X_1 - X_2| = \tfrac{1}{2}\lambda, \tfrac{3}{2}\lambda, \tfrac{5}{2}\lambda, \ldots$$
$$= (n + \tfrac{1}{2})\lambda \qquad n = 0, 1, 2, \ldots \qquad (3.12)$$ **Double-slit minima**

We can find the locations on the screen where the interference maxima occur in the following way. Let d be the separation of the slits, and let D be the distance from the slits to the screen. If y_n is the distance from the center of the screen to the nth maximum, then from the geometry of Figure 3.3 we find (assuming $X_1 > X_2$)

$$X_1^2 = D^2 + \left(\frac{d}{2} + y_n\right)^2$$

$$X_2^2 = D^2 + \left(\frac{d}{2} - y_n\right)^2 \tag{3.13}$$

Subtracting,

$$X_1^2 - X_2^2 = 2y_n d \tag{3.14}$$

and

$$y_n = \frac{(X_1 + X_2)(X_1 - X_2)}{2d} \tag{3.15}$$

In experiments with light, D is of order 1 m and y_n and d are at most 1 mm; thus $X_1 \cong D$ and $X_2 \cong D$, so $X_1 + X_2 \cong 2D$, and to a good approximation

$$y_n = (X_1 - X_2)\frac{D}{d} \tag{3.16}$$

Using Equation (3.11) for the values of $(X_1 - X_2)$ at the maxima,

$$y_n = n\frac{\lambda D}{d} \tag{3.17}$$

Another device for observing the interference of light waves is the *diffraction grating*, which is a sort of multiple-slit device for producing the interference of light waves. The operation of this device is illustrated in Figure 3.4; interference maxima corresponding to different wavelengths appear at different angles θ, according to

$$d \sin \theta = n\lambda \tag{3.18}$$

where d is the slit spacing and n is the order number of the maximum $(n = 1, 2, 3, \ldots)$. This relationship is derived in many introductory physics textbooks.

The advantage of the diffraction grating is its superior resolution—it enables us to get very good separation of wavelengths that are close to one another, and thus it is a very useful device for measuring wavelengths. Notice, however, that in order to get reasonable values of the angle θ, for example $\sin \theta$ in the range of 0.3 to 0.5, we must have d of the order of a few times the wavelength. For visible light this is not particularly difficult, but for very short wavelength radiations, it is not possible to construct a grating with such small values of d. For example, for X rays with a wavelength of the order of 0.1 nm, we would need to construct a grating in which the slits were less than 1 nm apart, which is about the same spacing as the atoms of most materials. The solution to this problem has been known since the pioneering experiments of Laue and Bragg—use the atoms themselves as a diffraction grating! A beam of X rays sees the regular spacings of the atoms in a crystal as a sort of three-dimensional diffraction grating. Consider the set of atoms shown in Figure 3.5, which represents a small portion of a two-dimensional slice of the crystal. The X rays are reflected from indi-

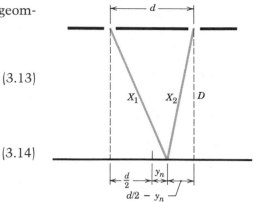

FIGURE 3.3 The geometry of the double-slit experiment.

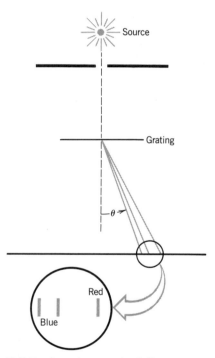

FIGURE 3.4 The use of a diffraction grating to break light into its constituent wavelengths.

X-ray diffraction

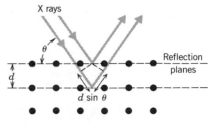

vidual atoms, in all directions, but in only one direction will the scattered "wavelets" constructively interfere to produce a reflected beam, and in this case we can regard the reflection as occurring from a plane drawn through the row of atoms. (This situation is identical with the reflection of light from a mirror—only in one direction will there be a beam of reflected light, and in that direction we can regard the reflection as occurring on a plane with the angle of incidence equal to the angle of reflection.) Suppose the rows of atoms are a distance d apart in the crystal. Then a portion of the beam is reflected from the front plane, and a portion is reflected from the second plane, and so forth. The wave fronts of the beam reflected from the second plane lag behind those reflected from the front plane, since the wave reflected from the second plane must travel an additional distance of $2d \sin \theta$, where θ is the angle of incidence as *measured from the face of the crystal* (in optics we have previously always measured angles with respect to the *normal* to the surface). If this path difference is a whole number of wavelengths, the reflected beams will interfere constructively and will give an intensity maximum; thus the basic expression for the interference maxima in X-ray diffraction from a crystal is

FIGURE 3.5 A beam of X rays reflected from a set of crystal planes of spacing d. The beam reflected from the second plane travels a distance of $2d \sin \theta$ greater than the beam reflected from the first plane.

$$2d \sin \theta = n\lambda \qquad n = 1, 2, 3, \ldots \qquad (3.19)$$ **Bragg's law**

This result is known as *Bragg's law* for X-ray diffraction. Notice the factor of 2 that appears in Equation (3.19) but does *not* appear in the otherwise similar expression of Equation (3.18) for the ordinary diffraction grating.

A single crystal of table salt (NaCl) is irradiated with a beam of X rays of wavelength 0.250 nm, and the first Bragg reflection is observed at an angle of 26.3°. What is the atomic spacing of NaCl?   **EXAMPLE 3.1**

Solving Bragg's law for the spacing d, we have   ***Solution***

$$d = \frac{n\lambda}{2 \sin \theta} = \frac{0.250 \text{ nm}}{2 \sin (26.3°)}$$

$$d = 0.282 \text{ nm}$$

Our drawing of Figure 3.5 was very arbitrary—we had no basis for choosing which set of atoms to draw the reflecting planes through. Figure 3.6 shows a larger section of the crystal. As you can see, there are many possible reflecting planes, each with a different value of θ and d. (Of course, d_i and θ_i are related and cannot be varied independently.) If we used a beam of X rays of a single wavelength, it might be difficult to find the proper angle and set of planes to observe the interference. However, if we use a beam of X rays of a continuous range of wavelengths, for each d_i and θ_i interference will occur for a certain wavelength λ_i and so there will be a pattern of interference maxima appearing at different angles of reflection as shown in Figure 3.6. The pattern of interference maxima depends not on the incident wavelengths (which form a continuous distribution) but rather on the spacing and the type of arrangement of the atoms in the crystal.

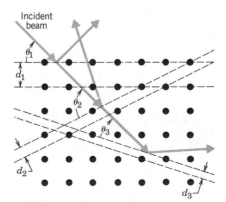

FIGURE 3.6 An incident beam of X rays can be reflected from many different crystal planes.

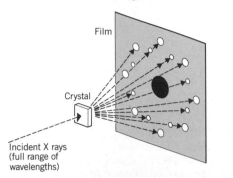

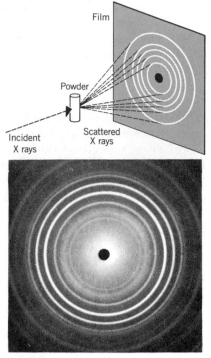

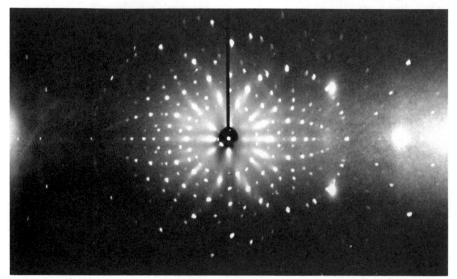

FIGURE 3.7 (Top) Apparatus for observing X-ray scattering by a crystal. An interference maximum (dot) appears on the film whenever a set of crystal planes happens to satisfy the Bragg condition for a particular wavelength. (Bottom) Laue pattern of NaCl crystal. (*Source:* Amoros, Buerger, and de Amoros, *The Laue Method*, Academic Press, 1975.)

FIGURE 3.9 (Top) Apparatus for observing X-ray scattering from a powdered sample. Because the many crystals in a powder have all possible different orientations, each scattered ray of Figure 3.7 becomes a cone which forms a circle on the film. (Bottom) Diffraction pattern (known as *Debye-Scherrer* pattern) of a powder sample. (*Source:* Eisberg & Resnick, *Quantum Physics*, John Wiley & Sons, 1974.)

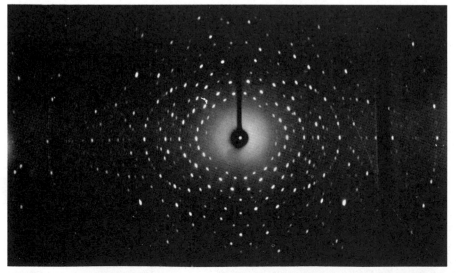

FIGURE 3.8 Laue pattern of a quartz crystal. The difference in crystal structure and spacing between quartz and NaCl makes this pattern look different from Figure 3.7.

Figures 3.7 and 3.8 show sample patterns (called *Laue patterns*) that are obtained from X-ray scattering from two different crystals. The bright dots correspond to interference maxima for wavelengths from the range of incident wavelengths that happen to satisfy Equation (3.19). Of course the three-dimensional pattern is more complicated than our two-dimensional drawings, but the individual dots have the same interpretation. Figure 3.9 shows the pattern obtained from a sample that consists of many tiny crystals, rather than one single crystal. (It looks like Figure 3.7 or 3.8 rotated rapidly about its center.) From such pictures it is also possible to deduce crystal structures and lattice spacing.

All of the examples we have discussed in this section depend on the wave properties of electromagnetic radiation. We now discuss some experiments with light and other electromagnetic radiation, which indicate that the wave interpretation is not a complete description of the properties of electromagnetic radiation.

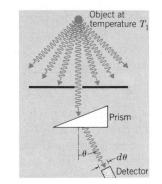

FIGURE 3.10 Measurement of the spectrum of thermal radiation. A device such as a prism is used to separate the wavelengths emitted by the object.

The first indication that all was not well with the classical wave picture of electromagnetic radiation (which worked so well to explain the nineteenth-century experiments of Young and of Hertz and which could be analyzed so precisely with the equations of Maxwell) followed from the failure of the wave theory to explain the observed spectrum of *thermal radiation*—that type of electromagnetic radiation emitted by all objects merely because of their temperature. Other experiments soon followed, including the studies of the electrons emitted when light strikes the surface of a metal (*the photoelectric effect*) and the scattering of light by electrons (*the Compton effect*), in which the wave theory similarly failed to explain the results of experiments. We discuss these other two experiments in the next two sections; here we discuss only thermal radiation.

A typical experimental arrangement is shown in Figure 3.10. An object is maintained at a temperature T_1. The radiation emitted by the object is detected by an apparatus that is sensitive to the wavelength of the radiation. For example, a dispersive medium such as a prism can be used so that different wavelengths will appear at different angles θ. By moving the radiation detector to different angles θ we can measure the intensity of the radiation at a specific wavelength. Since the detector is not a geometrical point (hardly an efficient detector!), but instead subtends a small range of angles $d\theta$, what we really measure is the amount of radiation in some range $d\theta$ at θ, or equivalently, in some range $d\lambda$ at λ. We call this quantity the *radiant intensity R*, and the result of the experiment is a series of values of $R\,d\lambda$ for as many different values of λ as we choose to measure. When we have finished, we plot the data as a function of λ, and the result might be similar to Figure 3.11. Increasing the temperature to T_2, we repeat the measurement, and obtain the result shown in the figure. Repeating the measurement a number of times, we learn two important details of the properties of thermal radiation:

3.2 BLACKBODY RADIATION

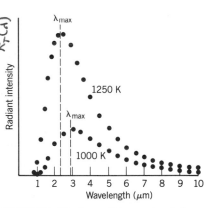

FIGURE 3.11 A possible result of the measurement of the radiant intensity over many different wavelengths. Each different temperature of the emitting body gives a different peak λ_{max}.

1. The total radiant intensity over all wavelengths increases as the fourth power of the temperature T; since the total intensity is just the area under either of the curves of Figure 3.11, we can write:

$$\int_0^\infty R\,d\lambda = \sigma T^4 \qquad (3.20) \quad \textbf{Stefan's law}$$

where we have introduced the proportionality constant σ. Equation (3.20) is called *Stefan's law* and σ is known as the *Stefan-Boltzmann constant*. Its value has been determined from experiments such as those illustrated in Figure 3.11:

$$\sigma = 5.6703 \times 10^{-8} \text{ W/m}^2\cdot\text{K}^4$$

2. The wavelength at which each curve has its peak, which we call λ_{max} (even though it is *not* a maximum wavelength) decreases as we raise the temperature, and in fact it decreases exactly in inverse proportion to the temperature, so that $\lambda_{max} \propto 1/T$. Again from experiments we learn the value of the proportionality constant, and we find

$$\lambda_{max} T = 2.898 \times 10^{-3} \text{ m}\cdot\text{K} \tag{3.21}$$

Wien's displacement law

This result is known as *Wien's displacement law*; the term "displacement" refers to the way the peak is moved or displaced as the temperature is varied.

EXAMPLE 3.2

(a) At what wavelength does a room-temperature $(T = 20°C)$ object emit the maximum thermal radiation? (b) To what temperature must we heat it until its peak thermal radiation is in the red region of the spectrum? (c) How many times as much thermal radiation does it emit at the higher temperature?

Solution

(a) Converting to absolute temperature, $T = 293$ K, and from Wien's displacement law,

$$\lambda_{max} = \frac{2.898 \times 10^{-3} \text{ m}\cdot\text{K}}{293 \text{ K}} = 9.89 \ \mu\text{m}$$

(b) Taking the wavelength of red light to be $\lambda \cong 650$ nm, we again use Wien's displacement law to find T:

$$T = \frac{2.898 \times 10^{-3} \text{ m}\cdot\text{K}}{650 \times 10^{-9} \text{ m}} = 4460 \text{ K}$$

(c) Since the total intensity of radiation is proportional to T^4, the ratio of the total thermal emissions will be

$$\frac{T_2^4}{T_1^4} = \frac{(4460)^4}{(293)^4} = 5.37 \times 10^4$$

Be sure to notice the use of absolute (Kelvin) temperatures in this example.

It is now time to try to analyze and understand these results (the dependence of R on λ, Stefan's law, Wien's law) based on theories of thermodynamics and electromagnetism. We will not give the complete discussion here, but only a brief outline of the theory. At the end of this chapter you will find references to more detailed (and advanced) discussions of this topic.

Ordinary objects are seen because of the light that is reflected by them. At room temperature the thermal radiation is mostly in the *infrared* region of the spectrum $(\lambda_{max} \cong 10 \ \mu\text{m})$, where our eyes are not sensitive. As we heat such objects, they begin to emit visible light; according to Equation (3.21), as T increases, λ_{max} decreases, and for relatively modest temperatures λ_{max} will move down to the visible region. For example, a piece of metal first begins to glow with a deep red color, and as the temperature is increased the color becomes more yellow.

Unfortunately, the radiation emitted by ordinary objects depends not only on the temperature, but also on other properties, such as the shape of the object, its surface properties, the material of which it is constructed, and so forth. It also depends on whether or not it reflects the radiation that falls on it from its surroundings. To eliminate some of these difficulties we consider not an arbitrary object, but rather one whose surface is completely black (*a blackbody*). If an object is completely black, it reflects none of the radiation that strikes it, and its surface properties can then not be seen. However, this generalization still does not simplify the problem sufficiently for us to be able to calculate the spectrum of emitted radiation, so we generalize still further to a special type of blackbody—a cavity, such as the inside of a metal box, with a small hole cut in one wall. *It is the hole, and not the box itself, that is the blackbody.* Radiation from outside that strikes the hole gets lost inside the box and has a negligible chance of reemerging from the hole; thus no reflections occur from the blackbody (the hole). Since the radiation that emerges from the hole is just a sample of the radiation inside the box, understanding the nature of the radiation inside the box allows us to understand the radiation that leaves through the hole.

Cavity radiation

The *classical* calculation of the radiant energy emitted at each wavelength now breaks down into several steps. Without proof, here are the essential parts of the derivation, which involves first computing the amount of radiation (number of waves) at each wavelength, then the contribution of each wave to the total energy in the box, and finally the radiant intensity corresponding to that energy.

1. *The box is filled with electromagnetic standing waves.* If the walls of the box are metal, radiation is reflected back and forth with a node of the electric field at each wall (the electric field must vanish inside a conductor). This is the same condition that applies to other standing waves, like those on a stretched string or a column of air in an organ pipe.

2. *The number of standing waves with wavelengths between λ and $\lambda + d\lambda$ is*

$$N(\lambda)\, d\lambda = \frac{8\pi V}{\lambda^4}\, d\lambda \qquad (3.22)$$

where V is the volume of the box. For one-dimensional standing waves, as on a stretched string of length L, the allowed wavelengths are $\lambda = 2L/n$, $(n = 1, 2, 3, \ldots)$. The number of possible standing waves with wavelengths between λ_1 and λ_2 is $n_2 - n_1 = 2L(1/\lambda_2 - 1/\lambda_1)$, and in the interval from λ to $\lambda + d\lambda$ there would be $N(\lambda)\, d\lambda = (2L/\lambda^2)\, d\lambda$ different waves. Extending this result to three dimensional electromagnetic waves, Equation (3.22) is obtained.

3. *Each individual wave contributes an energy of* kT *to the radiation in the box.* This result follows from classical thermodynamics. The radiation in the box is in thermal equilibrium with the walls at temperature T. Radiation is reflected from the walls because it is absorbed and quickly reemitted by the atoms of the walls, which in the process oscillate at the frequency of the radiation. At temperature T,

When Planck announced his results at a meeting of the German Physical Society in 1900, no great shock waves were felt in the world of accepted physical theories. Even Planck himself did not believe he had done anything more than to find an *ad hoc* explanation for a single physical phenomenon. It was another five years until Einstein, in his analysis of the photoelectric effect, showed that Planck's result was not just a curiosity associated with cavity radiation but was, in fact, a fundamental property of electromagnetic waves that led to a new and unexpected way of looking at the physical world.

In the photoelectric effect, a metal surface is illuminated with a beam of light, and electrons are emitted from the surface. In experimental studies of the photoelectric effect, we measure how the rate and kinetic energy of electron emission depend on the intensity and wavelength of the source of light. The experiment must be done in vacuum, so that the electrons do not lose energy in collisions with molecules of the air.

The experimental arrangement is shown in Figure 3.14. The rate of electron emission is measured as an electric current using an ammeter in the external circuit, and the kinetic energy of the electrons is determined by applying a retarding potential to the anode so that the electrons do not have sufficient kinetic energy to "climb" the potential hill. Experimentally, the retarding voltage is increased until the current reading on the ammeter falls to zero. This voltage is called the *stopping potential* V_s. The most energetic electrons cannot overcome the stopping potential, so measuring V_s is a way of determining the maximum electron kinetic energy K_{max}:

$$K_{max} = eV_s \qquad (3.28)$$

where e is the charge of the electron. Typical values of V_s are a few volts.

From such experiments, we learn the following details about the photoelectric effect:

1. The rate at which electrons are emitted depends on the intensity of the light.

2. The rate at which electrons are emitted is independent of the wavelength below a certain wavelength; above that value the current gradually falls and reaches zero at a cutoff wavelength λ_c. Typically λ_c is in the blue or ultraviolet region of the spectrum.

3. The value of λ_c depends on the kind of metal used for the photosensitive surface, but *not on the intensity of the light source*. Below λ_c, any light source, no matter how weak, will cause the emission of photoelectrons; above λ_c, no light source, no matter how strong, will cause the emission of photoelectrons.

4. The maximum kinetic energy of the emitted electrons depends on the wavelength of the light, but not at all on its intensity; the kinetic energy is found to increase linearly with the frequency of the light source.

5. When the light source is first turned on, current begins to flow essentially immediately (within 10^{-9} s).

3.3 THE PHOTOELECTRIC EFFECT

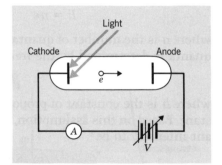

FIGURE 3.14 Apparatus for observing the photoelectric effect. Light shines on the metal surface (the cathode) and electrons are knocked loose. When they travel to the anode, they appear in the external circuit as a current, measured with the ammeter A.

Let us first see how the wave theory fails in its analysis of the photo-electric effect. According to the wave theory an atom will absorb energy from an incident electromagnetic wave in proportion to the area that it presents to that wave. The electrons oscillate in response to the electric field of the wave, until enough energy is absorbed to break an electron free from an atom. Increasing the brightness of the light source increases the rate of energy absorption, since the electric field is increased, and so the rate of electron emission should increase, in agreement with the experimental result. This absorption, however, should occur at *any* wavelength, so the existence of a cutoff wavelength is completely contrary to the wave picture. Even at wavelengths greater than λ_c, the wave theory suggests that it should still be possible for an electromagnetic wave to deliver enough energy to release electrons.

We can roughly estimate the time necessary for an atom to absorb enough energy to release an electron. For a light source we choose a modest laser, such as the helium-neon lasers you have probably seen in the laboratory. Such a laser might have an output of, at most, 10^{-3} W, spread over an area a few millimeters square (10^{-5} m^2). A typical atom has a diameter of the order of 10^{-10} m, and thus an area of the order of 10^{-20} m^2. The fraction of the laser intensity that falls on the atom is only 10^{-20} m$^2/10^{-5}$ m$^2 \cong 10^{-15}$. Thus only 10^{-18} W $= 10^{-18}$ J/s $\cong 6$ eV/s can be absorbed by the atom, and to absorb a few eV of energy takes about a second. According to the wave theory, we therefore expect to see no photoelectrons emitted until seconds after the light source is turned on; in practice we find the first photoelectrons emitted within 10^{-9} s.

In these two predictions (cutoff wavelength and time delay) the wave theory is therefore unable to account for the experimental observations.

Einstein put forth the correct theory of the photoelectric effect in 1905. He based his theory on Planck's idea of the *quantum of energy*, but he carried it a step further. Einstein assumed that the quantum of energy was not a special property of the atoms of the walls of the cavity radiator, but was a property of the *radiation itself*. The energy of electromagnetic radiation is absorbed not in a continuous stream of waves but in discrete little bundles or quanta, which we call *photons*. A photon is one emitted or absorbed quantum of electromagnetic energy, and in analogy with Planck's suggestion, each photon of the radiation of frequency ν has an energy

Photons

$$E = h\nu \tag{3.29}$$

where h is Planck's constant. Higher-frequency photons have more energy—a photon of blue light is more energetic than a photon of red light.

Since a classical electromagnetic wave of energy U carries a momentum $p = U/c$, the photons must also carry momentum, and in analogy with the classical wave, the momentum of a single photon of energy E is

$$p = \frac{E}{c} \tag{3.30}$$

From Equation (2.14) it must follow that $m_0 = 0$ for a photon—a photon behaves like a "particle" with no rest mass! Of course, Einstein began by assuming this was true from the beginning; special relativity does not permit us to "catch up" with a light beam, and therefore photons can never be brought to rest. Equation (2.10) also requires that m_0 must be zero for a photon or for any particle that travels at the speed of light; otherwise the energy mc^2 would become infinite.

Combining Equations (3.29) and (3.30) we find a direct relationship between the momentum and wavelength of a photon:

$$p = \frac{h}{\lambda} \qquad (3.31)$$

Einstein's theory is immediately able to explain all of the observed features of the photoelectric effect. Let us suppose an electron is bound in a metal with an energy W, known as the *work function*. Different metals have different work functions; a list is shown in Table 3.1. To remove the electron from the surface we must supply at least an energy of W. Electrons that are more tightly bound within the metal require more energy to remove. An electron is knocked loose whenever an atom absorbs a photon that delivers an energy of at least W. That is, if $h\nu < W$, there is no photoelectric effect; if $h\nu > W$, the electron is knocked free and the excess energy appears as its kinetic energy. The maximum kinetic energy K_{max} corresponds to those electrons knocked loose from the surface:

$$K_{max} = h\nu - W \qquad (3.32)$$

Those below the surface require an energy greater than W and so come off with less kinetic energy.

A photon which supplies an energy of W, just exactly what is needed to remove an electron, corresponds to light of a wavelength equal to the cutoff wavelength λ_c. At this wavelength there is no excess energy for kinetic energy, and so Equation (3.32) reduces to

$$W = h\nu = \frac{hc}{\lambda_c} \qquad (3.33)$$

and so

$$\lambda_c = \frac{hc}{W} \qquad (3.34)$$

Since we get one photoelectron for each absorbed photon, increasing the intensity of the light source causes more photoelectrons to be emitted, but they will have the same kinetic energy, since the photons all have the same energy.

Finally, the time delay before photoemission is expected to be small—as soon as the first photon is absorbed, the photoelectric current will begin to flow.

Thus all of the experimental features of the photoelectric effect are in agreement with the quantum behavior of electromagnetic radiation. Robert Millikan gave the most convincing demonstration of this agreement in a careful series of experiments in 1915. A sample of his results is shown in Figure 3.15. From the slope of the line, which is really just a plot of Equation (3.32), the value of Planck's constant can be derived:

$$h = 6.57 \times 10^{-34} \text{ J·s}$$

Table 3.1 Some Photoelectric Work Functions

Material	W(eV)
Na	2.28
Al	4.08
Co	3.90
Cu	4.70
Zn	4.31
Ag	4.73
Pt	6.35
Pb	4.14

Kinetic energy of photoelectrons

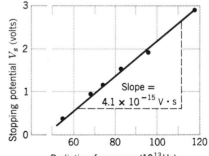

FIGURE 3.15 Millikan's results for the photoelectric effect in sodium. The slope of the line is just h/e; the experimental determination of the slope gives a way of determining Planck's constant. The intercept should give the cutoff frequency; however, in Millikan's time the contact potentials of the electrodes were not known precisely and so the vertical scale is displaced by a few tenths of a volt. The slope is not affected by this correction.

This value is in excellent agreement with the value derived from the measured Stefan-Boltzmann constant, as in Equation (3.27). This good agreement, derived from two very different kinds of experiments, one involving the absorption and the other the emission of electromagnetic radiation, shows that Planck's constant has an importance that goes beyond the mere explanation of any one experiment. It is today regarded as one of the universal constants of nature, and has been measured with high precision in a variety of experiments. The presently accepted value is

$$h = 6.62618 \times 10^{-34} \text{ J·s}$$

EXAMPLE 3.3

(a) What are the energy and momentum of a photon of red light of wavelength 650 nm? (b) What is the wavelength of a photon of energy 2.40 eV?

Solution

(a) Usually we want to know the relationship between the wavelength of light and the energy of its photons. Let us therefore rewrite Equation (3.29), recalling that for light, $\nu = c/\lambda$:

$$E = h\nu = \frac{hc}{\lambda}$$

Thus we can solve this problem:

$$E = \frac{(6.63 \times 10^{-34} \text{ J·s})(3.00 \times 10^8 \text{ m/s})}{(650 \times 10^{-9} \text{ m})}$$

$$= 3.06 \times 10^{-19} \text{ J}$$

Converting to electron-volts, we have

$$E = \frac{3.06 \times 10^{-19} \text{ J}}{1.60 \times 10^{-19} \text{ J/eV}} = 1.91 \text{ eV}$$

Problems of this sort are simplified if we express the combination hc in units of eV·nm, as we did in Chapter 1.

$$E = \frac{hc}{\lambda} = \frac{1240 \text{ eV·nm}}{650 \text{ nm}} = 1.91 \text{ eV}$$

The momentum is found in a similar way:

$$p = \frac{h}{\lambda} = \frac{1}{c}\frac{hc}{\lambda} = \frac{1}{c}\frac{1240 \text{ eV·nm}}{650 \text{ nm}} = 1.91 \text{ eV}/c$$

The momentum could also be found directly from the energy:

$$p = \frac{E}{c} = \frac{1.91 \text{ eV}}{c} = 1.91 \text{ eV}/c$$

(It may be helpful to review the discussion in Example 2.16 about these units of momentum.)

(b) Solving for λ,

$$\lambda = \frac{hc}{E} = \frac{1240 \text{ eV·nm}}{2.40 \text{ eV}} = 517 \text{ nm}$$

Robert A. Millikan (1868–1953, United States). Perhaps the best experimentalist of his era, his work included the precise determination of Planck's constant using the photoelectric effect (for which he received the 1923 Nobel prize) and the measurement of the charge of the electron (using his famous "oil-drop" apparatus).

EXAMPLE 3.4

The work function for tungsten metal is 4.52 eV. (a) What is the cutoff wavelength λ_c for tungsten? (b) What is the maximum kinetic energy of the electrons when radiation of wavelength 200.0 nm is used? (c) What is the stopping potential in this case?

(a) From Equation (3.34)

$$\lambda_c = \frac{hc}{W} = \frac{1240 \text{ eV·nm}}{4.52 \text{ eV}} = 274 \text{ nm}$$

in the ultraviolet region.

(b) At the shorter wavelength,

$$K_{max} = h\nu - W = \frac{hc}{\lambda} - W$$

$$= \frac{1240 \text{ eV·nm}}{200 \text{ nm}} - 4.52 \text{ eV}$$

$$= 1.68 \text{ eV}$$

(c) The stopping potential is just the voltage corresponding to K_{max}:

$$V_s = \frac{K_{max}}{e} = \frac{1.68 \text{ eV}}{e} = 1.68 \text{ V}$$

Another way for radiation to interact with atoms is by means of the Compton effect, in which radiation scatters from loosely bound, nearly free electrons. Part of the energy of the radiation is given to the electron, which is released from the atom; the remainder of the energy is reradiated as electromagnetic radiation. According to the wave picture, the scattered radiation is less energetic than the incident radiation (the difference going into the kinetic energy of the electron) but has the same wavelength. As we will see, the photon concept leads to a very different prediction for the scattered radiation.

The scattering process is analyzed simply as an interaction (a "collision" in the classical sense of particles) between a single photon and an electron, which we assume to be at rest. Figure 3.16 shows the collision. Initially, the photon has energy E given by

3.4 THE COMPTON EFFECT

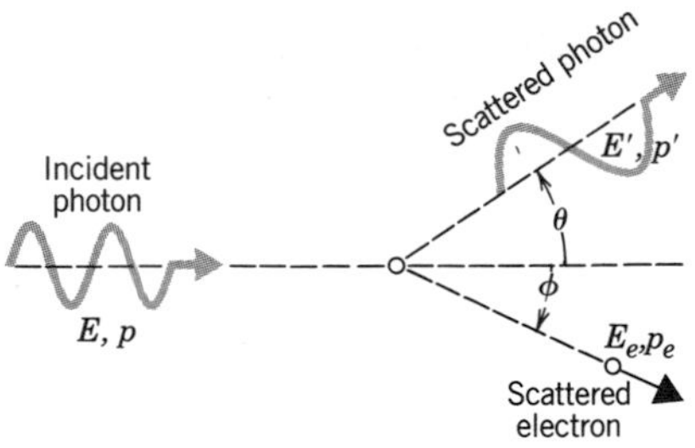

FIGURE 3.16 The geometry of Compton scattering.

$$E = h\nu = \frac{hc}{\lambda} \tag{3.35}$$

and linear momentum p,

$$p = \frac{E}{c} \tag{3.36}$$

The electron, at rest, has rest energy $m_e c^2$. After the scattering, the photon has energy E' and momentum p', and moves in a direction at an angle θ with respect to the direction of the incident photon. The electron has total energy E_e and momentum p_e and moves in a direction at an angle ϕ with respect to the initial photon. (To allow for the possibility of high-energy incident photons giving energetic scattered electrons, we use relativistic kinematics for the electron.) The usual conditions of conservation of energy and momentum are then applied:

$$E_{initial} = E_{final}$$

$$E + m_e c^2 = E' + E_e \tag{3.37a}$$

$$(p_x)_{\text{initial}} = (p_x)_{\text{final}}$$

$$p = p_e \cos\phi + p' \cos\theta \tag{3.37b}$$

$$(p_y)_{\text{initial}} = (p_y)_{\text{final}}$$

$$0 = p_e \sin\phi - p' \sin\theta \tag{3.37c}$$

We have three equations with four unknowns (θ, ϕ, E_e, E'; p_e and p' are not independent unknowns) that cannot be solved uniquely, but we can eliminate any two of the four unknowns by solving the equations simultaneously. If we choose to measure the energy and direction of the scattered photon, we eliminate E_e and ϕ. The angle ϕ is eliminated by combining the momentum equations:

$$p_e \cos\phi = p - p' \cos\theta$$

$$p_e \sin\phi = p' \sin\theta$$

Squaring and adding,

$$p_e^2 = p^2 - 2pp' \cos\theta + p'^2 \tag{3.38}$$

Recalling from Chapter 2 that the relativistic relationship between energy and momentum is, according to Equation (2.14),

$$E_e^2 = c^2 p_e^2 + m_e^2 c^4$$

we can substitute for E_e and p_e,

$$(E + m_e c^2 - E')^2 = c^2(p^2 - 2pp' \cos\theta + p'^2) + m_e^2 c^4 \tag{3.39}$$

and after a bit of algebra we find

$$\frac{1}{E'} - \frac{1}{E} = \frac{1}{m_e c^2}(1 - \cos\theta) \tag{3.40}$$ **Compton scattering formula**

This equation can also be written as

$$\lambda' - \lambda = \frac{h}{m_e c}(1 - \cos\theta) \tag{3.41}$$

where λ is the wavelength of the incident photon and λ' is the wavelength of the scattered photon. The quantity $h/m_e c$ is known as the *Compton wavelength of the electron* and has a value of 0.002426 nm; however keep in mind that it is *not* a true wavelength but rather is a *change* of wavelength.

Equations (3.40) and (3.41) give the change in energy or wavelength of the photon, as a function of the *scattering angle* θ. Since the quantity on the right hand side is never negative, E' is always less than E—the scattered photon has less energy than the original incident photon; the difference $E - E'$ is just the kinetic energy given to the electron, $(E_e - m_e c^2)$. Similarly, λ' is always greater than λ—the scattered photon has a longer wavelength than the incident photon; the change in wavelength ranges from 0 at $\theta = 0°$ to twice the Compton wavelength at $\theta = 180°$. Of course the descriptions in terms of energy and wavelength are equivalent, and the choice of which to use is merely a matter of convenience.

X rays of wavelength 0.2400 nm are Compton scattered and the scattered beam is observed at an angle of 60.0° relative to the incident beam. Find: (a) the wavelength of the scattered X rays, (b) the energy of the scattered X ray photons, (c) the kinetic energy of the scattered electrons, and (d) the direction of travel of the scattered electrons.

EXAMPLE 3.5

Solution

(a) λ' can be found immediately from Equation (3.41):

$$\lambda' = \lambda + \frac{h}{m_e c} (1 - \cos \theta)$$

$$= 0.2400 \text{ nm} + (0.00243 \text{ nm})(1 - \cos 60°)$$

$$= 0.2412 \text{ nm}$$

(b) The energy E' can be found directly from λ':

$$E' = \frac{hc}{\lambda'} = \frac{1240 \text{ eV·nm}}{0.2412 \text{ nm}} = 5141 \text{ eV}$$

(c) From Equation (3.37a) for conservation of energy, we have

$$E_e = (E - E') + m_e c^2 = K_e + m_e c^2$$

$$K_e = E - E'$$

The initial photon energy E is $hc/\lambda = 5167$ eV, and so

$$K_e = 5167 \text{ eV} - 5141 \text{ eV} = 26 \text{ eV}$$

(d) Solving Equations (3.37b) and (3.37c) for $p_e \cos \phi$ and $p_e \sin \phi$ as we did in deriving Equation (3.38), we divide (instead of adding and squaring):

$$\tan \phi = \frac{p' \sin \theta}{p - p' \cos \theta}$$

Multiplying top and bottom by c, and recalling that $E = pc$ and $E' = p'c$, we have

$$\tan \phi = \frac{E' \sin \theta}{E - E' \cos \theta} = \frac{(5141 \text{ eV})(\sin 60°)}{(5167 \text{ eV}) - (5141 \text{ eV})(\cos 60°)}$$

$$= 1.715$$

$$\phi = 59.7°$$

The first experimental demonstration of this type of scattering was done by Arthur Compton in 1923. A diagram of his experimental arrangement is shown in Figure 3.17. A beam of X rays is incident on a scattering target, for which Compton used carbon. (Although no scattering target contains actual "free" electrons, the outer or valence electrons in many materials are very weakly attached to the atom and behave like nearly free electrons. The kinetic energies of these electrons in the atom are very small compared with the kinetic energies K_e the electrons acquire in the scattering process.) A movable detector measured the energy of the scattered X rays at various angles θ.

Compton's original results are illustrated in Figure 3.18. At each angle, two peaks appear, corresponding to scattered X-ray photons with two different energies or wavelengths. The wavelength of one peak does not change as the angle is varied; this peak corresponds to scattering from the tightly bound "inner" electrons of the atom. These electrons

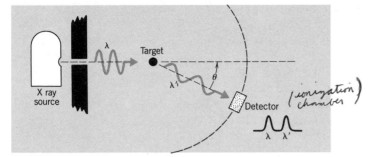

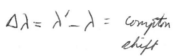

$$\Delta\lambda = \lambda' - \lambda = \text{Compton shift}$$

FIGURE 3.17 Schematic diagram of Compton scattering apparatus. The wavelength λ' of the scattered X rays is measured by the detector, which can be moved to different positions θ. The wavelength difference $\lambda' - \lambda$ varies with θ.

Arthur H. Compton (1892–1962, United States). His work on X-ray scattering verified Einstein's photon theory and earned him the 1927 Nobel prize. He was a pioneer in research with X rays and cosmic rays. During World War II he directed a portion of the U. S. atomic bomb research.

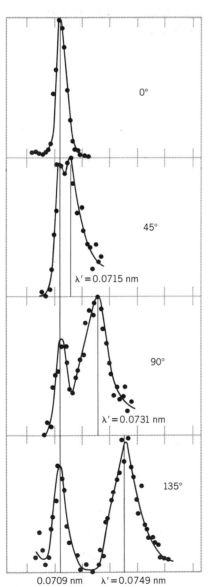

FIGURE 3.18 Compton's original results for X-ray scattering.

Classical model

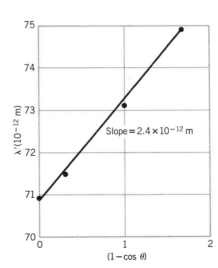

FIGURE 3.19 The scattered X-ray wavelengths λ', from Figure 3.18, for different scattering angles. The expected slope is 2.43×10^{-12} m, in agreement with the measured slope of Compton's data points.

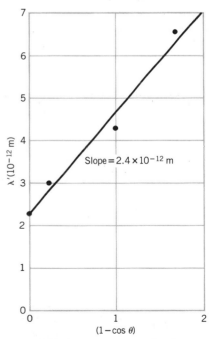

FIGURE 3.20 Compton's results for gamma-ray scattering. Even though the wavelengths are roughly two orders of magnitude smaller than the X-ray wavelengths, the slope is the same as in Figure 3.19, which the Compton formula, Equation (3.41), predicts.

are so tightly attached to the atoms that the photon scatters and loses no energy. The wavelength of the other peak, however, varies strongly with angle; as can be seen from Figure 3.19, this variation is exactly as the Compton formula predicts.

Similar results can be obtained for the scattering of gamma rays, which are higher-energy (shorter wavelength) photons emitted in various radioactive decays. Compton also measured the variation in wavelength of scattered gamma rays, as illustrated in Figure 3.20. The change in wavelength deduced from the gamma-ray measurements is identical with the change in wavelength deduced from the X-ray measurements; the Compton formula (3.41) leads us to expect this, since the change in wavelength does not depend on the incident wavelength.

Although Compton scattering and the photoelectric effect provided the earliest experimental evidence in support of the photon as a quantum of electromagnetic radiation, there are numerous other experiments that can also be interpreted correctly only if we assume the quantization (particlelike behavior) of electromagnetic radiation. In this section we will discuss some of these processes, which cannot be understood if we consider only the wave nature of electromagnetic radiation.

3.5 OTHER PHOTON PROCESSES

Bremsstrahlung and X-ray Production   When an electric charge, such as an electron, is accelerated or decelerated, it radiates electromagnetic energy; in our present framework, we would say that it emits photons. Suppose we have a beam of electrons, which has achieved an energy eV after having been accelerated through a potential V (Figure 3.21). When the electrons strike a target they are slowed down and even-

Bremsstrahlung

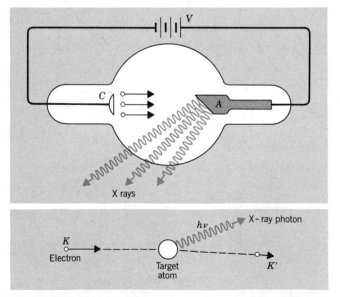

FIGURE 3.21 Apparatus for producing bremsstrahlung. Electrons from a cathode C are accelerated to the anode A through the potential difference V. When an electron strikes a target atom of the anode, it can slow down, with the emission of an X-ray photon.

tually come to rest, because they make collisions with the atoms of the target material. In such a collision, momentum is transferred to the atom, the electron slows down, and photons are emitted. The recoil kinetic energy of the atom is small (because the atom is so massive) and can safely be neglected. If the electron has a kinetic energy K before the encounter and if it leaves after the collision with a smaller kinetic energy K', then the photon energy is

$$h\nu = K - K' \qquad (3.42)$$

The amount of energy lost, and therefore the energy and wavelength of the emitted photon, are not uniquely determined, since K is the only known energy in Equation (3.42). An electron usually will make many collisions, and therefore emit many different photons, before it is brought to rest; the photons then will range all the way from very small energies (large wavelengths) corresponding to small energy losses, up to a maximum energy of K, corresponding to an electron which loses all of its energy in a single encounter. The smallest emitted wavelength is therefore determined by the maximum possible energy loss,

$$h\nu = K$$

$$\frac{hc}{\lambda_{\min}} = eV \qquad (3.43)$$

$$\lambda_{\min} = \frac{hc}{eV}$$

For typical accelerating voltages in the range of 10,000 V, $\lambda_{\min}$ is in the range of a few tenths of nm, which corresponds to the X-ray region of the spectrum. This *continuous* distribution of X rays (which is very different from the *discrete* X-ray energies that are emitted in atomic transitions; more about these in Chapter 8) is called *bremsstrahlung*, which is German for braking, or decelerating, radiation. Some sample bremsstrahlung spectra are illustrated in Figure 3.22.

Symbolically we can write the bremsstrahlung process as

$$\text{electron} \longrightarrow \text{electron} + \text{photon}$$

This is just the reverse process of the photoelectric effect, which is

$$\text{electron} + \text{photon} \longrightarrow \text{electron}$$

However, *neither* process will occur for free electrons. In both cases there must be a heavy atom in the neighborhood to take care of the recoil momentum.

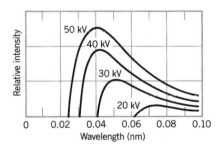

FIGURE 3.22 Some typical bremsstrahlung spectra. Each spectrum is labeled with the value of the accelerating voltage V.

Pair Production   Another process which can occur when photons encounter atoms is *pair production*, in which the photon loses all its energy and in the process two particles are created: an electron and a positron. (A positron is a particle that is identical in mass to the electron but which has a positive electric charge; more about *antiparticles* in Chapter 11.) Here we have an example of the creation of mass energy. The electron did not exist before the encounter of the photon with the atom (it was *not* an electron that was part of the atom). The photon energy is converted into the total relativistic energies E_+ and E_- of the positron and electron:

$$hv = E_+ + E_-$$
$$= (m_e c^2 + K_+) + (m_e c^2 + K_-)$$

(3.44)   **Pair production**

Since K_+ and K_- are always positive, the photon must have an energy of at least $2m_e c^2 = 1.02$ MeV in order for this process to occur; such high-energy photons are in the region of *nuclear gamma rays*. Symbolically,

photon $\longrightarrow$ electron + positron

This process, like brehmsstrahlung, will not occur unless there is an atom nearby to supply the necessary recoil momentum. The reverse process,

electron + positron $\longrightarrow$ photon

also occurs; this process is known as *positron annihilation** and can occur for free electrons and positrons as long as at least two photons are created. In this process the electron and positron disappear and are replaced by two photons. Conservation of energy requires that, if E_1 and E_2 are the photon energies,

$$(m_e c^2 + K_+) + (m_e c^2 + K_-) = E_1 + E_2$$

Positron annihilation

Usually K_+ and K_- are negligibly small, and we can assume the positron and electron to be essentially at rest. Momentum conservation then requires the two photons to have equal energies of $m_e c^2$ and to move in exactly opposite directions.

3.6 WHAT IS A PHOTON?

If one were asked "What is a tree?" an answer would not be difficult to give, but the type of answer to that question would depend on one's point of view. For example, most of us would probably describe some of the more obvious physical attributes of a tree—its size, mass, color, general construction, and so forth. A botanist, on the other hand, might describe how a tree originates, grows, and develops. A biochemist might give its chemical composition—cellulose, chlorophyll, and so forth. A farmer or a timber harvester might describe the products obtained from trees—fruit or lumber. All of these descriptions taken together go into our understanding of what a tree is, and if we were to try to describe a tree to someone who had never seen one, we would need to cover all of these descriptions.

The situation is similar for photons. If an answer were sought to the question "What is a photon?" we would need to answer from a variety of different perspectives. We could list its more obvious properties: it has no rest mass; it moves with the speed of light; it can interact as a particle but travel as a wave; it obeys the relationships $E = hv$, $p = h/\lambda$, and $E = pc$; it even feels the pull of gravity like other particles (even though its rest mass is zero; see Chapter 15). Alternatively, we could describe how photons originate (such as in bremsstrahlung processes) or

---

* An example of our cultural bias—on worlds made of antimatter, this process is called "electron annihilation"!

how they are guided as they travel from one place to another. We could even describe what their place is in fundamental physics—photons transmit the electromagnetic force; in this view, two electric charges interact by "exchanging" photons (photons are emitted by one charge and absorbed by the other). These photons are usually imaginary or "virtual" photons, which exist only in the mathematical framework of theoretical physics, but they have all the properties of real photons.

A question that has no known answer is that of the composition of a photon—what is the photon made of? As we will see in our discussion of elementary particles in Chapter 11, there are certain particles, such as the photon and electron, whose nature is such that we believe them to be "points" in the true mathematical sense—they have no physical size and they cannot be taken apart because they have no constituents.

The most difficult question to answer regarding the photon is whether it is a particle or a wave. Is the physical particle, with its list of properties, more real than the electromagnetic wave, with its *very different* list of properties?

We clearly have a paradox here. Certain experiments, such as those involving interference and diffraction effects, show that electromagnetic radiation interacts as waves; other experiments, which we have discussed in this chapter, show that electromagnetic radiation interacts as particlelike quanta known as photons. Surely the wave and particle interpretations are not consistent—particles deliver their energy in concentrated packets, while the energy of a wave is spread uniformly over the entire wave front. For example, if we think of light only as particles, it is difficult to explain the interference pattern observed in the double-slit experiment. A particle must go through one slit or the other; only a wave front can be split so that it passes through both slits and then recombines.

If we regard the wave and particle pictures as valid but exclusive alternatives, we must assume that the light emitted by a source must travel *either* as waves or as particles. How does the source know what sort of light (particles or waves) to emit? Suppose we place a double-slit apparatus on one side of the source and a photoelectric cell on the other side. Light emitted toward the double slit behaves like a wave and light emitted toward the photocell behaves like particles. How did the source know in which direction to aim the waves and in which direction to aim the particles?

Perhaps nature has a sort of "secret code" in which the kind of experiment we are doing is signaled back to the source so that it knows whether to emit particles or waves. Let us repeat our dual experiment with light from a distant galaxy, light which has been traveling toward us for a time roughly equal to the age of the universe (15×10^9 years). Surely the kind of experiment we are doing could not be signaled back to the limits of the known universe in the time it takes us to remove the double-slit apparatus from the laboratory table and replace it with the photoelectric apparatus. Yet we find that the starlight can produce both the double-slit interference and also the photoelectric effect.

We are therefore trapped into an uncomfortable conclusion: light is not *either* particles *or* waves; it is somehow *both* particles *and* waves, and only shows one or the other aspect, depending on the kind of experiment we are doing. A particle-type experiment shows the particle na-

Wave-particle duality

ture, while a wave-type experiment shows the wave nature. Our failure to classify light as *either* particle *or* wave is not so much a failure to understand the nature of light as it is a failure of our limited vocabulary (based on experiences with ordinary particles and waves) to describe a phenomenon that is more elegant and mysterious than either simple particles or waves.

The situation becomes even more difficult if we use our eye or a photographic film to observe this double-slit interference pattern. Our eye and the film both respond to individual photons. When a single photon is absorbed by a cell of the retina, an electrical impulse is produced that travels to the brain (of course, our vision is composed of many such impulses). When a single photon is absorbed by the film, a single grain of the photographic emulsion is darkened; a complete picture requires a large number of grains to be darkened.

Let us imagine for the moment that we could see individual grains of the film as they absorbed photons and darkened, and let us do the experiment with a light source that is so weak that there is a relatively long time interval between photons. We would see first one grain darken, then another, and so forth, until after a large number of photons we would see the interference pattern begin to emerge. Alternatively, the wave picture of the double-slit experiment suggests that we could find the net electric field of the wave that strikes the screen by superimposing the electric fields of the portions of the incident wave fronts that pass through the two slits; the intensity or power in that combined wave could then be found by a procedure similar to Equations (3.7) through (3.10), and we would expect that the resultant intensity should show maxima and minima just like the observed double-slit interference pattern.

In summary, the correct explanation of the origin and appearance of the interference pattern comes from the wave picture, and the correct interpretation of the evolution of the pattern on the film comes from the particle picture; the two explanations, which according to our limited vocabulary and common-sense experience cannot simultaneously be correct, must somehow be taken together to give a complete description of the properties of electromagnetic radiation.

This dilemma of wave-particle duality cannot be resolved with a simple explanation; physicists and philosophers have struggled with this problem ever since the quantum theory was introduced. The best we can do is to say that neither the wave nor the particle picture is wholly correct all of the time, that both are needed for a complete description of physical phenomena, and that in fact the two are *complementary* to one another. In the case of the double-slit experiment, we might reason as follows: the interaction between a "source" of radiation and the electromagnetic field is quantized, and we can think of the atoms of the source as emitting individual photons. The interaction at the opposite end of the experiment, the photographic film, is also quantized, and we have the similarly useful view of atoms absorbing individual photons. In between, the electromagnetic energy propagates smoothly and continuously as a wave and can show wave-type behavior (interference or diffraction). The effect of the double slit is to change the propagation of the wave (from a plane wave to the characteristic double-slit pattern, for example). Where the wave has large intensity, the film reveals the presence of many photons; where the wave has small intensity, few photons are observed. Recalling that the intensity of the wave is proportional to the square of its amplitude, we then have

probability to observe photons $\propto$ (electric field amplitude)^2

It is this expression that provides the ultimate connection between the wave behavior and the particle behavior, and we will see in the next two chapters that a similar expression connects the wave and particle aspects of those objects, like electrons, which have been previously considered to behave as classical particles.

SUGGESTIONS FOR FURTHER READING

Electromagnetic radiation and interference are discussed in most introductory physics texts; see for example the list at the end of Chapter 1. Another good reference is H. D. Young, *Fundamentals of Waves, Optics, and Modern Physics* (New York, McGraw-Hill, 1976). A collection of photographs illustrating the phenomena of classical optics can be found in M. Cagnet, M. Francon, and J. C. Thrierr, *Atlas of Optical Phenomena* (Berlin, Springer-Verlag, 1962).

For more complete discussions of blackbody radiation, including derivations of the Rayleigh-Jeans law, see the following:

R. Eisberg and R. Resnick, *Quantum Physics of Atoms, Molecules, Solids, Nuclei, and Particles* (New York, Wiley, 1974).
F. K. Richtmeyer, E. H. Kennard, and J. N. Cooper, *Introduction to Modern Physics*, 6th edition (New York, McGraw-Hill, 1969).

The properties of X rays, including diffraction and scattering, are discussed in the following:

L. Bragg, "X-Ray Crystallography," *Scientific American 219*, 58 (July 1968).
G. L. Clark, *Applied X Rays* (New York, McGraw-Hill, 1940).
A. H. Compton and S. K. Allison, *X Rays in Theory and Experiment* (New York, Van Nostrand, 1935).

QUESTIONS

1. The diameter of an atomic nucleus is about 10×10^{-15} m. Suppose you wanted to study the diffraction of photons by nuclei. What energy of photons would you choose? Why?

2. What are the fields of classical physics on which the classical theory of blackbody radiation is based? Why don't we believe that the "ultraviolet catastrophe" suggests that something is wrong with one of those classical theories?

3. In what region of the electromagnetic spectrum do room-temperature objects radiate? What problems would we have if our eyes were sensitive in that region?

4. How does the total intensity of thermal radiation vary when the temperature of an object is doubled?

5. How is the wave nature of light unable to account for the observed properties of the photoelectric effect?

6. In the photoelectric effect, why do some electrons have kinetic energies smaller than K_{max}?

7. Why doesn't the photoelectric effect work for free electrons?

8. What does the work function tell us about the properties of a metal? Of the metals listed in Table 3.1, which has the least tightly bound electrons? Which has the most tightly bound?

9. Electric current is charge per unit time. If we increase the kinetic energy of the photoelectrons (by increasing the energy of the incident photons), shouldn't the current increase, because the charge flows more rapidly? Why doesn't it?

10. What might be the effects on a photoelectric effect experiment if we were to double the frequency of the incident light? If we were to double the wavelength? If we were to double the intensity?

11. In the photoelectric effect, how can a photon moving in one direction eject an electron moving in a different direction? What happens to conservation of momentum?

12. Make a sketch showing how the photoelectric current might depend on the voltage V. Show voltages >0 and <0. Does the current drop suddenly to zero at V_s or does it drop gradually? What experimental difficulties does this suggest might arise in trying to measure V_s precisely?

13. Compton-scattered photons of wavelength λ' are observed at 90°. In terms of λ', what is the scattered wavelength observed at 180°?

14. The Compton-scattering formula suggests that objects viewed from different angles should reflect light of different wavelengths. Why don't we observe a change in color of objects as we change the viewing angle?

15. You have a monoenergetic source of X rays of energy 84 keV, but for an experiment you need 70 keV X rays. How would you convert the X-ray energy from 84 to 70 keV?

16. Often we read about the problems of X-ray emission by TV sets. What is the origin of these X rays? Estimate their wavelengths.

17. The X-ray peaks of Figure 3.18 are not sharp but are spread over a range of wavelengths. What reasons might account for that spreading?

18. A beam of photons passes through a block of matter. What are the three ways discussed in this chapter that the photons can lose energy in interacting with the material?

PROBLEMS

1. By differentiating Equation (3.26), show that $R(\lambda)$ has its maximum as expected according to Wien's displacement law, Equation (3.21).

2. Integrate Equation (3.26) to obtain Equation (3.20). Use the definite integral $\int_0^\infty x^3\, dx/(e^x - 1) = \pi^4/15$ to obtain Equation (3.27) relating Stefan's constant to Planck's constant.

3. Use the numerical value of Stefan's constant, $\sigma = 5.67 \times 10^{-8}$ W/m²·K⁴ to find the numerical value of Planck's constant from Equation (3.27).

4. At what wavelength does the sun emit its peak radiant intensity? The surface of the sun has a temperature of about 6000 K. How does this compare with the peak sensitivity of the human eye?

5. The universe is filled with thermal radiation, which has a blackbody spectrum at an effective temperature of 2.7 K (see Chapter 16). What is the peak wavelength of this radiation? What is the energy (in eV) of quanta at the peak wavelength? In what region of the electromagnetic spectrum is this peak wavelength?

6. Show how Equation (3.40) follows from Equation (3.39).

7. What is the cutoff wavelength for the photoelectric effect using an aluminum surface?

8. When sodium metal is illuminated with light of wavelength 4.20×10^2 nm, the stopping potential is found to be 0.65 V; when the wavelength is

changed to 3.10×10^2 nm, the stopping potential is 1.69 V. Using *only these data* and the values of the speed of light and the electronic charge, find the work function of sodium and a value of Planck's constant.

9. A metal surface has a photoelectric cutoff wavelength of 325.6 nm. It is illuminated with light of wavelength 259.8 nm. What is the stopping potential?

10. When light of wavelength λ illuminates a copper surface, the stopping potential is V. In terms of V, what will be the stopping potential if the same wavelength is used to illuminate a sodium surface?

11. The cutoff wavelength for the photoelectric effect in a certain metal is 254 nm. (a) What is the work function for that metal? (b) Will the photoelectric effect be observed for $\lambda > 254$ nm or for $\lambda < 254$ nm?

12. A surface of zinc is illuminated and photoelectrons are observed. (a) What is the largest wavelength that will cause photoelectrons to be emitted? (b) What is the stopping potential when light of wavelength 220.0 nm is used?

13. Incident photons of energy 10.39 keV are Compton scattered, and the scattered beam is observed at 45.00° relative to the incident beam. (a) What is the energy of the scattered photons at that angle? (b) How much kinetic energy is given to the scattered electron?

14. X-ray photons of wavelength 0.02480 nm are incident on a target and the Compton-scattered photons are observed at 90.0°. (a) What is the wavelength of the scattered photons? (b) What is the momentum of the incident photons? Of the scattered photons? (c) What is the kinetic energy of the scattered electrons? (d) What is the momentum (magnitude and direction) of the scattered electrons?

15. In Compton scattering, calculate the maximum kinetic energy given to the scattered electron for a given photon energy.

16. High-energy gamma rays can reach a radiation detector by Compton scattering from the surroundings, as shown in the diagram at right; this effect is known as backscattering. Show that, when $E \gg m_e c^2$, the backscattered photon has an energy of approximately 0.25 MeV, independent of the energy of the original photon, when the scattering angle is nearly 180°.

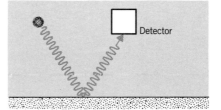

17. Gamma rays of energy 0.662 MeV are Compton scattered. (a) What is the energy of the scattered photon observed at a scattering angle of 60.0°? (b) What is the kinetic energy of the scattered electrons?

18. Assume that a 100-W light source delivers all its energy in the form of visible light, with average photon wavelength about 550 nm. How many photons per second strike a 20 cm × 30 cm sheet of paper 1 m from the light?

19. Prove that it is *not* possible to conserve both momentum and total relativistic energy in the following situation: A free electron moving at velocity **v** emits a photon and then moves at a slower velocity **v**′.

20. Find the momentum of (a) a 10.0-MeV gamma ray; (b) a 25-keV X ray; (c) a 1.0-μm infrared photon; (d) a 150-MHz radio-wave photon. Express the momentum in kg·m/s and eV/c.

21. Radio waves have a frequency of the order of 1 to 100 MHz. What is the range of energies of these photons? Our bodies are continuously bombarded by these photons. Why are they not dangerous to us?

22. (a) What is the wavelength of an X-ray photon of energy 10.0 keV? (b) What is the wavelength of a gamma-ray photon of energy 1.00 MeV?

23. What is the range of energies of photons of visible light with wavelengths 350 to 700 nm?

24. Suppose an atom of iron at rest emits an X-ray photon of energy 6.4 keV. Calculate the "recoil" momentum and kinetic energy of the atom. (*Hint.* Do you

expect to need classical or relativistic kinetic energy for the atom? Is the kinetic energy likely to be much smaller than the atom's rest energy?)

25. What is the minimum X-ray wavelength produced in bremsstrahlung by electrons that have been accelerated through 2.50×10^4 V?

26. A photon of energy E strikes an electron at rest and undergoes pair production, producing a positive electron (positron) and an electron:

$$\text{photon} + e^- \longrightarrow e^+ + e^- + e^-$$

The two electrons and the positron move off with identical momenta in the direction of the initial photon. Find the kinetic energy of the three final particles and find the energy E of the photon. (*Hint.* Conserve momentum and total relativistic energy.)

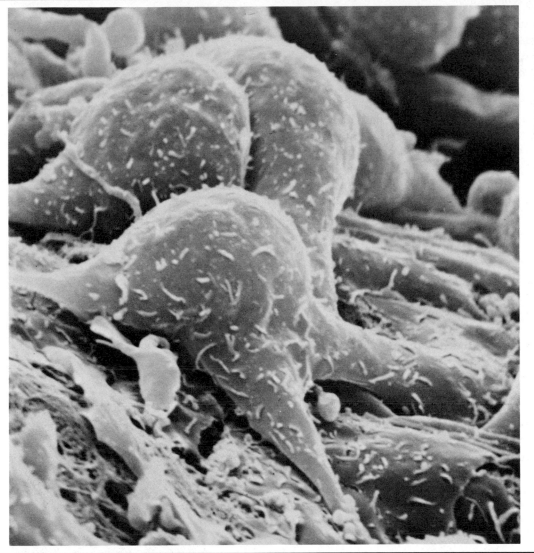

4

THE WAVELIKE PROPERTIES OF PARTICLES

The system of mechanics associated with quantum systems is sometimes called "wave mechanics." In this chapter we discuss the experimental evidence in support of the wave behavior of particles such as electrons. In classical physics, the laws governing the characteristics of waves and particles are fundamentally different. Projectiles obey particle-type laws, such as Newtonian mechanics. Waves undergo interference and diffraction, which cannot be explained by the Newtonian mechanics associated with particles. The energy carried by a particle (or projectile) is confined to a small region of space; a wave, on the other hand, distributes its energy throughout all space in ever-expanding wavefronts. In contrast to this clear distinction found in classical physics, the quantum theory requires that, in the microscopic domain, particles sometimes obey the rules that we have previously established for waves! We are going to be required to discard some of our classically inspired notions of the distinction between particles and waves. We have seen how electrons, when Compton scattered, behave like classical billiard balls, and we are led to believe that, given a sufficiently fine pair of forceps, we should be able to pick up an electron. If, however, an electron is a wave, we will certainly not be able to do so—imagine trying to grasp a sound wave or a water wave!

In attempting to supply a logical and mathematical system that resolves such dilemmas, we will refer to numerous axioms, analogies, and examples that have no counterparts in classical physics, and that may leave you doubting the logical foundation of quantum physics. You are not alone! Over the last 50 years, physicists have struggled with this same dilemma, and have not come any closer to an explanation of why this curious mix of wave and particle behavior should occur than they were when wave mechanics was first introduced. The important detail is that *it works!* The mathematical formulism enables us to compute the detailed properties of atoms and nuclei with incredible precision; when such computations are tested by experiment, they are found to be most accurate. Do not despair! You will, in this chapter and the next, be introduced to a physical theory of unmatched beauty and sophistication which has been tested in all aspects in many applications and found to give extremely accurate results; under such circumstances, the theory may be forgiven for its lack of *a priori* logical deduction.

As you study this chapter, notice the frequent references to such terms as the *probability* of the outcome of a measurement, the *average* of many repetitions of a measurement, and the *statistical* behavior of a system. These terms are a fundamental part of the quantum theory, and you cannot grasp the nature of the quantum theory until you feel comfortable with discarding such classical notions as fixed trajectories and certainty of outcome, while substituting the quantum mechanical notions of probability and statistically distributed outcomes.

Louis deBroglie (1892–   , France). A member of an aristocratic family, his work contributed substantially to the early development of the quantum theory.

4.1 DeBROGLIE'S HYPOTHESIS

Progress in physics can best be characterized by long periods of experimental and theoretical drudgery punctuated occasionally by flashes of insight that cause profound changes in the way we view the universe. Frequently the more profound the insight and the bolder the initial step, the simpler it seems in historical perspective, and the more likely we are to sit back and wonder "Why didn't I think of that?" Einstein's special theory of relativity is one example of such insight; the hypothesis of the Frenchman Louis deBroglie is another.

In the last chapter we discussed the double-slit experiment (which can be understood only if light behaves as a wave) and the photoelectric and Compton effects (which can be understood only if light behaves as a particle). Is this dual particle-wave nature a property only of light or of all material objects as well? In a bold and daring hypothesis in his doctoral dissertation, deBroglie chose the latter alternative. Examining Equation (3.29), $E = h\nu$, and Equation (3.31), $p = h/\lambda$, we find some difficulty in applying the first in the case of particles, for we cannot be sure whether E should be the kinetic energy, total energy, or total relativistic energy (all, of course, are identical for light). No such difficulties arise from the second relationship. DeBroglie suggested, lacking any experimental evidence in support of his hypothesis, that associated with any material particle moving with momentum p there is a wave of wavelength λ, related to p according to

$$\lambda = \frac{h}{p}$$

(4.1)   **DeBroglie wavelength**

The wavelength λ of a particle computed according to Equation (4.1) is called its *deBroglie wavelength*.

Compute the deBroglie wavelength of the following:

EXAMPLE 4.1

(a) A 1000-kg automobile traveling at 100 m/s (about 200 mi/h).
(b) A 10-g bullet traveling at 500 m/s.
(c) A smoke particle of mass 10^{-6} g moving at 1 cm/s.
(d) An electron with a kinetic energy of 1 eV.

(a) Using the classical relation between velocity and momentum,

Solution

$$\lambda = \frac{h}{p} = \frac{h}{mv} = \frac{6.6 \times 10^{-34}\ \text{J·s}}{(10^3\ \text{kg})(100\ \text{m/s})} = 6.6 \times 10^{-39}\ \text{m}$$

(b) As in part (a),

$$\lambda = \frac{h}{mv} = \frac{6.6 \times 10^{-34}\ \text{J·s}}{(10^{-2}\ \text{kg})(500\ \text{m/s})} = 1.3 \times 10^{-34}\ \text{m}$$

(c)

$$\lambda = \frac{h}{mv} = \frac{6.6 \times 10^{-34}\ \text{J·s}}{(10^{-9}\ \text{kg})(10^{-2}\ \text{m/s})} = 6.6 \times 10^{-23}\ \text{m}$$

(d) The rest energy $(m_0 c^2)$ of an electron is 5.1×10^5 eV. Since the kinetic energy (1 eV) is much less than the rest energy, we can use nonrelativistic kinematics.

$$p = \sqrt{2mK} = \sqrt{(2)(9.1 \times 10^{-31}\ \text{kg})(1\ \text{eV})(1.6 \times 10^{-19}\ \text{J/eV})}$$

$$= 5.4 \times 10^{-25}\ \text{kg·m/s}$$

Then,

$$\lambda = \frac{h}{p} = \frac{6.6 \times 10^{-34}\ \text{J·s}}{5.4 \times 10^{-25}\ \text{kg·m/s}} = 1.2 \times 10^{-9}\ \text{m} = 1.2\ \text{nm}$$

We can also find this solution in the following way, using $hc = 1240$ eV·nm.

$$p = \sqrt{2mK} = \sqrt{\frac{2(mc^2)K}{c^2}} = \frac{1}{c}\sqrt{2(mc^2)K}$$

$$cp = \sqrt{2(5.1 \times 10^5 \text{ eV})(1 \text{ eV})} = 1.0 \times 10^3 \text{ eV}$$

$$\lambda = \frac{h}{p} = \frac{hc}{pc} = \frac{1240 \text{ eV·nm}}{1.0 \times 10^3 \text{ eV}} = 1.2 \text{ nm}$$

This method may seem artificial at first, but with practice it becomes quite useful, especially since most energies are given in electron-volts in atomic and nuclear physics.

Note that the wavelengths computed in parts (a), (b), and (c) are far too small to be observed in the laboratory. Only in the last case, in which the wavelength is of the same order as the atomic spacing in a solid, do we have any chance of observing the wavelength. *Because of the smallness of* h, *only for particles of atomic or nuclear size will the wave behavior be observable.*

Two questions immediately follow. First, just what sort of wave is it which has this deBroglie wavelength? That is, what does the amplitude of the deBroglie wave measure? This is a question that has caused considerable difficulty for physicists over the last 50 years, and we defer a detailed answer until the last section of this chapter. For the present, we assume that, associated with the particle as it moves, there is a deBroglie wave of wavelength λ, which shows itself *when a wave-type experiment (such as diffraction) is performed on it.* The outcome of the wave-type experiment depends on this wavelength.

The second question then occurs: Why was this wavelength not directly observed before deBroglie's time? Suppose we try to observe the deBroglie wave that characterizes a marble. The classic means of observing wave behavior is by means of a double-slit experiment, so we erect a barrier, cut two slits to allow marbles to pass through, roll marbles through the slits, and allow them to leave marks when they strike a "screen." When we examine the screen, we expect the wave nature of the marble to be revealed by means of a pattern of fringes similar to that of Figure 3.2. If we did the experiment, we would alas find no fringes. The reason for our failure follows from the smallness of Planck's constant. The deBroglie wavelength of a marble (mass $\cong 1$ g, speed $\simeq$ 1 cm/s) is about 10^{-28} m, about 10^{18} times smaller than a single atom! The separation of the fringes is about the same size; of course, the separation of the fringes depends on how far the screen is from the slits—as we move the screen away, the separation of the fringes increases. If we moved the screen one *light-year* away, the spacing between the fringes would still be smaller than a single atom! There is no experiment we can do that will show the wave nature of macroscopic (laboratory sized) objects. It is only when we reach inside the atom to do experiments with atomic and nuclear particles that the deBroglie wavelength becomes observable.

Let us choose, instead of marbles, a beam of electrons. Beams of electrons can be prepared with any desired momentum by acceleration through a suitably chosen electric potential difference. Consequently, we can produce a beam of electrons in which we can vary the deBroglie wavelength over a wide range of values. The wave nature of the electron could be revealed by causing the beam of electrons to pass through a double slit. However, the construction of a suitable double slit for a beam of electrons is a difficult experimental problem that was not solved until many years after deBroglie's hypothesis had been supported by numerous other experiments. We will discuss the electron double-slit experiment a bit later on; for the present we will examine some other tests of the deBroglie hypothesis that demonstrate the wave behavior of particles.

Since interference and diffraction are the two indicators of wave behavior, the wave nature of electrons will be revealed only by doing those kinds of experiments. Two-slit interference was discussed in Chapter 3; diffraction is treated in most elementary physics texts. Figure 4.1 shows sample diffraction patterns obtained when light waves are incident on a single slit and a fine wire.

Atoms, with a size of the order of 10^{-10} m, are excellent diffracting objects for waves whose wavelength is also of the order of 10^{-10} m. Unfortunately, we cannot study diffraction by a single atom, but in Chapter 3 we discussed and illustrated the beautiful patterns produced when X rays are incident on a crystal, in which the regular spacing and order of the atoms causes readily identified interference maxima to appear.

To examine the wave nature of electrons we therefore adopt the following procedure. A beam of electrons is accelerated through a potential V, acquiring a nonrelativistic kinetic energy $K = eV$ and a momentum $p = \sqrt{2mK}$. Wave mechanics would describe the beam of electrons as a wave of wavelength $\lambda = h/p$. The beam strikes a crystal in exactly the

Electron diffraction

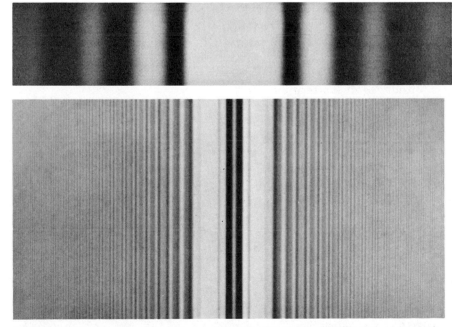

FIGURE 4.1 (Top) Diffraction pattern of a single slit. (Bottom) Diffraction pattern of a fine wire. (*Source:* Cagnet, et al., *Atlas of Optical Phenomena*, Springer-Verlag, 1971.)

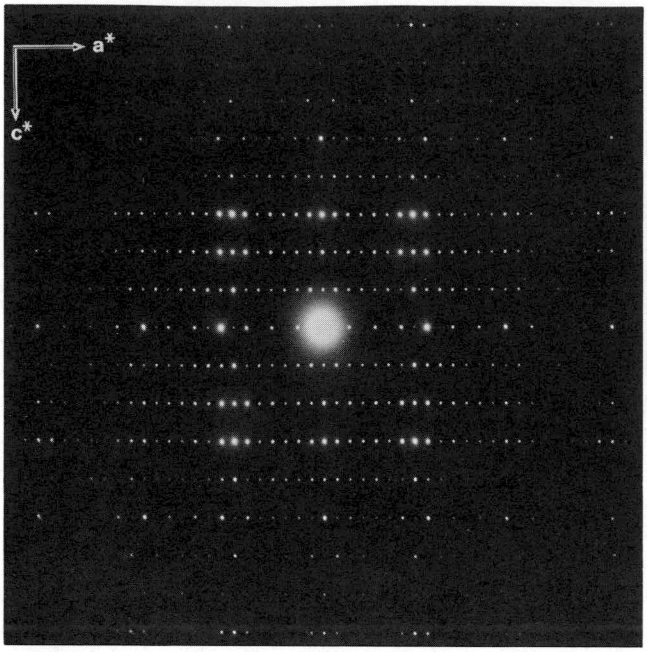

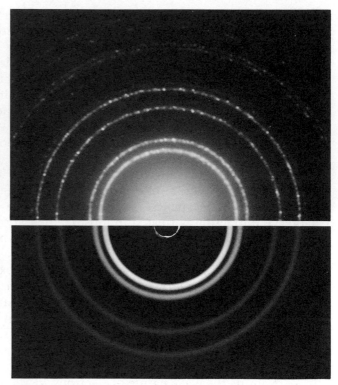

FIGURE 4.2 An electron diffraction pattern. Each bright dot is a region of constructive interference, as in the X-ray diffraction patterns of Figures 3.7 and 3.8. The target is a crystal of $Ti_2Nb_{10}O_{29}$. (Courtesy of Sumio Iijima, Arizona State University.)

FIGURE 4.3 Comparison of X-ray diffraction and electron diffraction. The upper half of the figure shows the result of scattering of 0.071 nm X rays by an aluminum foil, and the lower half shows the result of scattering of 600 eV electrons by aluminum. (The wavelengths are different so the scales of the two halves have been adjusted.) (*Source:* Education Development Center, Newton, MA.)

same way as the beam of X rays in Figure 3.7, and the scattered beam is photographed. Such a photograph is shown in Figure 4.2. The similarity between Figure 4.2 (electron diffraction) and Figures 3.7 and 3.8 (X-ray diffraction) strongly suggests that the electrons are behaving as waves.

A direct comparison of the "rings" produced in scattering by polycrystalline materials (see Figure 3.9) is shown in Figure 4.3. The effect of the comparison between electron scattering and X-ray scattering is striking, and there is again strong evidence for the similarity in the wave behavior of electrons and X rays.

The wave nature of particles is not exclusive to electrons; *any* particle with momentum p has deBroglie wavelength h/p. Neutrons are produced in nuclear reactors with kinetic energies corresponding to wavelengths of roughly 0.1 nm; these also should be suitable for diffraction by crystals. Figure 4.4 shows that diffraction of neutrons by a salt crystal produces the same characteristic patterns as the diffraction of electrons or X rays.

To study the nuclei of atoms, much smaller wavelengths are needed, of the order of 10^{-15} m. Figure 4.5 shows the diffraction pattern produced by the scattering of 1-GeV kinetic energy protons by oxygen nuclei. The maxima and minima of the diffraction pattern are similar to those of single-slit diffraction as shown in Figure 4.1. (The intensity at

FIGURE 4.4 Diffraction of neutrons by a sodium chloride crystal. (*Source:* Eisberg & Resnick, *Quantum Physics*, John Wiley & Sons, 1974.)

the minima does not fall to zero because nuclei do not have a sharp boundary. The determination of nuclear sizes from such diffraction patterns is discussed in Chapter 9.)

The interference and diffraction effects of Figures 4.2 through 4.5 are not representative merely of one special type of particle or of one special variety of target. They are examples of a *general* phenomenon, the wave behavior of particles, unobserved before 1920 because the proper experiments had not yet been done. Today this wave nature is used regularly as a standard tool by atomic physicists in studying atomic properties, by nuclear physicists in studying nuclear properties, by solid-state physicists, physical chemists, and all manner of other materials scientists in studying the properties of matter, by biologists and biochemists in studying microscopic life with the electron microscope, and by astrophysicists in attempting to explain many of the curious objects found in our universe.

The first experimental confirmation of the wave nature of electrons (and the *qualitative* confirmation of the deBroglie relationship $\lambda = h/p$) followed soon after deBroglie's original hypothesis. In 1926, at the Bell Telephone Laboratories, Clinton Davisson and Lester Germer were investigating the reflection of electron beams from the surface of nickel crystals. A schematic view of their apparatus is shown in Figure 4.6. A beam of electrons from a heated filament is accelerated through a potential difference V. Passing through a small aperture, the beam strikes a single crystal of nickel. Electrons are scattered in all directions by the

Davisson-Germer experiment

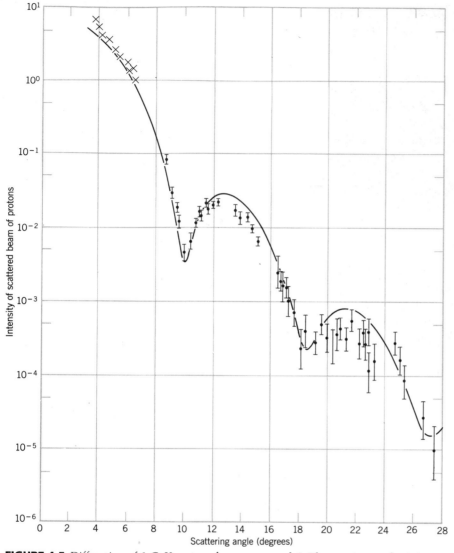

FIGURE 4.5 Diffraction of 1 GeV protons by oxygen nuclei. The maxima and minima are similar to those of single-slit diffraction of light waves.

atoms of the crystal, some of them striking a detector, which can be moved to any angle ϕ relative to the incident beam and which measures the intensity of the electron beam scattered at that angle.

If we assume that each of the atoms of the crystal can act as a scatterer, then the scattered *electron waves* can interfere, and we have a sort of crystal diffraction grating for the electron waves. Any plane of atoms in the crystal has the regularly spaced scattering centers that can produce an interference pattern; the scattering from one certain set of such planes is shown in Figure 4.7. The scattering angle θ, as we defined it in Chapter 3, is just $90° − \phi/2$.

An intense reflected beam will be observed at the angle ϕ whenever the *Bragg condition* (3.19) for constructive interference is satisfied. The

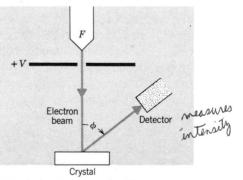

FIGURE 4.6 Apparatus used by Davisson and Germer to study electron diffraction. Electrons leave the filament F and are accelerated by the voltage V. The beam strikes a crystal and the scattered beam is detected at an angle ϕ relative to the incident beam. The detector can be moved in the range 0 to 90°.

atomic spacing a is related to the distance d according to

$$d = a \sin\left(\frac{\phi}{2}\right) \qquad (4.2)$$

The data recorded by Davisson and Germer are illustrated in Figure 4.8, which shows the intensity of the reflected beam at angles ϕ between 0 and 90°. The intense beam at $\phi = 50°$ occurs for $V = 54$ volts. Equations (4.2) and (3.19) then give us the value for the wavelength of the electron beam for scattering at 50°, since the lattice spacing of nickel atoms is known from other experiments to be $a = 0.215$ nm:

$$d = a \sin 25° = 0.0909 \text{ nm}$$

$$\lambda = 2d \sin \theta = 0.165 \text{ nm}$$

We can compare this value with that expected on the basis of the deBroglie theory. An electron accelerated through a potential difference of 54 V has a kinetic energy of 54 eV and therefore a momentum of

$$p = \sqrt{2mK} = \frac{1}{c}\sqrt{2mc^2K} = \frac{1}{c}(7430 \text{ eV})$$

The deBroglie wavelength is $\lambda = h/p = hc/pc$. Using $hc = 1240$ eV·nm,

$$\lambda = \frac{1240 \text{ eV·nm}}{7430 \text{ eV}} = 0.167 \text{ nm}$$

This is in excellent agreement with the value found from the diffraction maximum, and provides strong evidence in favor of the deBroglie theory.

For this experimental work, Davisson received the Nobel prize in 1937, in conjunction with G. P. Thomson, who obtained simultaneous proof of the wave nature of electrons by means of electron diffraction in thin metal foils, as shown in Figure 4.3.

The Davisson-Germer experiment demonstrates the diffraction of electron beams. It is also possible to demonstrate the wave nature of electrons by doing Young's double-slit experiment with an electron beam, but it is difficult to prepare a suitable double slit that can transmit an electron beam. This experiment was done in 1961 by Claus Jönsson, who accelerated a beam of electrons through 50,000 V and passed the beam through a double slit of separation 2.0×10^{-6} m and width 0.5×10^{-6} m. The double-slit interference pattern photographed by Jönsson is shown in Figure 4.9. The similarity with the double-slit patterns obtained with light sources (Figure 3.2) is striking, and again shows evidence for the wave nature of electrons.

When we do such a double-slit experiment with electrons, it is tempting to try to determine through which slit the electron passes. For example, we could surround each slit with a loop of wire, as in Figure 4.10. Since a moving electron behaves like an electric current, the passage of an electron through the plane of the loop is a changing current, and this gives rise to an induced current in the loop, which we could detect on the meter. With such apparatus we could (in principle, at least) observe the passage of a single electron through one or the other of the

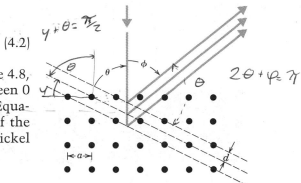

FIGURE 4.7 Detail of scattering from crystal planes. The atoms in the crystal are separated by a distance a, and at the Bragg angle θ the spacing of atomic planes is d. Constructive interference of the scattered beams occurs when the Bragg condition is satisfied.

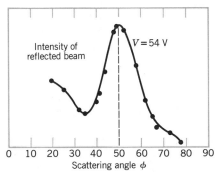

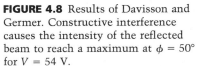

FIGURE 4.8 Results of Davisson and Germer. Constructive interference causes the intensity of the reflected beam to reach a maximum at $\phi = 50°$ for $V = 54$ V.

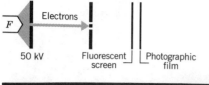

FIGURE 4.9 Double-slit diffraction with electrons. Electrons from the filament *F* are accelerated through 50 kV and pass through the double slit. They produce a visible pattern when they strike a fluorescent screen (like a TV screen) and the resulting visual pattern is photographed. The resulting interference pattern is shown. (*Source:* C. Jönsson, *Am. J. Phys. 42,* 4 (1974).)

slits. If we performed this experiment, we would indeed record the passage of electrons through one slit or the other, but in the process we would find that our interference pattern had been destroyed. Instead, we would record an intensity pattern similar to that shown in Figure 4.10, with "hits" in front of each of the slits, but no interference fringes. (The magnetic field of the induced current in the loop changes the motion of the electron just enough to destroy the interference pattern.) No matter what means we try in order to determine through which slit an electron

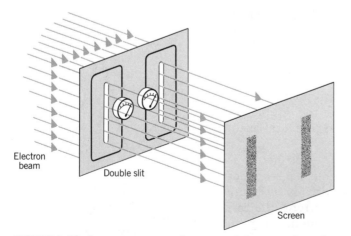

FIGURE 4.10 Apparatus to record passage of electrons through slits. Each slit is surrounded by a loop of wire with a current meter that signals the passage of an electron through the slit. No interference finges are seen on the screen.

passes, we find that the interference pattern is destroyed. If the electron is to behave like a classical *particle* it must pass through one slit or the other; only a *wave*, on the other hand, can pass through both slits. When we try to observe the electron passing through one slit or the other, we are investigating the *particle* aspects of its behavior, and therefore cannot see its *wave* nature (the interference pattern). The electron will behave like a particle *or* a wave, but we cannot observe both aspects of its behavior simultaneously. This is the basis of the *principle of complementarity*, which asserts that the complete description of a physical entity such as a photon or an electron cannot be done in terms of particles or waves exclusively, but that both aspects must be considered. Moreover, the particle and wave natures cannot be observed simultaneously, and the aspect of the behavior of the system that we observe depends on the kind of experiment we are performing.

Principle of complementarity

We explore in this section another important difference between classical particles and waves. Let us consider a wave of the form $y = y_1 \sin k_1 x$, as shown in Figure 4.11. This is a wave that repeats itself endlessly from $x = -\infty$ to $x = +\infty$. If we ask the question "Where is the wave located?" we cannot provide an answer—it is everywhere. (Its wavelength, on the other hand, is precisely determined and is equal to $2\pi/k_1$.) If we are to use a wave to represent a particle, the wave must have one of the important attributes of a particle—it must be *localized*, or able to be confined to a relatively small (atom sized or nucleus sized, for example) region of space. The pure sine wave is of no use in localizing particles.

Now consider what happens when we add to our original wave another wave of slightly different wavelength (i.e., different k), so that $y = y_1 \sin k_1 x + y_2 \sin k_2 x$. The characteristic pattern, known in the case of sound waves as "beats," is produced as shown in Figure 4.12. The pattern still repeats endlessly from $x = -\infty$ to $x = +\infty$, but we know a bit more about the "position" of the wave—at certain values of x the medium is less likely to be "waving" than at others (or at least it is "waving" with a smaller amplitude). In Figure 4.12, we would observe vibration at the point $x = x_A$, but not at $x = x_B$. The state of our knowledge of the "position" of the wave has improved, but it is at the expense of our knowledge of its wavelength—we added together two different wavelengths and so the wavelength is no longer precisely determined.

If we continued to add waves of different wavelengths (different wave numbers k), with properly chosen amplitudes and phases, we could eventually reach a situation similar to that illustrated in Figure 4.13. Such a wave has virtually no amplitude outside a rather narrow

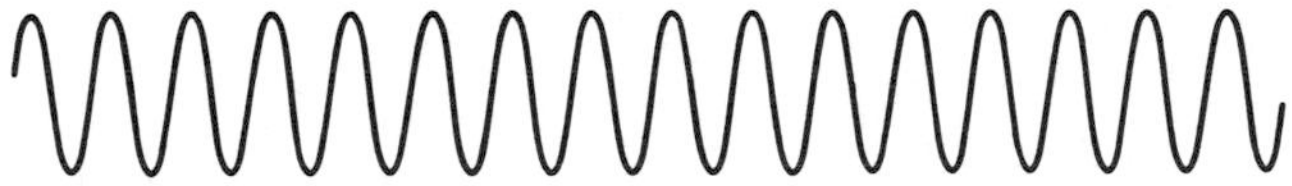

4.2 UNCERTAINTY RELATIONSHIPS FOR CLASSICAL WAVES

$y = y_1 \sin k_1 x$

FIGURE 4.11 A pure sine wave, which extends from $-\infty$ to $+\infty$.

Handwritten note (top right): $y = y_1 \sin k_1 x + y_2 \sin k_2 x$ "beats" occur

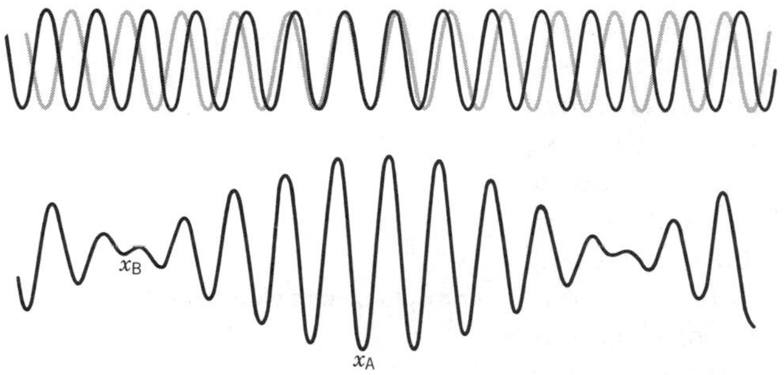

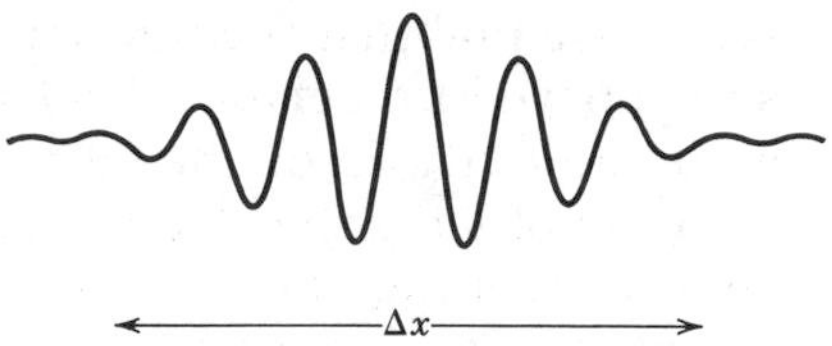

FIGURE 4.13 The resultant of the addition of many sine waves (of different wavelengths and possibly different amplitudes.)

FIGURE 4.12 The superposition of two sine waves of nearly equal wavelengths to give beats. The two sine waves differ in wavelength by 10 percent but have the same amplitude.

region of space Δx (Δx is not precisely defined, but is a rough measure of the region over which the wave has reasonably large amplitude). In order to achieve this situation, we had to add together a large number of waves of different wave numbers k. The wave thus represents a range of wave numbers (wavelengths) that we denote by Δk. When we had a single sine wave, Δk was zero (only one k) and Δx was infinite (the wave extended throughout all space). As we *increased* Δk (by adding more waves), we *decreased* Δx (the wave became more confined). We seem to have an inverse relationship between Δx and Δk; as one decreases, the other increases. An approximate mathematical relationship between Δx and Δk is

$$\Delta x \, \Delta k \sim 1 \qquad (4.3)$$

Wave number—position uncertainty relationship

Handwritten note (right): $\hbar = \dfrac{h}{2\pi}$ page 97

$\Delta x \, \Delta p_x \sim \hbar$

where the wavy equal sign is taken to mean "of order of magnitude." (Since Δx and Δk have not yet been precisely defined, they should be regarded as estimates, and Equation (4.3) is a rough indication of their relationship.) Equation (4.3) asserts that the product of Δx, the spatial extent of the wave, and Δk, the range of wave numbers it contains, is of order of magnitude one.

For any type of wave, the position can only be determined at the expense of our knowledge of its wavelength. This statement, and its mathematical representation given in Equation (4.3), are the first of our "uncertainty relationships" for classical waves. The position and wavelength (wave number) are mutually "uncertain" to the degree given by Equation (4.3).

We can interpret this relationship in a slightly different way. Suppose we attempt to measure the wavelength of a classical wave, such as a water wave. We can do this by measuring the distance between two adjacent wave crests. Suppose the wave is a very short pulse with only one wave crest (Figure 4.14). Measuring λ is very difficult, and we are likely to make a large error, perhaps of the order of one wavelength. That is, when the spatial extent of the wave $\Delta x \sim \lambda$, then $\Delta \lambda \sim \lambda$. (Remember, $\sim$ means "of the order of.") For this wave, we then have $\Delta x \, \Delta \lambda \sim \lambda^2$. Suppose now that the wave extends over many wavelengths, so that $\Delta x \sim N\lambda$. Now we can determine λ with much greater precision. Counting the number of wavelengths in Δx, we still may make an error of the order of

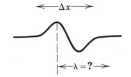

FIGURE 4.14 Two different groups of waves.

one wavelength (perhaps $\frac{1}{2}$ or $\frac{1}{3}$ or $\frac{1}{4}$, but still of order one) out of N, so now $\Delta\lambda \sim \lambda/N$, and once again $\Delta x\,\Delta\lambda \sim \lambda^2$. This uncertainty relationship, which connects the "size" of a wave with the uncertainty in determining its wavelength, can be shown to be equivalent to Equation (4.3).

Let us now try to measure the frequency of a wave (a sound wave, for example). Suppose we could observe the individual oscillations (on an oscilloscope, for example) and had a suitable apparatus for counting them. If we counted for 1 s and recorded 100 oscillations, we would then have a frequency of 100 Hz. However, we cannot be sure we counted *exactly* 100 oscillations. How far along the way was the one hundred and first oscillation when the 1-s interval ended? If half of one cycle had gone by, we still would have recorded only 100 oscillations, but the actual frequency would be 100.5 Hz, compared with our measured 100 Hz. There is an *uncertainty* in the frequency of perhaps 0.5 Hz. Suppose we now count for 2 s. We would record 200 oscillations, and would again have a deduced frequency of 100 Hz, but would again be unsure of how much of the two hundred and first oscillation had occurred. If half of one cycle had again gone by, we would have 200.5 oscillations in 2 s, or an actual frequency of 100.25 Hz. The uncertainty in the frequency (now 0.25 Hz) has been reduced by a factor of 2 when we doubled the time interval over which we measured. We therefore have the same sort of inverse relationship we deduced previously: the uncertainty in the frequency, $\Delta\nu$, is inversely proportional to the time, Δt, over which the measurement is made, and using the circular frequency $\omega = 2\pi\nu$ we can write

$$\Delta\omega\,\Delta t \sim 1 \tag{4.4}$$

Frequency—time uncertainty relationship

This is the second of the uncertainty relationships for classical waves, and is similar to Equation (4.3) in that it gives a rough relationship between estimates of uncertainties.

In a measurement of the wavelength of water waves, 10 wave crests are counted in a distance of 200 cm. Estimate the minimum uncertainty in the wavelength that might be obtained from this experiment.

EXAMPLE 4.2

In order to estimate $\Delta\lambda$, we need to relate it to Δk. We know k and λ are related by

$$k = \frac{2\pi}{\lambda}$$

and, after differentiating,

$$dk = \frac{-2\pi}{\lambda^2}\,d\lambda$$

Replacing the differentials with small intervals and taking magnitudes only, we find

$$\Delta k = \frac{2\pi}{\lambda^2}\,\Delta\lambda$$

(Think about how this is different from $\Delta k = 2\pi/\Delta\lambda$. Why is this latter equation incorrect? What happens when λ is known exactly?) From Equation (4.3), $\Delta x\,\Delta k \sim 1$, so

$$\Delta x\left(\frac{2\pi}{\lambda^2}\,\Delta\lambda\right) \sim 1$$

or

$$\Delta\lambda \sim \frac{1}{\Delta x}\frac{\lambda^2}{2\pi} = \frac{1}{200}\frac{(20)^2}{2\pi} \sim 0.3 \text{ cm}$$

An electronics salesman offers to sell you a frequency-measuring device. When hooked up to a sinusoidal signal, it automatically displays the frequency of the signal, and to account for frequency variations, the frequency is remeasured and the display updated once each second. The salesman claims the device to be accurate to 0.01 Hz. Is this claim valid?

EXAMPLE 4.3

Based on Equation (4.4), we know that a measurement of frequency in a time $\Delta t = 1$ s must have an associated uncertainty of $\Delta\omega \sim 1$ rad/s. The salesman's quoted accuracy is $\Delta\omega = 2\pi(\Delta\nu) = 0.06$ rad/s, and thus we have reason to doubt the quoted specification.

The uncertainty relationships discussed in the previous section apply to *all* waves, and we should therefore apply them to deBroglie waves. We can use the basic deBroglie relationship $p = h/\lambda$ along with the expression $k = 2\pi/\lambda$ to find $p = hk/2\pi$, which relates the momentum of a particle to the wave number of its deBroglie wave. The combination $h/2\pi$ occurs frequently in wave mechanics and is given the special symbol $\hbar$ ("h-bar")

4.3 HEISENBERG UNCERTAINTY RELATIONSHIPS

$$\hbar = \frac{h}{2\pi} = 1.05 \times 10^{-34} \text{ J·s}$$

$$= 6.58 \times 10^{-16} \text{ eV·s}$$

In terms of $\hbar$,

$$p = \hbar k \qquad (4.5)$$

and so $\Delta k = \Delta p/\hbar$. From the uncertainty relationship (4.3) we then have

Heisenberg uncertainty relationships

$$\Delta x\,\Delta p_x \sim \hbar \qquad (4.6)$$

The x subscript has been added to the momentum to remind us that Equation (4.6) applies to motion in a given direction and relates the uncertainties in position and momentum in that direction only. Similar and independent relationships can be applied in the other directions as necessary; thus $\Delta y \, \Delta p_y \sim \hbar$ or $\Delta z \, \Delta p_z \sim \hbar$.

The deBroglie relationship $E = h\nu$ can be written as $E = \hbar\omega$. Thus $\Delta\omega = \Delta E/\hbar$, and the uncertainty relationship (4.4) becomes

$$\Delta E \, \Delta t \sim \hbar \qquad (4.7)$$

Equations (4.6) and (4.7) are the *Heisenberg uncertainty relationships,* and they are the mathematical representation of the Heisenberg *uncertainty principle,* which states that no experiment can ever be performed to give uncertainties below the limits expressed by Equations (4.6) and (4.7).

These relationships give an estimate of the minimum uncertainty that can result from any experiment; measurement of the position and momentum of a particle will give a spread of values of widths Δx and Δp_x. We may, for other reasons, do much worse than (4.6) and (4.7), but *we can do no better.* (Occasionally, you may see these relationships written with $\hbar/2$ or h, rather than $\hbar$, on the right-hand side, or else with $>$ rather than $\sim$ showing the equality. This difference is not very important, since (4.6) and (4.7) give us only estimates. The actual uncertainties Δx and Δp_x depend on the distribution of different wave numbers (or wavelengths) we use to restrict the wave to the region Δx; the most compact distribution gives $\Delta x \, \Delta p_x = \hbar/2$ and all other distributions will have $\Delta x \, \Delta p_x > \hbar/2$. We are therefore safe in using $\hbar$ as an estimate.)

These relationships have a profound impact on our view of nature. It is quite acceptable to say that there is an uncertainty in locating the position of a water wave. It is quite another matter to make the same statement about a deBroglie wave, since *there is an implied corresponding uncertainty in the position of the particle.* Equations (4.6) and (4.7) say that *nature imposes a limit on the accuracy with which we can do experiments;* no matter how well designed our measuring apparatus might be, we still can do no better than Equations (4.6) or (4.7). To determine the momentum accurately, we must measure over a long distance Δx; if we wish to confine a particle to a small region of space Δx, we lose our ability to measure its momentum accurately. To measure an energy with small uncertainty takes a long time Δt; if a particle only lives for a short time, its energy uncertainty will be large.

The following examples give applications of these uncertainty relationships.

Werner Heisenberg (1901–1976, Germany). Best known for the uncertainty principle, he also developed a complete formulation of the quantum theory based on matrices.

Suppose a particle of mass m is confined to move in one dimension (imagine a bead sliding along a wire) between walls with which the particle collides elastically and rebounds. Initially we move the reflecting walls to $\pm\infty$ and place the particle on the wire such that it is at rest. (a) What will be the likely outcome of a measurement of the particle's position? (b) Describe what happens as the walls are moved in from $\pm\infty$.

EXAMPLE 4.4

(a) If the particle is at rest, $p_x = 0$ exactly, and thus $\Delta p_x = 0$. (We know the momentum with no uncertainty.) Therefore, according to Equation (4.6), $\Delta x = \infty$. The location of the particle is *completely* uncertain! We are permitted no knowledge at all of the location of the particle. All values of x are equally likely, and a measurement of the particle's position could yield any result at all. (b) Suppose we move the walls in from $x = \pm\infty$ so that they are now at $x = \pm L/2$ (separated by a distance L). The particle must then be located somewhere in this interval of length L. If we have no other information about its location, our knowledge of its position is uncertain to within the distance L, and thus $\Delta x \sim L$. Equation (4.6) requires $\Delta p_x \sim \hbar/L$. Its momentum is now uncertain to within this amount. Before we began to move the walls, every measurement of the momentum of the particle yielded $p_x = 0$, and thus $\Delta p_x = 0$. Now, the measurements are still clustered about $p_x = 0$ but not all measurements yield exactly $p_x = 0$. The distribution of measurements of p_x might look something like Figure 4.15. We were initially sure of the particle's momentum—it was zero. After the walls are moved in, the *average* of many measurements of p_x is zero, but not every measurement gives zero, and the spread Δp_x of the distribution of p_x values increases as the walls come together according to $\Delta p_x \sim \hbar/L$.

Of course, a classical particle cannot go from a state of rest to a state of motion without the application of a force. How then can the particle have a momentum different from zero? The dilemma here results from using the word "particle." We have seen that the terms "particle" and "wave" are not exclusive in quantum physics, and the proper description of a physical system must include both aspects. It is the *wave* behavior that causes the spreading of the momentum distribution as the spacing L is reduced. (A classical wave analogy, which like all classical analogies is of limited scope and should not be taken too seriously: a guitar string is plucked and then its length is gradually reduced, by sliding a finger along the fingerboard. The string vibrates at increasingly higher frequencies and therefore more rapidly.) To localize a particle, we must localize the amplitude of its associated wave, which is done by

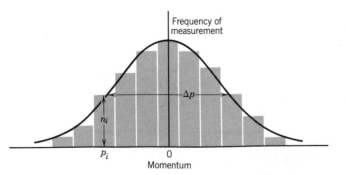

FIGURE 4.15 The momentum of a particle confined to a region Δx. Repeating the measurement many times, each value p_i is measured n_i times. The average momentum is zero, and the distribution has a width $\Delta p \sim \hbar/\Delta x$.

adding together many different component waves; the smaller L is made, the more waves must be added, in accordance with Equation (4.3). These various waves travel through the medium, in general at different speeds, reflecting back and forth between the walls. When the walls are at $\pm\infty$, only one wave is needed, no dispersion or reflection occurs, and the particle's behavior does not change with time. When the walls are moved in, more waves are needed, dispersion and reflection can occur, and it may occasionally happen that the phase and amplitude relationships between the different component waves combine to produce a momentary imbalance of waves moving to the right or to the left, which we observe as a nonzero value of p_x.

A large number of measurements will probably show the particle moving to the right as often as to the left, and so its *average* momentum p_{av} is zero, since the opposite momenta cancel. The average *magnitude* of the momentum $|p|_{av}$ is not zero. ($|p|_{av}$ is zero only when all p's are zero.) The closer the walls become, the more reflection occurs, and the greater the chance for several momentum components to interfere constructively and give a large momentum in a single direction. Consequently, the particle begins to "move" more rapidly; even though p_{av} is still zero, $|p|_{av}$ has become larger. It therefore seems that Δp is related to $|p|_{av}$, which is related to $(p^2)_{av}$. A rigorous definition of Δp is

$$\Delta p = \sqrt{(p^2)_{av} - (p_{av})^2} \tag{4.8}$$

Note the similarity of this definition to the statistical concept of the standard deviation of a quantity x that has an average value $\bar{x}$

$$\sigma_x = \sqrt{\frac{1}{N}\sum_{i=1}^{N}(x_i - \bar{x})^2}$$

$$= \sqrt{\frac{1}{N}\sum_{i=1}^{N}x_i^2 - (\bar{x})^2} = \sqrt{(x^2)_{av} - (x_{av})^2}$$

EXAMPLE 4.5

A beam of monoenergetic electrons (of momentum p_y) is moving in the y direction (i.e., $p_x = 0$). The beam passes through a narrow slit of width a parallel to the x axis. Find the uncertainty in the x component of the particle's momentum following its passage through the slit, and compare the interpretation with the conventional description of single-slit diffraction.

Solution

Since the beam was initially moving in the y direction, we know that it has no momentum in the x direction, so that p_x is exactly zero, and therefore $\Delta p_x = 0$. Thus, by Equation (4.6), $\Delta x = \infty$, and we have no knowledge of the location of the particles. Once they have passed through the slit, the location is no longer so uncertain—we have reduced Δx by the passage through the slit. Now $\Delta x \sim a$, so that, according to Equation (4.6), $\Delta p_x \sim \hbar/a$. Thus a measurement beyond the slit no longer shows the particle moving purely in the y direction. Even though p_y does not change in passing through the slit, p_x will no longer be measured to be exclusively zero, but will show a range of values about zero, distributed with a width of order $\hbar/a$. In passing through the slit, the typical particle acquires an

x component of momentum of roughly $\hbar/a$, according to the uncertainty principle.

In the conventional observation of single-slit diffraction, a screen placed beyond the slit shows a pattern similar to that of Figure 4.16. Let us define the most probable "location" of the particle at the screen to be between the first diffraction minima on either side of center. Wave theory gives the location of these minima at angles θ for which

$$a \sin \theta = \lambda \qquad (4.9)$$

The angle θ is small, and we can approximate $\sin \theta \cong \tan \theta = x/D$. Also, the wavelength of these "particle waves" is given by the deBroglie relationship, so that $\lambda = h/p_y$

$$a \left(\frac{x}{D} \right) = \frac{h}{p_y}$$

If the particle moves with an x component of momentum p_x, then

$$\frac{p_x}{p_y} = \frac{x}{D}$$

so that

$$p_x = p_y \frac{x}{D} = \frac{h}{a}$$

This result is consistent with the result $p_x \sim \hbar/a$ derived above using the uncertainty principle, and we conclude that the two descriptions are roughly equivalent.

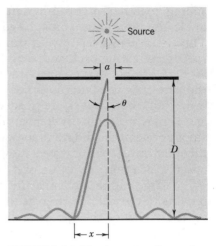

FIGURE 4.16 The intensity observed in single-slit diffraction. The slit has width a and the screen is at a distance D from the slit. The first diffraction minimum is a distance x from the center.

At $t = 0$, we have performed a measurement of the position of a particle, and we have found it to be somewhere in the interval $-a/2 \leq x \leq a/2$. Where will we be likely to find it at times $t > 0$?

EXAMPLE 4.6

Associated with our determination of the particle's position to within an uncertainty $\Delta x \sim a$ is a corresponding uncertainty in its momentum $\Delta p_x \sim \hbar/a$, and therefore in its velocity $\Delta v_x = \Delta p_x/m \sim \hbar/ma$. If the particle were moving with velocity v_x, then in a time t it would move from $x = 0$ to $x = v_x t$ if there were no uncertainty in its position or velocity. The uncertainty in its starting position $(-a/2$ to $a/2)$ suggests that, even if there were no uncertainty in its velocity, its position at t would be subject to the same uncertainty, and we could locate the particle only to within the region $(v_x t - a/2$ to $(v_x t + a/2)$. However, the uncertainty in v_x causes even more spreading—the particle might be moving with any speed in the range $(v_x - \Delta v_x$ to $(v_x + \Delta v_x)$. This uncertainty results in a range of possible positions at time t from $(v_x - \Delta v_x)t$ to $(v_x + \Delta v_x)t$, an uncertainty of $2(\Delta v_x)t$ or $2\hbar t/ma$. The uncertainty at time t thus consists of two parts, a contribution of constant magnitude a and a contribution $2\hbar t/ma$, which increases with time. Note that the rate at which this contribution spreads is inversely proportional to a—the more efficient we are in locating the particle at $t = 0$, the less we know about it at later times. Figure 4.17 shows a representation of this effect for several different values of a, assuming the two contributions add quadratically to give the total uncertainty. This phenomenon reminds us of the single-slit diffraction experiment, in which the narrower we make the slit, the more the wave spreads after passing through the slit.

Solution

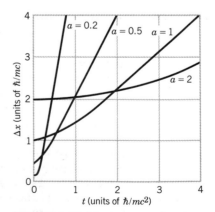

FIGURE 4.17 The increase of positional uncertainty with time. The smaller Δx is at $t = 0$, the faster it grows and the wider it is at large times.

EXAMPLE 4.7

In nuclear beta decay, electrons are observed to be ejected from the atomic nucleus. Suppose we assume that electrons are somehow trapped within the nucleus, and that occasionally one escapes and is observed in the laboratory. Take the diameter of a typical nucleus to be 10^{-14} m, and use the uncertainty principle to estimate the range of kinetic energies that such an electron must have.

Solution

To be trapped in a region $\Delta x \sim 10^{-14}$ m the electron must have a corresponding Δp_x of at least $\hbar/\Delta x = 1 \times 10^{-20}$ kg·m/s. That is, measurements of the momenta of electrons confined in such a small region of space would yield a distribution of values (around zero) with a range of about 1×10^{-20} kg·m/s. This corresponds to a range of kinetic energies of about 20 MeV, which is actually a lower limit, as determined by Equation (4.6). Electrons emitted in beta decay, however, have energies below about 1 MeV. This calculation suggests that electrons of such low energies cannot be confined within a region of space of dimension 10^{-14} m, and that we must find a different explanation for the source of beta-decay electrons. (As we will learn in Chapter 9, these electrons are "manufactured" by the nucleus at the instant of decay.)

[handwritten: pion — heavier than electron lighter than proton]

[handwritten: π^+, π^0, π^-]

EXAMPLE 4.8

(a) A charged pi meson has a rest energy of 140 MeV and a lifetime of 26 ns. Find the energy uncertainty of the pi meson, expressed in MeV and also as a fraction of its rest energy. (b) Repeat for the uncharged pi meson, with a rest energy of 135 MeV and a lifetime of 8.3×10^{-17} s. (c) Repeat for the rho meson, with a rest energy of 765 MeV and a lifetime of 4.4×10^{-24} s.

[handwritten: $E_0 = m_\pi c^2 = 140 MeV$]

Solution

[handwritten: $M_e c^2 = 0.511 MeV$]

(a) If the pi meson lives for 26 ns, we have only that much time in which to measure its rest energy, and Equation (4.7) tells us that *any* energy measurement done in a time Δt is uncertain by an amount of at least $\Delta E \sim \hbar/\Delta t$.

$$\Delta E = \frac{\hbar}{\Delta t} = \frac{6.58 \times 10^{-16} \text{ eV·s}}{26 \times 10^{-9} \text{ s}} = 2.5 \times 10^{-8} \text{ eV}$$

$$= 2.5 \times 10^{-14} \text{ MeV}$$

$$\frac{\Delta E}{E} = \frac{2.5 \times 10^{-14} \text{ MeV}}{140 \text{ MeV}} = 1.8 \times 10^{-16}$$

[handwritten: π^- rest mass can't be measured to the precision of 10^{-16}]

(b) In a similar way,

$$\Delta E = \frac{\hbar}{\Delta t} = \frac{6.58 \times 10^{-16} \text{ eV·s}}{8.3 \times 10^{-17} \text{ s}} = 7.9 \text{ eV}$$

$$= 7.9 \times 10^{-6} \text{ MeV}$$

$$\frac{\Delta E}{E} = \frac{7.9 \times 10^{-6} \text{ MeV}}{135 \text{ MeV}} = 5.9 \times 10^{-8}$$

(c) For the rho meson,

$$\Delta E = \frac{\hbar}{\Delta t} = \frac{6.58 \times 10^{-16} \text{ eV·s}}{4.4 \times 10^{-24} \text{ s}} = 1.5 \times 10^8 \text{ eV}$$

$$= 150 \text{ MeV}$$

$$\frac{\Delta E}{E} = \frac{150 \text{ MeV}}{765 \text{ MeV}} = 0.20$$

In the first case, the uncertainty principle does not give a large enough effect to be measured—particle rest masses cannot be measured to a precision of

10^{-16} (about 10^{-6} is the best precision that we can obtain). In the second example, the uncertainty principle contributes at about the level of 10^{-7}, which approaches the limit of our measuring instruments and therefore might be observable in the laboratory. In the third example, we see that the uncertainty principle can contribute substantially to the precision of our knowledge of the rest energy of the rho meson; measurements of its rest energy will cluster about 765 MeV with a spread of 150 MeV, and no matter how precise an instrument we use to measure the rest energy, we can never reduce that spread.

The lifetimes of very short-lived particles such as the rho meson cannot be measured directly. In practice we reverse the procedure of the calculation of this example—we measure its rest energy, which gives a distribution of the form of Figure 4.15, and from the "width" ΔE of the distribution we deduce the lifetime using Equation (4.7). This procedure is discussed in Chapter 11.

Estimate the minimum velocity of a billiard ball ($m \sim 100$ g) confined to a billiard table of dimension 1 m.

EXAMPLE 4.9

For $\Delta x \sim 1$ m, we have

Solution

$$\Delta p_x \sim \frac{\hbar}{\Delta x} = \frac{1.05 \times 10^{-34} \text{ J·s}}{1 \text{ m}} = 1 \times 10^{-34} \text{ kg·m/s}$$

so

$$\Delta v_x = \frac{\Delta p_x}{m} = \frac{1 \times 10^{-34} \text{ kg·m/s}}{0.1 \text{ kg}} = 1 \times 10^{-33} \text{ m/s}$$

Thus quantum effects "cause" motion of the billiard ball with a speed the order of 1×10^{-33} m/s. At this speed, the ball would move a distance of $\frac{1}{100}$ of the diameter of an atomic nucleus in a time equal to the age of the universe! Once again, we see that quantum effects are not observable with macroscopic objects.

A pure sine wave is completely unlocalized—it extends from $-\infty$ to $+\infty$. A classical particle, on the other hand, is completely localized. Our quantum description mixes particles and waves. The particles are approximately, but not completely, localized. An electron, for example, is bound to a specific atom. We know its position to within an uncertainty of the order of the diameter of the atom (10^{-10} m), but we don't know exactly where it is within that atom. The method used in physics to describe such a situation is that of a *wave packet*. A wave packet can be considered to be the superposition of a large number of waves, which interfere constructively in the vicinity of the particle, giving the resultant wave a large amplitude, and interfere destructively far from the particle, so that the resultant wave has a small amplitude in regions where we don't expect to find the particle. The exact interpretation of these large and small amplitudes is discussed in the next section. For the present we will establish the mathematical description of the wave packet and discuss some of its properties.

An ideal wave packet would be one such as is pictured in Figure 4.13. Its amplitude is negligibly small, except for a region of space of dimension Δx. This corresponds to a particle that is localized in the region

4.4 WAVE PACKETS

of dimension Δx. From our previous discussion, we know that such a situation results in a range of momenta Δp_x as specified by Equation (4.6); since each momentum corresponds to a unique deBroglie wavelength, a range of momenta Δp_x is equivalent to a range of wavelengths $\Delta\lambda$. Thus we expect that the mathematical description of the wave packet will be in terms of the addition (superposition) of a number of waves of varying wavelengths.

For simplicity, we will consider sinusoidal waves of the form $y = A \cos kx$, where k is the wave number $2\pi/\lambda$. We also assume that all of the waves have the same amplitude. In general this will *not* be true, but it simplifies our calculations.

Let us begin with a wave of wave number k_1 and add to it a wave of nearly equal wave number $k_2 = k_1 + \Delta k$. The resulting wave form, shown in Figure 4.12, illustrates the phenomenon of "beats." The component waves start out in phase at $x = 0$, so that the resultant has its maximum amplitude there. Further along, the slight difference in wavelength causes the two waves to become out of phase, and the resultant has zero amplitude. A bit of trigonometric manipulation yields the result

$$y(x) = A \cos k_1 x + A \cos k_2 x$$

$$= 2A \cos\left(\frac{\Delta k}{2}x\right) \cos\left(\frac{k_1 + k_2}{2}x\right) \quad (4.10)$$

The last term of Equation (4.10) gives the variation of the amplitude of the resultant within the envelope specified by the first cosine term.

Let us now consider these waves as traveling waves, whose mathematical description is obtained from Equation (4.10) by substituting $(kx - \omega t)$ for kx. The circular frequency of the oscillation is ω, and $v = \omega/k$ is the *phase velocity* of the wave—the speed with which a single component wave moves through the medium. The two waves are shown again in Figure 4.18 at $t = 0$ and at a later time t. In general, the phase velocities $v_1 = \omega_1/k_1$ and $v_2 = \omega_2/k_2$ may be different. Note that the envelope moves at a speed different from that of the individual waves. Again, we can derive an explicit expression for the resultant with a bit of trigonometric manipulation:

$$y(x, t) = A \cos (k_1 x - \omega_1 t) + A \cos (k_2 x - \omega_2 t)$$

$$= 2A \cos\left(\frac{\Delta k}{2}x - \frac{\Delta\omega}{2}t\right) \cos\left(\frac{k_1 + k_2}{2}x - \frac{\omega_1 + \omega_2}{2}t\right) \quad (4.11)$$

where $\Delta\omega = \omega_2 - \omega_1$. The envelope thus moves along with a speed $v = \Delta\omega/\Delta k$, while the wave within the envelope moves with speed $(\omega_1 + \omega_2)/(k_1 + k_2)$ which, if $\Delta\omega$ and Δk are small, does not differ greatly from v_1 or v_2.

The superposition of only two waves doesn't resemble the wave packet of Figure 4.13. We can make a better approximation by adding together more sine waves of different wave numbers k_i and possibly different amplitudes $A(k_i)$:

$$y(x) = \sum_{\substack{\text{many} \\ k_i}} A(k_i) \cos k_i x \quad (4.12)$$

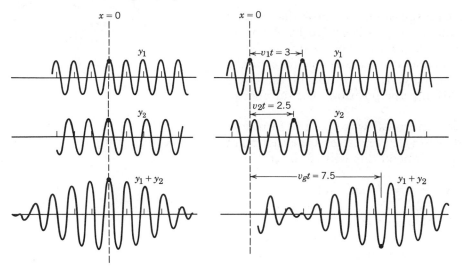

FIGURE 4.18 The group speed of a wave packet. At left is shown a "snapshot" at $t = 0$ of the waves y_1 and y_2 and their sum (y_1 has wavelength 1 unit and y_2 has wavelength $\frac{10}{9}$ unit.) Wave 1 moves with speed 3 units per second and wave 2 with speed 2.5 units per second. A snapshot at $t = 1$ s is shown at right. The two waves are not in phase until the point at 7.5 units, so the midpoint of the "beat" moves at a group speed of 7.5 units per second, in this case much greater than v_1 or v_2.

If there are many different wave numbers and if they are very close together, the sum in Equation (4.12) can be replaced with an integral:

$$y(x) = \int A(k) \cos kx \, dk \qquad (4.13)$$

where the integral is carried out over whatever range of wave numbers is permitted (possibly 0 to ∞).

For example, suppose we have a range of wave numbers from $k_0 - \Delta k/2$ to $k_0 + \Delta k/2$. If all of the waves have the same amplitude A, then from Equation (4.13) the form of the wave packet can be shown to be (see Problem 26 at the end of the chapter)

$$y(x) = \frac{2A}{x} \sin \left(\frac{\Delta k}{2} x \right) \cos k_0 x \qquad (4.14)$$

The function $\cos k_0 x$ oscillates within the envelope $(2A/x) \sin (\Delta k x/2)$. This function is illustrated in Figure 4.19 and looks much more like the finite wave packet we are after. The wave has large amplitude only in a region Δx, but we have achieved this only by adding together many different wave numbers. [This is just as required by the uncertainty principle in the form of Equation (4.3)—to make Δx small, Δk must be large.]

An even better approximation of the shape of the wave packet can be found by letting $A(k)$ vary; for example, the *Gaussian* $A(k) = e^{-(k-k_0)^2/2 (\Delta k)^2}$ gives (see Problem 27)

$$y(x) \propto e^{-(\Delta k \, x)^2/2} \cos k_0 x \qquad (4.15)$$

Once again, there is an envelope that modulates the cosine and reduces

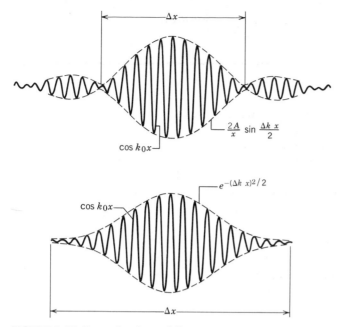

$\frac{2A}{x} \sin \frac{\Delta k \, x}{2}$

$\cos k_0 x$

$e^{-(\Delta k \, x)^2/2}$

$\cos k_0 x$

Δx

FIGURE 4.19 Example of two different wave packets. In each case there is a modulating function that reduces the amplitude of the cosine beyond the region Δx.

its amplitude outside a region of width Δx, as shown in Figure 4.19. To restrict the wave to this small region, we have once again used a large range of wave numbers.

Figure 4.19 should be regarded as a snapshot of the wave packet at a specific time, such as $t = 0$. Similarly, Equations (4.14) and (4.15) represent the waves only at $t = 0$. To convert to traveling waves we must replace kx with $kx - \omega t$, as we did in Equation (4.11). In the case of the two waves used for Equation (4.11), we found that the envelope moved with speed $\Delta\omega/\Delta k$. We generalize this to the case in which there are many different wave numbers by defining the *group velocity* as follows:

$$v_{\text{group}} = \frac{d\omega}{dk} \qquad (4.16)$$

The envelope of the wave packet moves at the group velocity, while within the envelope each individual component wave moves with its *phase velocity*

$$v_{\text{phase}} = \frac{\omega}{k} \qquad (4.17)$$

The phase velocity cannot be defined for the wave packet, but is meaningful only for a single component wave.

Group and phase velocities

(handwritten margin notes:) non dispersion $v_g = c$ $v_{phase} < c$

Certain ocean waves travel with a phase velocity $v_{\text{phase}} = \sqrt{g\lambda/2\pi}$, where g is the acceleration due to gravity. What is the group velocity of a "wave packet" of these waves? Express the result in terms of the phase velocity.

EXAMPLE 4.10

The group velocity is found from Equation (4.16). Since $k = 2\pi/\lambda$, $v_{\text{phase}} = \sqrt{g/k}$. But with $v_{\text{phase}} = \omega/k$, we have $\omega/k = \sqrt{g/k}$, so $\omega = \sqrt{gk}$ and $d\omega = \sqrt{g}(\frac{1}{2}k^{-1/2}) \, dk$. Therefore $d\omega/dk = \frac{1}{2}\sqrt{g/k}$, so $v_{\text{group}} = \frac{1}{2}v_{\text{phase}}$.

Solution

[handwritten: $w = \sqrt{gk}$]

Summarizing, a particle that is localized in a certain region of space must be represented not by a single deBroglie wave of definite frequency and wavelength, but by a wave packet representing the superposition of a large number of waves. The envelope of the wave packet moves at the group velocity $d\omega/dk$.

Our discussion is not complete unless we can provide a physical interpretation of the group velocity. Suppose we have a localized particle, represented by a group of deBroglie waves. For each component wave, the energy and frequency are related by the deBroglie relationship $E = \hbar\omega$, and similarly, the momentum and wave number are related by $p = \hbar k$. Thus the group velocity $v_g = d\omega/dk$ can also be represented in the following way:

$$v_g = \frac{d\omega}{dk} = \left(\frac{d\omega}{dE}\right)\left(\frac{dE}{dp}\right)\left(\frac{dp}{dk}\right) = \left(\frac{1}{\hbar}\right)\left(\frac{dE}{dp}\right)(\hbar) \quad (4.18)$$

[handwritten: ← wave packet]

$$v_g = \frac{dE}{dp} \quad \text{[handwritten: particle]}$$

The group velocity is *not* so much a property of the *component waves* as it is of the *medium* through which the wave packet moves. We now make the following assumption, which is central to the fundamental mechanics of the quantum theory. We assume that the response of the medium to the passage of the wave packet, given by dE/dp, is identical with the response of the medium to the passage of a particle. That is,

$$\left(\frac{dE}{dp}\right)_{\text{wave packet}} = \left(\frac{dE}{dp}\right)_{\text{particle}} \quad (4.19)$$

In the expression for the energy of a particle, only the kinetic energy K depends on momentum, so that $dE/dp = dK/dp$; since $K = p^2/2m$ for a nonrelativistic particle, $dK/dp = p/m$, which is just the velocity of the particle. Thus the right side of Equation (4.19) is just the velocity of the particle, while the left side is the group velocity of the wave packet. We have obtained an important result. *The velocity of a material particle is equal to the group velocity of the corresponding wave packet.*

Our discussion may then be summarized as follows. A particle confined to move in a certain region of space is described by a wave packet, a superposition of deBroglie waves. The wave packet moves with a speed equal to the group velocity of the medium, which is equal to the speed of the particle. The proper interpretation of the amplitudes of the deBroglie waves is discussed in the next sections.

Any single measurement of the position or momentum of a particle can be made with as much precision as our experimental skill permits. How

4.5 PROBABILITY AND RANDOMNESS

[handwritten margin notes: $k = \frac{2\pi}{\lambda}$ $p = \frac{h}{2\pi} k = \hbar k$]

then does the wavelike behavior of a particle become observable? How does the uncertainty in position or momentum affect our experiment?

Suppose we prepare an atom by attaching an electron to a nucleus. (For this example we regard the nucleus as being fixed in space.) Some time after preparing our atom, we measure the position of the electron. We then repeat the procedure, preparing the atom *in an identical way*, and find that a remeasurement of the position of the electron yields a value different from that found in our first measurement. In fact, each time we repeat the measurement, we may obtain a different outcome. If we repeat the measurement a large number of times, we find ourselves led to a conclusion that runs counter to a basic notion of classical physics—*systems that are prepared in identical ways do not show identical subsequent behavior.* What hope do we then have of constructing a mathematical theory that has any usefulness at all in predicting the outcome of a measurement, if that outcome is completely random? The solution to this dilemma lies in the consideration of the *probability* of obtaining any given result from an experiment whose possible results are subject to the laws of statistics. We cannot predict the outcome of a single flip of a coin or roll of the dice, because any *single result* is as likely as any other *single result*. We can, however, predict the *distribution* of a large number of individual measurements. For example, on a single flip of a coin, we cannot predict whether the outcome will be "heads" or "tails"; the two are equally likely. If we make a large number of trials, we expect that approximately 50 percent will turn up "heads" and 50 percent will yield "tails"; even though we cannot predict the result of any single toss of the coin, we can predict reasonably well the result of a large number of tosses.

Our study of systems governed by the laws of quantum physics leads us to a similar situation. We cannot predict the outcome of any *single* measurement of the position of the electron in the atom we prepared, but if we do a large number of measurements, we ought to find a statistical distribution of results. If we cannot produce a mathematical theory that will predict the result of a single measurement, we can attempt to obtain a mathematical theory that predicts the statistical behavior of a system (or of a large number of identical systems). The quantum theory provides such a mathematical procedure; although we shall not go into the details of this procedure, it is important to note that the mathematical theory of quantum mechanics enables us to calculate the average or probable outcome of measurements and the distribution of individual outcomes about the average. This is not such a disadvantage as it may seem, for in the realm of quantum physics, we seldom do measurements with, for example, a single atom. If we were studying the emission of light by a radiant system or the properties of a solid or the scattering of nuclear particles, we would be dealing with a large number of atoms, and so our concept of statistical averages is really quite useful.

In fact, such concepts are not as far removed from our daily lives as we might think. For example, what is meant when the TV weather forecaster "predicts" a 50 percent chance of rain tomorrow? Will it rain 50 percent of the time, or over 50 percent of the city? The proper interpretation of the forecast is that the existing set of atmospheric conditions will, in a large number of similar cases, result in rain in about half the cases. A surgeon who asserts that a patient has a 50 percent chance of surviving an operation means exactly the same thing—experience with a large number of similar cases suggests recovery in about half. Quantum physics predicts, for example, that for identically prepared hydrogen atoms, the probability of later finding the electron revolving clockwise is 50 percent. Of course a *single measurement* shows *either* clockwise or counterclockwise motion, not a combination of the two. (Similarly, it *either* rains or it doesn't; the patient *either* lives or dies.) It is a sometimes confusing characteristic of the mathematics of wave mechanics that, since we don't know the outcome of a measurement in advance, the complete description of the system (the atom, for example) must include both possible outcomes.

Of course, one could argue that the flip of a coin or the roll of the dice is not a random process, but that the apparently random nature of the outcome simply reflects our lack of knowledge of the state of the system. For example, if we knew exactly how the dice were thrown (magnitude and direction of initial velocity, initial orientation, rotational speed) and precisely what the laws are that govern their bouncing on the table, we should be able to predict exactly how they will land. (Similarly, if we knew a great deal more about atmospheric physics or physiology, we could predict with certainty whether or not it will rain tomorrow or an individual patient will survive.) When we instead analyze the outcomes in terms of probabilities, we are really admitting our inability to do the analysis exactly. There is a school of thought that asserts that the same situation exists in quantum physics. According to this interpretation, we could predict *exactly* the behavior of the electron in our atom if only we knew the nature of a set of so-called "hidden variables" that determine its motion. However, in the absence of evidence to confirm such a theory, we must conclude that the random behavior of a system governed by the laws of quantum physics is a fundamental aspect of nature and not a result of our limited knowledge of the properties of the system.

4.6 THE PROBABILITY AMPLITUDE

One final problem remains to be discussed. What does the amplitude of the deBroglie wave represent? In any wave phenomenon, a physical quantity such as displacement or pressure varies with distance and time. What is the physical property that varies as the deBroglie wave propagates?

In a previous section, we discussed the representation of a localized particle by a wave packet. If a particle is confined to a region of space of dimension Δx, the wave packet that represents the particle has large amplitude only in a region of space of dimension Δx and has small amplitude elsewhere. That is, the amplitude is large where the particle is likely to be found and small where the particle is less likely to be found. *The amplitude of the deBroglie wave at any point is related to the probability of finding the particle at that point.* In analogy with classical physics, in which the intensity of any wave is proportional to the square of its amplitude, the probability is proportional to the square of the amplitude of the deBroglie wave. In the next chapter, we discuss the mathematical framework for computing the wave amplitudes for a particle in various situations, and we also discuss a more mathematical definition of the probability. Our difficulty in giving a precise interpretation to the wave amplitude arises partly from the *complex* nature of the wave amplitude. (A complex variable, such as the probability amplitude, is one which contains an *imaginary* part, proportional to the square root of -1, denoted by i; see Section 5.6.) We cannot represent such variables in our world of real (nonimaginary) numbers, and hence cannot interpret or directly measure the wave amplitude. The probability, however, is defined in terms of the absolute squared magnitude of the amplitude; since this is always a real number, no difficulty in its interpretation is encountered.

Even though the amplitudes of these waves cannot be simply interpreted, the waves themselves have all the characteristics of well-behaved classical waves. For example, they can be reflected and refracted, they obey the principle of superposition, and waves traveling in opposite directions can combine to produce standing waves.

SUGGESTIONS FOR FURTHER READING

Many of the descriptive references listed in Chapter 1 are useful background reading for this chapter as well. For a delightful account of a world in which Planck's constant is so large that quantum effects are ordinary, see G. Gamow, *Mr. Tompkins in Paperback* (Cambridge, Cambridge University Press, 1967). Another nonmathematical discussion of quantum theory is B. Hoffmann, *The Strange Story of the Quantum* (New York, Dover, 1959). An imaginary dialogue, in which the protagonists of Galileo's dialogues are reunited to discuss quantum theory, measurement, and uncertainty, is in J. M. Jauch, *Are Quanta Real?* (Bloomington, Indiana University Press, 1973). The paradox of Schrödinger's cat is discussed in this last reference.

Other references, in which the philosophy of quantum theory is mixed with mathematics at about the same level as this text, are as follows:

R. P. Feynman, R. B. Leighton, and M. Sands, *The Feynman Lectures on Physics* (Reading, Addison-Wesley, 1965). Chapters 1 to 3 of Volume 3 are particularly good introductions to quantum waves and the philosophy of measurement.

R. Resnick, *Basic Concepts in Relativity and Early Quantum Theory* (New York, Wiley, 1972). Chapter 6 discusses the wave nature of particles and the uncertainty principle.

E. H. Wichmann, *Quantum Physics, Volume 4 of the Berkeley Physics Course* (New York, McGraw-Hill, 1971).

A more advanced work that is particularly careful about the philosophical background of quantum theory is:

D. Bohm, *Quantum Theory* (Englewood Cliffs, Prentice-Hall, 1951). See Chapters 5 and 6.

For a discussion of the uncertainty principle, see:

G. Gamow, "The Principle of Uncertainty," *Scientific American 198*, 51 (January 1958).

The time development of the electron interference pattern is discussed in:

P. G. Merli, G. F. Missiroli, and G. Pozzi, *American Journal of Physics 44*, 306 (1976).

A translation of Claus Jönsson's article on the electron double-slit experiment is given in *American Journal of Physics 42*, 4 (1974). This short, clearly written paper is very readable and is highly recommended as an example of the careful experimental technique that is necessary in doing interference experiments to illustrate the wave nature of particles.

QUESTIONS

1. When an electron moves with a certain deBroglie wavelength, does the electron shake back and forth or up and down at that wavelength?

2. Imagine a different world in which the laws of quantum physics still apply, but which has $h = 1$ J·s. What might be some of the difficulties of life in such a world? (See Gamow, *Mr. Tompkins in Paperback*, for a fanciful account of such a world.)

3. Suppose we try to measure an unknown frequency ν by listening for beats between ν and a known (and controllable) frequency ν'. (We assume ν' is known to arbitrarily small uncertainty.) The beat frequency is $|\nu' - \nu|$. If we hear no beats, then we conclude $\nu = \nu'$. (a) How long must we listen to hear "no" beats? (b) If we hear no beats in one second, how accurately have we determined ν? (c) If we hear no beats in 10 s, how accurately? In 100 s? (d) How is this experiment related to Equation (4.4)?

4. What difficulties does the uncertainty principle cause in trying to pick up an electron with a pair of forceps?

5. Does the uncertainty principle apply to nature itself or only to the results of experiments? That is, is it the position and momentum that are *really* uncertain, or merely our knowledge of them? Does it make any difference?

6. The uncertainty principle states in effect that the more we try to confine an object, the faster we are likely to find it moving. Is this why you can't seem to keep money in your pocket or purse for long? Make a numerical estimate.

7. Consider a collection of gas molecules trapped in a container. As we move the walls of the container closer together (compressing the gas) the molecules move faster (the temperature increases). This description is similar to Example 4.4. Does the gas behave this way because of the uncertainty principle? Justify your answer with some numerical estimates.

8. Many atoms are unstable and undergo radioactive decay to other atoms. The lifetimes for these decays are typically of the order of days to years. Do you expect that the uncertainty principle will cause a measurable effect in the precision to which we can measure the masses of these atoms?

9. Often it happens in physics that great discoveries are made inadvertently. What would have happened if Davisson and Germer had their accelerating voltage set below 32 V?

10. Suppose we cover one slit in the two-slit electron experiment with a very thin sheet of fluorescent material that emits a photon of light whenever an electron passes through. We then fire electrons one at a time at the double slit; whether or not we see a flash of light tells us which slit the electron went through. What effect does this have on the interference pattern? Why?

11. In another attempt to determine through which slit the electron passes, we suspend the double slit itself from a very fine spring balance and measure the "recoil" momentum of the slit as a result of the passage of the electron.

Electrons that strike the screen near the center must cause recoils in opposite directions depending on which slit they pass through. Sketch such an apparatus and describe its effect on the interference pattern. (*Hint*. Consider the uncertainty principle $\Delta p\,\Delta x \sim \hbar$ as applied to the motion of the slits suspended from the spring. How precisely do we know the position of the slit?)

12. Is it possible for v_{phase} to be greater than c? Can v_{group} be greater than c?

13. In a nondispersive medium, $v_{\text{group}} = v_{\text{phase}}$; this is another way of saying that all waves travel with the same phase velocity, no matter what their wavelengths. Is this true for (a) deBroglie waves? (b) Light waves in glass? (c) Light waves in vacuum? (d) Sound waves in air? What difficulties would be encountered in attempting communication (by speech or by radio signals for example) in a strongly dispersive medium?

PROBLEMS

1. Find the deBroglie wavelength of (a) a nitrogen molecule ($m = 28$ u) in air at room temperature. (b) A 5-MeV proton. (c) A 50-GeV electron. (d) An electron moving at $v = 10^6$ m/s.

2. The neutrons produced in a reactor are known as *thermal neutrons*, because their kinetic energies have been reduced (by collisions) until $K \cong \frac{3}{2}kT$ where T is room temperature. (a) What is the kinetic energy of such neutrons? (b) What is their deBroglie wavelength? Because this wavelength is of the same order as the lattice spacings of the atoms of a solid, neutron diffraction (like X-ray and electron diffraction) is a useful means of studying solid lattices.

3. To what voltages must we accelerate electrons (as in an electron microscope, for example) if we wish to resolve a virus of diameter 12 nm? An atom of diameter 0.12 nm? A proton of diameter 1.2 fm?

4. In an electron microscope we wish to study particles of diameter about 0.10 μm (about 1000 times the size of a single atom). (a) What should be the deBroglie wavelength of the electrons? (b) Through what voltage should the electrons be accelerated to have that deBroglie wavelength?

5. In order to study the atomic nucleus, we would like to observe the diffraction of particles whose deBroglie wavelength is about the same size as the nuclear diameter, about 14 fm for a heavy nucleus such as lead. What kinetic energy should we use if the diffracted particles are (a) electrons? (b) Neutrons? (c) Alpha particles ($m = 4$ u)?

6. A free electron bounces elastically back and forth in one dimension between two walls that are $L = 0.50$ nm apart. (a) Assuming that the electron is represented by a deBroglie standing wave with a node at each wall, show that the permitted deBroglie wavelengths are $\lambda = 2L/n$ ($n = 1, 2, 3, \ldots$). (b) Find the values of the kinetic energy of the electron for $n = 1, 2,$ and 3.

7. In the Davisson-Germer experiment, at what accelerating voltage would the second-order reflection ($n = 2$ in the Bragg equation) begin to appear? The third-order reflection? (Choose the proper set of reflecting lattice planes.)

8. In the Davisson-Germer experiment, at what angle would the reflected electron beam appear if the accelerating voltage had been 105 V instead of 54 V?

9. Show that the relationship $\Delta x\,\Delta \lambda \sim \lambda^2$ is essentially the same as Equation (4.3).

10. Suppose a traveling wave has a speed v (where $v = \lambda \nu$). Instead of measuring waves over a distance Δx, we stay in one place and count the number of

wave crests which pass in a time Δt. Show that $\Delta x \, \Delta \lambda \sim \lambda^2$ is equivalent to $\Delta \omega \, \Delta t \sim 1$ for this case.

11. The speed of an electron is measured to within an uncertainty of 2.0×10^4 m/s. In how large a region of space is the electron likely to be found?

12. An electron is confined to a region of space of the size of an atom (0.1 nm). (a) What is the uncertainty in the momentum of the electron? (b) What is the kinetic energy of an electron with a momentum equal to Δp? (c) Is this a reasonable value for the kinetic energy of an electron in an atom?

13. The Σ^* particle has a rest energy of 1385 MeV and a lifetime of 2.0×10^{-23} s. What would be a typical range of outcomes of measurements of the Σ^* rest energy?

14. The Δ particle has a rest energy of 1236 MeV with an experimental spread of 120 MeV. What is the minmum lifetime of the Δ?

15. A nucleus emits a gamma ray of energy 1.0 MeV from a state that has a lifetime of 1.2 ns. (a) What is the uncertainty in the energy of the gamma ray? (b) How does this compare with the experimental precision with which the best gamma-ray detectors can measure gamma-ray energies, which is of the order of several eV? Will this uncertainty be directly measurable?

16. A nucleus of helium with mass 5 u breaks up from rest into a nucleus of ordinary helium (mass = 4 u) plus a neutron (mass = 1 u). The mass energy liberated in the break-up is 0.89 MeV, which is shared (*not* equally) by the products. (a) Using momentum conservation, find the kinetic energy of the neutron. (b) The lifetime of the original nucleus is 1.0×10^{-21} s. What range of values of the neutron kinetic energy might we typically observe in the laboratory?

17. Particle A, with a rest energy $m_A c^2$ and a lifetime t_A, decays into particles B and C. Particle B has a rest energy $m_B c^2$ and a lifetime t_B; particle C has a rest energy $m_C c^2$ and is stable ($t_C = \infty$). In an experiment, particle A at rest decays into B + C, and we determine the total final kinetic energy $K_B + K_C$. Show that:

$$K_B + K_C = (m_A - m_B - m_C)c^2 \pm \sqrt{\left(\frac{\hbar}{t_A}\right)^2 + \left(\frac{\hbar}{t_B}\right)^2}$$

18. In a metal, the conduction electrons are not attached to any one atom, but are relatively free to move throughout the entire metal. Consider a 1 cm × 1 cm × 1 cm piece of copper. (a) What is the uncertainty in any one component of the momentum of an electron confined to the metal? (b) What is the resulting estimate of the typical kinetic energy of an electron in the metal? (Assume $\Delta p = [(\Delta p_x)^2 + (\Delta p_y)^2 + (\Delta p_z)^2]^{1/2}$.) (c) Assuming the heat capacity of copper to be 24.5 J/mole·K, would the contribution of this motion to the internal energy of the copper be important at room temperature? What do you conclude from this? (See also Problem 20.)

19. A proton or a neutron can sometimes "violate" conservation of energy by emitting and then reabsorbing a pi meson, which has a mass of 135 MeV/c^2. This is possible as long as the pi meson is reabsorbed within a short enough time Δt consistent with the uncertainty principle. (a) Consider $p \rightarrow p + \pi$. By what amount ΔE is energy conservation violated? (Ignore any kinetic energies.) (b) For how long a time Δt can the pi meson exist? (c) Assuming the pi meson to travel at very nearly the speed of light, how far from the proton can it go? (This procedure, as we discuss in Chapter 9, gives us an estimate of the *range* of the nuclear force, because we believe that protons and neutrons are held together in the nucleus by exchanging pi mesons.)

20. In a crystal, the atoms are a distance L apart; that is, each atom must be localized to within a distance of at most L. (a) What is the minimum uncer-

tainty in the momentum of the atoms of a solid that are 0.20 nm apart? (b) What is the typical kinetic energy of such an atom of mass 65 u? (c) What would a collection of such atoms contribute to the internal energy of a typical solid, such as copper? Is this contribution important at room temperature? (See also Problem 18.)

21. An apparatus is used to prepare an atomic beam by heating a collection of atoms to a temperature T and allowing the beam to emerge through a hole of diameter d in one side of the oven. The beam then travels through a straight path of length L. Show that the diameter of the beam at the end of the path is larger than d by an amount of order $L\hbar/d\sqrt{3mkT}$ where m is the mass of an atom. Make a numerical estimate for typical values of $T = 1500$ K, $m = 7$ u (lithium atoms), $d = 3$ mm, $L = 2$ m.

22. Do the trigonometric manipulation necessary to obtain Equation (4.11).

23. Sound waves travel through air at a speed of 330 m/s. A whistle blast at a frequency of about 1.0 kHz lasts for 2.0 s. (a) Over what distance in space does the "wave train" representing the sound extend? (b) What is the wavelength of the sound? (c) Estimate the precision with which an observer could measure the wavelength. (d) Estimate the precision with which an observer could measure the frequency.

24. A stone tossed into a body of water creates a disturbance at the point of impact which lasts for 4.0 s. The wave speed is 25 cm/s. (a) Over what distance on the surface of the water does the group of waves extend? (b) An observer counts 12 wave crests in the group. Estimate the precision with which the wavelength can be determined.

25. Show that the data used in Figure 4.18 are consistent with Equation (4.11), that is, use $\lambda_1 = 1$ and $\lambda_2 = 10/9$, $v_1 = 3$ and $v_2 = 2.5$ to show $v_{group} = 7.5$.

26. (a) Use a distribution of wave numbers of constant amplitude in a range Δk about k_0:

$$A(k) = A \qquad k_0 - \frac{\Delta k}{2} \le k \le k_0 + \frac{\Delta k}{2}$$

$$= 0 \qquad \text{otherwise}$$

and obtain Equation (4.14) from Equation (4.13). (b) Make a convenient choice of the width Δx of the wave packet, and show that $\Delta x\, \Delta k \sim 1$.

27. Use the distribution of wave numbers $A(k) = e^{-(k-k_0)^2/2(\Delta k)^2}$ for $k = -\infty$ to $+\infty$ to derive Equation (4.15). Ignore any multiplicative constant in $y(x)$.

28. (a) Show that the group velocity and phase velocity are related by:

$$v_{group} = v_{phase} - \lambda \frac{dv_{phase}}{d\lambda}$$

(b) When white light travels through glass, the phase velocity of each wavelength depends on the wavelength. (This is the origin of dispersion and the breaking up of white light into its component colors—different wavelengths travel at different speeds and have different indices of refraction.) How does v_{phase} depend on λ? Is $dv_{phase}/d\lambda$ positive or negative? Therefore, is $v_{group} > v_{phase}$ or $< v_{phase}$?

29. Certain surface waves in a fluid travel with phase speed $\sqrt{b/\lambda}$, where b is a constant. Find the group velocity of a packet of surface waves, in terms of the phase velocity.

30. Show that $dE/dp = v$ for a particle when (a) E is the classical kinetic energy, and (b) E is the relativistic total energy.

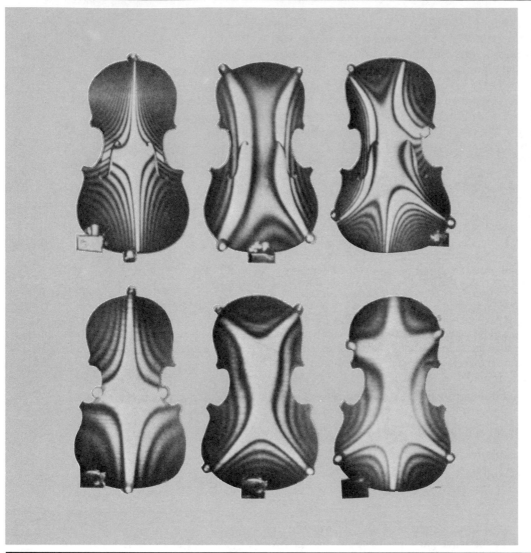

The mathematical techniques of quantum mechanics are formally similar to the techniques used to analyze classical waves. The resulting solutions of the quantum problems, such as that for a particle trapped in a two-dimensional region, often look similar to solutions of familiar classical wave problems, such as the vibrations of two-dimensional surfaces. The photograph shows the classical vibrations of violin plates, made visible using laser holograms.

5

THE SCHRÖDINGER EQUATION

Given a particle in a classical (nonrelativistic, nonquantum) situation, its future behavior may be predicted with absolute certainty by means of Newton's laws. That is, given a force $\mathbf{F}$ (derived from a potential V) acting on a particle initially at $\mathbf{x}_0$ moving with speed $\mathbf{v}_0$, by doing the mathematics necessary to solve Newton's second law, $\mathbf{F} = d\mathbf{p}/dt$ (a second-order, linear differential equation), we can find $\mathbf{x}(t)$ and $\mathbf{v}(t)$ at all future times t. The mathematics may be difficult, and in fact solution may not be possible in closed form (in which case an approximate solution can be obtained numerically with the aid of a computer); however, these are merely mathematical difficulties, and the physics consists of writing down the original equation $\mathbf{F} = d\mathbf{p}/dt$ and interpreting its solutions $\mathbf{x}(t)$ and $\mathbf{v}(t)$. For example, a satellite or planet moving under the influence of a gravitational force can be shown, after considerable manipulation of the equations, to follow exactly an elliptical path. Similarly, the electric and magnetic fields associated with any static distribution of charges and currents can be found by solving the first-order differential equations known as Maxwell's equations. The physics of the problem, as in the case of Newton's laws, consists of writing down the initial equation and in interpreting the solution; the act of obtaining the solution is mathematical manipulation. In the case of nonrelativistic quantum physics, the basic equation to be solved is a second-order differential equation known as the *Schrödinger equation.* Like Newton's laws, we find a solution for a certain force, although we will be more concerned with the potential than with the force; unlike Newton's laws, the solution to the Schrödinger equation, called the *wave function,* gives information on the wavelike properties of the particle. In this chapter we introduce the Schrödinger equation, study some solutions of the equation, and learn how to interpret those solutions.

To recapitulate, in the case of classical mechanics we have a problem characterized by the presence of a certain force $\mathbf{F}$. Writing Newton's second law for that specific force, we turn the mathematical crank and out come the position and velocity of the particle. In the case of electromagnetism, we have a problem characterized by a collection of charges or currents; we write Maxwell's equations for the specific distribution under study, turn the mathematical crank, and out come the electric and magnetic fields. In the case of quantum physics, our problem is characterized by a specific potential function; we write the Schrödinger equation for that potential and crank out the solution. Of course, in each case the solution is valid only for that specific situation; each different situation requires a new solution to the appropriate equation.

Neither Newton's laws, nor Maxwell's equations, nor the Schrödinger equation can be derived from basic principles; they are, instead, mathematical equations that yield solutions consistent with experiments. The Schrödinger equation can be solved in a tidy mathematical form for only a few potentials; the simplest of these, the constant potential and the harmonic-oscillator potential, are instructive in the general technique of solution and will be discussed in this chapter; however, these cases are not "physical" in that the solutions cannot be checked against experiment—we find in nature no examples of a particle trapped in a

Erwin Schrödinger (1887–1961, Austria). Although he disagreed with the probabilistic interpretation that was later given to his work, he developed the mathematical theory of wave mechanics that for the first time permitted the wave behavior of physical systems to be calculated.

5.1 JUSTIFICATION OF THE SCHRÖDINGER EQUATION

"one-dimensional box" or of an ideal quantum-mechanical harmonic oscillator (although these are often good approximations to physical situations).

If we for a moment try to imagine that we are Erwin Schrödinger and are searching for a differential equation that will yield the proper solution to problems of quantum physics, we find that we are stymied by the lack of experiments of a simple nature, the results of which could be compared with the solution to the equation. We must therefore be content with the following—let us write down the properties that we expect our equation to have, and then examine the kinds of equations that satisfy those criteria.

1. We must conserve energy. While we are willing to sacrifice a great deal of the framework of classical physics, conservation of energy is one principle that we expect to remain valid. We therefore take

$$K + V = E \tag{5.1}$$

   where K, V, and E are the kinetic, potential, and total energies. (We will be dealing exclusively with nonrelativistic situations in our study of quantum physics, and thus $K = \frac{1}{2}mv^2 = p^2/2m$; E represents only the sum of kinetic and potential energies, and *not* the relativistic mass energy.)

2. The differential equation, whatever its form, must be consistent with the deBroglie hypothesis—if we turn the mathematical crank for a free particle with momentum p, we must find as a solution a wave of wavelength λ equal to h/p. Using Equation (4.5), the kinetic energy of the free-particle deBroglie wave must be $K = p^2/2m = \hbar^2 k^2/2m$.

3. The equation must be "well behaved," in the mathematical sense. We expect our solution to tell us something about the probability of finding the particle; we should be very surprised to find, for example, the probability changing discontinuously, with the particle suddenly disappearing at one point in space and reappearing at another. It must be *single valued*—there ought not to be two different probabilities for finding the particle at the same point in space. It must also be *linear*, so that the waves have the *superposition* property we expect of well-behaved waves.

Working backward, we begin with a solution to the equation for which we are searching. You have previously studied waves on a stretched string, which have the mathematical form $y(x, t) = A \sin(kx - \omega t)$, and electromagnetic waves, which have a similar form $\mathbf{E}(x, t) = \mathbf{E}_0 \sin(kx - \omega t)$ and $\mathbf{B}(x, t) = \mathbf{B}_0 \sin(kx - \omega t)$. We therefore postulate that the free-particle deBroglie wave $\Psi(x,t)$ has mathematical form similar to $A \sin(kx-\omega t)$, which is the basic form of a wave of amplitude A traveling in the positive x direction. The wave has wavelength $\lambda = 2\pi/k$ and frequency $\nu = \omega/2\pi$. For the moment we will ignore the time dependence, and deal with a snapshot of the wave at a specific time, say $t = 0$. Thus, defining $\psi(x)$ to be $\Psi(x, t = 0)$,

$$\psi(x) = A \sin kx \tag{5.2}$$

The differential equation, of which $\Psi(x, t)$ is a solution, may contain many derivatives of $\Psi(x, t)$ with respect to x or t; however, it may depend on Ψ and its derivatives only to the first power, so that terms such as Ψ^2 or $(\partial\Psi/\partial t)^2$ may not appear. (This follows from our assumption of a linear single-valued equation.) Our equation must involve the potential V; if V appears to the first power, then in order to be consistent with conservation of energy $(V + K = E)$, K also must appear to the first power. Previously, we found $K = \hbar^2 k^2/2m$, and the only way to obtain a term in k^2 is to take two derivatives of $\psi(x) = A \sin kx$ with respect to x.

$$\frac{d^2\psi}{dx^2} = -k^2\psi = -\frac{2m}{\hbar^2}K\psi = -\frac{2m}{\hbar^2}(E - V)\psi$$

$$-\frac{\hbar^2}{2m}\frac{d^2\psi}{dx^2} + V\psi = E\psi \tag{5.3}$$

One-dimensional, time-independent Schrödinger equation

It must be emphasized that what we have done here is not a derivation; we have merely constructed a differential equation with three properties: (1) it is consistent with energy conservation; (2) it is linear and single valued; (3) it gives a free-particle solution consistent with a single deBroglie wave. Other equations could be constructed with the same properties, but only Equation (5.3) passes the stringent test of being *consistent with experimental results* in many different physical situations. Equation (5.3) is the *time-independent Schrödinger equation* in one dimension. Although real waves depend on the time as well as on the spatial coordinates, and although our world is three dimensional rather than one dimensional, we can learn much about the mathematics and physics of quantum mechanics by studying the solutions of Equation (5.3). Later in the chapter we discuss the extension to two and three dimensions and the inclusion of time dependence.

5.2 THE SCHRÖDINGER RECIPE

The techniques of solving Equation (5.3) are sufficiently similar, no matter what the form of the potential V (which is in general a function of x), that we may list a series of steps to be followed in obtaining the solutions. We assume that we are given a certain potential $V(x)$, and we wish to obtain the wave function $\psi(x)$ and the energy E. This is a general example of a type of equation known as an *eigenvalue equation*; we will find that it is possible to obtain solutions to the equation only for certain values of E, which are known as the *energy eigenvalues*.

1. Begin by writing Equation (5.3) with the appropriate $V(x)$. Note that if the potential changes discontinuously [$V(x)$ may be discontinuous; $\psi(x)$ may *not*], we may need to write different equations for different regions of space. Examples of this sort are given in Section 5.4.

2. Using general mathematical techniques suited to the form of the equation, find a mathematical function $\psi(x)$, which is a solution to the differential equation. Since there is no one specific technique for solving differential equations, we can only learn from example how to find solutions.

3. In general, several solutions may be found. By applying boundary conditions some of these may be eliminated and some arbitrary constants may be determined. It is usually the application of the boundary conditions that selects out the energy eigenvalues.

4. If you are seeking solutions for a potential that changes discon-

tinuously, you must apply the continuity conditions on ψ (and usually on $d\psi/dx$) at the boundary between different regions.

   5. Evaluate all undetermined constants, for example, A in Equation (5.2). The method for doing this is discussed in the following section.

   We consider now an example from classical physics that requires many of the same techniques of solution as do typical problems in quantum physics. The condition of continuity at the boundary between two regions is one that must frequently be applied in classical problems. By way of illustration, we study the following classical problem.

A mass m is dropped from rest at height H above a tank of water. On entering the water, it is subject to a buoyant force B greater than the weight of the object. (We neglect the viscous force exerted by the water.) Find the displacement and velocity of the object from the time it is released until it rises to the surface of the water.

EXAMPLE 5.1

We choose a coordinate system in which y is positive upward and take $y = 0$ at the surface of the water. During the time the object is initially in free fall, it is subject only to the force of gravity. Then in region 1 (above the water), Newton's second law gives

Solution

$$-mg = m\frac{d^2y_1}{dt^2}$$

which has the solutions

$$v_1(t) = v_{0_1} - gt$$

$$y_1(t) = y_{0_1} + v_{0_1}t - \tfrac{1}{2}gt^2$$

where v_{0_1} and y_{0_1} are the initial velocity and height at $t = 0$.

   When it enters the water (region 2) the force becomes $B - mg$, so Newton's second law becomes

$$B - mg = m\frac{d^2y_2}{dt^2}$$

which has the solutions

$$v_2(t) = v_{0_2} + \left(\frac{B}{m} - g\right) t$$

$$y_2(t) = y_{0_2} + v_{0_2}t + \frac{1}{2}\left(\frac{B}{m} - g\right) t^2$$

These solutions have four undetermined coefficients: y_{0_1}, v_{0_1}, y_{0_2}, and v_{0_2}. (Note that y_{0_2} and v_{0_2} are *not* the values at $t = 0$; they are merely constants to be determined.) The first two constants are found by applying the *initial conditions*—at $t = 0$ (when the object is released) $y_{0_1} = H$ and $v_{0_1} = 0$, since it is dropped from rest. The solutions in region 1 are therefore

$$v_1(t) = -gt$$

$$y_1(t) = H - \tfrac{1}{2}gt^2$$

The next step is to apply the *boundary conditions* at the surface of the water. Let t_1 be the time that the object enters the water. The boundary conditions require that v and y be continuous across the boundary between air and water:

$$y_1(t_1) = y_2(t_1)$$

and

$$v_1(t_1) = v_2(t_1)$$

The first condition states that the object does not disappear at one instant and reappear at a different point in space at the next instant. The second condition is equivalent to requiring that the speed changes smoothly at the water's surface. [If this were *not* so, then $v_1(t_1 - \Delta t) \neq v_2(t_1 + \Delta t)$ even as $\Delta t \to 0$, and the acceleration would be infinite.] To apply the boundary conditions, we must first find t_1, which is obtained by finding the time at which y_1 becomes zero.

$$y_1(t_1) = H - \tfrac{1}{2}gt_1^2 = 0$$

so

$$t_1 = \sqrt{\frac{2H}{g}}$$

We can then find the speed at which the object enters the water, $v_1(t_1)$:

$$v_1(t_1) = -gt_1 = -g\sqrt{\frac{2H}{g}} = -\sqrt{2gH}$$

The boundary conditions then give

$$y_2(t_1) = y_{0_2} + v_{0_2}\sqrt{\frac{2H}{g}} + \frac{1}{2}\left(\frac{B}{m} - g\right)\left(\frac{2H}{g}\right) = 0$$

and

$$v_2(t_1) = v_{0_2} + \left(\frac{B}{m} - g\right)\sqrt{\frac{2H}{g}} = -\sqrt{2gH}$$

These two equations may be solved simultaneously for y_{0_2} and v_{0_2}, yielding $v_{0_2} = -(B/m)\sqrt{2H/g}$ and $y_{0_2} = H(1 + B/mg)$. The complete solutions in region 2 are thus

$$v_2(t) = -\frac{B}{m}\sqrt{\frac{2H}{g}} + \left(\frac{B}{m} - g\right)t$$

$$y_2(t) = H + \frac{HB}{mg} - \frac{B}{m}\sqrt{\frac{2H}{g}}\,t + \frac{1}{2}\left(\frac{B}{m} - g\right)t^2$$

The equations for v_1, y_1, v_2, and y_2 give the behavior of the object from $t = 0$ until it rises to the surface of the water.

We can *apply* these results to calculating other properties of the motion; for example, we can find the maximum depth reached by the object, which occurs when $v_2 = 0$. If we let t_2 be the time at which this occurs, then

$$v_2(t_2) = -\frac{B}{m}\sqrt{\frac{2H}{g}} + \left(\frac{B}{m} - g\right)t_2 = 0$$

$$t_2 = \frac{B}{B - mg}\sqrt{\frac{2H}{g}}$$

The depth D is just the value of y_2 at this time t_2:

$$D = y_2(t_2) = \left(H + \frac{HB}{mg}\right) - \frac{B}{m}\sqrt{\frac{2H}{g}}\,t_2 + \frac{1}{2}\left(\frac{B}{m} - g\right)t_2^2$$

$$= \frac{-mgH}{B - mg}$$

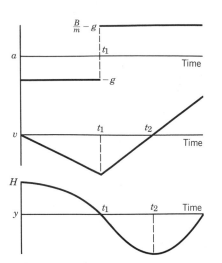

FIGURE 5.1 The acceleration, velocity, and position of the mass described in Example 5.1. Notice that the acceleration is discontinuous, but the position and velocity are continuous.

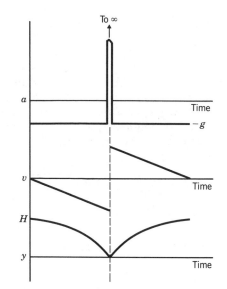

FIGURE 5.2 The acceleration, velocity, and position of a mass rebounding elastically from a hard surface. An infinite acceleration acts for an infinitesimally short time. The velocity is discontinuous (it changes instantly from $-v$ to $+v$) but the position is continuous.

In summary, in this example we have used the equations of motion to find a solution, we have evaluated the undetermined constants in our solution by applying initial and boundary conditions, and we have applied our resulting solution to calculate a feature of the future behavior of the object (in this case, the maximum depth D). *The same basic procedure should be applied to problems in quantum physics.*

The behavior of the object is illustrated in Figure 5.1, which shows the acceleration, velocity, and position as functions of the time. Note that $v(t)$ and $y(t)$ are both continuous, as we have demanded by our application of the boundary conditions.

Suppose we replaced the water by a rigid surface from which the (likewise rigid) object rebounded elastically. Then in an idealized representation, we might picture this situation in Figure 5.2. Notice that in this case the object is subject to an infinite force during the instant it is in contact with the surface, so that its velocity changes discontinuously, but its position changes continuously (it still does not disappear and reappear somewhere else).

Our conclusions for the classical problem can be summarized as follows; the equivalent wording for applications to quantum mechanics is in brackets. When an object moves across the boundary between two regions in which it is subject to different $\begin{Bmatrix} \text{forces} \\ \text{potentials} \end{Bmatrix}$, the basic behavior of the object is found by solving $\begin{Bmatrix} \text{Newton's second law} \\ \text{the Schrödinger equation} \end{Bmatrix}$. The $\begin{Bmatrix} \text{displacement} \\ \text{wave function} \end{Bmatrix}$ is always continuous across the boundary, and the

$\left\{\begin{array}{l}\text{velocity}\\\text{derivative } d\psi/dx\end{array}\right\}$ is also continuous as long as the $\left\{\begin{array}{l}\text{force}\\\text{change in potential}\end{array}\right\}$ remains finite.

Just as in the case of classical physics, each problem may require somewhat different techniques, and so it is difficult to establish a general procedure. The steps listed in this section should, however, suggest to you the general direction to take in seeking solutions. The best way to learn the techniques is by studying the examples given in this chapter. The recipe is incomplete at this point; we have discussed the mathematical technique for finding the solution $\psi(x)$, but we have not discussed the interpretation of the solution or its application to physical situations. These are discussed in the following sections.

5.3 PROBABILITIES AND NORMALIZATION

The remaining steps in the Schrödinger recipe depend on the physical interpretation of the solution to the differential equation. The meaning of the *wave function* $\psi(x)$ is not at all obvious, and this subject has produced a great quantity of debate in the physics literature over the past five decades. The function $\psi(x)$ represents a wave in the sense with which we are all familiar—it has a well-defined wavelength and moves with a well-defined phase velocity. The dilemma arises when we try to interpret the amplitude of the wave. What does the amplitude of $\psi(x)$ represent, and what is the physical variable that is waving? It is certainly not a displacement, as in the case of a water wave or a wave on a stretched piano wire, nor is it a pressure wave, as in the case of sound. *It is in fact a very different kind of wave, whose absolute squared amplitude gives the probability for finding the particle at a given point in space.* More precisely $|\psi|^2\, dx$ gives the probability of finding the particle in the infinitesimal interval dx at x (i.e., between x and $x + dx$). In one dimension, the distinction between "finding the particle at x" and "finding the particle in the infinitesimal interval dx at x" may not be important, but when we move to two and three dimensions the difference becomes apparent. For now, you may wish to think of this rule in terms of the lack of physical dimension of a single point in space; since a point in space has no dimension, the probability of finding a particle *at a point* is always zero, but there is a nonzero probability of finding the particle in an *interval dx*. If we define $P(x)$ as the *probability density* (probability per unit length, in one dimension), then the interpretation of $\psi(x)$ according to the Schrödinger recipe is

Probability density

$$P(x)\, dx = |\psi(x)|^2\, dx \qquad (5.4)$$

(The reason for the absolute magnitude bars will be explained in Section 5.6.)

This interpretation of $|\psi|^2$ helps us to understand the continuity condition of $\psi(x)$; we must not allow the probability to change discontinuously but, like any well-behaved wave, the amplitude varies slowly and continuously.

Our interpretation of $\psi(x)$ now permits us to complete the Schrödinger recipe and to illustrate how to use the wave function to calculate those quantities that we can measure in the laboratory. Steps 1 through 5 were given in the previous section; the recipe continues.

6. The probability for finding the particle between the points x_1 and x_2 is just the sum of all the probabilities $P(x)\,dx$ in the infinitesimal intervals between x_1 and x_2, which is of course an integral:

Probability of finding the particle between x_1 and x_2 $= \displaystyle\int_{x_1}^{x_2} P(x)\,dx$

$$= \int_{x_1}^{x_2} |\psi(x)|^2\,dx \quad (5.5) \quad \textbf{Probability}$$

As a corollary to this rule, we require a 100 percent probability of finding the particle *somewhere* along the x axis, and thus

$$\int_{-\infty}^{+\infty} |\psi(x)|^2\,dx = 1 \qquad (5.6) \quad \textbf{Normalization}$$

Equation (5.6) is known as the *normalization* condition, and shows us how to find the constant A discussed in step 5 of the recipe. Note that the constant A does not come out of the solution to the differential equation; in fact, as long as the Schrödinger equation is linear, if $\psi(x)$ is a solution, then *any* constant times $\psi(x)$ is also a solution. A wave function in which the arbitrary multiplicative constant is determined according to Equation (5.6) is said to be *normalized*; otherwise, it is unnormalized. Only a properly normalized wave function may be used for physically meaningful calculations. If the normalization has been done correctly, Equation (5.5) will always yield a probability that lies between 0 and 1.

7. Any solution to the Schrödinger equation for which $|\psi(x)|^2$ becomes infinite must be discarded—there can never be an infinite probability of finding the particle at any point. In practice, we "discard" a solution by setting its multiplicative constant equal to zero. For example, if the mathematical solution to the differential equation yields $\psi(x) = Ae^{kx} + Be^{-kx}$ for the *entire* region $x > 0$, then we must require $A = 0$ for the solution to be physically meaningful; otherwise $|\psi(x)|^2$ would become infinite as x goes to infinity. (However, if the solution is only to be valid in the region $0 < x < L$, then we cannot set $A = 0$.) If the solution is to be valid in the *entire* region $x < 0$, then we must set $B = 0$.

8. Since we can no longer speak with certainty about the position of the particle, we can no longer guarantee the outcome of a single measurement of any physical quantity that depends on its position. However, if we can calculate the probability associated with any coordinate, we can find the *probable* outcome of any single measurement or (equivalently) the *average* outcome of a large number of measurements. For example, suppose we wish to find the average location of a particle by measuring its coordinate x. Performing a large number of measurements, we find that we measure the value x_1 a certain number of times n_1, x_2 a number of times n_2, etc., and in the usual way we can find the average value

$$x_{\mathrm{av}} = \frac{n_1 x_1 + n_2 x_2 + \cdots}{n_1 + n_2 + \cdots} \qquad (5.7)$$

$$= \frac{\sum n_i x_i}{\sum n_i} \qquad (5.8)$$

If we know the probability for finding the particle at each x_i, then n_i is related to $P(x_i)$, and changing the sums to integrals, we have

$$X_{av} = \frac{\int_{-\infty}^{+\infty} P(x)\, x\, dx}{\int_{-\infty}^{+\infty} P(x)\, dx} \tag{5.9}$$

and thus

$$X_{av} = \int_{-\infty}^{+\infty} |\psi(x)|^2\, x\, dx \tag{5.10}$$

Average or expectation value

where the last step can be made if the wave function is normalized, since the denominator of (5.9) is then equal to one.

By analogy, the average value of any function of x can be found:

$$[f(x)]_{av} = \int_{-\infty}^{+\infty} |\psi(x)|^2\, f(x)\, dx \tag{5.11}$$

Average values calculated according to (5.10) or (5.11) are known as *expectation values*.

The Free Particle By a "free particle" we mean one moving with no forces acting on it in any region of space; that is, $F = 0$, and so $V(x) =$ constant for all x. We are free to choose the constant to be zero, since the potential is always determined only to within a constant of integration ($F = -dV/dx$ in one dimension).

We apply the recipe by writing Equation (5.3) with the appropriate potential ($V = 0$):

$$-\frac{\hbar^2}{2m}\frac{d^2\psi}{dx^2} = E\psi \tag{5.12}$$

or

$$\frac{d^2\psi}{dx^2} = -k^2\psi \tag{5.13}$$

where

$$k^2 = \frac{2mE}{\hbar^2} \tag{5.14}$$

Equation (5.13) is a familiar one; written in this form, with k^2 always positive, its solution is

$$\psi(x) = A \sin kx + B \cos kx \tag{5.15}$$

The allowed energy values can be found from Equation (5.14):

$$E = \frac{\hbar^2 k^2}{2m} \tag{5.16}$$

Since our solution has placed no restrictions on k, the energy is permitted to have any value (in the language of quantum physics, we say that the energy is *not* quantized). We note that Equation (5.16) is just the kinetic energy of a particle with momentum $p = \hbar k$ or, equivalently, $p = h/\lambda$; based on the discussion of Section 5.1, this is just what we

5.4 APPLICATIONS

$H = k$

$k^2 = \frac{2mE}{\hbar^2}$

would expect, since we have constructed the Schrödinger equation to yield the solution for the free particle corresponding to a single deBroglie wave.

Solving for A and B presents some difficulties because the normalization integral, Equation (5.6), cannot be evaluated from $-\infty$ to $+\infty$ for this wave function. (These difficulties would not occur if we made a linear superposition of many sine or cosine waves to form a wave packet, as we did in Section 4.4.)

Particle in a Box (One Dimension)

In this case we are concerned with a particle moving freely in a one-dimensional "box" of length L; the particle is completely trapped within the "box." (Imagine a bead sliding without friction along a wire stretched between two rigid walls and making perfectly elastic collisions with the walls.) This potential may be expressed as:

$$V(x) = 0 \qquad 0 \le x \le L$$
$$= \infty \qquad x < 0, x > L \tag{5.17}$$

The potential is shown in Figure 5.3 and is sometimes known as the *infinite square well* potential. Of course, we are free to choose any constant value for V in the region $0 \le x \le L$; we choose it to be zero for convenience.

The recipe must now be applied separately to the regions inside and outside the box. We can analyze the outside region in either of two ways. If we examine Equation (5.3) for the region outside the box, we find that the only way to keep the equation from becoming meaningless when $V \to \infty$ is to require $\psi = 0$, so that $V\psi$ will not become infinite. Alternatively, we can go back to the original statement of the problem. If the walls of the box are perfectly rigid, the particle must always be in the box, and the probability for finding it elsewhere must be zero. To make the probability zero everywhere outside the box, we must make $\psi = 0$ outside the box. Thus we have

$$\psi(x) = 0 \qquad x < 0, x > L \tag{5.18}$$

The Schrödinger equation for $0 \le x \le L$, when $V(x) = 0$, is identical with Equation (5.12) and has the same solution:

$$\psi(x) = A \sin kx + B \cos kx \qquad (0 \le x \le L) \tag{5.19}$$

with

$$k^2 = \frac{2mE}{\hbar^2} \tag{5.20}$$

Our solution is not yet complete, for we have not evaluated A or B, nor have we found the allowed values of the energy E. To do this, we must apply the requirement that $\psi(x)$ must be continuous across any boundary. In this case, we require that our solutions for $x < 0$ and $x > 0$ match up at $x = 0$; similarly, the solutions for $x > L$ and $x < L$ must match at $x = L$.

Let us begin at $x = 0$. At $x < 0$, we have found that $\psi = 0$, and so we must set $\psi(x)$ of Equation (5.19) to zero at $x = 0$.

$$\psi(0) = A \sin 0 + B \cos 0 = 0$$

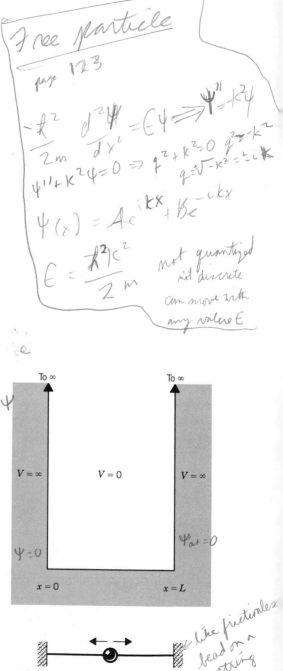

FIGURE 5.3 A particle moves freely in the one-dimensional region $0 \le x \le L$, but is excluded completely from $x < 0$ and $x > L$.

Thus

$$B = 0 \tag{5.21}$$

Since $\psi = 0$ for $x > L$, we must have $\psi(L) = 0$,

$$\psi(L) = A \sin kL + B \cos kL = 0 \tag{5.22}$$

Since we have already found $B = 0$, we must now have

$$A \sin kL = 0 \tag{5.23}$$

Either $A = 0$, in which case $\psi = 0$ *everywhere*, $\psi^2 = 0$ everywhere, and there is no particle (a meaningless solution) or else $\sin kL = 0$, which is true only when

$$kL = \pi, 2\pi, 3\pi, \ldots$$

or

$$kL = n\pi \quad n = 1, 2, 3, \ldots \tag{5.24}$$

Since $k = 2\pi/\lambda$, we have $\lambda = 2L/n$; this is identical with the result obtained in introductory mechanics for the wavelengths of the standing waves in a string of length L fixed at both ends. *Thus the solution to the Schrödinger equation for a particle trapped in a linear region of length* L *is just a series of standing deBroglie waves!* Not all wavelengths are permitted; only certain values, determined from Equation (5.24), may occur.

From Equation (5.20) we find that, since only certain values of k are permitted by Equation (5.24), only certain values of E may occur—*the energy is quantized!*

$$E = \frac{\hbar^2 k^2}{2m} = \frac{\hbar^2 \pi^2 n^2}{2mL^2} \tag{5.25}$$

Energy values in infinite, one-dimensional potential well

For convenience, let $E_0 = \hbar^2\pi^2/2mL^2$; this unit of energy is determined by the mass of the particle and the length of the box. Then $E_n = n^2 E_0$, and the particle can be found with an energy of E_0, $4E_0$, $9E_0$, $16E_0$, etc., but *never* with $3E_0$ or $6.2E_0$. Since the energy is purely kinetic in this case, our result means that only certain speeds are permitted for the particle. This is very different from the classical case, in which the bead (sliding without friction along the wire and colliding elastically with the walls) can be given any initial velocity and will move forever, back and forth, at the same speed. In the quantum case, this is not possible; only certain initial speeds can result in sustained states of motion; these special conditions are called "stationary states." (These states are "stationary" because, when the time dependence is included to make $\Psi(x, t)$, as in Section 5.6, $|\Psi(x, t)|^2$ is independent of time. Average values calculated according to Equation (5.11) likewise do not change with time. A particle initially in a pure stationary state remains in that state for all time.) The result of a measurement of the energy of a particle in a potential well *must* be one of the stationary state energies; no other result is possible.

Our solution for $\psi(x)$ is not yet complete, since we have not yet determined the constant A. To do this, we go back to the normalization condition, $\int_{-\infty}^{\infty} \psi^2 \, dx = 1$. Since $\psi = 0$ except for $0 \leq x \leq L$, the integral vanishes except inside that region, so that

$$\int_0^L A^2 \sin^2 \frac{n\pi x}{L} \, dx = 1 \tag{5.26}$$

from which we find $A = \sqrt{2/L}$. The complete wave function for $0 \le x \le L$ is then

$$\psi_n(x) = \sqrt{\frac{2}{L}} \sin \frac{n\pi x}{L} \qquad n = 1, 2, 3, \ldots \qquad (5.27)$$

[handwritten: $E_n = n^2 E_0$ $n = 1, 2, \ldots$]

Wave functions in infinite, one-dimensional potential well

In Figure 5.4, the allowed energy levels, wave functions, and probability densities ψ^2 are illustrated for the lowest several states. The lowest energy state, for which $n = 1$, is known as the *ground state,* and the states with higher energies $(n > 1)$ are known as *excited states.*

Let us try to interpret the results of our calculation. Suppose we carefully place a particle with energy E_0 into our region (our "wire") and subsequently measure its position. After repeating this measurement a

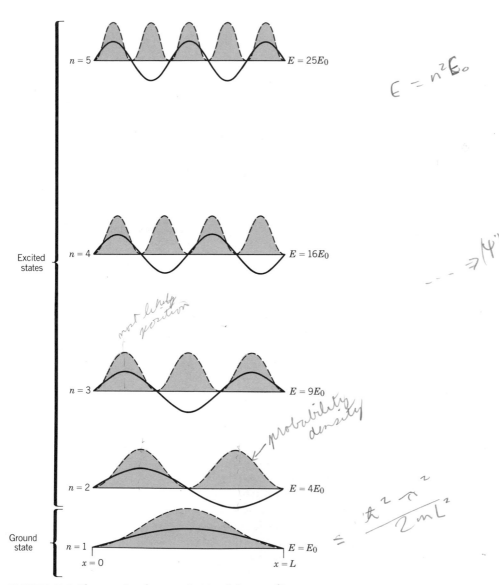

[handwritten: $E = n^2 E_0$]

[handwritten: $\Rightarrow |\psi|^2$]

[handwritten: most likely position]

[handwritten: probability density]

[handwritten: $\leq \dfrac{\hbar^2 \pi^2}{2mL^2}$]

FIGURE 5.4 The permitted energy levels of the one-dimensional infinite square well. The wave function for each level is shown by the solid curve, and the shaded region gives the probability density for each level.

large number of times, we should expect to find a distribution of results similar to ψ^2 for the $n = 1$ case—the probability is greatest at $x = L/2$, tapers off as we move off center and falls to zero at the edges. (If we used a classical, nonquantum particle, we expect to find the probability *constant* at all positions in the "box.") Suppose we repeat the measurement, except that now we prepare the particle with an energy of $4E_0$. We repeat our measurements of its position, and find a distribution of results in accordance with ψ^2 for $n = 2$; maxima in the probability at $x = L/4$ and $x = 3L/4$, and *zero* probability at $x = L/2$! The particle must travel so that it can be found occasionally at $L/4$ and at $3L/4$ without ever being found at $L/2$! Here we have a graphic illustration of the difference between classical and quantum physics. How can the particle get from $L/4$ to $3L/4$ without going through $L/2$? Our difficulty in answering this question comes from our desire to think in terms of particles, when quantum physics demands we think in terms of waves. The first overtone of a vibrating string of length L has a node in the center, and "information" travels from left to right and from right to left through the center, even though the midpoint does not move. When we speak of a position, we are referring to a *particle*; when we speak of motion from $L/4$ to $3L/4$, we are dealing with *waves*.

The calculation of probabilities and average values is illustrated by the following examples.

An electron is trapped in a one-dimensional region of length 1.0×10^{-10} m (a typical atomic diameter). (a) How much energy must be supplied to excite the electron from the ground state to the first excited state? (b) In the ground state, what is the probability of finding the electron in the region from $x = 0.090 \times 10^{-10}$ m to 0.110×10^{-10} m? (c) In the first excited state, what is the probability of finding the electron between $x = 0$ and $x = 0.250 \times 10^{-10}$ m?

EXAMPLE 5.2

(a) $\quad E_0 = \dfrac{\hbar^2 \pi^2}{2mL^2} = \dfrac{(1.05 \times 10^{-34}\ \text{J·s})^2 (3.14)^2}{2(9.1 \times 10^{-31}\ \text{kg})(10^{-10}\ \text{m})^2} = 6.0 \times 10^{-18}\ \text{J} = 37\ \text{eV}$

Solution

In the ground state, the energy is E_0. In the first excited state, the energy is $4E_0$. The difference, which must be supplied, is $3E_0$ or 111 eV.

(b) From Equation (5.5),

$$\text{probability} = \int_{x_1}^{x_2} \psi^2\ dx = \frac{2}{L} \int_{x_1}^{x_2} \sin^2 \frac{\pi x}{L}\ dx = \left(\frac{x}{L} - \frac{1}{2\pi} \sin \frac{2\pi x}{L} \right) \Bigg|_{x_1}^{x_2}$$

$$= 0.0038 = 0.38\ \text{percent}$$

(c)

$$\text{probability} = \int_{x_1}^{x_2} \left(\frac{2}{L} \right) \sin^2 \frac{2\pi x}{L}\ dx$$

$$= \left(\frac{x}{L} - \frac{1}{4\pi} \sin \frac{4\pi x}{L} \right) \Bigg|_{x_1}^{x_2}$$

$$= 0.25$$

(This result is of course what we would expect by inspection of the graph of ψ^2 for $n = 2$ in Figure 5.4. The interval from $x = 0$ to $x = L/4$ contains 25 percent of the total area under the ψ^2 curve.)

Show that the average value of x is $L/2$, independent of the quantum state.   **EXAMPLE 5.3**

We use Equation (5.10); since $\psi = 0$ except for $0 \leq x \leq L$, we use 0 and L as the   **Solution**
limits of integration

$$x_{\text{av}} = \frac{2}{L} \int_0^L \left(\sin^2 \frac{n\pi x}{L} \right) x \, dx$$

This can be integrated by parts or found in integral tables; the result is

$$x_{\text{av}} = \frac{L}{2}$$

Note that, as required, this result is independent of n. Thus a measurement of the average position of the particle yields no information about its quantum state.

Particle in a Box (Two Dimensional)

When we extend the previous situation to the more physical two- and three-dimensional cases, the principal features of the solution remain the same, but an important new feature is introduced. In this section we show how this occurs, since this new feature, known as *degeneracy*, will become very important in our study of atomic physics.

To begin with, we need a Schrödinger equation that is valid in more than one dimension; our previous version, Equation (5.3), was a one-dimensional version. We suspect immediately the following: if the potential is a function of x and y, ψ should also depend on x and y, and the derivatives with respect to x must be replaced by derivatives with respect to x and y. In two dimensions, we then have

$$-\frac{\hbar^2}{2m} \left(\frac{\partial^2 \psi(x,y)}{\partial x^2} + \frac{\partial^2 \psi(x,y)}{\partial y^2} \right) + V(x,y)\psi(x,y) = E\psi(x,y) \quad (5.28)$$

[The first two terms on the left side of this equation require *partial* derivatives; for well-behaved functions, these involve taking the derivative with respect to one variable while keeping the other constant. Thus if $f(x,y) = x^2 + xy + y^2$, $\partial f / \partial x = 2x + y$ and $\partial f / \partial y = 2y + x$.]

Our two-dimensional "box" can now be expressed as follows:

$$V(x,y) = 0 \qquad 0 \leq x \leq L, 0 \leq y \leq L$$
$$= \infty \qquad \text{otherwise} \qquad (5.29)$$

We picture a mass sliding without friction on a tabletop and colliding elastically with walls at $x = 0$, $x = L$, $y = 0$, and $y = L$, as in Figure 5.5. (For simplicity, we have made the box square; we could have made it rectangular by setting $V = 0$ when $0 \leq x \leq a$ and $0 \leq y \leq b$.)

Solving *partial* differential equations requires a technique more involved than we need to consider, so we will not give the details of the solution. We suspect that, as in the previous case, $\psi(x,y) = 0$ outside the box, in order to make the probability zero there. Inside the box, we consider solutions that are *separable*; that is, our function of x and y can be expressed as the product of one function that depends only on x and another that depends only on y:

$$\psi(x,y) = f(x)\,g(y) \quad \text{inside} \qquad (5.30)$$

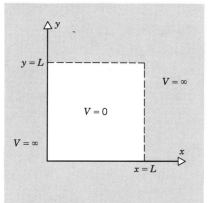

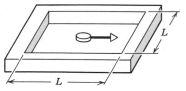

FIGURE 5.5 A particle moves freely in the two-dimensional region $0 \leq x \leq L$, $0 \leq y \leq L$.

to find it most often near the four points $(x, y) = (L/4, L/4)$, $(L/4, 3L/4)$, $(3L/4, L/4)$ and $(3L/4, 3L/4)$; we expect *never* to find it at $x = L/2$ or $y = L/2$. The *shape* of the probability density tells us something about the quantum numbers and therefore about the energy. Thus if we measured the probability density and found six maxima, as shown in Figure 5.7, we would deduce that the particle had an energy of $13E_0$ with $n_x = 2$ and $n_y = 3$, or else $n_x = 3$, $n_y = 2$.

Occasionally it happens that two different sets of quantum numbers n_x and n_y have exactly the same energy. This situation is known as *degeneracy*, and the energy levels are said to be *degenerate*. For example, the energy level at $E = 13E_0$ is degenerate, since both $n_x = 2$, $n_y = 3$ and $n_x = 3$, $n_y = 2$ have $E = 13E_0$. Since this degeneracy arises from interchanging n_x and n_y (which is the same as interchanging the x and y axes), the probability distributions in the two cases are not very different. However, consider the state with $E = 50E_0$, for which there are three sets of quantum numbers: $n_x = 7$, $n_y = 1$; $n_x = 1$, $n_y = 7$; and $n_x = 5$, $n_y = 5$. The first two result from the interchange of n_x and n_y and so have similar probability distributions, but the third represents a *very* different state of motion, as shown in Figure 5.8. The level at $E = 13E_0$ is said to be *two-fold* degenerate, while the level at $E = 50E_0$ is *three-fold* degenerate; we could also say that one level has a degeneracy of 2, while the other has a degeneracy of 3.

Degeneracy occurs in general whenever a system is labeled by two or more quantum numbers; as we have seen in the above calculation,

Degeneracy

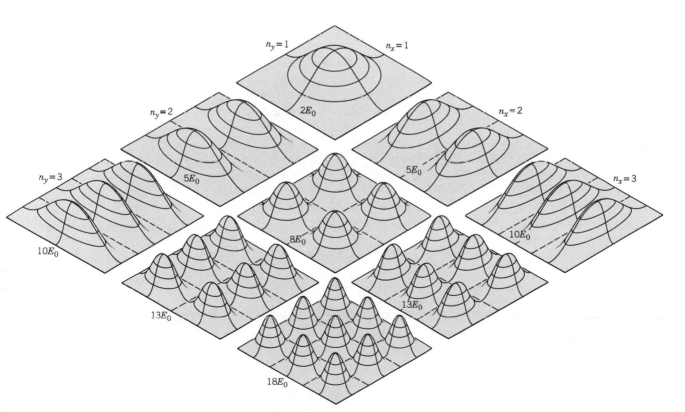

FIGURE 5.7 The probability density ψ^2 for some of the lower energy levels of the particle confined to the two-dimensional box.

where the functions f and g are similar to Equation (5.15)

$$f(x) = A \sin k_x x + B \cos k_x x$$
$$g(y) = C \sin k_y y + D \cos k_y y$$

(5.31)

The wave number k of the previous problem has become the separate wave numbers k_x for $f(x)$ and k_y for $g(y)$. We show later how these are related. (See also Problem 16 at the end of this chapter.)

The continuity condition on $\psi(x, y)$ requires that the solutions inside and outside match at the boundary; thus $\psi = 0$ at $x = 0$ and $x = L$ (for all y) and $\psi = 0$ at $y = 0$ and $y = L$ (for all x). The conditions at $x = 0$ and $y = 0$ require, in analogy with the previous problem, $B = 0$ and $D = 0$. The condition at $x = L$ requires that $\sin k_x L = 0$, and thus that $k_x L$ be an integer multiple of π; the condition at $y = L$ similarly requires that $k_y L$ be an integer multiple of π. These integers do not necessarily need to be the same, so we call them n_x and n_y to help keep track of them. We thus have:

$$\psi(x, y) = A' \sin \frac{n_x \pi x}{L} \sin \frac{n_y \pi y}{L}$$

(5.32)

where we have combined A and C into A'. The coefficient A' is once again found by the normalization condition, which in two dimensions becomes

$$\iint \psi^2 \, dx \, dy = 1$$

(5.33)

For our case this gives

$$\int_0^L dy \int_0^L A'^2 \sin^2 \frac{n_x \pi x}{L} \sin^2 \frac{n_y \pi y}{L} \, dx = 1$$

(5.34)

from which follows

$$A' = \frac{2}{L}$$

(5.35)

(The solutions to this problem, which are standing deBroglie waves on a two-dimensional surface, are similar to the solutions of the classical problem of the vibrations of a stretched membrane like a drumhead.)

Finally, we can plug our solution for $\psi(x, y)$ back into Equation (5.28) to find the energy:

$$E = \frac{\hbar^2 \pi^2}{2mL^2} (n_x^2 + n_y^2)$$

(5.36)

Compare this result with Equation (5.25). Once again we let $E_0 = \hbar^2 \pi^2 / 2mL^2$ so that $E = E_0(n_x^2 + n_y^2)$. In Figure 5.6 the energies of the excited states are shown. You can see how different the energies are from those of the one-dimensional case shown in Figure 5.4.

Figure 5.7 shows the probability density ψ^2 for several different combinations of the *quantum numbers* n_x and n_y. The probability has maxima and minima, just like the probability in the one-dimensional problem. For example, if we gave the particle an energy of $8E_0$ and then made a large number of measurements of its position, we would expect

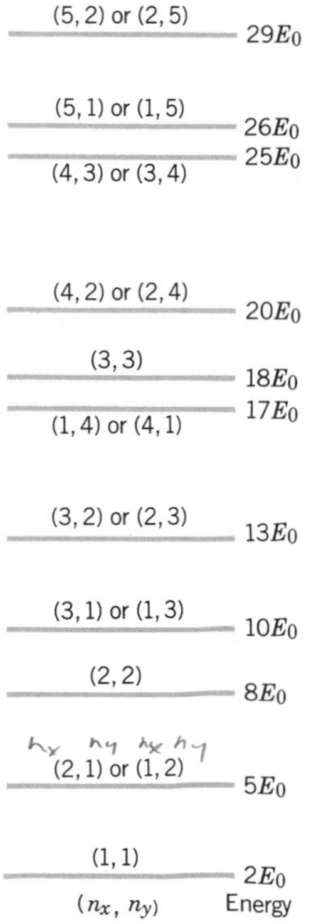

(n_x, n_y)	Energy
(5, 2) or (2, 5)	$29E_0$
(5, 1) or (1, 5)	$26E_0$
(4, 3) or (3, 4)	$25E_0$
(4, 2) or (2, 4)	$20E_0$
(3, 3)	$18E_0$
(1, 4) or (4, 1)	$17E_0$
(3, 2) or (2, 3)	$13E_0$
(3, 1) or (1, 3)	$10E_0$
(2, 2)	$8E_0$
(2, 1) or (1, 2)	$5E_0$
(1, 1)	$2E_0$

FIGURE 5.6 The lower permit energy levels of the particle co the two-dimensional box.

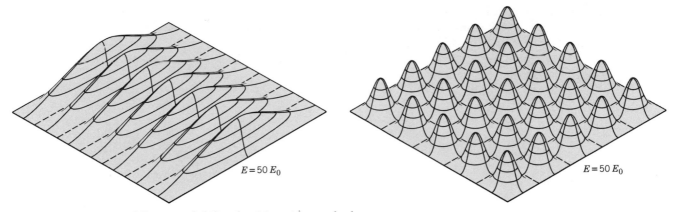

$E = 50 E_0$ $E = 50 E_0$

FIGURE 5.8 Two very different probability densities with exactly the same energy.

different combinations of quantum numbers often can give the same value of the energy. The number of different quantum numbers required by a given physical problem turns out to be exactly equal to the number of dimensions in which the problem is being solved—one-dimensional problems need only one quantum number, two-dimensional problems need two, and so forth. When we get to three dimensions, as in some problems at the end of this chapter and especially in the hydrogen atom in Chapter 7, we find that the effects of degeneracy become more significant; in the case of atomic physics, the degeneracy is a major contributor to the structure and properties of atoms.

Another abstraction that can be handled easily using the Schrödinger equation is the one-dimensional simple harmonic oscillator. The classical oscillator we can consider is the mass m suspended from a spring of force constant k and thus subject to a restoring force, $F = -kx$, where x is the displacement from equilibrium. Such an oscillator can be analyzed using Newton's laws to have a frequency $\omega_0 = \sqrt{k/m}$ and a period $T = 2\pi\sqrt{m/k}$. The oscillator has its maximum kinetic energy at $x = 0$; its kinetic energy vanishes at the *turning points* $x = \pm A_0$, where A_0 is the amplitude of the motion. At the turning points the oscillator comes to rest for an instant and then reverses its direction of motion. The motion is, of course, confined to the region $-A_0 \leq x \leq +A_0$.

Why analyze the motion of such a system using quantum mechanics? Although we never find in nature an example of a one-dimensional quantum oscillator, there are systems that behave approximately as one—a vibrating diatomic molecule, for example. In fact, any system in a potential minimum behaves, to lowest order, like a simple harmonic oscillator.

A force $F = -kx$ has the associated potential $V = \frac{1}{2}kx^2$, and so we have the Schrödinger equation:

$$-\frac{\hbar^2}{2m}\frac{d^2\psi}{dx^2} + \frac{1}{2}kx^2\psi = E\psi \qquad (5.37)$$

(Since we are working in one dimension, V and ψ are functions only of x.) This differential equation is difficult to solve directly, and so we

5.5 THE SIMPLE HARMONIC OSCILLATOR

guess at its solutions. The solution must approach zero as $x \to \pm\infty$, and in the limit $x \to \pm\infty$, the solutions of Equation (5.37) behave like exponentials of $-x^2$. We therefore try $\psi(x) = Ae^{-ax^2}$, where A and a are constants that are determined by evaluating Equation (5.37) for this choice of $\psi(x)$. We begin by evaluating $d^2\psi/dx^2$.

$$\frac{d\psi}{dx} = -2ax(Ae^{-ax^2})$$

$$\frac{d^2\psi}{dx^2} = -2a(Ae^{-ax^2}) - 2ax(-2ax)Ae^{-ax^2}$$

don't want functions to blow up

and we plug $\psi(x)$ and $d^2\psi/dx^2$ into (5.37) to see if there is a solution.

$$-\frac{\hbar^2}{2m}(-2aAe^{-ax^2} + 4a^2x^2Ae^{-ax^2}) + \frac{1}{2}kx^2(Ae^{-ax^2}) = EAe^{-ax^2} \quad (5.38)$$

Canceling the common factor Ae^{-ax^2} yields

must hold for any value of x

$$\frac{\hbar^2 a}{m} - \frac{2a^2\hbar^2}{m}x^2 + \frac{1}{2}kx^2 = E \quad (5.39)$$

Equation (5.39) is *not* an equation to be solved for x, because we are looking for a solution which is valid for *any* x, not just for one specific value. In order for this to hold for *any* x, the coefficients of x^2 must cancel and the remaining constants must be equal. (That is, consider the equation $ax + b = 0$. This equation is of course valid for $x = -b/a$, but if we wish it to be true for *any* and *all* x, we must require that $a = 0$ and $b = 0$.) Thus

$$-\frac{2a^2\hbar^2}{m} + \frac{1}{2}k = 0 \quad (5.40)$$

and

$$\frac{\hbar^2 a}{m} = E \quad (5.41)$$

yielding

$$a = \frac{\sqrt{km}}{2\hbar} \quad (5.42)$$

$= \frac{1}{2}\frac{m\omega_0}{\hbar}$ *(margin note)*

and

$$E = \tfrac{1}{2}\hbar\sqrt{k/m} \quad (5.43)$$

We can also write the energy in terms of the classical frequency $\omega_0 = \sqrt{k/m}$ as

$$E = \tfrac{1}{2}\hbar\omega_0 \quad (5.44)$$

margin notes: $m\omega_0^2 = k$ $\omega_0 = \sqrt{\frac{k}{m}}$

$E = \frac{\hbar a}{m}$

The coefficient A must be found from the normalization condition (see Problem 18 at the end of the chapter).

The solution we have just found is illustrated in Figure 5.9. One feature of the solution is striking—there is a nonzero probability of finding the particle beyond the classical turning points $x = \pm A_0$. The total energy E is constant, and beyond $x = \pm A_0$, the potential energy is greater than E, so that the kinetic energy would become *negative*. This is an impossibility in the realm of classical physics, and so the classical *particle* can never be found at $|x| > A_0$. It is, however, possible for the quantum *wave* to penetrate into the classically forbidden region. We shall pursue this question when we discuss barriers in the next section.

The solution we have found corresponds to the *ground state* of the

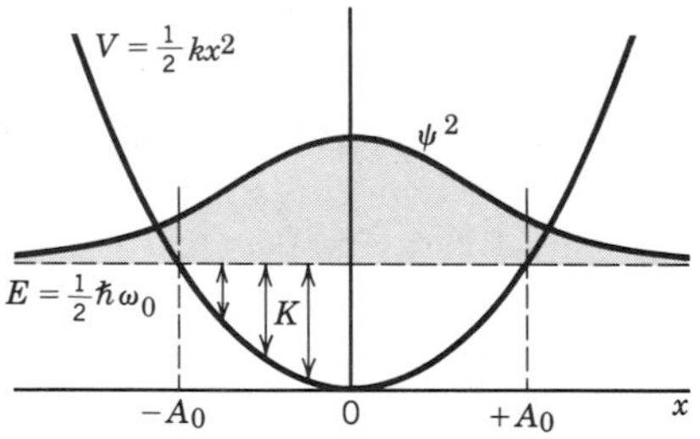

FIGURE 5.9 The ground state of the one-dimensional harmonic oscillator. The kinetic energy K is the difference between the total energy E and the potential energy $V = \tfrac{1}{2}kx^2$. Classical physics does not permit the particle to move beyond the classical turning points $x = \pm A_0$, where its kinetic energy would be negative. The probability density ψ^2 extends beyond the classical turning points, so there is according to quantum physics some probability for the particle to enter the classically forbidden region.

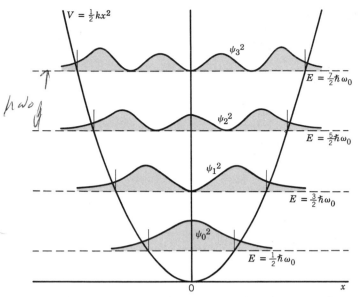

FIGURE 5.10 The lowest few energy levels and corresponding probability densities of the harmonic oscillator.

oscillator. The mathematically difficult general solution is of the form $\psi_n(x) = Af_n(x)e^{-ax^2}$, where $f_n(x)$ is a polynomial in which the highest power of x is x^n. The corresponding energies are

$$E_n = (n + \tfrac{1}{2})\hbar\omega_0 \qquad (5.45)$$

Harmonic oscillator energy levels

where n is an integer 0, 1, 2, . . . Note that these levels, illustrated along with their probability densities in Figure 5.10, are *uniformly spaced,* in contrast to the particle in a box. All of the solutions have the property of penetration of probability density into the forbidden region beyond the classical turning points. The probability density oscillates, somewhat like a sine wave, between the turning points, and decreases like e^{-2ax^2} to zero beyond the turning points.

We have thus far not considered the time dependence of the Schrödinger equation or of its solutions. We will not consider the method of solution in detail, but will merely state the result: given a time-independent solution $\psi(x)$ of Equation (5.3) corresponding to the energy E, the *time-dependent wave function* $\Psi(x, t)$ is found according to*

$$\Psi(x, t) = \psi(x)e^{-i\omega t} \qquad (5.46)$$

5.6 TIME DEPENDENCE

---

* The imaginary number i is defined as $\sqrt{-1}$. A *complex number* can be represented as having a real part, which does not depend on i, and an imaginary part, which depends on i. We require no manipulations with complex numbers for this chapter, and need only the following relationship: the complex exponential $e^{i\theta}$ can be represented in terms of real trigonometric functions as

$$e^{i\theta} = \cos\theta + i\sin\theta$$

and

$$e^{-i\theta} = \cos\theta - i\sin\theta$$

Thus

$$\sin\theta = \frac{1}{2i}(e^{i\theta} - e^{-i\theta})$$

and

$$\cos\theta = \tfrac{1}{2}(e^{i\theta} + e^{-i\theta})$$

where the frequency ω is given by the deBroglie relationship

$$\omega = \frac{E}{\hbar} \qquad (5.47)$$

As mentioned in Section 4.1, it is not clear whether the energy E in the deBroglie relationship should be the classical total energy or the relativistic total energy, since we get no clue from the corresponding relationship $E = h\nu$ for photons. In this chapter we have assumed the classical relationship $E = V + K$ and neglected the rest energy contribution to E. To be strictly correct, we should write $E = V + K + m_0 c^2$ (but we still consider only cases where $v \ll c$ so that the classical $\frac{1}{2}mv^2$ is acceptable for K). The addition of the rest energy changes Equation (5.46) by introducing a factor $e^{-im_0 c^2 t/\hbar}$, but since all of the measurable properties of $\Psi(x, t)$ depend on $\Psi^*\Psi$, the product of Ψ and its *complex conjugate* obtained by replacing i with $-i$, this additional factor has no observable consequences and can safely be ignored. As in Problem 30 of Chapter 4, the group velocity depends on dE/dp and the addition of constants like $m_0 c^2$ to E does not affect the motion of the wave packet.

In order to see how multiplying by $e^{-i\omega t}$ gives a wave, we consider how the wave function of the free particle, Equation (5.15), gives the wave function $\Psi(x, t)$. This process is simplified if we first rewrite Equation (5.15) in terms of the complex exponentials e^{ikx} and e^{-ikx}, so that

$$\psi(x) = A'e^{ikx} + B'e^{-ikx} \qquad (5.48)$$

The constants A' and B' can be found from the constants A and B. Now we have, for the time-dependent wave function,

$$\Psi(x, t) = (A'e^{ikx} + B'e^{-ikx})e^{-i\omega t}$$
$$= A'e^{i(kx-\omega t)} + B'e^{-i(kx+\omega t)} \qquad (5.49)$$

The first term on the right represents a trigonometric function with phase $(kx - \omega t)$, and thus is a wave moving in the *positive* x direction; the second term corresponds to a wave moving in the *negative* x direction. The squared magnitudes of the coefficients give the intensities of the waves; thus the wave moving in the positive x direction has intensity $|A'|^2$ and the wave moving in the negative x direction has intensity $|B'|^2$.

Suppose we have a beam of monoenergetic particles moving in the positive x direction, which are represented by a wave function in the form of the *first* term of Equation (5.49). The probability for locating a particle is then given by $|A'|^2$. This is a constant, independent of the position x—a particle is equally likely to be found at any location along the x axis. (Of course, this is consistent with the uncertainty relationship. If the beam is monoenergetic, $\Delta p = 0$ and therefore $\Delta x = \infty$.)

If the wave function contains equal amplitudes of waves moving in both directions (i.e., if $|A'| = |B'|$), certain locations can be found at which the probability density $\Psi^*\Psi$ is equal to zero. *There can be points at which there is zero probability of finding the particle!* In analogy with classical physics, when we add two waves of equal amplitude traveling in opposite directions, we obtain a *standing wave,* in which there are certain points (known as "nodes") at which the combined wave amplitude vanishes at all times.

In this general type of problem, we will analyze what happens when a particle moving (again in one dimension) in a region of constant potential suddenly moves into a region of different, but also constant, potential. We will not discuss in detail the solutions to these problems, but the methods of solution of each are so similar that we can outline the steps to take in the solution. In this discussion, we will let E be the (fixed) total energy of the particle and V_0 will be the value of the constant potential.

1. Whenever E is greater than V_0, the solution to the Schrödinger equation is of the form

$$\psi(x) = A \sin kx + B \cos kx \qquad (5.50)$$

where

$$k = \sqrt{\frac{2m}{\hbar^2}(E - V_0)} \qquad (5.51)$$

A and B are constants to be found from the continuity and normalization conditions. For example, consider the step potential shown in Figure 5.11:

$$V(x) = 0 \qquad x < 0$$
$$\quad = V_0 \qquad x \geq 0$$

If E is the total energy and is greater than V_0, then we can simply write down the solutions to the Schrödinger equation in the two regions:

5.7 STEPS AND BARRIERS

$$\psi_0(x) = A \sin k_0 x + B \cos k_0 x \qquad k_0 = \sqrt{\frac{2mE}{\hbar^2}} \qquad x < 0 \quad (5.52a)$$

Step potential, $E > V_0$

$$\psi_1(x) = C \sin k_1 x + D \cos k_1 x \qquad k_1 = \sqrt{\frac{2m}{\hbar^2}(E - V_0)} \quad x > 0 \quad (5.52b)$$

Relationships among the four coefficients, A, B, C, and D, may be found by applying the condition that $\psi(x)$ and $\psi'(x) = d\psi/dx$ must be continuous at the boundary; thus $\psi_0(0) = \psi_1(0)$, $\psi_0'(0) = \psi_1'(0)$. A typical solution might look like that sketched in Figure 5.12. Note that the application of the continuity condition guarantees a smooth transition from one wave to the next at the boundary.

Once again, we may use the equation $e^{i\theta} = \cos\theta + i\sin\theta$ to transform these solutions from sines and cosines to complex exponentials:

$$\psi_0(x) = A'e^{ik_0 x} + B'e^{-ik_0 x} \qquad x < 0 \qquad (5.53a)$$
$$\psi_1(x) = C'e^{ik_1 x} + D'e^{-ik_1 x} \qquad x > 0 \qquad (5.53b)$$

When the time dependence has been added by multiplying each term by $e^{-i\omega t}$, we can then make the following identification of the component

FIGURE 5.11 A step of height V_0.

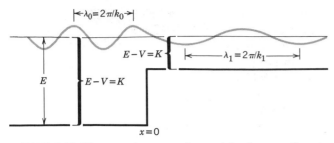

FIGURE 5.12 The wave function of a particle of energy E encountering a step of height V_0, for the case $E > V_0$. The deBroglie wavelength changes from λ_0 to λ_1 when the particle crosses the step, but ψ and $d\psi/dx$ are continuous at $x = 0$.

waves, recalling that $(kx - \omega t)$ is the phase of a wave moving in the positive x direction, while $(kx + \omega t)$ is the phase of a wave moving in the negative x direction, and assuming that *the squared magnitude of each coefficient gives the intensity of the corresponding component wave.* In the region $x < 0$, Equation (5.53a) describes the superposition of a wave of intensity $|A'|^2$ moving in the positive x direction (from $-\infty$ to 0) and a wave of intensity $|B'|^2$ moving in the negative x direction. Suppose we had intended our solution to describe particles that are incident from the left on this potential. Then $|A'|^2$ gives the intensity of the *incident* wave (more exactly, the deBroglie wave describing the incident beam of particles) and $|B'|^2$ gives the intensity of the *reflected* wave. The ratio $|B'|^2/|A'|^2$ tells us the reflected fraction of the incident wave intensity. In the region $x > 0$, the wave of intensity $|D'|^2$ moving in the negative x direction (from $x = +\infty$ to $x = 0$) cannot exist if we are firing particles from the negative x axis, so *in this particular experimental situation* we would set D' to zero. The intensity of the *transmitted* wave is then $|C'|^2$.

We can also analyze the solutions in terms of the kinetic energy in each region. Where the kinetic energy is greatest, the linear momentum p $(= \sqrt{2mK})$ will be greatest, and the deBroglie wavelength λ $(= h/p)$ will be smallest. Thus in the region $x < 0$, the wavelength is smaller than in the region $x > 0$.

2. Whenever E becomes less than V_0, a different solution results:

$$\psi(x) = Ae^{kx} + Be^{-kx} \qquad (5.54)$$

where

$$k = \sqrt{\frac{2m}{\hbar^2}(V_0 - E)} \qquad (5.55)$$

If the region in which this solution is to be valid extends to $+\infty$ or $-\infty$, we must keep ψ from becoming infinite by setting A or B to zero; if the region contains only finite x, this need not be done.

As an example, in the previous problem, if E were less than V_0, then the solution for ψ_0 (for $x < 0$) would still be given by Equation (5.52a) or (5.53a), but the solution ψ_1 (for $x > 0$) becomes

$$\psi_1(x) = Ce^{k_1x} + De^{-k_1x} \qquad k_1 = \sqrt{\frac{2m}{\hbar^2}(V_0 - E)} \qquad (5.56)$$ **Step potential, $E < V_0$**

Again, we must be sure that the solutions match smoothly at the boundaries; the application of the boundary conditions proceeds as before. (We set $C = 0$ to keep $\psi_1(x)$ from becoming infinite as $x \to +\infty$.) A possible solution might look like the one shown in Figure 5.13.

This solution illustrates an important difference between classical and quantum mechanics. Classically, the particle can *never* be found at $x > 0$, since its total energy is not sufficient to overcome the potential energy step. However, quantum mechanics allows the wave function, and therefore the particle, to penetrate the classically forbidden region. (As we discuss below, the particle can never be *observed* in the forbidden region.)

The probability density in the $x > 0$ region is $|\psi_1|^2$, which according to Equation (5.56) is proportional to $e^{-2k_1 x}$. If we define a representative penetration distance Δx to be the distance from $x = 0$ to the point at which the probability drops by $1/e$, then

$$e^{-2k_1 \Delta x} = e^{-1}$$

$$\Delta x = \frac{1}{2k_1} = \frac{1}{2} \frac{\hbar}{\sqrt{2m(V_0 - E)}} \tag{5.57}$$

If it were to enter the region with $x > 0$, the *particle* must gain an energy of at least $V_0 - E$ in order to get over the potential step; it must in addition gain some kinetic energy if it is to move in the region $x > 0$. Of course, it is a violation of conservation of energy for the particle to spontaneously gain *any* amount of energy, but according to the uncertainty relationship $\Delta E \, \Delta t \sim \hbar$ conservation of energy does not apply at times smaller than Δt except to within an amount $\Delta E \sim h/\Delta t$. That is, if the particle "borrows" an amount of energy ΔE and "returns" the borrowed energy within a time $\Delta t \sim \hbar/\Delta E$, we observers will still believe energy is conserved. Suppose we borrow an energy sufficient to give the particle a kinetic energy of K in the forbidden region. How far into the forbidden region does it penetrate?

The "borrowed" energy is $(V_0 - E) + K$; the term $(V_0 - E)$ gets the particle to the top of the step, and the extra K gives it its motion. We must return this energy within a time

$$\Delta t = \frac{\hbar}{V_0 - E + K} \tag{5.58}$$

The particle moves with speed $v = \sqrt{2K/m}$, and so the distance it can travel is

$$\Delta x = \tfrac{1}{2} v \, \Delta t$$

$$= \frac{1}{2} \sqrt{\frac{2K}{m}} \frac{\hbar}{V_0 - E + K} \tag{5.59}$$

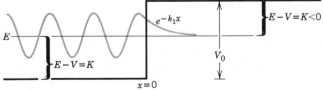

FIGURE 5.13 The wave function of a particle of energy E encountering a step of height V_0, for the case $E < V_0$. The wave function decreases exponentially in the classically forbidden region, where the classical kinetic energy would be negative. At $x = 0$, ψ and $d\psi/dx$ are continuous.

(The factor of $\frac{1}{2}$ is present because in the time Δt the particle must penetrate the distance Δx and return.)

In the limit $K \to 0$, the penetration distance Δx goes to 0 according to Equation (5.59) because the particle has zero velocity; similarly, $\Delta x \to 0$ in the limit $K \to \infty$, because it moves for a vanishing time interval Δt. In between those limits, there must be a maximum value of Δx for some particular K. Differentiating Equation (5.59) we can find the maximum value

$$\Delta x_{\max} = \frac{1}{2} \frac{\hbar}{\sqrt{2m(V_0 - E)}} \tag{5.60}$$

This value of Δx is identical with Equation (5.57)! This demonstrates that the penetration into the forbidden region given by the solution to the Schrödinger equation is entirely consistent with the uncertainty relationship. (The agreement between Equations (5.57) and (5.60) is really somewhat accidental, since the factor $1/e$ used to obtain (5.57) was chosen arbitrarily. What we have really demonstrated is that the Schrödinger equation gives the same estimates of uncertainty as the Heisenberg relationships.)

Consider now the barrier potential shown in Figure 5.14.

$$V(x) = 0 \qquad x < 0$$
$$= V_0 \qquad 0 \leq x \leq a$$
$$= 0 \qquad x > a$$

Particles with energy E less than V_0 are incident from the left. Our experience then leads us to expect solutions of the form shown in Figure 5.15—sinusoidal oscillation in the region $x < 0$ (an incident wave and a reflected wave), exponentials in the region $0 \leq x \leq a$, and sinusoidal oscillations in the region $x > a$ (the transmitted wave). The intensity of the transmitted wave can be found by proper application of the continuity conditions, which we will not discuss; the intensity is found to depend on the energy of the particle and on the height and thickness of the barrier. Classically, the particles should never appear at $x > a$, since they have not sufficient energy to overcome the barrier; this situation is an example of *barrier penetration*, sometimes called quantum mechanical *tunneling*. The particles can never be *observed* in the classically forbidden region $0 \leq x \leq a$, but can "tunnel" *through* that region and be observed at $x > a$.

Barrier penetration (tunneling)

Although the potential of Figure 5.14 is quite schematic and hypothetical, there are many examples in nature of such tunneling. We consider three such examples.

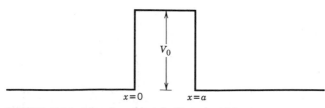

FIGURE 5.14 A barrier of height V_0 and width a.

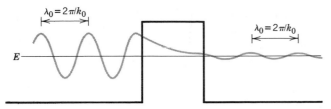

$\lambda_0 = 2\pi/k_0$

$\lambda_0 = 2\pi/k_0$

E

FIGURE 5.15 The wave function of a particle of energy $E < V_0$ encountering a barrier potential (the particle would be incident from the left in the figure). The wavelength λ_0 is the same on both sides of the barrier, but the amplitude beyond the barrier is much less than the original amplitude. The particle can never be observed inside the barrier (where it would have negative kinetic energy) but it can be observed *beyond* the barrier.

1. *Alpha decay* An atomic nucleus consists of protons and neutrons in a constant state of motion; occasionally these particles form themselves into an aggregate of two protons and two neutrons, called an alpha particle. In one form of radioactive decay, the nucleus can emit an alpha particle, which can be detected in the laboratory. However, in order to escape from the nucleus the alpha particle must penetrate a potential of the form shown in Figure 5.16. The probability for the alpha particle to penetrate the barrier, and be detected in the laboratory, can be computed based on the energy of the alpha particle, and the height and thickness of the barrier. These decay probabilities can be measured in the laboratory and are found to be in excellent agreement with those predicted on the basis of a quantum-mechanical calculation based on barrier penetration.

2. *Ammonia inversion* Figure 5.17 is a representation of the ammonia molecule NH_3. If we were to try to move the nitrogen atom along the axis of the molecule, toward the plane of the hydrogen atoms, we would find repulsion caused by the three hydrogen atoms, which pro-

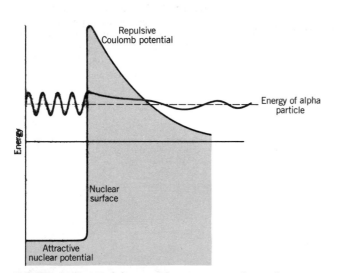

Repulsive Coulomb potential

Energy of alpha particle

Energy

Nuclear surface

Attractive nuclear potential

FIGURE 5.16 An alpha particle penetrating the nuclear potential barrier. The probability to penetrate the barrier depends on its thickness and its height.

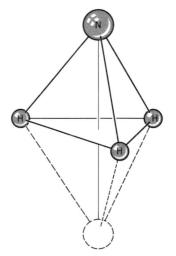

FIGURE 5.17 A schematic diagram of the ammonia molecule. The Coulomb repulsion of the three hydrogens establishes a barrier against the nitrogen atom moving to a symmetric position (shown in dashed lines) on the opposite side of the plane of hydrogens.

duces a potential of the form shown in Figure 5.18. According to classical mechanics, unless we give the nitrogen atom sufficient energy, it should not be able to surmount the barrier and appear on the other side of the plane of hydrogens. According to quantum mechanics, the nitrogen can tunnel through the barrier and appear on the other side of the molecule. In fact, the N atom actually tunnels back and forth with a frequency in excess of 10^{10} oscillations per second.

3. **The tunnel diode**   An electronic device that uses the phenomenon of tunneling is the tunnel diode. We reserve a detailed discussion of the properties of semiconductor devices to Chapter 14. Schematically, the potential "seen" by an electron in a tunnel diode can be represented in a form similar to that of Figure 5.19. The current that flows through such a device is produced by electrons tunneling through from one side to the other. The rate of tunneling, and therefore the current, can be regulated merely by changing the height of the barrier, which can be done with an applied voltage. This can be done rapidly, so that switching frequencies in excess of 10^9 Hz can be obtained. Ordinary semiconductor diodes depend on the diffusion of electrons across a junction, and therefore operate on much longer time scales (at lower frequencies).

Before concluding this discussion, we return briefly to the question of particle-wave duality. The *particle* is never observed in the forbidden region; to find it there would violate energy conservation. Every particle that is incident from the left on the step potential of Figure 5.13 is reflected and ends up moving back in the negative x direction. Some are reflected at $x = 0$, while others may penetrate a distance Δx before turning around and reemerging. Every particle incident on the barrier of Figure 5.15 is either reflected or transmitted; the number of incident particles is equal to the number reflected back to $x < 0$ plus the number transmitted to $x > a$. None are "trapped" or ever seen in the forbidden region $0 < x < a$. How can the incident *particle* get from $x < 0$ to $x > a$? As a particle, *it can't!* Only the *wave* can penetrate to $x > a$, but of course wherever the wave is, particles must be also.

You must keep in mind that all of these unusual phenomena we are

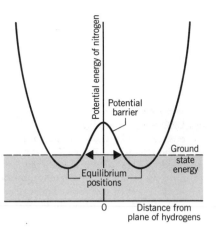

FIGURE 5.18 The potential energy seen by the nitrogen atom in an ammonia molecule. The nitrogen can penetrate the barrier and move from one equilibrium position to another.

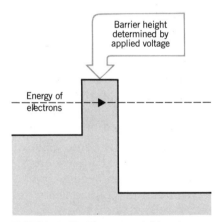

FIGURE 5.19 The potential barrier seen by an electron in a tunnel diode. The conductivity of the device is determined by the electron's probability to penetrate the barrier, which depends on the height of the barrier.

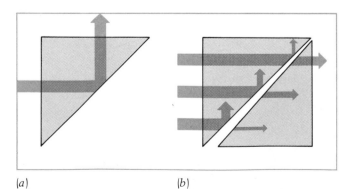

(a)   (b)

FIGURE 5.20 (a) Total internal reflection of light waves at a glass-air boundary. (b) Frustrated total internal reflection. The thicker the air gap, the smaller the probability to penetrate. Note that the light beam does not appear *in* the gap.

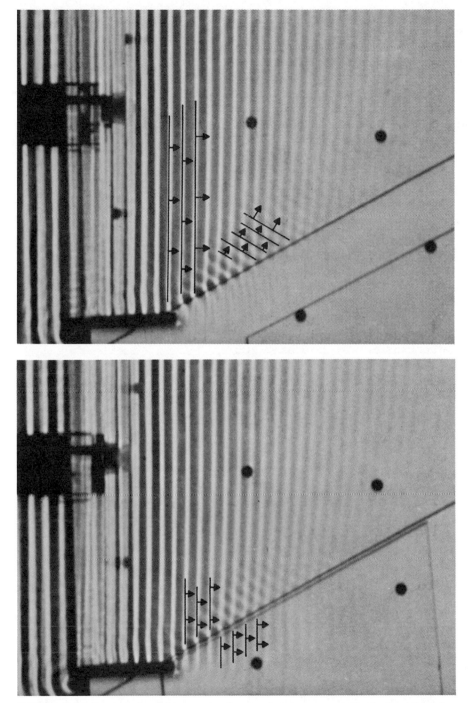

FIGURE 5.21 Frustrated total internal reflection for water waves. At the boundary the depth increases suddenly and the waves are totally reflected. When the gap is made narrow, the waves can penetrate and appear on the other side. (Courtesy of Education Development Center, Inc., Newton, MA.)

discussing are well-known properties of waves (even classical waves). The only new ingredient provided by quantum physics is to demand that these wave properties be applied to deBroglie waves, which means that the particles they represent must have the same properties.

The property of penetration into a forbidden region is a general property of classical waves. Consider the case of total internal reflection* of light waves, as shown in Figure 5.20a. How does the light wave "know" that there is air on the other side of the boundary? As long as the light wave is entirely inside the glass, it cannot "know" what is beyond the glass. To find out, it must penetrate into the forbidden air a short distance, perhaps a few wavelengths, before it realizes that entry into the air is forbidden and that it must return to the glass. It is, of course, never observed in the air; the laws of reflection and refraction forbid its presence there. If, however, a second piece of glass is placed within the penetration distance, the beam can reappear in the second piece of glass, as shown in Figure 5.20b. This phenomenon is called *frustrated total internal reflection.* We can imagine that the beam sends out "feelers" into the forbidden air gap, finds the other glass, realizes that it is not forbidden from entering the glass, and proceeds on its way. Just as with the barrier of Figure 5.15, the probability to penetrate the air gap decreases as the thickness of the gap increases.

The same behavior is shown by other classical waves, such as water waves as in Figure 5.21.

Although the situations we have discussed in this chapter are schematic and nonphysical, the principles involved are general and apply to all systems governed by wave mechanics. We see in Chapter 7 how atomic properties are subject to the same effects of wave mechanics, and in later chapters we study the wave-mechanical properties of molecules and solids.

SUGGESTIONS FOR FURTHER READING

An introduction to the formalism of quantum physics and its historical development is found in W. H. Cropper, *The Quantum Physicists and an Introduction to Their Physics* (New York, Oxford University Press, 1970).

Other discussions of the Schrödinger equation and examples of its use are given in the following books, in approximately increasing order of difficulty beginning at the level of this book:

N. Ashby and S. C. Miller, *Principles of Modern Physics* (San Francisco, Holden-Day, 1970).

P. A. Lindsay, *Introduction to Quantum Mechanics for Electrical Engineers* (New York, McGraw-Hill, 1967).

E. H. Wichmann, *Quantum Physics, Volume 4 of the Berkeley Physics Course* (New York, McGraw-Hill, 1971).

J. Norwood, *Twentieth Century Physics* (Englewood Cliffs, Prentice-Hall, 1976).

R. Eisberg and R. Resnick, *Quantum Physics of Atoms, Molecules, Solids, Nuclei, and Particles* (New York, Wiley, 1974).

R. B. Leighton, *Principles of Modern Physics* (New York, McGraw-Hill, 1959).

* When a beam of light moves from a medium of high index of refraction (glass, for example) to a medium of low index of refraction (air), the beam can be totally reflected back into the glass if the angle of incidence exceeds the critical angle $\sin^{-1}(1/n)$, where n is the index of refraction of the glass. This topic is discussed in introductory physics texts.

QUESTIONS

1. Newton's laws can be solved to give the future behavior of a particle. In what sense does the Schrödinger equation also do this? In what sense does it not?

2. Why is it important for a wave function to be normalized? Is an unnormalized wave function a solution to the Schrödinger equation?

3. What is the physical meaning of $\int_{-\infty}^{\infty} |\psi|^2 \, dx = 1$?

4. What are the dimensions of $\psi(x)$? Of $\psi(x, y)$?

5. None of the following are permitted as solutions of the Schrödinger equation. Give the reasons in each case.
   (a) $\psi(x) = A \cos kx \qquad x < 0$
       $\psi(x) = B \sin kx \qquad x > 0$
   (b) $\psi(x) = \dfrac{Ae^{-kx}}{x} \qquad -L \leq x \leq L$
   (c) $\psi(x) = A \sin^{-1} kx$
   (d) $\psi(x) = A \tan kx \qquad x > 0$

6. What happens to the probability density in the infinite well when $n \to \infty$? Is this consistent with classical physics?

7. How would the solution to the infinite potential well be different if the well extended from $x = x_0$ to $x = x_0 + L$, where x_0 is a nonzero value of x? Would any of the measurable properties be different?

8. How would the solution to the one-dimensional infinite potential well be different if the potential were not zero for $0 \leq x \leq L$ but instead had a constant value V_0? What would be the energies of the excited states? What would be the wavelengths of the standing deBroglie waves? Sketch the behavior of the lowest two wave functions.

9. Assuming a pendulum to behave like a quantum oscillator, what are the energy differences between the quantum states of a pendulum of length 1 m? Are such differences observable?

10. For the barrier potential (Figure 5.14), is the wavelength for $x > a$ the same as the wavelength for $x < 0$? Is the amplitude the same?

11. Suppose particles were incident on the step potential from the *positive x* direction. Which of the four coefficients of Equation (5.53) would be set to zero? Why?

12. The energies of the excited states of the systems we have discussed in this chapter have been exact—there is no energy uncertainty. What does this suggest about the lifetime of particles in those excited states? Left on its own, will a particle really make transitions from one state to another?

PROBLEMS

1. A classical particle moves freely in the positive x direction with speed v_0. When it crosses the origin (at $t = 0$) it enters region 1 in which it feels a deceleration a_1. At time t_1 and position x_1 it leaves region 1 and enters region 2, where it feels a deceleration $a_2 = \frac{1}{2}a_1$. It leaves region 2 at time t_2 and coordinate x_2, when its velocity is $v_0/2$, and once again moves freely. Without setting up the equations of motion, make a sketch similar to Figure 5.1 showing the acceleration, velocity, and position from times less than zero to times greater than t_2. Think carefully about whether a, v, and x (and their slopes!) are continuous.

2. Suppose that a particle in the one-dimensional infinite well emitted photons in jumping from one state to a lower one, with no restrictions on the change of n. In terms of E_0, list all possible photon energies that could be

emitted when a particle in a one-dimensional infinite potential well goes from the $n = 4$ state to the ground state.

3. Suppose that a particle in the one-dimensional infinite well obeyed a rule that n could change by only one unit as the particle made transitions among the excited states. Show that the photons emitted in those transitions have energies $3E_0$, $5E_0$, $7E_0$, . . . $(2n - 1)E_0$, . . . where E_0 is the ground-state energy.

4. An electron is confined to a one-dimensional region in which its minimum energy is 1.0 eV. (a) What is the size of the region? (b) How much energy must be supplied to move the electron to the first excited state? (c) From a certain excited state, the electron gives up 24.0 eV to reach the ground state. What was the quantum number n of the excited state?

5. Show that Equation (5.26) gives the value $A = \sqrt{2/L}$.

6. (a) A particle is trapped in a one-dimensional region of dimension L. In its *second* excited state $(n = 3)$, show that the probability to find it between $x = 0$ and $x = L/3$ is $\frac{1}{3}$. (b) Show that the probability is $1/n$ to find the particle between $x = 0$ and $x = L/n$ in the state with quantum number n.

7. What is the minimum energy of an electron trapped in a one-dimensional region the size of an atomic nucleus $(1.0 \times 10^{-14}$ m)? Compare this result with the equivalent value found in Example 4.7.

8. What is the minimum energy of a proton or neutron $(mc^2 \cong 940$ MeV) confined to a region of space of nuclear dimensions $(1.0 \times 10^{-14}$ m)?

9. For a classical particle trapped in a one-dimensional potential well, the probability to find it in any interval of width w is (a) independent of the location in the well and (b) equal to w/L. Use the general wave function of the particle in a one-dimensional well to evaluate the probability to find the particle between x and $x + w$. What happens as $n \to \infty$? Is this consistent with the expected classical result?

10. Show that the average value of x^2 in the one-dimensional potential well is $(x^2)_{av} = L^2(\frac{1}{3} - 1/2n^2\pi^2)$.

11. Use the result of Problem 10 to show that, for the infinite one-dimensional well, defining $\Delta x = \sqrt{(x^2)_{av} - (x_{av})^2}$ gives

$$\Delta x = L \sqrt{\frac{1}{12} - \frac{1}{2\pi^2 n^2}}$$

12. (a) In the infinite one-dimensional potential well, what is $(p)_{av}$? (Use a symmetry argument.) (b) What is $(p^2)_{av}$? [*Hint.* What is $(p^2/2m)_{av}$?] (c) What is $\Delta p = \sqrt{(p^2)_{av} - (p_{av})^2}$? (d) Use the result of the previous problem to evaluate the smallest possible value of $\Delta x \, \Delta p$. How does this compare with the Heisenberg uncertainty relationship?

13. The problem of the particle in the *finite* potential well is similar to that of the infinite well, but the potential has the value V_0 for $x < 0$ and $x > L$. (a) Including six undetermined constants, write down the wave functions for the three regions $x < 0$, $0 < x < L$, and $x > L$, if $E < V_0$. (b) Two of the coefficients must be set to zero. Which ones? Why? (c) Keeping in mind the boundary conditions at $x = 0$ and $x = L$, sketch the three lowest-energy wave functions and probability densities. Do not attempt to apply the boundary conditions explicitly; just show the expected form of the wave function. Let the solutions for the infinite potential well guide you in deciding on the form of the wave functions inside the well.

14. What is the next level (above $E = 50E_0$) of the two-dimensional particle in a box in which the degeneracy is greater than 2?

15. A particle is confined to a two-dimensional box of length L and width $2L$. The energy values are $E = (\hbar^2\pi^2/2mL^2)(n_x^2 + n_y^2/4)$. Find the two lowest degenerate levels.

16. Show by direct substitution that Equation (5.31) gives a solution to the two-dimensional Schrödinger equation, Equation (5.28). Find the relationship between k_x, k_y, and E.

17. A particle is confined to a three-dimensional region of space of dimensions L by L by L. The energy levels are $(\hbar^2\pi^2/2mL^2)(n_x^2 + n_y^2 + n_z^2)$. Sketch an energy level diagram, showing the energies, quantum numbers, and degeneracies for the lowest 10 energy levels.

18. Using the normalization condition, show that the constant A has the value $(m\omega_0/\hbar\pi)^{1/4}$ for the one-dimensional simple harmonic oscillator.

19. Use the ground-state wave function of the simple harmonic oscillator to find x_{av}, $(x^2)_{av}$, and Δx. Use the normalization constant $A = (m\omega_0/\hbar\pi)^{1/4}$.

20. (a) What value do you expect for $(p)_{av}$ for the simple harmonic oscillator? Use a symmetry argument rather than a calculation. (b) Conservation of energy for the harmonic oscillator can be used to relate p^2 to x^2. Use this relation, along with the value of $(x^2)_{av}$ from Problem 19, to find $(p^2)_{av}$. (c) Evaluate Δp, using the results of (a) and (b).

21. From the results of Problems 19 and 20, evaluate $\Delta x\ \Delta p$ for the harmonic oscillator. Is the result consistent with the uncertainty relationship?

22. The first excited state of the harmonic oscillator has a wave function of the form $\psi(x) = Axe^{-ax^2}$. Follow the method outlined in Section 5.5 to find a and the energy E. Find the constant A from the normalization condition.

23. Assuming there is a rule that allows transitions between excited states of the harmonic oscillator *only* if n changes by one unit, what are the possible transition energies?

24. A two-dimensional harmonic oscillator has energy $E = \hbar\omega_0(n_x + n_y + 1)$, where n_x and n_y are integers beginning with zero. (a) Justify this result based on the energy of the one-dimensional oscillator. (b) Sketch an energy-level diagram similar to Figure 5.6, showing the values of E and the quantum numbers n_x and n_y. (c) Show that each level is degenerate, with degeneracy equal to $n_x + n_y + 1$.

25. Find the value of K at which Equation (5.59) has its maximum value, and show that Equation (5.60) is the maximum value of Δx.

26. Sketch a possible solution to the Schrödinger equation for each of the potentials shown below. In each case show several cycles of the wave function.

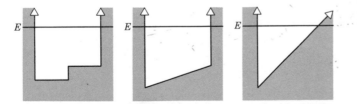

27. For a particle with energy $E < V_0$ incident on the step potential, use ψ_0 from Equation (5.52a) and ψ_1 from Equation (5.56), and evaluate the constants B and D in terms of A by applying the boundary conditions at $x = 0$.

28. Using the wave functions of Equation (5.53) for the step potential, apply the boundary conditions on ψ and $d\psi/dx$ to find B' and C' in terms of A', for the step potential when particles are incident from the negative x direction. Evaluate the ratios $|B'|^2/|A'|^2$ and $|C'|^2/|A'|^2$ and interpret.

29. (a) Write down the wave functions for the three regions of the barrier potential (Figure 5.14) for $E < V_0$. You will need six coefficients in all. Use complex exponential notation. (b) Use the boundary conditions at $x = 0$ and at $x = a$ to find four relationships among the six coefficients. (Do not try to solve these relationships.) (c) Suppose particles are incident on the barrier from the left. Which coefficient should be set to zero? Why?

30. Repeat Problem 29 for the barrier potential when $E > V_0$, and sketch the solutions.

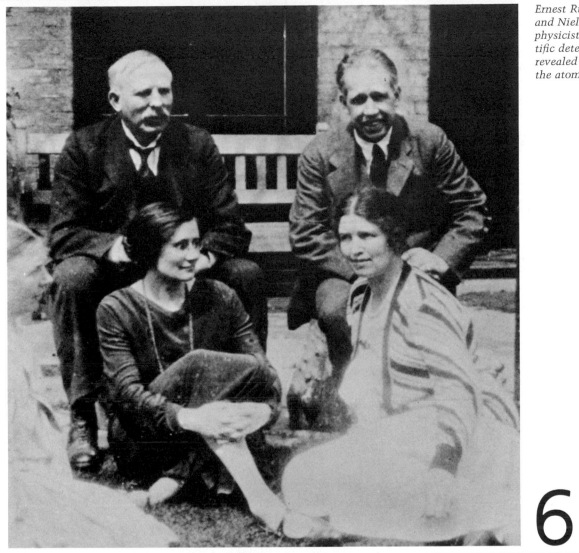

Ernest Rutherford (left) and Niels Bohr, two physicists whose scientific detective work first revealed the structure of the atom.

6

THE RUTHERFORD-BOHR MODEL OF THE ATOM

Our goal in this chapter is to understand some of the details of atomic structure, based on what we can learn of atoms by doing experiments. In the usual sense, we "learn" about the operation of a mechanical system by (1) understanding the properties of the individual parts of the system and (2) observing the parts acting together to make the system operate.

For example, suppose our goal were to "understand" the operation of a transmission. We would first learn all about nuts, bolts, gears and axles, and then we would study the operation of the transmission. When we were finished with our observations, we would know some gear ratios, rotary speeds, mechanical advantages, and so forth, and we could then say that we "understand" the operation of the transmission. In fact, we could have carried out much the same procedure without making any *visual* observations at all—a good technician could have obtained all the same information, and achieved the same level of understanding, using only the sense of touch; this is possible because the parts of the transmission and the irregularities in the parts (size of gear teeth, size of nuts and bolts) are all of roughly the same size as our fingers, so that we can "feel" the structure of the parts and the operation of the system. We have considerably more difficulty trying to understand the operation of a mechanical watch.* Without our sense of sight, we would be unable to "feel" the operation of the watch or to identify its components, simply because our fingers are too large to sense the variations and irregularities of the components. On the other hand, the wavelength of light is much larger than the size of the parts of the watch, so we can see the parts and their interactions. *The smaller the system, the smaller the probe we must use to study it.*

Returning for the moment to our study of the transmission, we would find that when we completed our investigations, we would have a "picture" of the operation of a transmission. (Such a picture can be a diagram on paper or a mental image.) We may regard this picture as a "model" of the operation of a transmission. Our model *can* be no more detailed than the experiments that we performed, and it *need* be no more detailed than the use for which it is intended. In *driving* a car, the "model" of the operation of a transmission need be no more complex than the use of the shift lever to put the car in gear; in *designing* a car, we would need a more detailed model.

How does all this apply to the study of atoms? We use knowledge gained through experiments to build an *atomic model,* a mental image that helps us to understand the structure of atoms. We will study a few basic experiments and show how our model follows from these experiments. Unfortunately, as with our study of the transmission, our model may help us to understand some structural features but it may not be sufficiently detailed to explain *all* of the possible experiments we can do with atoms. For this we need to explore the way in which the *wave behavior* of electrons changes our model for the simplest atom (which we do in Chapter 7) and the way more complicated atoms are constructed of the same building blocks (which follows in Chapter 8).

---

* The author is of a generation whose watches all had moving parts.

Before we begin to construct a model of the atom, let us summarize some of the basic properties of atoms.

1. *Atoms are very small*, about 0.1 nm (0.1×10^{-9} m) in radius. Thus any effort to "see" an atom using visible light ($\lambda \cong 500$ nm) is hopeless owing to diffraction effects. In effect, our clumsy "fingers" of visible light are too large to probe the structure of our fine atomic "watch." (We can make a crude estimate of the *maximum* size of an atom in the following way. Consider a cube of elemental matter, for example iron. Iron has a density of about 8 g/cm³ and an atomic weight of about 50. One mole of iron (50 g) contains Avogadro's number of atoms, about 6×10^{23}. Thus 6×10^{23} atoms occupy about 6 cm³ and so 1 atom occupies about 10^{-23} cm³. If we assume the atoms of a solid are packed together in the most efficient possible way, like hard spheres in contact, then the diameter of one atom would be about $\sqrt[3]{10^{-23} \text{ cm}^3} \cong 2 \times 10^{-8}$ cm = 0.2 nm.)

2. *Atoms are stable*—they do not spontaneously break apart into smaller pieces or collapse; therefore the internal forces that hold the atom together must be in equilibrium. This immediately tells us that the forces that pull the parts of an atom together must be opposed in some way; otherwise all the atoms in the universe would collapse.

3. *Atoms contain negatively charged electrons, but are electrically neutral.* If we shake or tickle an atom or collection of atoms with sufficient force, electrons are emitted. We learned this fact from studying the Compton effect and the photoelectric effect. (We also learned in Chapter 4 that even though electrons were emitted from the nuclei of atoms in certain radioactive decay processes, they don't "exist" in those nuclei but are manufactured there by some process. Electrons were excluded from the nucleus based on the uncertainty relationship, which forbids emitted electrons of the energies observed in the laboratory from existing in the nucleus. The uncertainty principle places no such restriction on atoms, so we assume the electron to be present in atoms.) We can also easily observe that bulk matter is electrically neutral, and we assume that this is likewise a property of the atoms. Experiments with beams of individual atoms support this assumption.

   From these experimental facts we deduce that an atom with Z negatively charged electrons must also contain a net positive charge of Ze.

4. *Atoms emit and absorb electromagnetic radiation.* This radiation may take many forms—visible light ($\lambda \sim 500$ nm), X rays ($\lambda \sim 1$ nm), ultraviolet rays ($\lambda \sim 10$ nm), infrared rays ($\lambda \sim 0.1$ μm), and so forth. In fact it is from observation of these emitted and absorbed radiations that we learn most of what we know about atoms. In a typical emission measurement, an electric current is passed through a tube containing a small sample of the gas phase of the element under study, and radiation is emitted when an excited atom returns to its ground state. The wavelengths of the many different emitted radiations can be measured with great precision, such as with a diffraction grating in the case of visible light. The absorption wavelengths can be measured by passing a beam of white light through a sample of the

gas and noting which colors are removed from the white light by absorption in the gas. One particularly curious feature of the atomic radiations is that atoms don't always emit and absorb radiations at the same wavelengths—some wavelengths present in the *emission* experiment do not also appear in the *absorption* experiment. Any successful theory of atomic structure must be able to account for these emission and absorption wavelengths.

*6.2 THE THOMSON MODEL

An early model of the structure of the atom was proposed by J. J. Thomson, who was known for his previous identification of the electron and measurement of its charge-to-mass ratio e/m. The Thomson model incorporates many of the known properties of atoms: size, mass, number of electrons, and electrical neutrality. In this model, an atom contains Z electrons which are imbedded in a uniform sphere of positive charge. The total positive charge of the sphere is Ze, the mass of the sphere is essentially the mass of the atom (the electrons don't contribute significantly to the total mass), and the radius R of the sphere is the radius of the atom. (This model is sometimes known as the "plum-pudding" model, since the electrons are distributed throughout the atom like raisins in a plum pudding.)

The force on an electron at a distance r from the center of a uniformly charged sphere of radius R can be computed using some basic results from electrostatics. According to Figure 6.1, a sphere of radius r will contain a fraction of the volume of the entire sphere of radius R and the same fraction of the total charge Ze. Thus

$$q_{\text{enclosed}} = Ze\,\frac{\frac{4}{3}\pi r^3}{\frac{4}{3}\pi R^3} = Ze\,\frac{r^3}{R^3} \tag{6.1}$$

According to Gauss's law, the electric field at the radius r can be found from the total charge enclosed by the sphere:

$$\int \mathbf{E}\cdot d\mathbf{S} = \frac{1}{\varepsilon_0}\,q_{\text{enclosed}} \tag{6.2}$$

Owing to the spherical symmetry, the electric field $\mathbf{E}$ is constant over the spherical surface, and in the usual way the integral reduces to $E\cdot 4\pi r^2$, so that

$$E = \frac{1}{4\pi\varepsilon_0}\,\frac{q_{\text{enclosed}}}{r^2}$$

and using Equation (6.1) for the enclosed charge,

$$E = \frac{1}{4\pi\varepsilon_0}\,\frac{Ze}{R^3}\,r \tag{6.3}$$

An electron with charge e will experience a force of magnitude $F = eE$, and so

$$F = \frac{Ze^2}{4\pi\varepsilon_0 R^3}\,r = kr \tag{6.4}$$

where we define $k = Ze^2/4\pi\varepsilon_0 R^3$.

This force tends to draw the electron toward the center of the atom, resulting in a collapse of the atom. We must therefore provide another

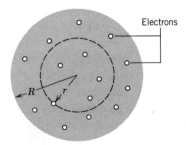

FIGURE 6.1 The Thomson model of the atom. Z tiny electrons are imbedded in a uniform sphere of positive charge Ze and radius R. An imaginary spherical surface of radius r contains a fraction r^3/R^3 of the charge.

force, which opposes the electrical attraction and keeps the electrons in equilibrium at the radius r. The mutual repulsion of the electrons supplies this additional force and maintains the electrons in stable equilibrium.

The resulting situation is similar to a mass m suspended from a spring of force constant k, in the Earth's gravity. The restoring force exerted on the mass by the spring, of magnitude $F = kx$, is opposed by the pull of the Earth's gravity, which exerts a force $F = mg$. The mass is in equilibrium under the influence of these two forces. When the mass is displaced a small distance from equilibrium, it oscillates with a frequency $\nu = (1/2\pi)\sqrt{k/m}$.

We therefore expect the electrons in the Thomson atom to oscillate about their equilibrium positions with a similar frequency $\nu = (1/2\pi) \times \sqrt{k/m}$, where k is the constant defined by Equation (6.4). Since an oscillating electric charge radiates electromagnetic waves whose frequency is identical to the oscillation frequency, we might expect, based on the Thomson model, that the radiation emitted by atoms would show this characteristic frequency. As you will show in Problem 1 at the end of this chapter, the frequencies thus obtained do not correspond to the observed frequencies of radiation emitted by atoms.

Another difficulty with this interpretation arises when we consider the absorption of radiation by atoms. We would expect that an electron in the Thomson atom could either *emit* radiation at its oscillation frequency, with an accompanying decrease in the amplitude of oscillation, or else *absorb* radiation at the same frequency, with an increase in the amplitude. However, as mentioned previously, atoms sometimes do not emit and absorb radiation at the same frequency. This feature is difficult for the Thomson model to explain.

The most serious failure of the Thomson model arises from the scattering of charged particles by atoms. Consider the passage of a single positively charged particle through the atom. The particle is deflected somewhat from its original trajectory owing to the electrical forces exerted on the particle by the atom. These forces are (1) a repulsive force due to the positive charge of the atom, and (2) an attractive force due to the negatively charged electrons. We assume that the mass of the deflected particle is both much greater than the mass of an electron and yet much less than the mass of the atom. In the encounter between the projectile and an electron, the forces exerted on each by the other are of course equal and opposite (by Newton's third law), and so the principal victim of the encounter is the much less massive electron; the effect on the projectile is negligible. (Imagine rolling a bowling ball through a field of Ping-Pong balls!) We thus need only consider the positively charged atom as a cause of the deflection of the particle. By the same argument, we neglect any possible motion of the more massive atom caused by the passage of the projectile. Our experiment, then, is the scattering of a positively charged projectile by the stationary positively charged massive part of the atom.

Figure 6.2 shows a representation of the deflection of the projectile, which is moving at speed v (we assume $v \ll c$ and use nonrelativistic mechanics, so $K = \frac{1}{2}mv^2$) along a line that would pass a distance b from the center of the atom if the projectile were not deflected. (The distance b is called the *impact parameter*.) The electrical repulsion causes a

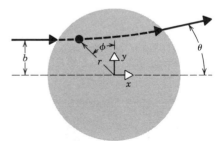

FIGURE 6.2 A positively charged alpha particle is deflected by an angle θ as it passes through a Thomson-model atom. The coordinates r and ϕ locate the alpha particle while it is inside the atom.

small deflection, and so the particle leaves the atom moving in a slightly different direction, at an angle θ with respect to the original direction.

We can calculate the deflection angle θ by considering the *impulse* received by the projectile, which gives it some momentum in the y direction

$$\Delta p_y = \int F_y \, dt \qquad (6.5)$$

At an arbitrary point along the trajectory,

$$F_y = F \cos \phi$$

Assuming the projectile to have a charge $q = ze$, the force F is just qE, where E is given by Equation (6.3)

$$F = \frac{1}{4\pi\varepsilon_0} \frac{zZe^2}{R^3} r = zkr \qquad (6.6)$$

where k is the same constant defined by Equation (6.4). Since $\cos \phi \cong b/r$, we have

$$\Delta p_y \cong \int zkr \cdot \frac{b}{r} \cdot dt = zkb \int dt$$

$$= zkbT \qquad (6.7)$$

where T is the total time it takes for the projectile to travel through the atom, which is just the total distance traveled in the atom divided by the average speed. Since the deflection is small, the path can be approximated as a straight line, as shown in Figure 6.3, and the average speed is very nearly equal to v, so

$$T \cong \frac{2\sqrt{R^2 - b^2}}{v} \qquad (6.8)$$

and

$$\Delta p_y \cong \frac{2zkb}{v} \sqrt{R^2 - b^2} \qquad (6.9)$$

Assuming that p_x doesn't change

$$\tan \theta = \frac{p_y}{p_x} \cong \frac{\Delta p_y}{p} \qquad (6.10)$$

and if θ is small, $\tan \theta \cong \theta$, so

$$\theta \cong \frac{\Delta p_y}{p} = \frac{2zkb}{mv^2} \sqrt{R^2 - b^2} \qquad (6.11)$$

When we do this kind of experiment, known as a *scattering* experiment, we can't shoot a single projectile at a single atom, and we can't control or determine the impact parameter b. A beam of particles incident on a thin foil of material will perhaps be deflected as shown in Figure 6.4, and we can determine the average scattering angle Θ_{avg} or the maximum scattering angle Θ_{max}. (Θ represents the measured scattering angle; θ is the result of scattering by one atom.)

Differentiating Equation (6.11) with respect to b, we can find θ_{max},

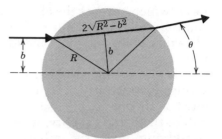

FIGURE 6.3 An approximate geometry of the deflection of an alpha particle by a Thomson atom. The scattering angle, whose maximum value is about 0.01°, has been greatly exaggerated.

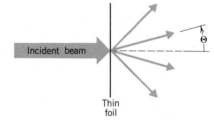

FIGURE 6.4 A typical scattering experiment. An incident beam is scattered by a thin foil; scattered particles are observed at all possible values of Θ in the laboratory.

Scattering angle for Thomson atom

the *maximum* scattering angle for a *single collision*:

$$\theta_{max} = \frac{zkR^2}{mv^2} \qquad (6.12)$$

In order to find the average scattering angle, we view the atom from the perspective of the projectile, and imagine the resulting circular disc to be divided into concentric rings. Each time the projectile enters a ring at a radius b with a width db, it is scattered into a range of angles $d\theta$ about θ (see Figure 6.5). The *fraction* of projectiles incident on the atom that enter that ring (and therefore are scattered at θ) is just the fraction of the target area occupied by that ring, or $2\pi b \, db/\pi R^2$. Averaging over all possible values of b, we have the *average* scattering angle for a *single collision*:

$$\theta_{avg} = \int_0^R \frac{2\pi b \, db}{\pi R^2} \theta \qquad (6.13)$$

and carrying out the integration, with θ from Equation (6.11),

$$\theta_{avg} = \frac{\pi}{4} \frac{zkR^2}{mv^2} \qquad (6.14)$$

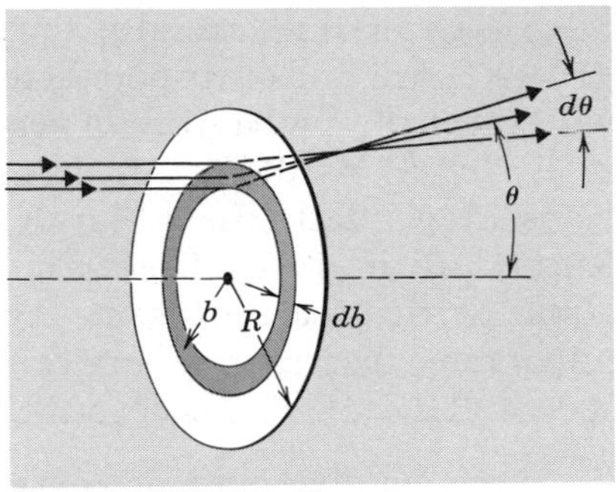

FIGURE 6.5 Scattering geometry for one atom. Particles entering a circular ring of radius b and width db are scattered into a range of angles $d\theta$ at θ.

EXAMPLE 6.1

Using the Thomson model with $R = 0.1$ nm (a typical atomic radius), calculate the average deflection angle per collision when 5 MeV alpha particles ($z = 2$) are scattered from gold ($Z = 79$).

Solution

The quantity zkR^2 can be computed to be

$$zkR^2 = 2\left(\frac{Ze^2}{4\pi\varepsilon_0 R^3}\right) R^2 = 2 \cdot 79 \cdot \frac{1.44 \text{ eV·nm}}{0.1 \text{ nm}} \cong 2.3 \text{ keV}$$

(Recall that $e^2/4\pi\varepsilon_0 = 1.44$ eV·nm.) Also, $mv^2 = 2K = 10$ MeV, so

$$\theta_{avg} \cong \frac{\pi}{4} \frac{2.3 \text{ keV}}{10 \text{ MeV}} \cong 2 \times 10^{-4} \text{ rad}$$

This is an extremely small angle, and it is questionable whether such a deflection could be measured in the laboratory. What we have failed to consider is that, in passing through a certain thickness of material, the projectile will undergo *many* collisions with atoms, and *each collision* will deflect the projectile by an angle whose average value is θ_{avg}. Some of these collisions will result in larger total deflection angles, and some will result in smaller deflection angles (see Figure 6.6). The value of the total observed scattering angle Θ is governed by the laws of statistics; in particular if there are N collisions, then $\Theta_{avg} = \sqrt{N} \, \theta_{avg}$, and the probability of scattering at any angle greater than some Θ is $e^{-(\Theta/\Theta_{avg})^2}$.

In traveling through a gold foil of thickness 1 μm (10^{-6} m), the projectile will encounter about 10^4 atoms (since each atom has a diameter of about 0.1 nm), so the average laboratory scattering angle Θ_{avg} will be about $\sqrt{10^4} \, \theta_{avg}$, or about 1°. This is in reasonable agreement with what one actually observes in such experiments.

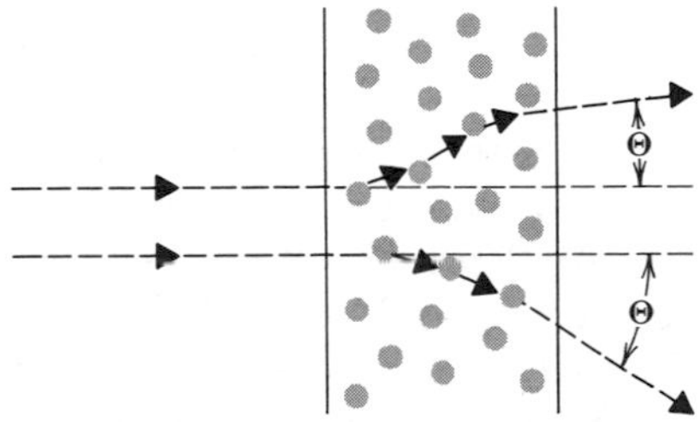

FIGURE 6.6 A microscopic representation of the scattering. Some individual scatterings tend to increase Θ, while others tend to decrease Θ.

However, if we examine the probability for scattering at large angles ($\Theta > 90°$), we find quite poor agreement with experiment. For $\Theta_{avg} \cong 1°$, we expect the probability of scattering at angles in excess of 90° to be $e^{-90^2} = e^{-8100} = 10^{-3500}$. An experiment of this sort was performed by Hans Geiger and Ernest Marsden in the laboratory of Professor Ernest Rutherford in 1910. Their results showed that the probability of an alpha particle scattering at angles beyond 90° was about 10^{-4}. This remarkable discrepancy between the expected value (10^{-3500}) and the observed value (10^{-4}) was described by Prof. Rutherford in this way:

> It was quite the most incredible event that ever happened to me in my life. It was as incredible as if you fired a 15-inch shell at a piece of tissue paper and it came back and hit you.

The analysis of the results of such scattering experiments led Rutherford to propose that the mass and positive charge of the atom were not distributed uniformly over the volume of the atom, but instead were concentrated in an extremely small region, about 10^{-14} m in diameter, at the center of the atom. In the next section we will see how this proposal is consistent with the large-angle scattering results.

6.3 THE RUTHERFORD NUCLEAR ATOM

In the previous section we have learned that, when a beam of projectiles such as alpha particles is incident on a thin target such as a gold foil, the *average* angle through which the beam is scattered is small (of the order of 1°). Furthermore, we learned that in such a circumstance the probability of many small deflections adding to give an observed large deflection was extremely small (10^{-3500}) and totally in disagreement with the experimental result (about 10^{-4}). The most likely way that an alpha particle ($m = 4$ u) can be deflected to large angles is by a *single* collision with a more massive object. Rutherford therefore proposed that the charge and mass of the atom were concentrated at its center, in a region called the *nucleus*. Figure 6.7 illustrates the scattering geometry in this case. The projectile, of charge ze, experiences a repulsive force due to the positively charged nucleus:

$$F = \frac{(ze)(Ze)}{4\pi\varepsilon_0 r^2} \tag{6.15}$$

(We assume that the projectile is always outside the nucleus, so it feels the full nuclear charge Ze.) The atomic electrons, with their small mass, do not appreciably affect the path of the projectile and we neglect their effect on the scattering. We also assume that the nucleus is so much more massive than the projectile that it does not move during the scattering process; since no recoil motion is given to the nucleus, the initial and final kinetic energies K of the projectile are equal.

As Figure 6.7 shows, for each impact parameter b, there is a certain scattering angle θ, and we need the relationship between b and θ. As derived in several texts*, the projectile follows a hyperbolic path; in polar coordinates r and ϕ, the equation of the hyperbola is

---

* See, for example, R. M. Eisberg and R. Resnick, *Quantum Physics of Atoms, Molecules, Solids, Nuclei, and Particles* (New York, Wiley, 1974).

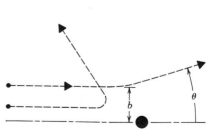

FIGURE 6.7 Scattering by a nuclear atom. The path of the scattered particle is a hyperbola. Smaller impact parameters give larger scattering angles.

$$\frac{1}{r} = \frac{1}{b} \sin \phi + \frac{zZe^2}{8\pi\varepsilon_0 b^2 K} \left(\cos \phi - 1\right) \tag{6.16}$$

As shown in Figure 6.8, the initial position of the particle is $\phi = 0$, $r \to \infty$, and the final position is $\phi = \pi - \theta$, $r \to \infty$. Using the coordinates at the final position, Equation (6.16) reduces to

$$b = \frac{zZe^2}{8\pi\varepsilon_0 K} \cot \tfrac{1}{2}\theta = \frac{zZ}{2K} \frac{e^2}{4\pi\varepsilon_0} \cot \tfrac{1}{2}\theta \tag{6.17}$$

Rutherford scattering angle

(This result is written in this form so that $e^2/4\pi\varepsilon_0 = 1.44$ eV·nm can be easily inserted.) A projectile that approaches the nucleus with impact parameter b will be scattered at an angle θ; projectiles approaching with smaller values of b will be scattered through larger angles, as shown in Figure 6.7.

We divide our study of the scattering of charged projectiles by nuclei (which is commonly called *Rutherford scattering*) into three parts: (1) calculation of the fraction of projectiles scattered at angles greater than some value of θ, (2) the Rutherford scattering formula and its experimental verification, and (3) the closest approach of a projectile to the nucleus.

1. *The fraction of projectiles scattered at angles greater than θ.* From Figure 6.7 we see immediately that every projectile with impact parameters less than a given value of b will be scattered at angles greater than its corresponding θ. What is the chance of a projectile having an impact parameter less than b? Suppose the foil were one atom thick — a single layer of atoms packed tightly together, as in Figure 6.9. Each *atom* looks like a circular disc, of area πR^2. If the foil contains N atoms, its total area is $N\pi R^2$. For scattering at angles greater than θ, the impact parameter must fall between zero and b—that is, the projectile must approach the atom within a circular disc of area πb^2. If the projectiles are spread uniformly over the area of the foil, then the fraction of projectiles that fall within that area is just $\pi b^2/\pi R^2$.

A real scattering foil may be thousands or tens of thousands of atoms thick. Let t be the thickness of the foil and A its area, and let ρ and M be the density and molecular weight of the material of which the foil is made. The volume of the foil is then just At, its mass is ρAt, the number of moles is $\rho At/M$ and the number of atoms or nuclei per unit volume is

$$n = N_A \frac{\rho At}{M} \frac{1}{At} = \frac{N_A \rho}{M} \tag{6.18}$$

where N_A is Avogadro's number (the number of atoms per gram-molecular weight). As seen by an incident projectile, the number of nuclei per unit area is $nt = N_A \rho t/M$; that is, on the average, each nucleus contributes an area $(N_A \rho t/M)^{-1}$ to the field of view of the projectile. For scattering at angles greater than θ, it must once again be true that the projectile must fall within an area πb^2 of the center of an atom; the fraction scattered at angles greater than θ is just the fraction that approaches an atom within the area πb^2:

$$f_{<b} = f_{>\theta} = nt\pi b^2 \tag{6.19}$$

Fraction scattered at angles greater than θ

assuming that the incident particles are spread uniformly over the area of the foil.

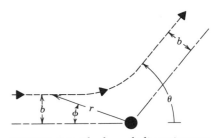

FIGURE 6.8 The hyperbolic trajectory of a scattered particle.

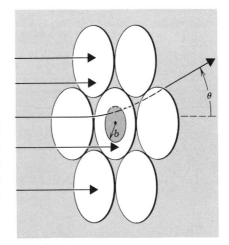

FIGURE 6.9 Scattering geometry for many atoms. For impact parameter b, the scattering angle is θ. If the particle enters the atom within the disc of area πb^2, its scattering angle will be larger than θ.

A gold foil (ρ = 19.3 g/cm³, M = 197 g/mole) has a thickness of 2.0×10^{-4} cm. It is used to scatter alpha particles of kinetic energy 8.0 MeV. (a) What fraction of the alpha particles is scattered at angles greater than 90°? (b) What fraction of the alpha particles is scattered at angles between 90° and 45°?

EXAMPLE 6.2

(a) For this case the number of nuclei per unit volume can be computed as

Solution

$$n = \frac{N_A \rho}{M} = \frac{(6.02 \times 10^{23} \text{ atoms/mole})(19.3 \text{ g/cm}^3)}{197 \text{ g/mole}}$$

$$= 5.9 \times 10^{22} \text{ atoms/cm}^3$$

$$= 5.9 \times 10^{28} \text{ atoms/m}^3$$

For scattering at 90°, the impact parameter b can be found from Equation (6.17):

$$b = \frac{(2)(79)}{2(8.0 \times 10^6 \text{ eV})}(1.44 \text{ eV·nm}) \cot 45° = 1.4 \times 10^{-14} \text{ m}$$

so $\pi b^2 = 6.4 \times 10^{-28}$ m²/nucleus and we then have

$$f_{>90°} = (5.9 \times 10^{28} \text{ nuclei/m}^3)(2.0 \times 10^{-6} \text{ m})(6.4 \times 10^{-28} \text{ m}^2\text{/nucleus})$$

$$= 7.5 \times 10^{-5}$$

(b) Repeating the calculation for $\theta = 45°$,

$$b = \frac{(2)(79)}{2(8.0 \times 10^6 \text{ eV})}(1.44 \text{ eV·nm}) \cot 22.5° = 3.4 \times 10^{-14} \text{ m}$$

and $f_{>45°} = 4.4 \times 10^{-4}$. If a total fraction of 4.4×10^{-4} is scattered at angles greater than 45°, and of that, 7.5×10^{-5} is scattered at angles greater than 90°, the fraction scattered *between* 45° and 90° must be $4.4 \times 10^{-4} - 7.5 \times 10^{-5} = 3.6 \times 10^{-4}$.

2. *The Rutherford scattering formula and its experimental verification.* In order to find the probability that a projectile be scattered into a small angular range at θ (between θ and $\theta + d\theta$), we require that the impact parameter lie within a small range of values db at b (see Figure 6.10). The fraction, df, is then

$$df = nt(2\pi b \ db)$$

from Equation (6.19). Differentiating Equation (6.17) we find db in terms of $d\theta$:

$$db = \frac{zZ}{2K}\frac{e^2}{4\pi\varepsilon_0}(-\csc^2 \tfrac{1}{2}\theta)(\tfrac{1}{2} \ d\theta) \qquad (6.20)$$

and so

$$|df| = \pi nt \left(\frac{zZ}{2K}\right)^2 \left(\frac{e^2}{4\pi\varepsilon_0}\right)^2 \csc^2 \tfrac{1}{2}\theta \cot \tfrac{1}{2}\theta \ d\theta \qquad (6.21)$$

[The minus sign in Equation (6.20) is not important—it just tells us that θ increases as b decreases.] Suppose we place a detector for the scattered projectiles at the angle θ a distance r from the nucleus. The probability for a projectile to be scattered into the detector depends on df; however, df merely gives the chance for those projectiles to be scattered at θ into

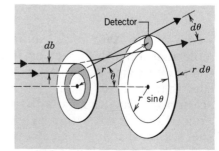

FIGURE 6.10 Particles entering the ring between b and $b + db$ are distributed uniformly along a ring of angular width $d\theta$. A detector is at a distance r from the scattering foil.

$d\theta$, and you can see that these particles will be uniformly distributed around a ring of radius $r \sin \theta$ and thickness $r\, d\theta$. The area of the ring is $dA = (2\pi r \sin \theta) r\, d\theta$. In order to calculate the rate at which projectiles are scattered *into our detector* we must know the probability *per unit area* for scattering into the ring. This is $|df|/dA$, which we shall call $N(\theta)$, and, after some manipulation, we find:

$$N(\theta) = \frac{nt}{4r^2} \left(\frac{zZ}{2K}\right)^2 \left(\frac{e^2}{4\pi\varepsilon_0}\right)^2 \frac{1}{\sin^4 \frac{1}{2}\theta} \qquad (6.22)$$

Rutherford scattering formula

This is the *Rutherford scattering formula*.

In Rutherford's laboratory, Geiger and Marsden tested the predictions of this formula, in a remarkable series of experiments requiring great care and experimental skill. They used *alpha particles* $(z = 2)$ to scatter from nuclei in a variety of thin metal foils. In those days before electronic recording and processing equipment was available, Geiger and Marsden observed and recorded the alpha particles by counting the scintillations (flashes of light) produced when the alpha particles struck a zinc sulfide screen. A schematic view of their apparatus is shown in Figure 6.11. In all, four predictions of the Rutherford scattering formula were tested:

(a) $N(\theta) \propto t$. With a source of 8 MeV alpha particles from radioactive decay, Geiger and Marsden used scattering foils of varying thicknesses t while keeping the scattering angle θ fixed at about 25°. Their results are summarized in Figure 6.12, and the linear dependence of $N(\theta)$ on t is apparent. This is also evidence that, even at this moderate scattering angle, *single* scattering is much more important than *multiple* scattering. (In a random statistical theory of multiple scattering, the probability for scattering at a large angle would be proportional to the square root of the number of single scatterings, and we would expect $N(\theta) \propto t^{1/2}$. Figure 6.12 shows clearly that this is not true.)

(b) $N(\theta) \propto Z^2$. In this experiment, Geiger and Marsden used a variety of different scattering materials, of approximately (*but not exactly*) the same thickness. This proportionality is therefore much more difficult to test than the previous one, since it involves the comparison of

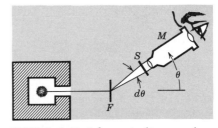

FIGURE 6.11 Schematic diagram of alpha-particle scattering experiment. A radioactive source of alpha particles is in a shield with a small hole. Alpha particles strike the foil F and are scattered into the angular range $d\theta$. Each time a scattered particle strikes the screen S a flash of light is emitted and observed with the movable microscope M.

Ernest Rutherford (1871–1937, England). Founder of nuclear physics, he is known for his pioneering work on alpha-particle scattering and radioactive decays. His inspiring leadership influenced a generation of British nuclear and atomic scientists.

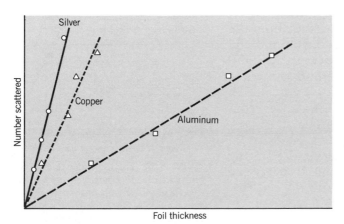

FIGURE 6.12 The dependence of scattering rate on foil thickness for three different scattering foils.

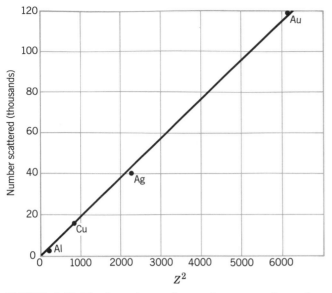

FIGURE 6.13 The dependence of scattering rate on the nuclear charge Z for foils of different materials. The data are plotted against Z^2.

different thicknesses of *different materials.* However, as shown in Figure 6.13, the results are consistent with the proportionality of $N(\theta)$ to Z^2.

(c) $N(\theta) \propto 1/K^2$. In order to test this prediction of the Rutherford scattering formula, Geiger and Marsden kept the thickness of the scattering foil constant and varied the speed of the alpha particles. They accomplished this by slowing down the alpha particles emitted from the radioactive source, using thin sheets of mica. From independent measurements they knew the effect of different thicknesses of mica on the velocity of the alpha particles. The results of the experiment are shown in Figure 6.14; once again we see excellent agreement with the expected relationship.

(d) $N(\theta) \propto 1/sin^4 \frac{1}{2}\theta$. This dependence of N on θ is perhaps the most important and distinctive feature of the Rutherford scattering formula. It also produces the largest variation in N over the range accessible by experiment. In the previous tests, N varies by perhaps an order of magnitude; in this case N varies by about *five* orders of magnitude from the smaller to the larger angles. Geiger and Marsden used a gold foil and varied θ from 5 to 150°, to obtain the relationship between N and θ plotted in Figure 6.15. The agreement with the Rutherford formula is again very good.

Thus all predictions of the Rutherford scattering formula were confirmed by experiment, and the "nuclear atom" was verified.

3. *The closest approach of a projectile to the nucleus.* A positively charged projectile slows down as it approaches a nucleus, exchanging part of its initial kinetic energy for the electrostatic potential energy due to the nuclear repulsion. The closer the projectile gets to the nucleus, the more potential energy it gains, since

$$V = \frac{1}{4\pi\varepsilon_0} \frac{zZe^2}{r}$$

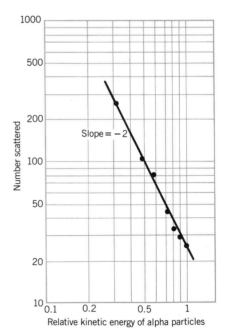

FIGURE 6.14 The dependence of scattering rate on the kinetic energy of the incident alpha particles for scattering by a single foil. Note the log-log scale; the slope of -2 shows that $\log N \propto -2 \log K$, or $N \propto K^{-2}$, as expected from the Rutherford formula.

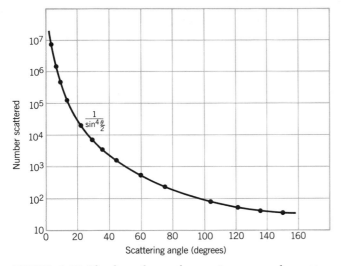

FIGURE 6.15 The dependence of scattering rate on the scattering angle θ, using a gold foil. The $\sin^{-4}(\theta/2)$ dependence is exactly as predicted by the Rutherford formula.

The maximum potential energy, and thus the minimum kinetic energy, occurs at the minimum value of r. We assume that $V = 0$ when the projectile is far from the nucleus, where it has total energy $E = K = \frac{1}{2}mv^2$. As the projectile approaches the nucleus, K decreases and V increases, but $V + K$ remains constant. At the distance r_{min}, the speed is v_{min} and:

$$E = \frac{1}{2}mv_{min}^2 + \frac{1}{4\pi\varepsilon_0}\frac{zZe^2}{r_{min}} = \frac{1}{2}mv^2 \qquad (6.23a)$$

(See Figure 6.16.)

Angular momentum is also conserved. Far from the nucleus, the angular momentum is mvb, and at r_{min}, the angular momentum is $mv_{min}r_{min}$, so:

$$mvb = mv_{min}r_{min}$$

or

$$v_{min} = \frac{b}{r_{min}}v \qquad (6.23b)$$

Combining Equations (6.23a) and (6.23b), we find

$$\frac{1}{2}mv^2 = \frac{1}{2}m\left(\frac{b^2v^2}{r_{min}^2}\right) + \frac{1}{4\pi\varepsilon_0}\frac{zZe^2}{r_{min}} \qquad (6.24)$$

This expression can be solved for the value of r_{min}.

Notice that the kinetic energy of the projectile is not zero at r_{min}, *unless* $b = 0$. (See Figure 6.16.) In this case, the projectile would lose all of its kinetic energy, and thus get closest to the nucleus. At this point its distance from the nucleus is d, *the distance of closest approach*. We find this distance by solving Equation (6.24) for r_{min} when $b = 0$, and find

$$d = \frac{1}{4\pi\varepsilon_0}\frac{zZe^2}{K} \qquad (6.25) \qquad \textbf{Distance of closest approach}$$

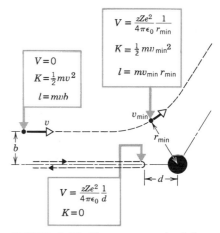

FIGURE 6.16 Closest approach of the projectile to the nucleus.

Find the distance of closest approach of an 8.0-MeV alpha particle incident on a gold foil.

EXAMPLE 6.3

$$d = \frac{zZe^2}{4\pi\varepsilon_0} \frac{1}{K} = (2)(79)(1.44 \text{ eV·nm}) \frac{1}{8 \times 10^6 \text{ eV}}$$

Solution

$$= 28 \times 10^{-6} \text{ nm}$$

$$= 2.8 \times 10^{-14} \text{ m}$$

Although this distance is very small (much less than an atomic radius, for example) it is larger than the nuclear radius of gold (about 7×10^{-15} m). Thus the projectile is always *outside* of the nuclear charge distribution, and we expect that the Rutherford scattering law, which was derived assuming the projectile to remain outside the nucleus, will describe the scattering. If we increase the kinetic energy of the projectile, or decrease the electrostatic repulsion by using a target nucleus with low Z, this may not be the case. Under certain circumstances, the distance of closest approach can be less than the nuclear radius. When this happens, the projectile no longer feels the full nuclear charge, and the Rutherford scattering law no longer holds. In fact, as we will see in Chapter 9, this gives us a convenient way of measuring the size of the nucleus.

6.4 LINE SPECTRA

The radiation from atoms can be classified into continuous spectra and discrete or line spectra. In a continuous spectrum, all wavelengths from some minimum, perhaps 0, to some maximum, perhaps approaching ∞, are emitted. The radiation from a hot glowing object is an example of this category. You know that white light is a mixture of all of the different colors of visible light; an object that glows white hot is emitting

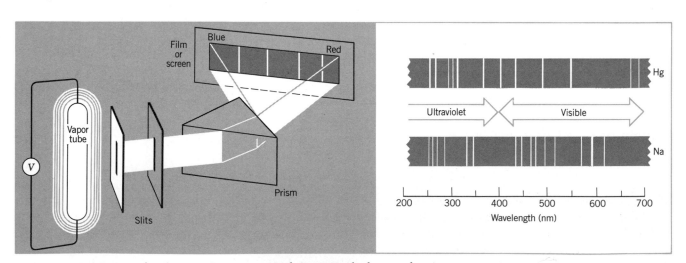

FIGURE 6.17 Apparatus for observing line spectra. Light is emitted when an electric discharge is created in a tube containing a vapor of an element. The light passes through a dispersive medium, such as a prism or a diffraction grating, which displays the individual component wavelengths at different positions. Sample line spectra are shown for mercury and sodium in the visible and near ultraviolet.

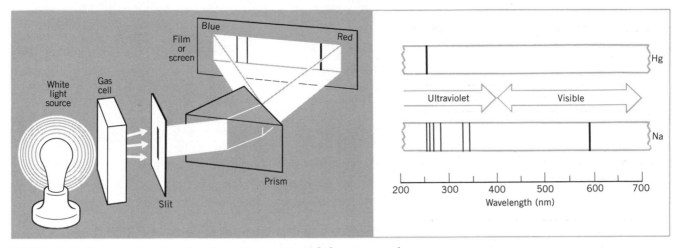

FIGURE 6.18 Apparatus for observing absorption spectra. A light source produces a continuous range of wavelengths, some of which are absorbed by a gaseous element. The light is dispersed, as in Figure 6.17. The result is a continuous "rainbow" spectrum, with dark lines at wavelengths where the light was absorbed by the gas.

light at all frequencies of the visible spectrum. If, on the other hand, we force an electric discharge in a tube containing a small amount of the gas or vapor of a certain element, such as mercury, sodium, or neon, light is emitted at a few discrete wavelengths and not at any others. Examples of such "line" spectra are shown in Figure 6.17. The strong 436 nm (blue) and 546 nm (green) lines in the mercury spectrum give mercury-vapor street lights their blue-green tint; the strong yellow line at 590 nm in the sodium spectrum (which is actually a *doublet*—two very closely spaced lines) gives sodium-vapor street lights a softer, yellowish color. The strong red line of neon is responsible for the red color of "neon signs."

Another possible experiment is to pass a beam of white light, containing all wavelengths, through a sample of a gas. When we do so, we find that certain wavelengths have been absorbed from the light, and again a line spectrum results. These wavelengths correspond to many (*but not all*) of the wavelengths seen in the emission spectrum. Examples are shown in Figure 6.18.

In general, the interpretation of line spectra is very difficult in complex atoms, and so we will deal exclusively with the study of the line spectrum of the simplest atom, hydrogen. This spectrum, owing to the simplicity of the atom with its single electron, exhibits regularities in the emission spectrum and absorption spectrum which are shown in Figure 6.19. Notice that, as with the other spectra, not all of the lines present in the emission spectrum are present in the absorption spectrum.

In our discussion of blackbody radiation, we found an example of the "reverse scientific method," in which, in the absence of a theory that explains the data, we try to find a function that fits the data, and then try to find a theory that explains the derived function. A Swiss schoolteacher named Johannes Balmer noticed that the wavelengths of the group of emission lines of hydrogen in the visible region could be

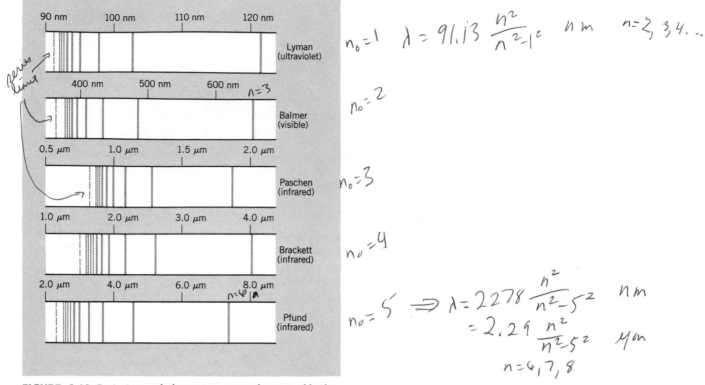

FIGURE 6.19 Emission and absorption spectral series of hydrogen. Note the regularities in the spacing of the spectral lines. The lines get closer together as the limit of each series (dashed line) is approached. Only the Lyman series appears in the absorption spectrum; all series are present in the emission spectrum.

very accurately fitted by the formula

$$\lambda = 364.5 \frac{n^2}{n^2 - 4}$$ (6.26)   **Balmer formula**

where λ is in units of nm and where n can take integer values beginning with 3. For example, for $n = 3$, $\lambda = 656.1$ nm. This formula is now known as the Balmer formula and the series of lines that it fits is called the Balmer series. The wavelength 364.5 nm, corresponding to $n \to \infty$, is called the *series limit*. It was soon discovered that all of the groupings of lines in the hydrogen spectrum could be fit with a similar formula of the form

$$\lambda = \lambda_{\text{limit}} \frac{n^2}{n^2 - n_0^2}$$ (6.27)

where λ_{limit} is the wavelength of the appropriate series limit and where n takes integer values beginning with $n_0 + 1$. (Obviously, $n_0 = 2$ for the Balmer series.) The other series are today known as Lyman ($n_0 = 1$), Paschen ($n_0 = 3$), Brackett ($n_0 = 4$), and Pfund ($n_0 = 5$).

Another interesting property of the hydrogen wavelengths is summarized in the *Ritz combination principle*. If we convert the hydrogen emission wavelengths to frequencies, we find the curious property that

Ritz combination principle

certain pairs of frequencies added together give other frequencies which appear in the spectrum.

Any successful model of the hydrogen atom must be able to explain the occurrence of these interesting arithmetic regularities in the emission spectra.

The series limit of the Paschen series $(n_0 = 3)$ is 820.1 nm. What are the three longest wavelengths of the Paschen series?

From Equation (6.27),

$$\lambda = 820.1 \frac{n^2}{n^2 - 3^2} \qquad n = 4, 5, 6, \ldots$$

The three longest wavelengths are:

$$n = 4: \qquad \lambda = 820.1 \frac{4^2}{4^2 - 3^2} = 1875 \text{ nm}$$

$$n = 5: \qquad \lambda = 820.1 \frac{5^2}{5^2 - 3^2} = 1281 \text{ nm}$$

$$n = 6: \qquad \lambda = 820.1 \frac{6^2}{6^2 - 3^2} = 1094 \text{ nm}$$

These transitions are in the infrared region of the electromagnetic spectrum.

Show that the longest wavelength of the Balmer series and the longest *two* wavelengths of the Lyman series satisfy the *Ritz combination principle*. For the Lyman series, $\lambda_{\text{limit}} = 91.13$ nm.

The longest wavelength of the Balmer series was found previously to be 656.1 nm, using Equation (6.26) with $n = 3$. Converting this to a frequency, we find $\nu = 4.57 \times 10^{14}$ Hz. Using Equation (6.27) for $n = 2$ and 3 with $n_0 = 1$, we find the longest two wavelengths of the Lyman series to be 121.5 nm and 102.5 nm, corresponding to frequencies of 24.67×10^{14} Hz and 29.24×10^{14} Hz. Adding the smallest frequency of the Lyman series to the smallest frequency of the Balmer series gives the next smallest Lyman frequency:

$$24.67 \times 10^{14} \text{ Hz} + 4.57 \times 10^{14} \text{ Hz} = 29.24 \times 10^{14} \text{ Hz}$$

demonstrating the Ritz combination principle.

EXAMPLE 6.4

Solution

EXAMPLE 6.5

Solution

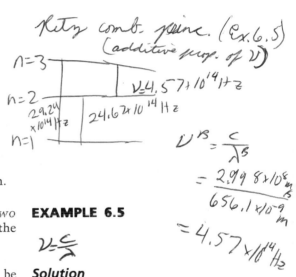

6.5 THE BOHR MODEL

Following Rutherford's proposal that the mass and positive charge are concentrated in a very small region at the center of the atom, the Danish physicist Niels Bohr in 1913 suggested that the atom was in fact like a miniature planetary system, with the electrons circulating about the nucleus like planets circulating about the sun. The atom thus doesn't collapse under the influence of the electrostatic Coulomb attraction between nucleus and electrons for the same reason that the solar system doesn't collapse under the influence of the gravitational attraction between sun and planets. In both cases, the attractive force provides the centripetal acceleration necessary to maintain the orbital motion.

We consider for simplicity the hydrogen atom, with a single electron circulating about a nucleus with a single positive charge, as in Figure 6.20. The radius of the circular orbit is r, and the electron (of mass m)

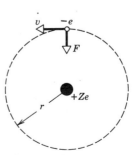

FIGURE 6.20 The Bohr model of the atom $(Z = 1$ for hydrogen).

moves with constant tangential speed v. The attractive Coulomb force provides the centripetal acceleration v^2/r, so

Coulomb force = centripetal force

$$F = \frac{1}{4\pi\varepsilon_0}\frac{q_1 q_2}{r^2} = \frac{1}{4\pi\varepsilon_0}\frac{e^2}{r^2} = \frac{mv^2}{r} \qquad (6.28)$$

Manipulating this equation, we can find the kinetic energy of the electron (we are assuming the nucleus to remain at rest—more about this later)

$$K = \frac{1}{2}mv^2 = \frac{1}{8\pi\varepsilon_0}\frac{e^2}{r} \qquad (6.29)$$

The potential energy of the electron-nucleus system is just the Coulomb potential energy:

$$V = -\frac{1}{4\pi\varepsilon_0}\frac{e^2}{r} \qquad (6.30)$$

The total energy of the system is thus

$$E = K + V = \frac{1}{8\pi\varepsilon_0}\frac{e^2}{r} - \frac{1}{4\pi\varepsilon_0}\frac{e^2}{r} \qquad \text{when } r \to \infty$$
$$E = 0$$
$$E = -\frac{1}{8\pi\varepsilon_0}\frac{e^2}{r} \qquad (6.31)$$

We have ignored one serious difficulty with this model thus far. Classical physics requires that an accelerated electric charge, such as our orbiting electron, must continuously radiate electromagnetic energy. As it radiates this energy, its total energy decreases, the electron spirals in toward the nucleus and the atom collapses. Bohr proposed to circumvent this difficulty by postulating *"stationary states"*—certain states of motion in which the electron would not radiate electromagnetic energy. Bohr deduced that *these are states in which the orbital angular momentum of the electron takes values that are integer multiples of $\hbar$.*

The angular momentum vector was defined in classical physics as $\mathbf{l} = \mathbf{r} \times \mathbf{p}$. Taking the angular momentum of the electron about the nucleus, $\mathbf{r}$ is perpendicular to $\mathbf{p}$, and so we may simplify to $l = rp = mvr$. Thus Bohr's postulate is

$$l = \mathbf{r} \times \mathbf{p} = \boxed{mvr = n\hbar} \qquad \textit{orbital } \angle \textit{ momentum} \qquad (6.32)$$

where n is an integer ($n = 1, 2, 3, \ldots$). We can use this expression with the relationship (6.29) for the kinetic energy

$$\frac{1}{2}mv^2 = \frac{1}{2}m\left(\frac{n\hbar}{mr}\right)^2 = \frac{1}{8\pi\varepsilon_0}\frac{e^2}{r} \qquad (6.33)$$

to find a series of allowed values of the radius r:

$$r_n = \frac{4\pi\varepsilon_0\hbar^2}{me^2}n^2 = a_0 n^2 \qquad (6.34)$$

where the *Bohr radius* a_0 is defined as

$$a_0 = \frac{4\pi\varepsilon_0\hbar^2}{me^2} = 0.0529 \text{ nm} \qquad (6.35)$$

Niels Bohr (1885–1962, Denmark). He developed a successful theory of the radiation spectrum of atomic hydrogen and also contributed the concepts of stationary states and complementarity to quantum mechanics. Later he developed the theory of nuclear fission. His institute of theoretical physics in Copenhagen still attracts scholars from throughout the world.

$$f(v) = \frac{1}{4\pi\varepsilon_0}\frac{(-e)(+e)}{r^2}$$

total energy is negative in a bound situation

Quantization of angular momentum

$$r \cdot \frac{h}{\lambda} = n\frac{h}{2\pi}$$

$$n = \frac{2\pi r}{\lambda}$$

Hydrogen atom radii

out of straightened electron orbit

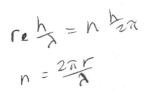

This important result is very different from what we expect from classical physics. A satellite may be placed into Earth orbit at any desired radius by boosting it to the appropriate altitude and then supplying the proper tangential speed. This is not true for an electron's orbit—only certain radii are allowed by the Bohr model. The radius of the electron's orbit may be a_0, $4a_0$, $9a_0$, $16a_0$, and so forth, but *never* $3a_0$ or $5.3a_0$.

Our expression for r may be combined with Equation (6.31) to give

$$E_n = -\frac{me^4}{32\pi^2\varepsilon_0^2\hbar^2}\frac{1}{n^2} \tag{6.36}$$

We have added a subscript n to the energy to serve as an index to identify the energy levels. The constants may be evaluated, yielding

$$E_n = \frac{-13.6\text{ eV}}{n^2} \tag{6.37}$$

The *energy levels* are indicated schematically in Figure 6.21. The electron's energy is *quantized*—only certain energy values are possible, as shown in Figure 6.21. In its lowest level, with $n = 1$, the electron has energy $E_1 = -13.6$ eV and orbits with a radius of 0.0529 nm. This state is the *ground state*. The higher states ($n = 2$ with $E_2 = -3.4$ eV, $n = 3$ with $E_3 = -1.5$ eV, etc.) are the *excited states*.

When the electron and nucleus are separated by an infinite distance, corresponding to $n = \infty$, we have $E = 0$. We might begin with the electron and nucleus separated by an infinite distance and then bring the electron closer to the nucleus until it is in orbit in a particular state n. Since this state has less energy than the $E = 0$ with which we began, we have "gained" an amount of energy equal to E_n. Conversely, if we have an electron in a state n, we can "take the atom apart" by supplying an energy E_n. This energy is known as the *binding energy* of the state n. If we supply more energy than E_n to the electron, the excess will appear as kinetic energy of the now free electron.

The *excitation energy* of an excited state n is just the energy above the ground state, $E_n - E_1$. Thus the first excited state ($n = 2$) has excitation energy -3.4 eV $- (-13.6$ eV) or 10.2 eV, the second excited state has excitation energy 12.1 eV, and so forth.

We previously discussed the emission and absorption spectra of atomic hydrogen, and our discussion of the Bohr model is not complete without an understanding of the origin of these spectra. Bohr postulated that, even though the electron doesn't radiate when it orbits in a certain level, it may make transitions from one level to a lower level. In this lower level, the electron has less energy than in the original level, and the energy difference appears as a quantum of radiation whose energy $h\nu$ is equal to the energy difference between the levels. That is, if the electron jumps from $n = n_1$ to $n = n_2$, as in Figure 6.22, a photon appears with energy

$$h\nu = E_{n_1} - E_{n_2} \tag{6.38}$$

or

$$\nu = \frac{me^4}{64\pi^3\varepsilon_0^2\hbar^3}\left(\frac{1}{n_2^2} - \frac{1}{n_1^2}\right) \tag{6.39}$$

Hydrogen atom energy levels

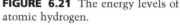

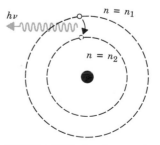

FIGURE 6.21 The energy levels of atomic hydrogen.

FIGURE 6.22 An electron jumps from the state n_1 to the state n_2 and emits a photon.

and thus the wavelength of the emitted radiation is

$$\lambda = \frac{c}{\nu} = \frac{64\pi^3 \varepsilon_0^2 \hbar^3 c}{me^4} \left(\frac{n_1^2 n_2^2}{n_1^2 - n_2^2} \right)$$

$$= \frac{1}{R_\infty} \left(\frac{n_1^2 n_2^2}{n_1^2 - n_2^2} \right)$$

(6.40)

The constant R_∞, called the Rydberg constant, has the value $1.0973731 \times 10^7 \text{ m}^{-1}$.

Find the wavelengths of the transitions from $n_1 = 3$ to $n_2 = 2$ and from $n_1 = 4$ to $n_2 = 2$.

EXAMPLE 6.6

Equation (6.40) gives

Solution

$$\lambda = \frac{1}{1.0973731 \times 10^7} \left(\frac{3^2 2^2}{3^2 - 2^2} \right) = 656.1 \text{ nm}$$

and

$$\lambda = \frac{1}{1.0973731 \times 10^7} \left(\frac{4^2 2^2}{4^2 - 2^2} \right) = 486.0 \text{ nm}$$

These wavelengths are remarkably close to the values of the two longest wavelengths of the Balmer series. In fact, if we calculate the wavelengths for transitions from any state n_1 to $n_2 = 2$, we find

$$\lambda = (364.5 \text{ nm}) \left(\frac{n_1^2}{n_1^2 - 4} \right)$$

This is identical with Equation (6.26) for the Balmer series. Thus we see that the transitions identified as the Balmer series are those from higher levels which terminate on the $n = 2$ level. Similar identifications can be made for other series of transitions, as shown in Figure 6.23.

The Bohr formulas also explain the Ritz combination principle, according to which certain frequencies in the emission spectrum could be summed to give other frequencies. Let us consider a transition from a state n_3 to a state n_2, which is followed by a transition from n_2 to n_1. Equation (6.39) can be used for this case to give

$$\nu_{n_3 \to n_2} = cR_\infty \left(\frac{1}{n_3^2} - \frac{1}{n_2^2} \right)$$

$$\nu_{n_2 \to n_1} = cR_\infty \left(\frac{1}{n_2^2} - \frac{1}{n_1^2} \right)$$

Thus

$$\nu_{n_3 \to n_2} + \nu_{n_2 \to n_1} = cR_\infty \left(\frac{1}{n_3^2} - \frac{1}{n_2^2} \right) + cR_\infty \left(\frac{1}{n_2^2} - \frac{1}{n_1^2} \right)$$

$$= cR_\infty \left(\frac{1}{n_3^2} - \frac{1}{n_1^2} \right)$$

But this is just equal to the frequency of the single photon emitted in a

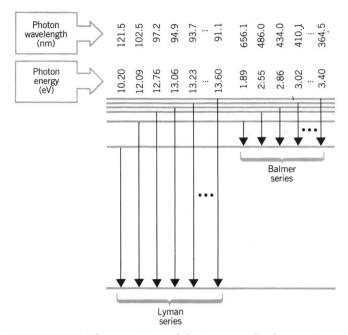

FIGURE 6.23 The transitions of the Lyman and Balmer series in hydrogen.

short wavelengths ⟹ large energy difference

direct transition from n_3 to n_1, so

$$\nu_{n_3 \to n_2} + \nu_{n_2 \to n_1} = \nu_{n_3 \to n_1}$$

The Bohr model is thus entirely consistent with the Ritz combination principle. (Since the frequency of an emitted photon is simply related to its energy by $E = h\nu$, the summing of frequencies is equivalent to the summing of energies. We may thus restate the Ritz combination principle in terms of energy. The energy of a photon emitted in a transition that skips or crosses over one or more states is simply equal to the step-by-step sum of the energies of the transitions connecting all of the individual states; see Problem 22 at the end of the chapter.)

Finally, we examine how the Bohr model helps us to understand why the absorption spectrum doesn't contain all of the lines present in the emission spectrum. Consider a hydrogen atom in its ground state, with the electron in the $n = 1$ level. Let us shine a beam of photons of a continuous range of energies on the hydrogen atom. If we hit the atom with a photon of energy 13.6 eV, the atom will absorb the photon, yielding a free electron. An atom in its ground state can also absorb a photon of energy 10.2 eV, corresponding to the energy difference between the $n = 1$ and $n = 2$ levels. In the process of this absorption, the atom will be excited to the $n = 2$ level, from which it rapidly decays back down to the ground state. If we shine a beam of radiation initially of continuous energy distribution through a vessel of atomic hydrogen, the transmitted beam will show a reduced intensity at $E = 10.2$ eV and at $E = 13.6$ eV, among others. Since the $n = 2$ level decays rapidly back down to the ground state, the electron doesn't exist in the $n = 2$ level long enough to absorb another photon and jump to the $n = 3$ level. Thus the absorption spectrum shows no line corresponding to the $n = 2$ to $n = 3$ transition (which is, of course, the longest wavelength of the

"e" jumps up from lower state n_1 to higher state n_2 by absorbing a photon of energy

$$h\nu = E_{n_2} - E_{n_1}$$

Balmer series). *The absorption spectrum contains only those transitions involving the* n = 1 *level.*

In a way, the good agreement between our calculated values of the Balmer series wavelengths and the experimental values was fortuitous, since we made two errors. The first error is related to the finite mass of the nucleus. We have neglected the mass of the nucleus in our calculations. The electron does not orbit about the nucleus, but instead the electron and the nucleus *both* orbit about their common center of mass. Since the nucleus is about 2000 times as massive as the electron, the center of mass of the system lies very close to the center of the nucleus. The kinetic energy should thus include an additional term due to the nuclear motion, and the effect of this inclusion is to reduce the Rydberg constant from the value R_∞ (so called because it would be correct if the nuclear mass were infinite) to the value $R = R_\infty/(1 + m/M)$, where M is the mass of the nucleus. This effect would tend to *increase* the calculated wavelengths. The second mistake we made was in converting the frequencies into wavelengths. The frequency of the emitted photon is directly related to the energy difference of the atomic levels; to find the wavelength, we use the expression $c = \lambda\nu$, which is correct *only in vacuum*. Since the experiment is usually performed in laboratories in air, the correct expression would be $\lambda\nu = v_{\text{air}}$, where v_{air} is the speed of light in air and is equal to $c/1.0003$. This effect tends to *decrease* the calculated wavelengths somewhat. Thus our two mistakes offset one another to some extent.

The same procedure we used to obtain the Bohr theory for hydrogen can be used for any atom with a single electron, even though the nuclear charge may be greater than 1. For example, we can calculate the energy levels of singly ionized helium (helium with one electron removed), doubly ionized lithium, etc. He^+

Reviewing the derivation of the Bohr theory, we find that the nuclear electric charge enters only in one place—in the expression for the electrostatic force between nucleus and electron, Equation (6.28). If the nucleus has a charge of Ze, the Coulomb force acting on the electron is

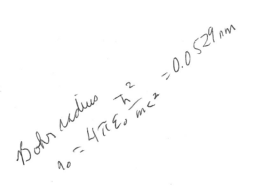

$$F = \frac{1}{4\pi\varepsilon_0}\frac{Ze^2}{r^2} \tag{6.41}$$

That is, where we had e^2 previously, we now have Ze^2. Making the same substitution in the final results, we can find the allowed radii:

$$r_n = \frac{4\pi\varepsilon_0\hbar^2}{Ze^2m}n^2 = \frac{a_0n^2}{Z} \tag{6.42}$$

Radii and energies of one-electron atoms

and the energies become

$$E_n = -\frac{m(Ze^2)^2}{32\pi^2\varepsilon_0^2\hbar^2}\frac{1}{n^2} = -(13.6\text{ eV})\frac{Z^2}{n^2} \tag{6.43}$$

The orbits in the higher-Z atoms are closer to the nucleus and have larger (negative) energies; that is, the electron is more tightly bound to the nucleus.

Calculate the two longest wavelengths of the Balmer series of triply ionized beryllium ($Z = 4$).   **EXAMPLE 6.7**

Since the radiations of the Balmer series are those that end with the $n = 2$ level, the two longest wavelengths are the radiations corresponding to $n = 3 \rightarrow n = 2$ and $n = 4 \rightarrow n = 2$. The energies of the radiations and their corresponding wavelengths are

$$E_3 - E_2 = -(13.6 \text{ eV})(4^2)\left(\frac{1}{9} - \frac{1}{4}\right) = 30.2 \text{ eV}$$

$$\lambda = \frac{hc}{E} = \frac{1240 \text{ eV·nm}}{30.2 \text{ eV}} = 41.0 \text{ nm}$$

$$E_4 - E_2 = -(13.6 \text{ eV})(4^2)\left(\frac{1}{16} - \frac{1}{4}\right) = 40.8 \text{ eV}$$

$$\lambda = \frac{hc}{E} = \frac{1240 \text{ eV·nm}}{40.8 \text{ eV}} = 30.4 \text{ nm}$$

These radiations are in the ultraviolet region.

Solution

$$F = \frac{1}{4\pi \varepsilon_0} \frac{Z_e^{2}}{r^2}$$

$$e^2 \rightarrow Z_e^{2}$$

$$r_n = \frac{4\pi \varepsilon_0 \hbar^2}{Z e^2 m} n^2$$

$$r_n = \frac{a_0 n^2}{Z}$$

$$E_n = -\frac{m(Z e^2)^2}{32 \pi^2 \varepsilon^2 \hbar^2} \frac{1}{n^2}$$

6.6 THE FRANCK-HERTZ EXPERIMENT

Let us imagine the following experiment, performed with the apparatus shown schematically in Figure 6.24. Electrons leave the cathode, which is heated by a filament heater. These electrons are accelerated toward the grid by the potential difference V, which we control. Electrons with energies V electron-volts pass through the grid and reach the plate if V exceeds V_0, a small retarding voltage between the grid and the plate. The current of electrons reaching the plate is measured using the ammeter A.

Now suppose the tube is filled with atomic hydrogen gas. As the voltage is increased from zero, more and more electrons reach the plate, and the current rises accordingly. The electrons inside the tube may make collisions with atoms of hydrogen, *but lose no energy in these collisions*—the collisions are perfectly elastic. The only way the electron can give up energy in a collision is if the electron has enough energy to cause the hydrogen atom to make a transition to an excited state. Thus, when the energy of the electrons reaches and barely exceeds 10.2 eV (or when the voltage reaches 10.2 V), the electrons can make *inelastic* collisions, leaving 10.2 eV of energy with the atom (now in the

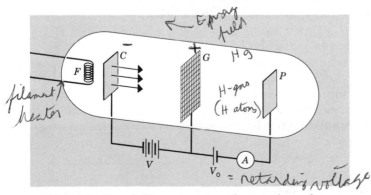

FIGURE 6.24 Franck-Hertz apparatus. Electrons leave the cathode C, are accelerated by the voltage V toward the grid G, and reach the plate P where they are recorded on the ammeter A.

To reach P, "e" must have KE_i eV where $V > V_0$

$n = 2$ level), and the original electron moves off with very little energy. If it should pass through the grid, the electron might not have sufficient energy to overcome the small retarding potential and reach the plate. Thus when $V = 10.2$ V, a drop in the current will be observed. As V is increased further, the current rises again, and would then drop again at $V = 12.1$ V, where an inelastic collision would leave the atom in the $n = 3$ state. This process would continue until $V = 13.6$ V, at which point the collision would ionize the atom. As V is increased further, we would begin to see the effects of multiple collisions. That is, when $V = 20.4$ V, an electron might make an inelastic collision, leaving the atom in the $n = 2$ state. The electron loses 10.2 eV of energy in this process, and so it moves off after the collision with a remaining 10.2 eV of energy, which is sufficient to excite a second hydrogen atom in an inelastic collision. Thus, if a drop in the current is observed at V, similar drops would be observed at $2V$, $3V$, ... Moreover, if drops are observed at V_1 and V_2, similar drops would be observed at $V_1 + V_2$, $2V_1 + V_2$, $V_1 + 2V_2$, etc.

This experiment should thus give rather direct evidence for the existence of atomic excited states. Unfortunately, it is not easy to do this experiment with hydrogen, because hydrogen occurs naturally in the molecular form H_2, rather than in atomic form. The molecules can absorb energy in a variety of ways, which would confuse the interpretation of the experiment. A similar experiment was done in 1914 by Franck and Hertz, using a tube filled with mercury vapor. Their results are shown in Figure 6.25, which shows clear evidence for an excited state at 4.9 eV; whenever the voltage is a multiple of 4.9 V, a drop in the current appears. Coincidentally, the *emission* spectrum of mercury shows an intense ultraviolet line of wavelength 254 nm, which corresponds to an energy of 4.9 eV; this could result from a transition between the same 4.9 eV excited state and the ground state. This early evidence for the discrete energies of atomic states thus not only confirmed the general principles of Bohr's atomic model but also showed in a direct way the energy quantization of physical systems.

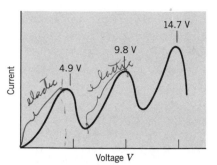

FIGURE 6.25 Result of Franck-Hertz experiment using mercury vapor. The current drops at $V = 4.9$ V, $V = 9.8$ V ($= 2 \times 4.9$ volts), $V = 14.7$ V ($= 3 \times 4.9$ V).

*6.7 THE CORRESPONDENCE PRINCIPLE

We have seen how Bohr's model permits calculations of transition wavelengths in atomic hydrogen that are in excellent agreement with the wavelengths observed in the emission and absorption spectra. However, in order to obtain this agreement, Bohr had to introduce two postulates which are radical departures from classical physics. In particular, an accelerated charged particle radiates electromagnetic energy according to classical physics, but an electron in Bohr's atomic model, accelerated to move in a circular orbit, does not radiate (unless of course it jumps to another orbit). Here we have a very different case than we did in our study of special relativity. You will recall, for example, that relativity gives us one expression for the kinetic energy, $K = E - E_0$, and classical physics gives us another, $K = \frac{1}{2}mv^2$; however, we showed that $E - E_0$ reduces to $\frac{1}{2}mv^2$ when $v \ll c$. Thus these two expressions are really not very different—one is merely a special case of the other. The dilemma associated with the accelerated electron is not simply a matter of atomic physics (as an example of quantum physics) being a special case of classical physics. Either the accelerated charge radiates, or it doesn't! Bohr's solution to this serious dilemma was to propose the *correspondence principle*, which simply states that the laws of classical physics are valid in the classical domain, while the laws of quantum physics are valid in the atomic domain; where the two domains overlap, both sets of laws must give the same result.

Let us see how we can apply this principle to the Bohr atom. According to classical physics, an electric charge moving in a circle radiates at a frequency equal to its frequency of rotation. For an atomic orbit, the period of revolution is just the distance traveled in one orbit, $2\pi r$, divided by the orbital speed $v = \sqrt{2K/m}$, where K is the kinetic energy. Thus, using our expression (6.29) for the kinetic energy, the orbital period T is

$$T = \frac{2\pi r}{\sqrt{2K/m}} = \frac{\pi r\sqrt{2m}\ \sqrt{8\pi\varepsilon_0 r}}{e} \tag{6.44}$$

The frequency ν is just the inverse of the period:

$$\nu = \frac{1}{T} = \frac{e}{\sqrt{16\pi^3\varepsilon_0 m r^3}} \tag{6.45}$$

Using the expression (6.34) for the allowed orbits, we find

$$\nu_n = \frac{me^4}{32\pi^3\varepsilon_0^2\hbar^3}\frac{1}{n^3} \tag{6.46}$$

A "classical" electron moving in an orbit of radius r_n would radiate at this frequency ν_n.

If we made the radius of the Bohr atom so large that it went from a quantum-sized object (10^{-10} m) to a laboratory-sized object (10^{-3} m), we might expect the atom to behave classically. Since the radius increases with increasing n like n^2, we expect that for n in the range $10^3 - 10^4$, the atom would behave classically. Let us then calculate the frequency of the radiation emitted by such an atom when the electron drops from the orbit n to the orbit $n - 1$. According to Equation (6.39), the frequency is

$$\begin{aligned}\nu &= \frac{me^4}{64\pi^3\varepsilon_0^2\hbar^3}\left(\frac{1}{(n-1)^2} - \frac{1}{n^2}\right) \\ &= \frac{me^4}{64\pi^3\varepsilon_0^2\hbar^3}\frac{2n-1}{n^2(n-1)^2}\end{aligned} \tag{6.47}$$

If n is very large, then we can approximate $n - 1$ by n and $2n - 1$ by $2n$, giving

$$\begin{aligned}\nu &\cong \frac{me^4}{64\pi^3\varepsilon_0^2\hbar^3}\frac{2n}{n^4} \\ &= \frac{me^4}{32\pi^3\varepsilon_0^2\hbar^3}\frac{1}{n^3}\end{aligned} \tag{6.48}$$

This is identical with Equation (6.46) for the "classical" frequency. The "classical" electron spirals slowly in toward the nucleus, radiating at the frequency given by (6.46), while the "quantum" electron jumps from the orbit n to the orbit $n - 1$ and then to the orbit $n - 2$, and so forth, radiating at the frequency given by the identical expression (6.48). (When the circular orbits are very large, this jumping from one circular

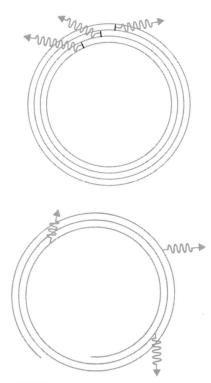

FIGURE 6.26 (Top) A large quantum atom. Photons are emitted in discrete transitions as the electron jumps to lower states. (Bottom) A classical atom. Photons are emitted continuously by the accelerated electron.

orbit to the next smaller one looks very much like a spiral, as in Figure 6.26.)

In the region of large n, where classical and quantum physics overlap, the classical and quantum expressions for the radiation frequencies are identical. This is an example of an application of Bohr's correspondence principle. The applications of the correspondence principle go far beyond the Bohr atom, and this principle is important to us in understanding how we get from the domain in which the laws of classical physics are valid to the domain in which the laws of quantum physics are valid.

6.8 DEFICIENCIES OF THE BOHR MODEL

The Bohr model gives us a clear picture of how electrons move about the nucleus, and many of our attempts to explain the behavior of atoms refer to this picture, even though it is not strictly correct. It is remarkable (perhaps even fortunate or coincidental) that the successes of the model, with its new concepts of discrete energy levels and stationary states, occurred a full decade before deBroglie's work and the birth of wave mechanics.

Still, in spite of its successes, it is at best an incomplete model. It is only useful for atoms that contain one electron (hydrogen, singly ionized helium, doubly ionized lithium, etc.), but not for atoms with two or more electrons, since we have considered only the force between electron and nucleus, and not the force between the electrons themselves. Furthermore, if we look very carefully at the emission spectrum, we find that many lines are in fact not single lines, but very closely spaced combinations of two or more lines; the Bohr model is unable to account for these *doublets* of spectral lines. The model is also limited in its usefulness as a basis from which to calculate other properties of the atom; although we can accurately calculate the energies of the spectral lines, we cannot calculate their intensities. For example, how often will an electron in the $n = 3$ state jump directly to the $n = 1$ state, emitting the corresponding photon, and how often will it jump first to the $n = 2$ state and then to the $n = 1$ state, emitting two photons? A complete theory should provide a way to calculate this property.

A more serious deficiency of the model is that it completely violates the uncertainty relationship. (In Bohr's defense, remember that this was a decade before the introduction of wave mechanics, with its accompanying ideas of uncertainty.) The uncertainty relationship $\Delta x \, \Delta p_x \gtrsim \hbar$ is valid for any direction in space. If we choose the radial direction, then $\Delta r \, \Delta p_r \gtrsim \hbar$. For an electron moving in a circular orbit, we know the value of r exactly, and thus $\Delta r = 0$. If it is moving in a circle we also know p_r exactly (in fact it is exactly zero), and so $\Delta p_r = 0$. This simultaneous exact knowledge of r and p_r violates the uncertainty principle.

We do not wish, however, to discard the model completely. The Bohr model gives a useful mental picture of the structure of an atom. There are many atomic properties, especially those associated with magnetism, which *can* be understood on the basis of Bohr orbits. When we treat the problem correctly in the next chapter, we will find that the energy levels of hydrogen calculated by solving the Schrödinger equation are in fact identical with those of the Bohr atom. We have in this chapter considered two problems of atomic physics—Rutherford scattering and the hydrogen emission spectrum. We have seen how both could be understood based on calculations done without reference to wave mechanics, even though we expect wave mechanics to be an important consideration on the atomic scale. In fact, these two examples are unique in that the "correct" wave-mechanical calculations of the Rutherford scattering formula and the hydrogen emission wavelengths give the same results as our classical calculation! This was not the case with many other phenomena we have studied, including blackbody radiation, Compton scattering, and the photoelectric effect. It is interesting to speculate on the development of physics in the era of Bohr and Rutherford if their classical calculations had not yielded correct results.

SUGGESTIONS FOR FURTHER READING

A discussion of many of the basic properties of atoms may be found in:

M. R. Wehr, J. A. Richards, and T. W. Adair, *Physics of the Atom* (Reading, Addison-Wesley, 1978).

For a historical perspective on the development of atomic theory, see:

H. A. Boorse and L. Motz, editors, *The World of the Atom* (New York, Basic Books, 1966).
G. K. T. Conn and H. D. Turner, *The Evolution of the Nuclear Atom* (London, Iliffe Books, 1965).
F. Friedman and L. Sartori, *The Classical Atom* (Reading, Addison-Wesley, 1965).

For a popular summary of Rutherford's work, see:

E. N. da C. Andrade, "The Birth of the Nuclear Atom," *Scientific American 195*, 93 (November 1956).

The early papers on the Rutherford model and its experimental confirmation illustrate the difficulty of the experiments and the care and abilities of the experimenters. They are easily readable and require no mathematics beyond the present level.

E. Rutherford, *Philosophical Magazine 21*, 669 (1911).
H. Geiger, *Proceedings of the Royal Society of London A83*, 492 (1910).
H. Geiger and E. Marsden, *Philosophical Magazine 25*, 604 (1913).

QUESTIONS

1. Does the Thomson model fail at large scattering angles or at small scattering angles? Why?
2. What principles of physics would be violated if we scattered a beam of alpha particles with a single impact parameter from a single target atom at rest?
3. Could we use the Rutherford scattering formula to analyze the scattering of:
   (a) Protons incident on iron?
   (b) Alpha particles incident on lithium $(Z = 3)$?
   (c) Silver nuclei incident on gold?
   (d) Hydrogen *atoms* incident on gold?
   (e) Electrons incident on gold?
4. What determines the angular range $d\theta$ in the alpha-particle scattering experiment (Figure 6.11)?
5. Why didn't Bohr use the concept of deBroglie waves in his theory?

6. In which Bohr orbit does the electron have the largest velocity? Are we justified in treating the electron nonrelativistically in that case?

7. How does an electron in hydrogen get from $4a_0$ to a_0 without being anywhere in between?

8. How is the quantization of the energy in the hydrogen atom similar to the quantization of the systems discussed in Chapter 5? How is it different? Do the quantizations originate from similar causes?

9. In a Bohr atom, an electron jumps from state n_1, with angular momentum $n_1\hbar$, to state n_2, with angular momentum $n_2\hbar$. How can an isolated system change its angular momentum? (In classical physics, a change in angular momentum requires an external torque.) Can the photon carry away the difference in angular momentum? Estimate the maximum angular momentum, relative to the center of the atom, which the photon can have. Does this suggest another failure of the Bohr model?

10. The product $E_n r_n$ for the hydrogen atom is (1) independent of Planck's constant and (2) independent of the quantum number n. Does this observation have any significance? Is this a classical or a quantum effect?

11. (a) How does a Bohr atom violate the $\Delta x\, \Delta p$ uncertainty relationship? (b) How does a Bohr atom violate the $\Delta E\, \Delta t$ uncertainty relationship? (What is ΔE? What does this imply about Δt? What do you conclude about transitions between levels?)

12. List the assumptions made in deriving the Bohr theory. Which of these are a result of neglecting small quantities? Which of these violate basic principles of relativity or quantum physics?

13. List the assumptions made in deriving the Rutherford scattering formula. Which of these are a result of neglecting small quantities? Which of these violate basic principles of relativity or quantum physics?

14. In both the Rutherford theory and the Bohr theory, we used the classical expression for the kinetic energy. Estimate the velocity of an electron in the Bohr atom and of an alpha particle in a typical scattering experiment, and decide if the use of the classical formula is justified.

15. In both the Rutherford theory and the Bohr theory, we neglected any wave properties of the particles. Estimate the deBroglie wavelength of an electron in a Bohr atom and compare it with the size of the atom. Estimate the deBroglie wavelength of an alpha particle and compare it with the size of the nucleus. Is the wave behavior expected to be important in either case?

16. Why are the decreases in current in the Franck-Hertz experiment not sharp?

17. As indicated by the Franck-Hertz experiment, the first excited state of mercury is at an energy of 4.9 eV. Do you expect mercury to show absorption lines in the visible spectrum?

18. Is the correspondence principle a necessary part of quantum physics or is it merely an accidental agreement of two formulas? Where do we draw the line between the world of quantum physics and the world of classical, nonquantum physics?

PROBLEMS

1. (a) Compute the oscillation frequency of the electron and the expected absorption or emission wavelength in a Thomson-model hydrogen atom. Use $R = 0.053$ nm. Compare with the observed wavelength of the strongest emission and absorption line in hydrogen, 122 nm. (b) Repeat for sodium $(Z = 11)$. Use $R = 0.18$ nm. Compare with the observed wavelength, 590 nm.

2. Derive Equation (6.12) from Equation (6.11).

3. Use Equation (6.13) to derive Equation (6.14).

4. Alpha particles of kinetic energy 5.00 MeV are scattered at 90° by a gold foil. (a) What is the impact parameter? (b) What is the minimum distance between alpha particles and gold nucleus? (c) Find the kinetic and potential energies at that minimum distance.

5. How much kinetic energy must an alpha particle have before its distance of closest approach to a gold nucleus is equal to the nuclear radius (7.0×10^{-15} m)?

6. What is the distance of closest approach when alpha particles of kinetic energy 6.0 MeV are scattered by a thin copper foil?

7. Protons of energy 5.0 MeV are incident on a silver foil of thickness 4.0×10^{-6} m. What fraction of the incident protons is scattered at angles: (a) Greater than 90°? (b) Greater than 10°? (c) Between 5° and 10°? (d) Less than 5°?

8. Protons are incident on a copper foil 12 μm thick. (a) What should the proton kinetic energy be in order that the distance of closest approach equal the nuclear radius (5.0 fm)? (b) If the proton energy were 7.5 MeV, what is the impact parameter for scattering at 120°? (c) What is the minimum distance between proton and nucleus for this case? (d) What fraction of the protons is scattered beyond 120°?

9. Alpha particles of kinetic energy K are scattered either from a gold foil or a silver foil of identical thickness. What is the ratio of the number of particles scattered at angles greater than 90° by the gold foil to the same number for the silver foil?

10. The maximum kinetic energy given to the target nucleus will occur in a head-on collision with $b = 0$. (Why?) Evaluate the maximum kinetic energy given to the target nucleus when 8.0 MeV alpha particles are incident on a gold foil. Are we justified in neglecting this energy?

11. The maximum kinetic energy that an alpha particle can transmit to an *electron* occurs during a head-on collision. Compute the kinetic energy lost by an alpha particle of kinetic energy 8.0 MeV in a head-on collision with an electron at rest. Are we justified in neglecting this energy in the Rutherford theory?

12. Alpha particles of energy 9.6 MeV are incident on a silver foil of thickness 7.0 μm. For a certain value of the impact parameter, the alpha particles lose exactly half their incident kinetic energy when they reach their minimum separation from the nucleus. Find the minimum separation, the impact parameter, and the scattering angle.

13. Alpha particles of kinetic energy 6.0 MeV are incident at a rate of 3.0×10^7 per second on a gold foil of thickness 3.0×10^{-6} m. A circular detector of diameter 1.0 cm is placed 12 cm from the foil at an angle of 30° with the direction of the incident alpha particles. At what rate does the detector measure scattered alpha particles?

14. In the $n = 3$ state of hydrogen, find the electron's velocity, kinetic energy, and potential energy.

15. Use the Bohr theory to find the series wavelength limits of the Lyman and Paschen series of hydrogen.

16. Show that the speed of an electron in the nth Bohr orbit of hydrogen is $\alpha c/n$, where α is the fine structure constant. What would be the speed in a hydrogenlike atom with a nuclear charge of Ze?

17. An electron is in the $n = 5$ state of hydrogen. To what states can the electron make transitions, and what are the energies of the emitted radiations?

18. A hydrogen atom is in the $n = 6$ state. (a) Counting all possible paths, how many different photon energies can be emitted if the atom ends up in the ground state? (b) Suppose only $\Delta n = 1$ transitions were allowed. How many different photon energies would be emitted? (c) How many different photon energies would occur in a Thomson-model hydrogen atom?

19. Continue Figure 6.23, showing the transitions of the Paschen series and computing their energies and wavelengths.

20. A collection of hydrogen atoms in the ground state is illuminated with ultraviolet light of wavelength 59.0 nm. Find the kinetic energy of the emitted electrons.

21. The *ionization energy* is the energy required to remove an electron from an atom. Find the ionization energy of: (a) The $n = 3$ level of hydrogen. (b) The $n = 2$ level of He^+ (singly ionized helium). (c) The $n = 4$ level of Li^{++} (doubly ionized lithium).

22. Use the Bohr formula to find the energy differences $E(n_1 \rightarrow n_2) = E_{n_1} - E_{n_2}$ and show that (a) $E(4 \rightarrow 2) = E(4 \rightarrow 3) + E(3 \rightarrow 2)$; (b) $E(4 \rightarrow 1) = E(4 \rightarrow 2) + E(2 \rightarrow 1)$. (c) Interpret these results based on the Ritz combination principle.

23. What is the difference in wavelength between the first line of the Balmer series in ordinary hydrogen $(M \cong 1.01$ u$)$ and in "heavy" hydrogen $(M \cong 2.01$ u$)$?

24. Find the shortest and the longest wavelengths of the Lyman series of singly ionized helium.

25. Draw an energy-level diagram showing the lowest four levels of singly ionized helium. Show all possible transitions from the levels and label each transition with its wavelength.

26. An electron is in the $n = 8$ level of ionized helium. (a) Find the three longest wavelengths that are emitted when the electron makes a transition from the $n = 8$ level to a lower level. (b) Find the shortest wavelength that can be emitted. (c) Find the three longest wavelengths at which the electron in the $n = 8$ level will *absorb* a photon and move to a higher state, if we could somehow keep it in that level long enough to absorb. (d) Find the shortest wavelength that can be absorbed.

27. The lifetimes of the levels in a hydrogen atom are of the order of 10^{-8} s. Find the energy uncertainty of the first excited state and compare it with the energy of the state.

28. The *Handbook of Chemistry and Physics* lists the following emission wavelengths (in nm) for ionized helium:

23.73	30.38	121.5	251.1	468.6	1012.4
24.30	102.5	164.0	273.4	541.1	1162.6
25.63	108.5	238.5	320.3	656.0	1863.7

Using the same values of n_0 as the hydrogen spectrum, group these spectral lines into series, showing the index n for each line identified. Give the series limit of each series. In what region of the electromagnetic spectrum is each series located?

29. The *Handbook of Chemistry and Physics* lists the following emission wavelengths (in nm) for doubly ionized lithium: 11.39, 13.50, 54.00, 72.91. Identify these lines with the proper spectral series (as in hydrogen), giving the index n for each line and the series limit for each series.

30. When an atom emits a photon in a transition from a state of energy E_1 to a state of energy E_2, the photon energy is not precisely equal to $E_1 - E_2$. Conservation of momentum requires that the atom must recoil, and so some

energy must go into recoil kinetic energy K_R. Show that $K_R \cong (E_1 - E_2)^2/2Mc^2$ where M is the mass of the atom. Evaluate this recoil energy for the $n = 2$ to $n = 1$ transition of hydrogen.

31. A long time ago, in a galaxy far, far away, electric charge had not yet been invented, and atoms were held together by gravitational forces. Compute the Bohr radius and the $n = 2$ to $n = 1$ transition energy in a gravitationally bound hydrogen atom.

32. (a) Find an expression for the Bohr radius a_0 in terms of the fine structure constant α (see Chapter 1), the rest energy of the electron, and the constant hc. (b) Do the same for the hydrogen ground-state energy E_1.

33. In a *muonic atom*, the electron is replaced by a negatively charged particle called the *muon*. The muon mass is 207 times the electron mass. What is the shortest wavelength of the Lyman series in a muonic hydrogen atom? In what region of the electromagnetic spectrum does this belong?

34. What is the radius of the first Bohr orbit of a muonic lead atom $(Z = 82)$? Compare with the nuclear radius of about 7 fm.

35. An alternative development of the Bohr theory begins by assuming that the stationary states are those for which the circumference of the orbit is an integral number of deBroglie wavelengths. (a) Show that this condition leads to standing deBroglie waves around the orbit. (b) Show that this condition gives the relationship (6.32) used in the Bohr theory.

36. Show that the energy of the photon emitted when a hydrogen atom makes a transition from state n to state $n - 1$ is, when n is very large, $\Delta E \cong \alpha^2(mc^2/n^3)$ where α is the fine structure constant.

37. Suppose all of the excited levels of hydrogen had lifetimes of 10^{-8} s. As we go to higher and higher excited states, they get closer and closer together, and soon they are so close in energy that the energy uncertainty of each state becomes as large as the energy spacing between states, and we can no longer resolve individual states. Use the result of Problem 36 for the energy spacing and find the value of n for which this occurs. What is the radius of such an atom?

38. Compare the frequency of revolution of an electron with the frequency of the photons emitted in transitions from n to $n - 1$ for (a) $n = 10$; (b) $n = 100$; (c) $n = 1000$; (d) $n = 10,000$.

39. A hypothetical atom has only two excited states, at 4.0 and 7.0 eV, and has a ground-state ionization energy of 9.0 eV. If we used a vapor of such atoms for the Franck-Hertz experiment, for what voltages would we expect to see decreases in the current? List all voltages up to 20 V.

40. The first excited state of sodium decays to the ground state by emitting a photon of wavelength 590 nm. If sodium vapor is used for the Franck-Hertz experiment, at what voltage will the first current drop be recorded?

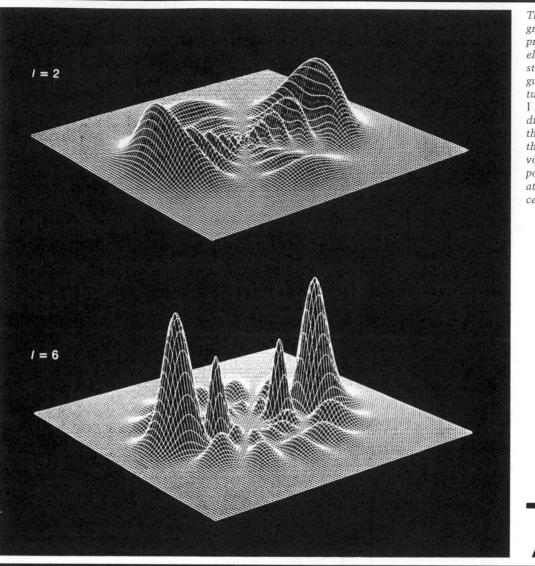

These computer-drawn graphs represent the probability to locate the electron in the n = 8 state of hydrogen for angular momentum quantum number l = 2 and l = 6; the vertical coordinate at any point gives the probability to find the electron in a small volume element at that point. The nucleus of the atom would be at the center of each graph.

7

THE HYDROGEN ATOM IN WAVE MECHANICS

In this chapter we study the solutions of the Schrödinger equation for the hydrogen atom. We will see how these solutions, although leading to the same energy levels calculated in the Bohr model, are consistent with the requirements of wave mechanics, especially the uncertainties in localizing the electron.

Two other deficiencies of the Bohr model are not so easily eliminated by solving the Schrödinger equation. The so-called "fine structure" splitting of the spectral lines, which appears when we examine the lines very carefully, cannot be explained by our solutions; the proper explanation of this effect requires the introduction of a new property of the electron, the *intrinsic spin*. Second, the mathematical difficulties of solving the Schrödinger equation for atoms containing two or more electrons are formidable, so we will restrict our discussion in this chapter to one-electron atoms, in order to see how wave mechanics enables us to understand some basic atomic properties. In the next chapter we discuss the structure of many-electron atoms.

7.1 THE SCHRÖDINGER EQUATION IN SPHERICAL COORDINATES

The Schrödinger equation in three dimensions has the following form:

$$-\frac{\hbar^2}{2m}\left(\frac{\partial^2\psi}{\partial x^2}+\frac{\partial^2\psi}{\partial y^2}+\frac{\partial^2\psi}{\partial z^2}\right)+V(x,y,z)\psi=E\psi \tag{7.1}$$

where ψ is a function of x, y, and z. The usual procedure for solving a partial differential equation of this type is to separate the variables. The potential for the force between the nucleus and electron is $V=-(1/4\pi\varepsilon_0)(e^2/r)$; since $r=\sqrt{x^2+y^2+z^2}$,

$$V(x,y,z)=-\frac{1}{4\pi\varepsilon_0}\frac{e^2}{\sqrt{x^2+y^2+z^2}} \tag{7.2}$$

The potential in this form does *not* lead to a separable equation, but if we work in the more convenient (at least for this calculation) spherical polar coordinates (r,θ,ϕ) instead of (x,y,z), we can separate the variables and find a set of solutions. The variables of spherical polar coordinates are illustrated in Figure 7.1. This simplification in the solution is at the expense of an increased complexity of the partial differential equation, which becomes:

$$-\frac{\hbar^2}{2m}\left[\frac{\partial^2\psi}{\partial r^2}+\frac{2}{r}\frac{\partial\psi}{\partial r}+\frac{1}{r^2\sin\theta}\frac{\partial}{\partial\theta}\left(\sin\theta\frac{\partial\psi}{\partial\theta}\right)+\frac{1}{r^2\sin^2\theta}\frac{\partial^2\psi}{\partial\phi^2}\right]$$
$$+V(r,\theta,\phi)\psi=E\psi \tag{7.3}$$

where now $\psi=\psi(r,\theta,\phi)$. We consider only those solutions that are *separable* and can be factored as

$$\psi(r,\theta,\phi)=R(r)\Theta(\theta)\Phi(\phi) \tag{7.4}$$

where $R(r)$, $\Theta(\theta)$, and $\Phi(\phi)$ are each functions of a single variable. This procedure gives three differential equations, each of a single variable (r, θ, or ϕ).

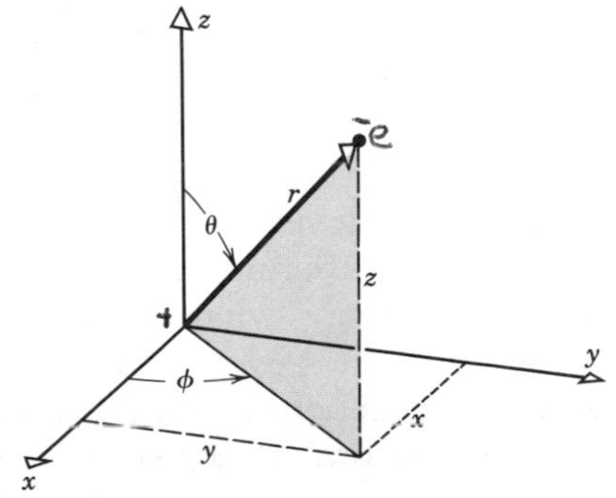

FIGURE 7.1 Spherical polar coordinates for the hydrogen atom. The proton is at the origin and the electron is at a radius r, in a direction determined by the polar angle θ and the azimuthal angle ϕ.

The mathematics of solving the Schrödinger equation in spherical coordinates (r, θ, ϕ) is rather difficult, so we will merely present the solutions and then discuss them.

As we discussed in our introduction to the Schrödinger equation, a three-dimensional problem requires three quantum numbers to identify the solutions. Three quantum numbers will therefore be necessary to describe the hydrogen atom wave functions. The first quantum number, n, is associated with the solution for the radial part of the equation, $R(r)$. It is the same n as the label of the levels in the Bohr model. The solution to the polar equation, $\Theta(\theta)$, gives a quantum number l, and the solution to the azimuthal equation, $\Phi(\phi)$, yields a third quantum number m_l.

The quantum number n, known as the *principal quantum number*, takes integer values 1, 2, 3, ... When we specify the number n, we choose a specific energy level, as in the Bohr model. Furthermore, when we solve the Schrödinger equation, we find that the quantized energy levels are, in exact agreement with the Bohr model,

$$E_n = -\frac{me^4}{32\pi^2\varepsilon_0^2\hbar^2}\frac{1}{n^2}$$

Note that the energy depends only on the quantum number n, and not on l and m_l.

The quantum numbers l and m_l take values which are limited by the value of n. The *angular momentum quantum number l* takes integer values from 0 to $n - 1$. For example, for $n = 1$, only $l = 0$ is permitted; for $n = 2$, $l = 0$ and $l = 1$ are permitted. For each l value, the *magnetic quantum number m_l* has the values 0, ± 1, ± 2, . . . , $\pm l$.

Let us see now how the levels are labeled with the three quantum numbers (n, l, m_l). The ground state has $n = 1$, and therefore $l = 0$. There is only one m_l value allowed, $m_l = 0$. Thus the ground state has the quantum numbers $(1, 0, 0)$. The first excited state has $n = 2$, and the allowed l values are $l = 0$ or $l = 1$. For $l = 0$, only $m_l = 0$ is permitted. For $l = 1$, the m_l values are -1, 0 or $+1$. The possible quantum numbers of this level are then $(2, 0, 0)$, $(2, 1, 1)$, $(2, 1, 0)$, and $(2, 1, -1)$. All of these levels have $n = 2$, and therefore all have the same energy, since the energy depends only on n. They are thus *degenerate*, and we would say that the $n = 2$ level is four-fold degenerate. If we listed all of the possible combinations of quantum numbers for the $n = 3$ level, we would find nine possible combinations, so that the level is nine-fold degenerate. In general, the nth level is n^2-fold degenerate. Figure 7.2 illustrates the labeling of the levels.

If these different combinations of quantum numbers have exactly the same energy, what is the purpose of listing them separately? First, we will find in the last section of this chapter that the sublevels are not precisely degenerate, but are separated by a very small energy (perhaps 10^{-5} eV). Second, in the study of the transitions between the levels, we find that the intensities of the individual transitions depend on the particular sublevel from which the transition originates. Third, and perhaps most important, *each of these sublevels has a very different wave function, and therefore represents a very different state of motion of the electron*. In understanding this last point, we must consider the geometrical interpretation of the quantum numbers, for which we will briefly return to the language of the Bohr model.

7.2 QUANTUM NUMBERS AND DEGENERACY

Principal quantum number n

$R_{n\ell} \quad \Theta_{\ell m_\ell} \quad \Phi_{m_\ell}$

$= -\frac{1}{2}\frac{1}{(4\pi\varepsilon_0)^2}\frac{me^4}{\hbar^2}$ ← from Bohr model

Angular momentum quantum number l, magnetic quantum number m_l

spin quantum # is the
The 4th quantum state

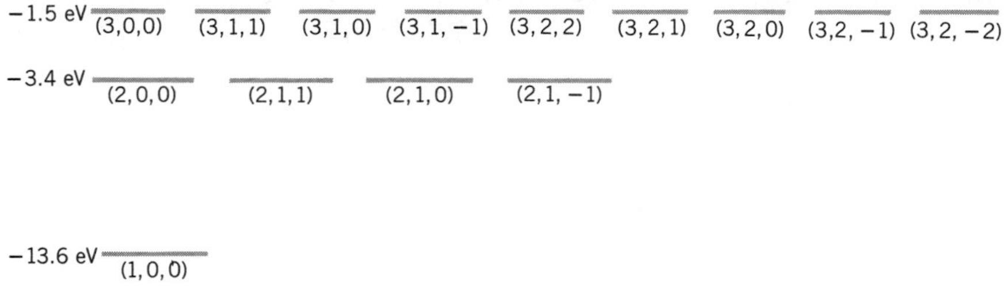

-1.5 eV $\overline{\underset{(3,0,0)}{\quad\quad}}$ $\overline{\underset{(3,1,1)}{\quad\quad}}$ $\overline{\underset{(3,1,0)}{\quad\quad}}$ $\overline{\underset{(3,1,-1)}{\quad\quad}}$ $\overline{\underset{(3,2,2)}{\quad\quad}}$ $\overline{\underset{(3,2,1)}{\quad\quad}}$ $\overline{\underset{(3,2,0)}{\quad\quad}}$ $\overline{\underset{(3,2,-1)}{\quad\quad}}$ $\overline{\underset{(3,2,-2)}{\quad\quad}}$

-3.4 eV $\overline{\underset{(2,0,0)}{\quad\quad}}$ $\overline{\underset{(2,1,1)}{\quad\quad}}$ $\overline{\underset{(2,1,0)}{\quad\quad}}$ $\overline{\underset{(2,1,-1)}{\quad\quad}}$

-13.6 eV $\overline{\underset{(1,0,0)}{\quad\quad}}$

FIGURE 7.2 The lower energy levels of hydrogen, labeled with the quantum numbers (n, l, m_l). The first excited state is four-fold degenerate and the second excited state is nine-fold degenerate.

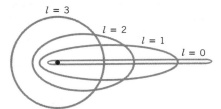

FIGURE 7.3 Electron orbits for $n = 4$. Note that (1) the average values of r are roughly the same; (2) in the orbits with smaller l-values, the electron spends more time *both* close to the nucleus and far away from the nucleus.

each sublevel has a different wave function. For a given n, $l = n-1 \Rightarrow$ circular orbit, other l's $\Rightarrow$ elliptical orbit

In the Bohr model, the value of n determined the radius of the orbit of the electron—the larger the value of n, the larger the radius. The quantum number l determines (in the context of the Bohr model) whether the orbit is circular or elliptical. Figure 7.3 illustrates the basic orbits for the different l values of the $n = 4$ level. In this interpretation of l, we can see why this quantum number is related to the angular momentum of the electron. The orbits with the largest possible value of l ($l = n - 1$) have the largest angular momentum about the nucleus, and are therefore circular. Smaller values of l give elliptical orbits, and the smallest value of l ($l = 0$) gives a completely flattened ellipse that passes through the nucleus. The quantum number m_l gives the orientation of the plane of the orbit relative to the x, y plane. Figure 7.4 illustrates two possible orientations of an electron's orbit. Once again, these geometrical interpretations are only meaningful within our schematic picture based on the Bohr model, and should not be taken too seriously; in fact, a fixed orbital plane violates the uncertainty principle.

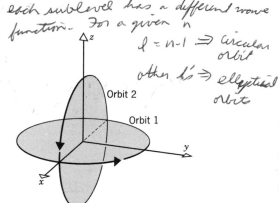

FIGURE 7.4 Two different orientations of the plane of an electron's orbit. Orbit 1 lies in the xy plane and represents $m_l = l$; orbit 2 lies in the xz plane and represents $m_l = 0$.

m_l gives the orientation of the plane of orbit rel. to the $x-y$ plane

7.3 THE VECTOR MODEL

For many purposes, the Bohr model will help us to understand the properties of atoms, and we have seen in the last section how the three quantum numbers (n, l, m_l) tell us about the "shape" of the electron's orbit. However, there are other properties of atoms, especially their behavior in magnetic fields, which are easier to understand if we use a model in which the angular momentum is considered to behave much like an ordinary vector (although the vector will have some special properties not found in "classical" vectors).

For each of the electron's possible orbits, the angular momentum l remains constant during the orbit. (Such is also true of objects in gravitational orbits; a comet speeds up as it passes near the sun, so that the decrease in its distance r from the sun is exactly balanced by an increase in its linear momentum p, and the product $\mathbf{r} \times \mathbf{p}$ remains constant.) We represent that angular momentum by a vector $\mathbf{l}$; in the classical sense this is a vector through the nucleus perpendicular to the plane of the electron's orbit. A careful calculation based on the solutions to the Schrödinger equation (but beyond the level of this text) gives the relationship between the length of the vector $\mathbf{l}$, which we denote by $|\mathbf{l}|$, and the quantum number l:

$$|\mathbf{l}| = \sqrt{l(l + 1)}\, \hbar \tag{7.6}$$

Length of l

Compute the length of the angular momentum vectors that represent the motion of an electron in a state with $l = 1$ and in another state with $l = 2$.

EXAMPLE 7.1

Equation (7.6) gives the relationship between the length of the vector and the associated quantum number l. For $l = 1$

$$|\mathbf{l}| = \sqrt{1(1 + 1)}\,\hbar = \sqrt{2}\,\hbar$$

and for $l = 2$

$$|\mathbf{l}| = \sqrt{2(2 + 1)}\,\hbar = \sqrt{6}\,\hbar$$

Note two important points here. First, the length of the vector $|\mathbf{l}|$ is always greater than $l\hbar$, since $\sqrt{l(l + 1)}$ is always greater than l. The importance of this point will be discussed later. Second, these values of $|\mathbf{l}|$, which we can interpret as the "magnitude" of the electron's angular momentum, are totally different from those found in the Bohr model. For example, an electron with $n = 3$ in the Bohr model has an angular momentum $|\mathbf{l}| = 3\hbar$ (see Section 6.5). In our quantum-mechanical vector model, an electron with $n = 3$ can have $l = 2$ (with $|\mathbf{l}| = \sqrt{6}\,\hbar$), or $l = 1$ (with $|\mathbf{l}| = \sqrt{2}\,\hbar$), or even $l = 0$ (with $|\mathbf{l}| = 0$).

Just like an ordinary classical vector, the vector $\mathbf{l}$ can have components along any axis in space. Once again, the wave functions deduced using the Schrödinger equation give us the rules for computing the components of $\mathbf{l}$. (We generally choose the z axis for special consideration, since it is an axis of reference in the spherical polar coordinate system.) The z component of $\mathbf{l}$, which we denote by l_z, is restricted to the values

$$l_z = m_l\hbar \qquad (7.7)$$

where m_l is just the magnetic quantum number, which takes values 0, ± 1, ± 2, . . . $\pm l$.

What are the possible z components of the vector $\mathbf{l}$ which represents the orbital angular momentum of a state with $l = 2$?

EXAMPLE 7.2

The possible m_l values for $l = 2$ are $+2$, $+1$, 0, -1, -2, and so the $\mathbf{l}$ vector can have any of five possible z components: $l_z = 2\hbar$, $1\hbar$, 0, $-1\hbar$, or $-2\hbar$. The length of the vector $\mathbf{l}$, as we found previously, is $\sqrt{6}\,\hbar$.

The components of the vector $\mathbf{l}$ for $l = 2$ are illustrated in Figure 7.5. Each different orientation in space of the vector $\mathbf{l}$ corresponds to a different m_l value. The polar angle θ that the vector $\mathbf{l}$ makes with the z axis can be easily found by referring to the figure. Since $l_z = |\mathbf{l}| \cos \theta$,

$$\cos \theta = \frac{l_z}{|\mathbf{l}|} = \frac{m_l\hbar}{\sqrt{l(l + 1)}\,\hbar}$$

so

$$\cos \theta = \frac{m_l}{\sqrt{l(l + 1)}} \qquad (7.8)$$

This behavior represents a curious aspect of quantum physics called *spatial quantization*, in which only certain orientations of angular momentum vectors are allowed. The number of these orientations is equal to $2l + 1$ (the number of different possible m_l values) and the magnitudes of their successive z components always differ by $\hbar$. As an example, suppose we could prepare a collection of hydrogen atoms in a state with $l = 1$. Choosing arbitrarily a z axis and using an appropriate experimental technique, we measure the z component of $\mathbf{l}$. From such a measurement we expect to find $l_z = \hbar$, 0, or $-\hbar$. Choosing a completely different z axis, we repeat the measurement and once again we find $l_z = \hbar$, 0, or $-\hbar$. This behavior is completely different from that of classical vectors. A classical vector of length 1.0 would have a z component of 1.0 if we choose our z axis along the vector, or -1.0 if we choose the opposite direction, or 0.5 if we choose our z axis at an angle of 60° to the vector, or 0.7 if we choose the z axis at 45° to the vector. A quantum-mechanical vector representing $l = 1$ has z components that are restricted to $+1.0$, 0, or -1.0. The curious fact is that we observe this result no matter which direction we choose for the z axis!

You may perhaps be wondering why we have singled out the z axis for special attention. Aside from its convenience in polar coordinates, there is an important reason. According to quantum physics, we can know exactly only *one* of the three components of $\mathbf{l}$ (by convention, we choose the z component); the other components of $\mathbf{l}$ are completely uncertain. This follows from an additional form of the uncertainty principle,

$$\Delta l_z \, \Delta \phi \gtrsim \hbar \qquad (7.9)$$

where ϕ is the azimuthal angle defined in Figure 7.1. If we know l_z exactly ($\Delta l_z = 0$), then we have no knowledge at all of the angle ϕ—all values are equally probable. This is equivalent to saying that we know nothing at all about l_x and l_y; whenever one component of $\mathbf{l}$ is determined, the other components are completely undetermined. Figure 7.6 gives a pictorial representation of the behavior of the $\mathbf{l}$ vector. We think of the vector as revolving, or *precessing*, about the z axis, so rapidly that we can never see the motion, all the while keeping l_z constant. In this interpretation, you can see how we can have no knowledge of the x and y components of $\mathbf{l}$. You should also see why it *must* be true that $|\mathbf{l}| > l\hbar$. If it were possible to have $|\mathbf{l}| = l\hbar$, then when m_l had its maximum value ($m_l = +l$), we would have $l_z = m_l\hbar = l\hbar$. Since the length of the vector would be equal to its z component, it must lie along the z axis, so that $l_x = l_y = 0$. However, this simultaneous exact knowledge of all three components of $\mathbf{l}$ violates the form of the uncertainty principle as expressed in Equation (7.9), and therefore this situation is not permitted to occur.

The quantum numbers (n, l, m_l) that label the states of the hydrogen atom have, as we have seen, two interpretations. The quantum numbers are labels that arise from the mathematical procedures involved in solving the Schrödinger equation, but they also have a geometrical interpretation. In this section we are more concerned with the mathematical

Spatial quantization

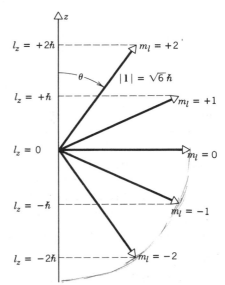

FIGURE 7.5 The orientations in space and z components of a vector with $l = 2$. There are five different possible orientations.

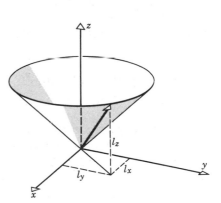

FIGURE 7.6 The vector $\mathbf{l}$ precesses rapidly about the z axis, so that l_z stays constant, but l_x and l_y are variable.

$$|\vec{l}| = \sqrt{\ell(\ell+1)}\ \hbar > \ell\hbar$$

$\vec{l}$ would not lie along the z axis

7.4 THE HYDROGEN ATOM WAVE FUNCTIONS

Table 7.1 Some Hydrogen Atom Wave Functions

n	l	m_l	$R(r)$	$\Theta(\theta)$	$\Phi(\phi)$
1	0	0	$\dfrac{2}{a_0^{3/2}}e^{-r/a_0}$	$\dfrac{1}{\sqrt{2}}$	$\dfrac{1}{\sqrt{2\pi}}$
2	0	0	$\dfrac{1}{(2a_0)^{3/2}}\left(2 - \dfrac{r}{a_0}\right)e^{-r/2a_0}$	$\dfrac{1}{\sqrt{2}}$	$\dfrac{1}{\sqrt{2\pi}}$
2	1	0	$\dfrac{1}{\sqrt{3}(2a_0)^{3/2}}\dfrac{r}{a_0}e^{-r/2a_0}$	$\sqrt{\dfrac{3}{2}}\cos\theta$	$\dfrac{1}{\sqrt{2\pi}}$
2	1	±1	$\dfrac{1}{\sqrt{3}(2a_0)^{3/2}}\dfrac{r}{a_0}e^{-r/2a_0}$	$\dfrac{\sqrt{3}}{2}\sin\theta$	$\dfrac{1}{\sqrt{2\pi}}e^{\pm i\phi}$

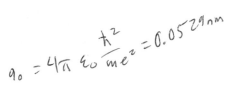

note: $m_l = 0 \Rightarrow \Phi(\phi) = $ constant

$l = 0 \Rightarrow$ constant

properties of the solutions, in which the quantum numbers are labels or indices of the different wave functions.

The wave functions $\psi(r, \theta, \phi)$ may be written as the product of three functions of a single variable

probability amplitude → *total wave function*
$$\psi_{n,l,m_l}(r, \theta, \phi) = R_{n,l}(r)\Theta_{l,m_l}(\theta)\Phi_m(\phi) \tag{7.10}$$

Different indices (n, l, m_l) give different wave functions, some of which are listed in Table 7.1 for a few values of the quantum numbers n, l, m_l. [The Bohr radius a_0 was defined in Equation (6.35).]

As we learned in Chapter 5, the probability of finding the electron is determined by the square of the wave function. More specifically, $|\psi(r, \theta, \phi)|^2$ gives the *probability density* (probability per unit volume) of finding the electron at the location (r, θ, ϕ). To compute the actual probability of finding the electron, we multiply the probability per unit volume by the volume element dV located at (r, θ, ϕ). In spherical polar coordinates (see Figure 7.7) the volume element is

$$dV = r^2 \sin\theta \, dr \, d\theta \, d\phi \tag{7.11}$$

of finding electron in the volume

and therefore the probability is

$$|\psi_{n,l,m_l}(r, \theta, \phi)|^2 \, dV = |R_{n,l}(r)|^2 \, |\Theta_{l,m_l}(\theta)|^2 \, |\Phi_{m_l}(\phi)|^2 \, r^2 \sin\theta \, dr \, d\theta \, d\phi \tag{7.12}$$

Using this expression for the probability *density*, we can calculate many features of the electron's spatial distribution. For example, we can find the radial probability $P(r) \, dr$ for locating the electron somewhere between r and $r + dr$ no matter what the values of θ and ϕ. To put it another way, we imagine a thin spherical shell of radius r and thickness dr, and ask what is the probability of the electron being within the volume of the shell. Since we are not interested in θ or ϕ, we integrate over all possible values of these variables:

$$P(r) \, dr = |R_{n,l}(r)|^2 \, r^2 \, dr \int_0^\pi |\Theta_{l,m_l}(\theta)|^2 \sin\theta \, d\theta \int_0^{2\pi} |\Phi_{m_l}(\phi)|^2 \, d\phi \tag{7.13}$$

The θ and ϕ integrals are each equal to unity, since each of the functions R, Θ, and Φ are individually *normalized*. Thus the *radial probability density* is

$$P(r) = r^2|R_{n,l}(r)|^2 \tag{7.14}$$

Figure 7.8 shows this function for several of the lowest levels of hydrogen.

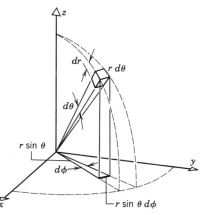

FIGURE 7.7 The volume element in spherical polar coordinates.

$a_0 = 4\pi\varepsilon_0\dfrac{\hbar^2}{me^2} = 0.0529$ nm

$r: 0 \rightarrow \infty$
$\theta: 0 \rightarrow \pi$
$\phi = 0 \rightarrow 2\pi$

Radial probability density

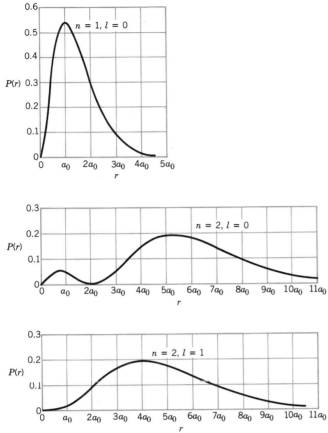

FIGURE 7.8 The radial probability density $P(r)$ for the three lowest states of hydrogen.

Prove that the most likely distance from the origin of an electron in the $n = 2$, $l = 1$ state is $4a_0$.   **EXAMPLE 7.3**

In the $n = 2$, $l = 1$ level, the radial probability density is   **Solution**

$$P(r) = r^2|R_{2,1}(r)|^2$$

$$= r^2 \frac{1}{24a_0^3}\frac{r^2}{a_0^2} e^{-r/a_0}$$

We wish to find where this function has its maximum; in the usual fashion, we take the first derivative of $P(r)$ and set it equal to zero:

$$\frac{dP(r)}{dr} = \frac{1}{24a_0^5}\frac{d}{dr}(r^4 e^{-r/a_0})$$

$$= \frac{1}{24a_0^5}\left[4r^3 e^{-r/a_0} + r^4\left(-\frac{1}{a_0}\right)e^{-r/a_0}\right] = 0$$

$$\frac{1}{24a_0^5}e^{-r/a_0}\left[4r^3 - \frac{r^4}{a_0}\right] = 0$$

The only solution that yields a maximum is $r = 4a_0$.

Notice that this is just the radius of the $n = 2$ level according to the Bohr model. As a general result, for each n, the most probable radius of the state with $l = n - 1$ (the maximum possible l for that n) is $n^2 a_0$, as given by the Bohr model. Other values of l for that same n will have different values for the radii at which $P(r)$ is maximum.

[handwritten margin notes:]
$r_n = n^2 a_0 \quad \text{for } n=2 \qquad r_2 = 4 a_0$
state w/ max $l = n-1$

An electron is in the $n = 1$, $l = 0$ state. What is the probability of finding the electron closer to the nucleus than the Bohr radius?   **EXAMPLE 7.4**

We are again interested in the radial probability density,   **Solution**

$$P(r)\, dr = r^2 |R_{1,0}(r)|^2\, dr$$

$$= r^2 \frac{4}{a_0^3} e^{-2r/a_0}\, dr$$

The total probability of finding the electron between $r = 0$ and $r = a_0$ is just

$$P = \int_0^{a_0} P(r)\, dr = \frac{4}{a_0^3} \int_0^{a_0} r^2 e^{-2r/a_0}\, dr$$

Letting $x = 2r/a_0$, we rewrite this as

$$P = \frac{1}{2} \int_0^2 x^2 e^{-x}\, dx$$

Evaluating the integral gives

$$P = 0.32$$

That is, 32 percent of the time the electron is closer than 1 Bohr radius to the nucleus.

Of the $n = 2$ states ($l = 0$ and $l = 1$), which one has the greater probability of being found inside the Bohr radius?   **EXAMPLE 7.5**

For the $n = 2$, $l = 0$ level, we have   **Solution**

$$P(r)\, dr = r^2 |R_{2,0}(r)|^2\, dr$$

$$= r^2 \frac{1}{8a_0^3} \left(2 - \frac{r}{a_0}\right)^2 e^{-r/a_0}\, dr$$

The total probability of finding the electron between $r = 0$ and $r = a_0$ is

$$P = \int_0^{a_0} P(r)\, dr = \frac{1}{8a_0^3} \int_0^{a_0} \left(4r^2 - \frac{4r^3}{a_0} + \frac{r^4}{a_0^2}\right) e^{-r/a_0}\, dr$$

and again, letting $x = r/a_0$,

$$P = \frac{1}{8} \int_0^1 (4x^2 - 4x^3 + x^4) e^{-x}\, dx$$

$$= \frac{1}{2} \int_0^1 x^2 e^{-x}\, dx - \frac{1}{2} \int_0^1 x^3 e^{-x}\, dx + \frac{1}{8} \int_0^1 x^4 e^{-x}\, dx$$

Evaluating the integrals,

$$P = 0.034.$$

For the $n = 2$, $l = 1$ level we have

$$P(r)\, dr = r^2 |R_{2,1}(r)|^2$$

$$= r^2 \frac{1}{24 a_0^3} \frac{r^2}{a_0^2} e^{-r/a_0}\, dr$$

The total probability between $r = 0$ and $r = a_0$ is

$$P = \int_0^{a_0} P(r)\, dr = \frac{1}{24 a_0^3} \int_0^{a_0} \frac{r^4}{a_0^2} e^{-r/a_0}\, dr$$

$$= \frac{1}{24} \int_0^1 x^4 e^{-x}\, dx$$

$$= 0.0037$$

The $n = 2$, $l = 1$ level has the smaller probability of being inside the Bohr radius. (In our simple geometrical picture, the $n = 2$, $l = 1$ orbit is circular with $r = 4a_0$ and is *never* inside the Bohr radius, while the $n = 2$, $l = 0$ orbit is a flattened ellipse that is *sometimes* within the Bohr radius.) This result, that the electrons in elliptical orbits ($l < n - 1$) spend more time close to the nucleus than do electrons in circular orbits ($l = n - 1$), will be important in the discussion of atomic structure in Chapter 8.

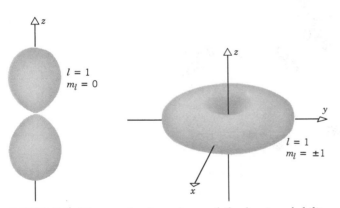

FIGURE 7.9 The angular dependence of the $l = 1$ probability density.

Figure 7.9 illustrates the angular dependence of the probability density. For $m_l = 0$, the electron is most likely to be found along the z axis; for $m_l = \pm 1$, it has the highest probability to be found in the xy plane. (Compare this with Figure 7.4 for the $m_l = 0$ and $m_l = l$ Bohr orbits.)

Of course, it is not possible to observe directly the motion of the electron in the hydrogen atom. All we can ever observe is the "smeared-out" distribution of electronic charge, with a spatial distribution given by the probability $|\psi|^2$. Figure 7.10 shows some representative examples of $|\psi|^2$. These probability distributions have important consequences for the joining of atoms in molecules, as we discuss in Chapter 13.

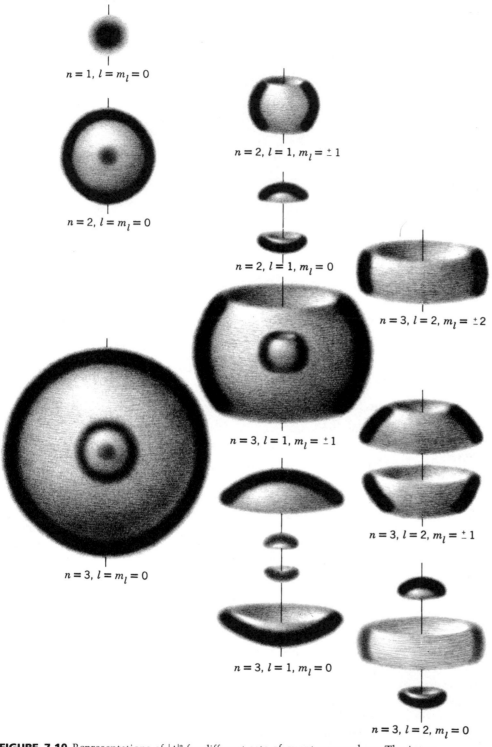

$n = 1, l = m_l = 0$

$n = 2, l = m_l = 0$

$n = 2, l = 1, m_l = \pm 1$

$n = 2, l = 1, m_l = 0$

$n = 3, l = 2, m_l = \pm 2$

$n = 3, l = m_l = 0$

$n = 3, l = 1, m_l = \pm 1$

$n = 3, l = 2, m_l = \pm 1$

$n = 3, l = 1, m_l = 0$

$n = 3, l = 2, m_l = 0$

FIGURE 7.10 Representations of $|\psi|^2$ for different sets of quantum numbers. The intensity of each diagram at any point is proportional to the probability of locating an electron in a small volume element at that point. [*Source.* R. Eisberg and R. Resnick, *Quantum Physics of Atoms, Molecules, Solids, Nuclei, and Particles* (New York, Wiley, 1974)].

Let us imagine a collection of identical, *noninteracting* small permanent magnets randomly arranged on a table, as shown in Figure 7.11. Each magnet has a strength that we characterize by its *magnetic dipole moment* μ, a vector that points from the S pole to the N pole of the magnet. An externally applied magnetic field **B** exerts a torque on each of the magnetic moments and tries to rotate them so that they line up with the field. Assuming there is a small frictional force between the magnets and the table, the field will not be completely successful in lining up all of the magnets; but the stronger we make the applied field, the more successful we are in lining up the magnets. As we increase the field, the magnets move smoothly and continuously toward alignment with the field. This is an example of the behavior of a classical system of magnets.

Figure 7.12 shows the corresponding experiment done with a collection of hydrogen atoms. Let us assume that each atom is in a state with $n = 2$ and $l = 1$. Using the Bohr model, the circulating electron looks like a circular loop carrying a current $i = dq/dt = q/T$, where q is the charge of the electron $(-e)$ and T is the time it takes the electron to complete one circuit of the loop. (If the electron moves with speed $v = p/m$ around a loop of radius r, then $T = 2\pi r/v = 2\pi rm/p$.) The magnetic moment of the loop is just the product of the current and the area of the loop, so

$$\mu = iA = \frac{q}{(2\pi rm/p)} \, \pi r^2 = \frac{q}{2m} \, rp = \frac{q}{2m} \, l \qquad (7.15)$$

since $l = rp$. Writing μ and l as vectors, and putting $-e$ for the electronic charge,

$$\boldsymbol{\mu} = \frac{-e}{2m} \, \mathbf{l} \qquad (7.16)$$

The negative sign, which is present because the electron has a negative charge, indicates that the vectors **l** and μ point in opposite directions. Before we turn on the field, the magnetic moments are randomly oriented. However, compare the random orientation of Figure 7.12 with that of the classical magnetic moments illustrated in Figure 7.11. Only three possible orientations of **l**, and therefore of μ, are possible; since $l = 1$, we can have $m_l = -1, 0,$ or $+1$. In the random situation, there are equal numbers of atoms with each m_l value. As we increase the strength of the field, the magnetic moments once again try to line up with the field, but they can't move continuously as the classical magnetic moments did—all the atom can do is jump from one m_l value to another.

Before we consider further the behavior of μ, we discuss the similar behavior of an *electric dipole*, which consists of two equal and opposite charges q separated by a distance r. The electric dipole moment **p** has magnitude qr and points from the negative charge to the positive charge. As shown in Figure 7.13, in a uniform *electric* field, the dipole experiences a torque that tends to rotate it into alignment with **E**. Suppose now that the field is not uniform—the field acting on the positive charge is *not* equal to the field that acts on the negative charge, as in Figure 7.14. There is still a net *torque* that tends to rotate the dipole, but there is also a net *force* that tends to move the dipole. Consider the two dipoles shown in Figure 7.15. Suppose the electric field near the bottom

7.5 INTRINSIC SPIN

Orbital magnetic moment

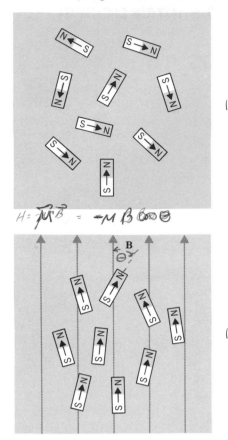

FIGURE 7.11 (*a*) A collection of small, noninteracting permanent magnets. The arrow shows the direction of the magnetic moments. (*b*) An applied magnetic field **B** rotates the magnetic moments into alignment with the field.

$$H = \vec{\mu} \cdot \vec{B} = -\mu B \cos\theta$$

$$N = \frac{-e}{2m} \vec{l}$$

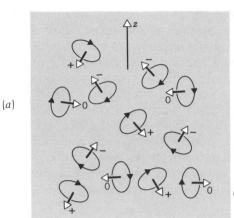

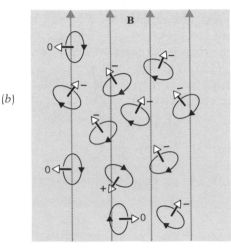

FIGURE 7.12 (*a*) A collection of non-interacting atomic magnetic moments. The arrow gives the direction of the magnetic moment. The *z* components of **l** are indicated as +, 0, and − for +1, 0, and −1. (Because of the minus sign in Equation (7.16), **μ** and **l** are in opposite directions.) Before the field is applied, there are equal numbers of atoms with each *z* component. (*b*) When the field is applied, the moments rotate around into alignment with the field, so that there are now many more atoms with $m_l = -1$.

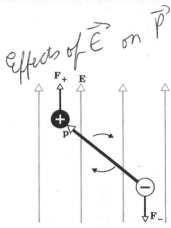

(1) FIGURE 7.13 An electric dipole in a uniform electric field **E**. A force F_+ on the positive charge and a force F_- on the negative charge produce a net torque on the dipole.

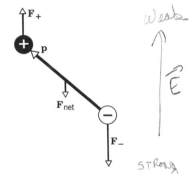

Weak

$\vec{E}$

Strong

(2) FIGURE 7.14 An electric dipole in a nonuniform field. The field decreases from the bottom to the top of the figure, so that the force F_- is greater than the force F_+. There is a net downward force on the dipole.

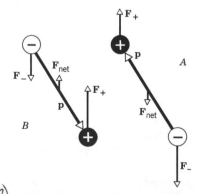

(3) FIGURE 7.15 Two dipoles with oppositely directed moments in a nonuniform field. The dipoles will move in opposite directions under the influence of the net force.

$$F = \rho E$$

effects of B on

of the figure is greater in magnitude than the field near the top, and further suppose that the field points upward. Dipole A, with its dipole moment **p** inclined upward, experiences a net downward force, F_{net}, since the (downward) force F_- on the negative charge is greater than the upward force F_+ on the positive charge. On the other hand, dipole B, with its dipole moment **p** inclined downward, experiences a net upward force, since now F_+ is greater than F_-. We can state this result in another way that will be more applicable to our discussion of *magnetic* dipole mo-

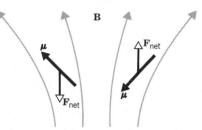

FIGURE 7.16 Two magnetic dipoles in a nonuniform magnetic field. Oppositely directed dipoles experience net forces in opposite directions.

ments. Let the field direction define the z axis. Then all dipoles with $p_z > 0$ (as dipole A) experience a net negative force and move in the negative z direction, and all dipoles with $p_z < 0$ (as dipole B) experience a net positive force and move in the positive z direction.

A magnetic dipole moment $\boldsymbol{\mu}$ behaves in an identical way. (In fact, if we imagine fictitious N and S poles, the behavior of a magnetic moment would be described by illustrations similar to Figures 7.13, 7.14, and 7.15.) A nonuniform *magnetic* field not only rotates the *magnetic* moments, but also gives an unbalanced force that causes a displacement. Figure 7.16 illustrates the behavior of magnetic moments having different orientations in a nonuniform field. The two different orientations give net forces in opposite directions.

Imagine the following experiment, illustrated schematically in Figure 7.17. A beam of hydrogen atoms is prepared in the $n = 2, l = 1$ state. The beam consists of equal numbers of atoms in the $m_l = -1, 0,$ and $+1$ states. (We will assume we can do the experiment so quickly that the $n = 2$ state doesn't decay to the $n = 1$ state.) The beam passes through a region in which there is a nonuniform magnetic field. The atoms with $m_l = +1$ experience a net upward force and are deflected upward, while the atoms with $m_l = -1$ are deflected downward. The atoms with $m_l = 0$ are undeflected. After passing through the field let the beam strike a screen where it makes a visible dot. When the field is off, we would expect to see one dot in the center of the screen, since there is no deflection at all. When the field is on, we would see three dots—one in the center (corresponding to $m_l = 0$), one above the center ($m_l = +1$) and one below the center ($m_l = -1$). If the atom were in the ground state ($l = 0$), we would expect to see one dot whether the field was off or on (recall that a $m_l = 0$ atom is not deflected). If we had prepared the beam in a state with $l = 2$, we would see five dots with the beam on. *The number of dots that appears is just the number of different* m_l *values,* and with a bit of thought, you can convince yourself that this is just $2l + 1$. Since l has possible values $0, 1, 2, 3, \ldots$, it follows that $2l + 1$ has the values $1, 3, 5, 7, \ldots$; that is, we should always see an *odd*

$$\text{current} = \frac{dq}{dt} = \Delta H = \frac{Q}{T}$$

$$M_z = -\mu_B m_l$$

$$l = 0 \implies m_l = 0$$
$$|\vec{l}| = \sqrt{l(l+1)}$$
$$|\vec{l}| = 0$$

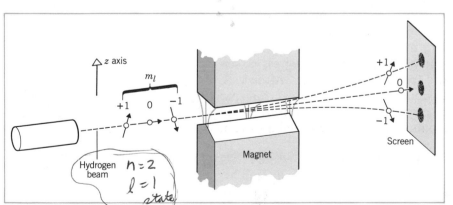

FIGURE 7.17 Schematic diagram of Stern-Gerlach experiment. A beam of atoms enters a region where there is a nonuniform magnetic field. Atoms with their magnetic dipole moments in opposite directions experience forces in opposite directions.

expect $B \neq 0 \implies 3$ dots

number of dots on the screen. However, if we were actually to perform the experiment with hydrogen in the $l = 1$ state, we would find not three but *four* dots on the screen! Even more confusing, when we do the experiment with hydrogen in the $l = 0$ state, we find not one but *two* dots on the screen, one representing an upward deflection and one a downward deflection! In the $l = 0$ state, the vector **l** has length zero, and so we expect that there is *no magnetic moment* for the magnetic field to deflect. We observe this not to be true—even when $l = 0$, the atom still has a magnetic moment, in contradiction to Equation (7.16).

This experiment was done by O. Stern and W. Gerlach in 1921. They used a beam of silver atoms; although the electronic structure of silver is much more complicated than that of hydrogen (as we will discuss in Chapter 8), the same basic principle applies—the silver must have $l = 0, 1, 2, 3, \ldots$, and so an *odd number* of dots is expected to appear on the screen. In fact, they observed the beam to split into *two* components, producing two dots on the screen.

The appearance of dots, rather than a smeared-out band, was the first conclusive evidence of *space quantization*; classical magnetic moments would have all possible orientations and would make a smeared-out pattern on the screen, but the observation of a number of discrete spots on the screen means that the atomic magnetic moments can only take certain discrete orientations in space. These correspond to the discrete orientations of the magnetic moment (or, equivalently, of the angular momentum).

If we believe our previous discussion of the number of spots on the screen, then for two spots to appear, we must have $2l + 1 = 2$, or $l = \frac{1}{2}$. But this is impossible, because the mathematics of the Schrödinger equation restricts l to integer values $0, 1, 2, \ldots$. The solution to this dilemma requires the introduction of *intrinsic angular momentum*.

Associated with the motion of the Earth are two types of angular momentum—the *orbital angular momentum* of the motion about the sun and the *intrinsic angular momentum* of the rotation of the Earth about its axis. Similarly the electron has an *orbital angular momentum* **l**, which characterizes the motion of the electron about the nucleus, and an *intrinsic angular momentum* **s**, which behaves as if the electron were spinning about its axis. For this reason, **s** is usually called the *intrinsic spin*. (The concept of the electron as a tiny ball of charge spinning on its axis is a useful one, just like the Bohr model. Unfortunately it is not a correct concept. However, it frequently happens in the progress of science that the right idea is introduced for the wrong reason. After S. A. Goudsmit and G. E. Uhlenbeck introduced the concept of electron spin in 1925, P. A. M. Dirac showed that a proper *relativistic* quantum theory of the electron gives the electron spin directly as an additional quantum number.)

In order to explain the result of the Stern-Gerlach experiment, we must assign to the electron an intrinsic spin s of $\frac{1}{2}$. The intrinsic spin behaves much like the orbital angular momentum; there is the quantum number s (which we can regard as a label arising from the mathematics), the angular momentum vector **s** (with a length $|\mathbf{s}| = \sqrt{s(s + 1)}\hbar$), an associated magnetic moment $\boldsymbol{\mu}_s = (-e/m)\mathbf{s}$, a z component $s_z = m_s\hbar$, and a spin magnetic quantum number m_s, which can have the values $+\frac{1}{2}$ or $-\frac{1}{2}$. Figure 7.18 illustrates the vector properties of **s**.

Stern-Gerlach experiment

(handwritten marginal notes:)
no orbital magnetic moment $(\ell = \mathbf{0})$ in ground state
expect 1 dot but got two
(or spin)
∴ intrinsic angular momentum

$2S + 1 = 2 \Rightarrow \boxed{S = \frac{1}{2}}$

orb ang mom ℓ 0, 1, 2, 3

intrinsic ang Mom $S = \frac{1}{2}$

$|S| = \sqrt{(S^2 + S)}\,\hbar = \sqrt{\frac{3}{4}}\,\hbar$

Intrinsic angular momentum (spin)

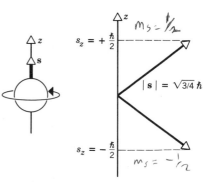

FIGURE 7.18 The spin angular momentum of an electron and the spatial orientation of the spin angular momentum vector.

We previously described all of the possible electronic states in hydrogen by three quantum numbers (n, l, m_l), but as we have seen, a fourth property of the electron, the intrinsic angular momentum or *spin*, will require the introduction of a fourth quantum number. We don't need to specify the spin s, since it is always $\frac{1}{2}$ (we regard it as a fundamental property of the electron, like its electric charge or its rest mass), but we do need to specify the value of the quantum number m_s ($+\frac{1}{2}$ or $-\frac{1}{2}$), which tells us about the z component of **s**. Thus the complete description of an electronic state requires the four quantum numbers (n, l, m_l, m_s).

For example, the ground state of hydrogen was previously labeled as $(n, l, m_l) = (1, 0, 0)$. With the addition of m_s, this would become either $(1, 0, 0, +\frac{1}{2})$ or $(1, 0, 0, -\frac{1}{2})$. The degeneracy of the ground state is now 2. The first excited state would have eight possible labels: $(2, 0, 0, +\frac{1}{2})$, $(2, 0, 0, -\frac{1}{2})$, $(2, 1, 1, +\frac{1}{2})$, $(2, 1, 1, -\frac{1}{2})$, $(2, 1, 0, +\frac{1}{2})$, $(2, 1, 0, -\frac{1}{2})$, $(2, 1, -1, +\frac{1}{2})$, and $(2, 1, -1, -\frac{1}{2})$. Since there are now two possible labels for each previous single label (each n, l, m_l becomes $n, l, m_l, +\frac{1}{2}$ and $n, l, m_l, -\frac{1}{2}$), the degeneracy of each level is $2n^2$.

It is only when we place an atom in a magnetic field that the distinction between different m_l or m_s values becomes important. For many applications, the values of m_l and m_s are of no importance, and it is cumbersome to write them each time we wish to refer to a certain level of an atom. We will therefore use a different notation, known as *spectroscopic notation*, to label the levels. In this system we use letters to stand for the different l values: for $l = 0$, we use the letter s (do not confuse this with the quantum number s), for $l = 1$, we use the letter p, and so on. The complete notation is as follows:

value of l	0	1	2	3	4	5	6
designation	s	p	d	f	g	h	i

(The first four letters stand for sharp, principal, diffuse, and fundamental, which were terms used to describe atomic spectra before atomic theory was developed.) In spectroscopic notation, the ground state of hydrogen would be labeled $1s$, where the value $n = 1$ is specified before the s. Figure 7.19 illustrates the labeling of the hydrogen atom levels in this notation.

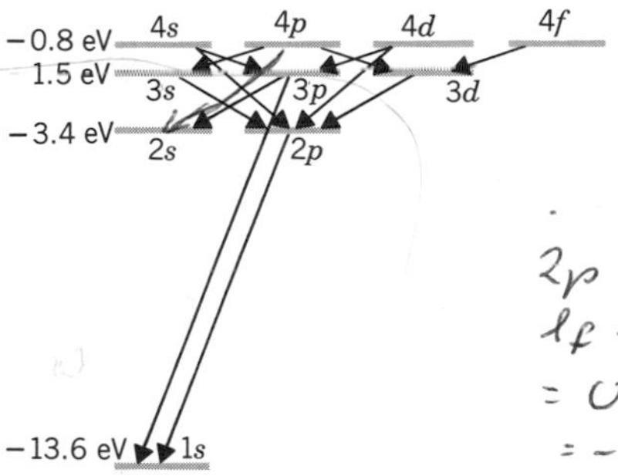

Also shown on Figure 7.19 are lines representing some different photons that can be emitted when the atom makes a transition from one state to a lower state. These lines indicate an additional feature of the level diagram, known as a *selection rule*. Not all transitions are allowed to occur. By solving the Schrödinger equation and using the solutions to compute *transition probabilities*, we find that the transitions most likely to occur are those that change l by one unit, and thus the selection rule is

$$\Delta l = \pm 1 \tag{7.17}$$

The $3s$ level cannot emit a photon in a transition to the $2s$ level ($\Delta l = 0$), but rather must go to the $2p$ level ($\Delta l = 1$). There is no selection rule for n, so that the $3p$ level could go to $2s$ or $1s$ (but not $2p$).

Spectroscopic notation

FIGURE 7.19 A partial energy level diagram of hydrogen, showing the spectroscopic notation of the levels and the transitions that satisfy the $\Delta l = \pm 1$ selection rule.

Selection rule

Let us consider for the moment a hypothetical (and less interesting) world in which the electron had no spin, and therefore no spin magnetic moment. Suppose we prepared a hydrogen atom in a $2p$ ($l = 1$) level and placed it in an external uniform magnetic field **B** (supplied by a laboratory electromagnet, for example). The magnetic moment $\boldsymbol{\mu}$ associated with the orbital angular momentum then interacts with the field, and the energy associated with this interaction is

$$V = -\boldsymbol{\mu}\cdot\mathbf{B} \tag{7.18}$$

That is, magnetic moments aligned in the direction of the field have less energy than those aligned oppositely to the field. Let us assume that the field is in the z direction. Using Equation (7.16) for the magnetic moment, we have

$$V = -\left(-\frac{e}{2m}\mathbf{l}\right)\cdot\mathbf{B}$$
$$= \frac{e}{2m}l_z B \tag{7.19}$$

where the latter result follows from choosing **B** in the z direction. Since l_z is just $m_l\hbar$,

$$V = m_l\left(\frac{e\hbar}{2m}\right)B$$
$$= m_l\mu_B B \tag{7.20}$$

The quantity $e\hbar/2m$ is known as the *Bohr magneton,* with the symbol μ_B and a value of 9.27×10^{-24} J/T. In the absence of a magnetic field, the $2p$ level has a certain energy $E_0(-3.4$ eV$)$. When the field is turned on, the energy becomes $E_0 + V = E_0 + m_l\mu_B B$; that is, there are now three different possible energies for the level, depending on the value of m_l. Figure 7.20 illustrates this situation.

Now suppose the atom emits a photon in a transition to the ground state (1s). In the absence of the magnetic field, a single photon is emitted with an energy of 10.2 eV and a corresponding wavelength of 122 nm. When the magnetic field is present, three photons can be emitted, with energies of 10.2 eV $+ \mu_B B$, 10.2 eV, and 10.2 eV $- \mu_B B$. The photon wavelength can be found according to $E = hc/\lambda$. Let us examine how a small change in energy ΔE (where ΔE is just $\mu_B B$) affects the wavelength. Differentiating, we have

$$dE = \frac{-hc}{\lambda^2}\,d\lambda \tag{7.21}$$

and taking small differentials with absolute magnitudes

$$\Delta\lambda = \frac{\lambda^2}{hc}\,\Delta E \tag{7.22}$$

Figure 7.21 illustrates the three transitions, and shows an example of the result of a measurement of the emitted wavelengths.

In analyzing transitions between different m_l states, often we need to use a second *selection rule:* the only transitions that occur are those that change m_l by 0, +1, or −1. Changes in m_l of two or more are not permitted.

*7.7 THE ZEEMAN EFFECT

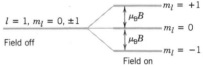

FIGURE 7.20 The Zeeman splitting of an $l = 1$ level in an external magnetic field. (The effects of the electron's spin angular momentum are ignored.) The energy in a magnetic field is different for different values of m_l.

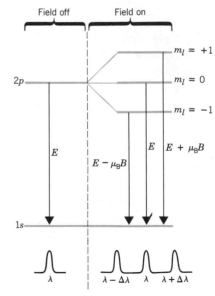

FIGURE 7.21 The normal Zeeman effect. When the field is turned on, the single wavelength λ becomes three separate wavelengths.

Compute the change in wavelength of the $2p - 1s$ photon when a hydrogen atom is placed in a magnetic field of 2.00 T.

EXAMPLE 7.6

The energy of the photon is

$$E = \frac{-13.6}{4} + 13.6 = 10.2 \text{ eV}$$

and its wavelength is, therefore,

$$\lambda = \frac{hc}{E} = \frac{1240 \text{ eV·nm}}{10.2 \text{ eV}} = 122 \text{ nm}$$

The energy change ΔE of the levels is

$$\Delta E = \mu_B B = (9.27 \times 10^{-24} \text{ J/T})(2 \text{ T})$$

$$= 18.5 \times 10^{-24} \text{ J}$$

$$= 11.6 \times 10^{-5} \text{ eV}$$

and so, from Equation (7.22),

$$\Delta\lambda = \frac{\lambda^2}{hc} \Delta E = \frac{(122 \text{ nm})^2}{1240 \text{ eV·nm}} 11.6 \times 10^{-5} \text{ eV}$$

$$= 0.00139 \text{ nm}$$

Even for such a relatively large laboratory magnetic field as 2 T, the change in wavelength is very small.

The experiment we have just considered is an example of the *Zeeman effect*—the splitting of a single wavelength into several different wavelengths when a magnetic field is applied. In the *normal Zeeman effect* a single spectral line splits into three components; this occurs only in atoms without spin. (All electrons of course have spin, unlike the hypothetical spinless electrons we considered; however, in certain atoms with several electrons, the spins can pair off and cancel, so that the atom behaves like a spinless one.) In our world, where spin *is* present, we must consider not only the effect of the orbital magnetic moment but also the spin magnetic moment. The resulting pattern of level splittings is more complicated, and spectral lines may split into more than three components. This case is known as the *anomalous Zeeman effect*.

In our discussion of the hydrogen spectrum in Chapter 6, it was mentioned that a careful inspection of the emission lines showed that many of them are in fact not single lines but very closely spaced combinations of two lines. In this section we examine the origin of that effect, known as *fine structure*.

For many centuries, it was believed that the Earth was the center of the solar system and that the sun moved around the Earth; today we know that the physical laws that govern the behavior of the solar system can best be interpreted if we choose a frame of reference in which the Earth orbits around the sun. There is no frame of reference that is "more

*7.8 FINE STRUCTURE

correct" than any other—the choice of a reference frame is merely a matter of convenience.

In this calculation it is more convenient for us to examine the hydrogen atom from the electron's frame of reference, in which the nucleus *appears* to travel around the electron, just as the sun *appears* to travel around the Earth. For convenience, we treat this problem in the context of the Bohr model; however, the calculation is correct in the quantum physics sense also.

Figure 7.22 shows the atom from the frame of reference of the electron. The motion of the proton in a circular orbit of radius r appears like a current loop, which causes a magnetic field **B** at the electron. This magnetic field interacts with the spin magnetic moment of the electron, $\boldsymbol{\mu}_s = (-e/m)\mathbf{s}$. The interaction energy of the magnetic moment $\boldsymbol{\mu}_s$ in a magnetic field is

$$V = -\boldsymbol{\mu}_s \cdot \mathbf{B} \tag{7.23}$$

That is, when $\boldsymbol{\mu}_s$ and **B** are parallel, the energy $(V = -\mu_s B)$ is lower than it is when $\boldsymbol{\mu}_s$ and **B** are antiparallel $(V = +\mu_s B)$. Let us define the z direction to be the direction of **B**; with $\boldsymbol{\mu}_s = (-e/m)\mathbf{s}$, we have

$$V = -\boldsymbol{\mu}_s \cdot \mathbf{B} = \frac{e}{m}\,\mathbf{s} \cdot \mathbf{B}$$

$$V = \frac{e}{m}\,s_z B \tag{7.24}$$

s_z can have only two possible values: $+\frac{1}{2}\hbar$ or $-\frac{1}{2}\hbar$. The states with $s_z = +\frac{1}{2}\hbar$ are shifted upward in energy by an amount $V = e\hbar B/2m = \mu_B B$; states with $s_z = -\frac{1}{2}\hbar$ are shifted down in energy by an equal amount.

At this point, the result looks rather similar to that of our previous discussion of the Zeeman effect, but it is important to note one significant difference: the magnetic field B in this case is *not* a field in the laboratory that can be turned on or off; it is, instead, a field produced by the apparent motion of the proton, that is *always* present.

Figure 7.22 shows an electron moving about the nucleus with **s** parallel to **l**. In the electron's frame of reference, the nucleus appears to

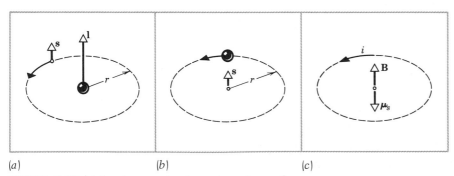

(a)   (b)   (c)

FIGURE 7.22 (a) An electron circulates about the nucleus with orbital angular momentum **l**. The spin of the electron is parallel to **l**. (b) From the point of view of the electron, the proton circulates as shown. (c) The apparently circulating proton is represented by the current i and causes a magnetic field **B** at the electron. The spin magnetic moment of the electron is opposite to its spin angular momentum.

move as shown. Since the nucleus has a positive charge, the equivalent current loop is a positive current, and thus the **B** field has the direction shown, determined using the right-hand rule. Since $\boldsymbol{\mu}_s = -(e/m)\mathbf{s}$, $\boldsymbol{\mu}_s$ and **s** are antiparallel, and so for this case **B** and $\boldsymbol{\mu}_s$ are antiparallel. We previously calculated that this state was shifted *upward,* and so when **s** and **l** are parallel, the energy of the state is shifted upward by an amount $\mu_B B$. We therefore expect to find each level split into two states, a higher energy state with **l** and **s** parallel, and a lower energy state with **l** and **s** antiparallel. Figure 7.23 is a representation of this.

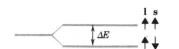

FIGURE 7.23 The fine-structure splitting in hydrogen. The state with **l** and **s** parallel is slightly higher in energy than the state with **l** and **s** antiparallel.

We can estimate the magnitude of this energy splitting again using the Bohr model. A circular loop of radius r carrying current i establishes at its center a magnetic field

$$B = \frac{\mu_0 i}{2r} \tag{7.25}$$

The current i is just the charge carried around the loop ($+e$ in this case) divided by the time T for one orbit. The time for one orbit is the distance traveled ($2\pi r$) divided by the speed v.

$$B = \frac{\mu_0}{2r}\frac{e}{T} = \frac{\mu_0}{2r}\frac{ev}{2\pi r} \tag{7.26}$$

Since the parallel state is shifted upward and the antiparallel state is shifted downward, the spacing ΔE between the states is

$$\Delta E = 2V = 2\mu_B B$$

$$= \frac{\mu_0 ev}{2\pi r^2}\mu_B \tag{7.27}$$

Since $l = mvr = n\hbar$, (remember, we are using the Bohr model), the speed v is $n\hbar/mr$, so

$$\Delta E = \frac{\mu_0 e^2 \hbar^2 n}{4\pi m^2 r^3} \tag{7.28}$$

In Chapter 6, we found

$$r_n = \frac{4\pi\varepsilon_0 \hbar^2}{me^2}n^2 \tag{7.29}$$

so

$$\Delta E = \frac{\mu_0 e^2 \hbar^2 n}{4\pi m^2}\frac{m^3 e^6}{(4\pi\varepsilon_0)^3\hbar^6}\frac{1}{n^6}$$

$$= \frac{\mu_0 m e^8}{256\pi^4\varepsilon_0^3\hbar^4}\frac{1}{n^5}. \tag{7.30}$$

We can rewrite this in a somewhat simpler form by recalling that $c^2 = 1/\varepsilon_0\mu_0$ and (from Chapter 1)

$$\alpha = \frac{e^2}{4\pi\varepsilon_0\hbar c} \tag{7.31}$$

which gives

$$\Delta E = (mc^2)\alpha^4\frac{1}{n^5} \tag{7.32}$$

Estimate of fine-structure splitting in hydrogen

α is known as the *fine structure constant* and is a dimensionless constant with a value very nearly equal to $\frac{1}{137}$. For the $n = 2$ state of hydrogen, we expect the state with **l** and **s** parallel to differ in energy from the state with **l** and **s** antiparallel by $(0.511 \text{ MeV}) \times \left(\frac{1}{137}\right)^4 \times \left(\frac{1}{32}\right) = 4.53 \times 10^{-5}$ eV. We can compare this estimate with the experimental value, based on the observed splitting of the first line of the Lyman series, which gives 4.54×10^{-5} eV. We see that in spite of the assumptions we have made, our use of the Bohr model, and our failure to use the hydrogen wave functions to do this calculation, the agreement with the experimental value is remarkably good. (In fact, the agreement is so good as to be embarrassing, for we neglected to consider the important *relativistic* effect of the motion of the electron, which contributes to the fine structure about equally as the *spin-orbit* interaction discussed in this section. We really should regard this calculation as an *order-of-magnitude* estimate, which happens by chance to give the correct numerical value.)

SUGGESTIONS FOR FURTHER READING

A more detailed treatment of the hydrogen atom, especially of the fine structure, is in the following:

J. Norwood, *Twentieth Century Physics* (Englewood Cliffs, Prentice-Hall, 1976).

An excellent and detailed full-scale treatment of the solutions of the Schrödinger equation for the hydrogen atom is Chapter V of the following:

L. Pauling and E. B. Wilson, *Introduction to Quantum Mechanics* (New York, McGraw-Hill, 1935).

Representing the three-dimensional probability distributions on a two-dimensional paper is a great challenge for illustrators, and the interpretation of such illustrations is often a similar challenge for students. It might be helpful to look at some other representations:

N. Ashby and S. C. Miller, *Principles of Modern Physics* (San Francisco, Holden-Day, 1970).
R. B. Leighton, *Principles of Modern Physics* (New York, McGraw-Hill, 1959).

A remarkable set of computer-drawn probability distributions can be found in the following:

D. Kleppner, M. G. Littman, and M. L. Zimmerman, "Highly Excited Atoms," *Scientific American 244*, 130 (May 1981).

QUESTIONS

1. How does the quantum-mechanical interpretation of the hydrogen atom differ from the Bohr model?
2. How does a quantized angular momentum vector differ from a classical angular momentum vector?
3. What are the meanings of the quantum numbers n, l, m_l according to (a) the quantum-mechanical calculation; (b) the vector model; (c) the Bohr (orbital) model?
4. List the dynamical quantities that are constant for a specific choice of n and l. List the dynamical quantities that are *not* constant. Compare these lists with the Bohr model.

5. How does the orbital angular momentum differ between the Bohr model and the quantum-mechanical calculation?

6. What does it mean that **l** precesses about the z axis? Can we observe the precession?

7. At the end of Section 7.2 the statement is made that "a fixed orbital plane violates the uncertainty principle." Explain this statement.

8. In the Bohr model, we calculated the total energy from the potential energy and kinetic energy for each orbit. In the quantum-mechanical calculation, is the potential energy constant for any set of quantum numbers? Is the kinetic energy? Is the total energy?

9. What is meant by the term *spatial quantization*? Is space really quantized?

10. In the questions to Chapter 6, it was pointed out that yet another deficiency of the Bohr model is the problem of angular momentum conservation in transitions between levels. Discuss this problem in relation to the quantum-mechanical angular momentum properties of the atom, especially the selection rule (7.17). The photon can be considered to carry angular momentum $\hbar$.

11. The $2s$ electron has a greater probability to be close to the nucleus than the $2p$ electron (see Example 7.5) and also a greater probability to be farther away (see Problem 11). How is this possible?

12. The probability density $\psi^*\psi$ does not depend on ϕ for the wave functions listed in Table 7.1. What is the significance of this?

13. How would the wave functions of Table 7.1 change if the nuclear charge were Ze instead of e? (Recall how we made the same change in the Bohr model in Section 6.5.) What effect would this have on the radial probability densities $P(r)$?

14. Can a hydrogen atom in its ground state absorb a photon (of the proper energy) and end up in the $3d$ state?

15. Is it *correct* to think of the electron as a tiny ball of charge spinning on its axis? Is it *useful*? Is this situation similar to using the Bohr model to represent the electron's orbital motion?

16. What are the differences between Zeeman splitting and fine-structure splitting?

17. How would the calculated fine structure be different in an atom with a single electron and a nuclear charge of Ze?

18. Does the fine structure, as we have calculated it, have any effect on the $n = 1$ level?

19. How would (a) the Zeeman effect and (b) the fine structure be different in a muonic hydrogen atom? (See Problems 33 and 34 in Chapter 6.) The muon has the same spin and spin magnetic moment as the electron, but is 207 times as massive.

20. Even though our calculation of the fine structure was based on a very simplified model, it does yield a result similar to the more correct calculation: the fine-structure splitting decreases as we go to higher excited states. Give at least two qualitative reasons for this.

PROBLEMS

1. List the 16 possible sets of quantum numbers of the $n = 4$ level of hydrogen without the inclusion of electron spin (as in Figure 7.2).

2. (a) What are the possible values of l for $n = 6$? (b) What are the possible values of m_l for $l = 6$? (c) What is the smallest possible value of n for which l can be 4? (d) What is the smallest possible l that can have a z component of $4\hbar$?

3. An electron is in the $n = 4, l = 3$ state of hydrogen. (a) What is the length of the electron's angular momentum vector? (b) How many different possible z components can the angular momentum vector have? List the possible z components. (c) What are the values of the angle that the $\mathbf{l}$ vector makes with the z axis? (d) Would your answers to (a), (b), or (c) change if the principal quantum number n were 5 instead of 4?

4. What angles does the $\mathbf{l}$ vector make with the z axis when $l = 2$?

5. Show that the (1, 0, 0) and (2, 0, 0) wave functions listed in Table 7.1 are properly normalized.

6. Show by direct substitution that the $n = 2, l = 0, m_l = 0$ and $n = 2, l = 1$, $m_l = 0$ wave functions of Table 7.1 are both solutions of Equation (7.3) corresponding to the energy of the first excited state of hydrogen.

7. Show by direct substitution that the wave function corresponding to $n = 1$, $l = 0, m_l = 0$ is a solution of Equation (7.3) corresponding to the ground-state energy of hydrogen.

8. Show that the radial probability density of the $1s$ level has its maximum value at $r = a_0$.

9. Find the values of the radius where the $n = 2, l = 0$ radial probability density has its maximum values.

10. What is the probability of finding a $n = 2, l = 1$ electron between a_0 and $2a_0$?

11. Find the probabilities for the $n = 2, l = 0$ and $n = 2, l = 1$ states to be further than $5a_0$ from the nucleus. Which has the greater probability to be far from the nucleus?

12. The mean or average value of the radius r can be found according to $r_{av} = \int_0^\infty rP(r)\, dr$. Show that the mean value of r for the $1s$ state of hydrogen is $\frac{3}{2}a_0$. Why is this greater than the Bohr radius?

13. Find the value of r_{av} (see Problem 12) for the $2s$ and $2p$ levels.

14. The mean or average value of the potential energy of the electron in a hydrogen atom can be found from $V_{av} = \int_0^\infty V(r)\, P(r)\, dr$. Find V_{av} in the $1s$ state and compare with the potential energy computed with the Bohr model when $n = 1$.

15. (a) Show that the *transverse force* on an atom in a Stern-Gerlach experiment is $F_z = \mu_z(dB_z/dz)$. (b) Suppose the source of atoms were an oven of temperature 1000 K. Assume the magnetic field gradient to be 10 T/m, and take the length of the magnetic field region and the field-free region between magnet and screen to be 1 m each. Make any other assumptions you may need and estimate the separation of the dots observed on the screen.

16. List the excited states (in spectroscopic notation) to which the $4p$ state can make downward transitions.

17. A hydrogen atom is in an excited $5g$ state, from which it makes a series of transitions, ending in the $1s$ state. Show, on a diagram similar to Figure 7.19, the sequence of transitions that can occur. (b) Repeat part (a) if the atom begins in the $5d$ state.

18. Consider the normal Zeeman effect applied to the $3d$ to $2p$ transition. (a) Sketch an energy-level diagram that shows the splitting of the $3d$ and $2p$ levels in an external magnetic field. Indicate all possible transitions from each m_l state of the $3d$ level to each m_l state of the $2p$ level. (b) Which transitions satisfy the $\Delta m_l = \pm 1$ or 0 selection rule? (c) Show that there are only three different transition energies emitted.

19. A collection of hydrogen atoms is placed in a magnetic field of 3.50 T. Ignoring the effects of electron spin, find the wavelengths of the three normal

Zeeman components (a) of the $3d$ to $2p$ transition; (b) of the $3s$ to $2p$ transition.

20. Calculate the wavelengths of the components of the first line of the Lyman series, taking the fine structure of the $2p$ level into account.

21. Calculate the energies and wavelengths of the $3d$ to $2p$ transition, taking into account the fine structure of *both* levels. How many component wavelengths might there be in the transition?

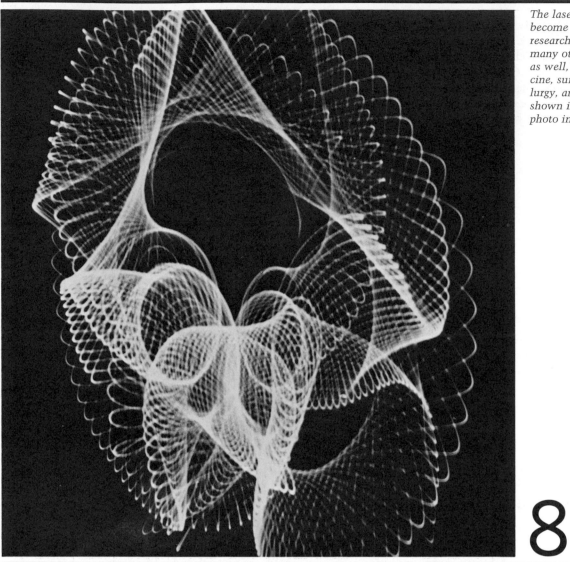

8

MANY-ELECTRON ATOMS

A number of difficulties arise when we attempt to construct a model that helps us to understand the structure and properties of atoms with more than one electron. For example, let's try to make an atom of helium. The nuclear charge is $Z = 2$, and we must therefore supply two electrons. As we bring in the first electron, the energy levels through which it must pass to reach the ground state are well described using our previous calculation, putting $Z = 2$. This electron would eventually occupy the $1s$ state. As we bring in the second electron, it feels the *attraction* of the nucleus with $Z = 2$ and also the electrical *repulsion* of the first electron. (The first $1s$ electron, you will recall, appears as a fuzzy ball of charge that completely surrounds the nucleus; see Figure 7.10.) This second electron thus feels a net attraction toward the nucleus with a force that is somewhat weaker than we would otherwise expect for $Z = 2$; there is an *effective value* of the nuclear charge seen by the second electron, with a contribution of $+2$ from the nucleus and a contribution of perhaps -1 from the first electron, giving $Z_{eff} \cong +1$. However, as the second electron is brought in, the *first* electron feels a repulsive force from the second electron, which changes the energy levels of the *first* electron. This problem of the mutual interactions of three or more objects is an example of what physicists call the *many-body problem*; it is not possible to solve a simple set of equations for the energy levels of an atom with two or more electrons. It is a problem that must be solved by approximation techniques; owing to the mathematical difficulty of obtaining solutions to this problem, we discuss only some very general results for the energy levels of many-electron atoms. Our goal in this chapter is to understand some of the properties of atoms (chemical, electrical, magnetic, optical, etc.) based on those energy levels.

Another problem that we must face is the question of which level an individual electron will occupy in a many-electron atom. We expect that each electron will end up in the lowest possible energy level. Based on our previous model, this would lead us to believe that all electrons will occupy the $1s$ level. From this kind of atom, we would expect a rather smooth or regular variation in the atomic properties between an atom having Z electrons and its neighbors having $Z \pm 1$ electrons. Indeed, certain of the physical properties, such as the energies of the X rays emitted by an atom, are consistent with this picture. However, many of the physical properties do not vary in this manner. For example, neon (with $Z = 10$) is an *inert gas*; it is practically unreactive and does not form chemical compounds under most conditions. Its neighbors, fluorine ($Z = 9$) and sodium ($Z = 11$), are among the most reactive of the elements and will under most conditions combine with other substances, sometimes violently. As another example, nickel ($Z = 28$) is strongly magnetic (ferromagnetic) and, for a metal, does not have a particularly large electrical conductivity. Copper ($Z = 29$) is an excellent electrical conductor but is not magnetic.

In this chapter, we try to understand these properties by studying the electronic energy levels of these atoms.

We first discuss the rule that prevents all of the electrons in an atom from falling into the $1s$ level. This rule was proposed by Wolfgang Pauli in 1925, based on a study of the transitions that are present, and those

8.1 THE PAULI EXCLUSION PRINCIPLE

that are expected but *not* present, in the emission spectra of atoms. Simply stated, the Pauli exclusion principle is as follows:

> No two electrons in a single atom can have the same set of quantum numbers (n, l, m_l, m_s).

The Pauli principle is the most important rule governing the structure of atoms, and no study of the properties of atoms can be attempted without a thorough understanding of this principle.

Let us illustrate how the Pauli principle works in the case of helium $(Z = 2)$. The first electron in helium, in the ground state, has quantum numbers $n = 1$, $l = 0$, $m_l = 0$, $m_s = +\frac{1}{2}$ or $-\frac{1}{2}$. The second electron can have the same n, l, and m_l, but it cannot have the same m_s, since then the exclusion principle would be violated. Thus if the first electron has $m_s = +\frac{1}{2}$, the second electron must have $m_s = -\frac{1}{2}$. Now suppose we are constructing an atom of lithium $(Z = 3)$. Just as with helium, the first two electrons will have quantum numbers $(n, l, m_l, m_s) = (1, 0, 0, +\frac{1}{2})$ and $(1, 0, 0, -\frac{1}{2})$. Since the third electron, according to the exclusion principle, cannot have the same set of quantum numbers as the first two, it *cannot go into the* $n = 1$ *level*, since there are only two different sets of quantum numbers available in the $n = 1$ level, and both of those sets have already been used. The third electron must therefore go into one of the $n = 2$ levels; experience indicates that the next level available of the two $n = 2$ levels ($2s$ or $2p$) is the $2s$ level, so the third electron might have quantum numbers $(n, l, m_l, m_s) = (2, 0, 0, +\frac{1}{2})$ or $(2, 0, 0, -\frac{1}{2})$. The fourth electron, in the case of beryllium $(Z = 4)$, would have the same n, l, and m_l, but the opposite m_s value as the third electron. When we reach boron, with $Z = 5$, the fifth electron cannot go into the $2s$ state, since we have already assigned both possible sets of quantum numbers in that level; the fifth electron then goes into one of the $2p$ levels. We might therefore expect that the properties of boron, with a $2p$ electron, would be different from the properties of lithium or beryllium, which have only $2s$ electrons.

It is this process of first using up all of the possible quantum numbers for one level, and then placing electrons in the next level, which accounts for the variations in the chemical and physical properties of the elements.

Wolfgang Pauli (1900–1958, Switzerland). His exclusion principle gave the basis for understanding atomic structure, but he also contributed to the development of quantum theory, to the theory of beta decay, and to the understanding of symmetry in physical laws.

8.2 ELECTRONIC STATES IN MANY-ELECTRON ATOMS

Figure 8.1 illustrates the ordering of the energy levels in many-electron atoms as the atomic number Z increases. The $1s$ level remains the lowest energy level, and the $2s$ and $2p$ levels remain fairly close in energy. The $2s$ level always lies a bit lower in energy than the $2p$ level. (The fine structure splitting is very small on the scale of this diagram.) We can understand why the $2s$ level lies lower in energy if we recall Example 7.5. An electron in the $2s$ level spends more of its time inside the Bohr radius than an electron in the "circular" $2p$ level does. The $2p$ electron sees the nuclear electric charge $+Ze$ shielded, or "screened," by the two $1s$ electrons. The $2s$ electron, however, occasionally is found closer to the nucleus than the Bohr radius, where it feels the attraction of the *entire* nuclear charge. The $2s$ electron feels, on the average, a slightly greater force of attraction to the nucleus than the $2p$ electron does; therefore the $2s$ electron is more tightly bound to the atom, and lies lower in energy.

A more extreme example of the *tighter binding* of the penetrating orbits can be found in the $n = 3$ and $n = 4$ levels. The $3s$ electron penetrates the inner electron orbits the most, and so is most tightly bound. The $3p$ electron penetrates almost as much, while the $3d$ electron (in the Bohr picture) moves in a circular orbit, which doesn't penetrate the inner orbits at all. The $4s$ electron again is in a highly eccentric elliptical orbit, and penetrates all of the inner orbits. This penetration causes such an extra tight binding that the $4s$ level is pulled all the way down to the energy of the $3d$ level. The small difference in energy of the $4s$ and $3d$ levels is an important contributing factor to the large electrical conductivity of copper, as we discuss later in this chapter. (Once again, keep in mind that the concept of Bohr orbits is often useful but ultimately *not* correct. The proper explanation of this tight binding of the s states is found in Figures 7.8 or 7.10, *not* in Figure 7.3. The s states have larger probability densities at smaller r than the states with $l > 0$.)

The levels with a certain value of n and l (for instance, $2s$ or $3d$) are known as *subshells*. The number of electrons that can be placed in each subshell is $2(2l + 1)$. The $(2l + 1)$ factor comes from the number of different m_l values for each l, since m_l can take the values $0, \pm 1, \pm 2, \pm 3, \ldots \pm l$. The extra factor of 2 comes from the two different m_s values; for each m_l, we can have $m_s = +\frac{1}{2}$ or $m_s = -\frac{1}{2}$. According to this scheme, the $1s$ subshell has a capacity of $2(2 \times 0 + 1) = 2$ electrons; the $3d$ subshell has a capacity of $2(2 \times 2 + 1) = 10$ electrons. (Note that this capacity doesn't depend on n; *any* d subshell has a capacity of 10 electrons.) Table 8.1 shows the ordering of the subshells.

As we saw in the case of the hydrogen atom, orbits with a certain n value all lie about the same average distance from the nucleus. (The electrons in the penetrating orbits spend some of their time closer to the

Penetrating orbits

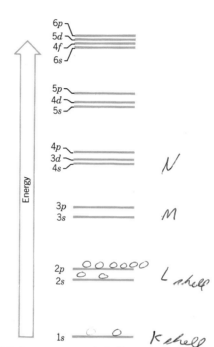

FIGURE 8.1 Atomic subshells, in order of increasing energy. The energy groupings are not to scale, but are merely representative of the relative energies of the subshells.

Table 8.1 Filling of Atomic Subshells

n	l	Subshell	Capacity $2(2l + 1)$
1	0	$1s$	2
2	0	$2s$	2
2	1	$2p$	6
3	0	$3s$	2
3	1	$3p$	6
4	0	$4s$	2
3	2	$3d$	10
4	1	$4p$	6
5	0	$5s$	2
4	2	$4d$	10
5	1	$5p$	6
6	0	$6s$	2
4	3	$4f$	14
5	2	$5d$	10
6	1	$6p$	6
7	0	$7s$	2
5	3	$5f$	14
6	2	$6d$	10

No 2 electrons in a single atom can have the same set of quantum nos (n, l, m_l, m_s) pauli

subshell with smaller value of l (same n) lies in lower in general

$4s$ is lower than $3d$
$5s$ " " " " $4d$
tighter binding " penetrating orbit

nucleus than the nonpenetrating orbits, but also some of their time further from the nucleus; the average distance from the nucleus of the penetrating orbits is then about the same as the average distance from the nucleus of the nonpenetrating orbits. See Problem 13 in Chapter 7 for a verification of this for the hydrogen atom.) The set of orbits with a certain value of n, with about the same average distance from the nucleus, is known as an atomic *shell.* The atomic shells are designated by letter, with K standing for the $n = 1$ shell, L for the $n = 2$ shell, M for the $n = 3$ shell, and so forth.

8.3 THE PERIODIC TABLE

Figure 8.2 shows the periodic table, which is merely an orderly array of the chemical elements, listed in order of increasing atomic number Z and arranged in such a way that the vertical columns, called *groups,* contain elements with rather similar physical and chemical properties. In this section we discuss the way in which the filling of electronic subshells helps us understand the arrangement of the periodic table. In the next section we examine some of the physical and chemical properties of the elements.

In attempting to understand the ordering of subshells and the periodic table, we must follow two rules for filling the electronic subshells:

Rules for filling electronic subshells

1. The capacity of each subshell is $2(2l + 1)$. (This is of course just another way of stating the Pauli exclusion principle.)

2. The electrons will occupy the lowest energy states available.

To indicate the electron configuration of each element, we use a notation in which the identity of the subshell and the number of electrons in it are listed. The identity of the subshell will be indicated in the usual

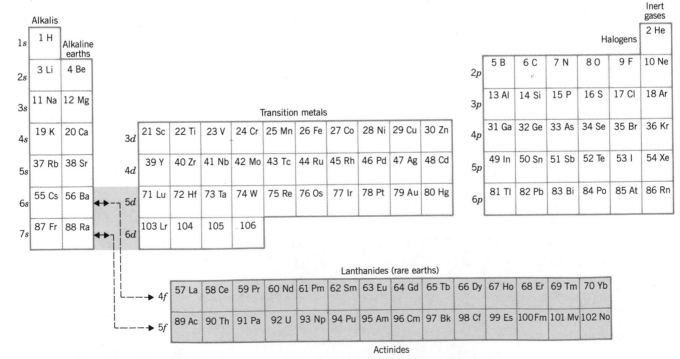

FIGURE 8.2 Periodic table of the elements.

way, and the number of electrons in that subshell is indicated by a superscript. Thus hydrogen has the configuration $1s^1$, for one electron in the $1s$ shell, and helium has the configuration $1s^2$. Helium has both a filled subshell (the $1s$) and a closed major shell (the K shell) and thus is an extraordinarily stable and inert element. With lithium ($Z = 3$), we begin to fill the $2s$ subshell; lithium has the configuration $1s^22s^1$. With beryllium ($Z = 4$, $1s^22s^2$) the $2s$ subshell is full, and the next element must begin filling the $2p$ subshell (boron, $Z = 5$, $1s^22s^22p^1$). The $2p$ subshell has a capacity of six electrons, and with neon ($Z = 10$, $1s^22s^22p^6$) both the $2p$ subshell and the L shell ($n = 2$) are complete. The next row (or *period*) begins with sodium ($Z = 11$, $1s^22s^22p^63s^1$), and the $3s$ and $3p$ subshells are filled in much the same way as the $2s$ and $2p$ subshells, ending with the inert gas argon ($Z = 18$, $1s^22s^22p^63s^23p^6$). The elements of the third row (period) are chemically similar to the corresponding elements of the second row (period), and so are written directly under them. The next electron would ordinarily go into the $3d$ level. However, the highly penetrating orbit of the $4s$ electron causes the $4s$ level to appear at roughly the same energy as the $3d$ level, and it turns out that the $4s$ level appears somewhat lower in energy and fills first. The configurations of potassium ($Z = 19$) and calcium ($Z = 20$) are therefore respectively $1s^22s^22p^63s^23p^64s^1$ and $1s^22s^22p^63s^23p^64s^2$. These elements have properties similar to, and therefore appear directly under, the corresponding elements with one and two s-subshell electrons in the second and third periods. We now begin to fill the $3d$ subshell. Since there is no $1d$ or $2d$ subshell (why not?), we would expect the first element with a d-subshell configuration to have rather different chemical properties than the elements we have placed previously; thus it should not appear in any of our previously occupied groups (columns), and so we begin a new group with scandium ($Z = 21$, $1s^22s^22p^63s^23p^64s^23d^1$) and the $3d$ subshell eventually closes with zinc ($Z = 30$, $1s^22s^22p^63s^23p^64s^23d^{10}$). (Along the way there are some minor variations; the most important is copper, with $Z = 29$. For this case the $3d$ level lies slightly lower than the $4s$ level, and so the $3d$ subshell fills before the $4s$, resulting in the configuration $1s^22s^22p^63s^23p^63d^{10}4s^1$. As we will see in the next section, this configuration is responsible for the large electrical conductivity of copper.) In the next series of elements, the $4p$ subshell is filled, from gallium ($Z = 31$) to the inert gas krypton ($Z = 36$). When we move to the next period, we again fill the $5s$ subshell before the $4d$ subshell, and the series of 10 elements corresponding to the filling of the $4d$ subshell is written directly under the series that had unfilled configurations in the $3d$ subshell. (Silver, with $Z = 47$, corresponds exactly to copper in the fourth period, with the $4d$ subshell filling before the $5s$.) After the completion of the $4d$ subshell, the $5p$ subshell is filled, ending with the inert gas xenon ($Z = 54$). The next period begins with cesium and barium filling the $6s$ subshell, and lanthanum beginning the $5d$ subshell. As was the case in the previous periods, the $5d$ and $6s$ lie at almost the same energy; since the $6s$ is at a slightly lower energy, it fills first. However, there is yet another subshell at about the same energy as the $6s$ and $5d$—the $4f$ subshell, which now begins to fill, from cerium to lutetium. This series of elements, called the lanthanides or rare earths, is usually written separately in the periodic table, since there have been no other f-subshell elements under which to write them. The $4f$ subshell has a

Table 8.2 Electronic Configurations of Some Elements

H	$1s^1$	Y	[Kr] $5s^2 4d^1$
He	$1s^2$	Mo	[Kr] $5s^1 4d^5$
Li	$1s^2 2s^1$	Ag	[Kr] $5s^1 4d^{10}$
Be	$1s^2 2s^2$	In	[Kr] $5s^2 4d^{10} 5p^1$
B	$1s^2 2s^2 2p^1$	Xe	[Kr] $5s^2 4d^{10} 5p^6$
Ne	$1s^2 2s^2 2p^6$	Cs	[Xe] $6s^1$
Na	[Ne] $3s^1$	La	[Xe] $6s^2 5d^1$
Al	[Ne] $3s^2 3p^1$	Ce	[Xe] $6s^2 5d^1 4f^1$
Ar	[Ne] $3s^2 3p^6$	Pr	[Xe] $6s^2 4f^3$
K	[Ar] $4s^1$	Gd	[Xe] $6s^2 5d^1 4f^7$
Sc	[Ar] $4s^2 3d^1$	Dy	[Xe] $6s^2 4f^{10}$
Cr	[Ar] $4s^1 3d^5$	Yb	[Xe] $6s^2 4f^{14}$
Mn	[Ar] $4s^2 3d^5$	Lu	[Xe] $6s^2 5d^1 4f^{14}$
Cu	[Ar] $4s^1 3d^{10}$	Re	[Xe] $6s^2 5d^5 4f^{14}$
Zn	[Ar] $4s^2 3d^{10}$	Au	[Xe] $6s^1 5d^{10} 4f^{14}$
Ga	[Ar] $4s^2 3d^{10} 4p^1$	Hg	[Xe] $6s^2 5d^{10} 4f^{14}$
Kr	[Ar] $4s^2 3d^{10} 4p^6$	Tl	[Xe] $6s^2 5d^{10} 4f^{14} 6p^1$
Rb	[Kr] $5s^1$	Rn	[Xe] $6s^2 5d^{10} 4f^{14} 6p^6$

A symbol in brackets [ ] means that the atom has the configuration of the previous inert gas plus the additional electrons listed.

capacity of 14 electrons, and so there are 14 elements in the lanthanide series. Once the 4f subshell is complete, we return to filling the 5d subshell, writing those elements in the groups under the corresponding 3d and 4d elements, and then complete the period with the filling of the 6p subshell, ending with the inert gas radon. The seventh period begins much like the sixth, with a series known as the actinides, written under the lanthanides, corresponding to the filling of the 5f subshell.

Table 8.2 lists the electronic configurations of some of the elements.

What is most remarkable about this scheme is that the arrangement of the periodic table was known well before the introduction of atomic theory. The elements were organized into groups and periods based on their physical and chemical properties by Mendeleev in 1859; understanding that organization in terms of atomic levels is a great triumph for the atomic theory. What remains is to interpret the chemical and physical properties based on the atomic levels.

Atomic theory and the periodic table

In this section we briefly study the way our knowledge of atomic structure helps us to understand the physical and chemical properties of the elements. Our discussion will be based on the following two precepts:

8.4 PROPERTIES OF THE ELEMENTS

1. Filled subshells are normally very stable configurations. An atom with one electron beyond a closed shell will readily give up that electron to another atom to form a chemical bond. Similarly, an atom lacking one electron from a closed shell will readily accept an additional electron from another atom in forming a chemical bond.

2. Filled subshells do not normally contribute to the chemical or physical properties of an atom. Only the electrons in the unfilled subshells need be considered.

We will consider a number of different physical properties of the elements, and will try to understand those properties based on atomic theory.

1. *Atomic Radii.* We have already learned that the radius of an atom is not a precisely defined quantity, since the electron probability density determines the "size" of an atom. The radii are similarly hard to define experimentally, and in fact different kinds of experiments may give different values for the radii. One way of defining the radius is by means of the spacing between the atoms in a crystal containing that element. However, if the element appears in different crystals with different valence states, different radii may be measured. Figure 8.3 shows how such typical atomic radii vary with Z.

2. *Ionization Energy.* The minimum energy needed to remove an electron from an atom is known as its *ionization energy*. For example hydrogen has an ionization energy of 13.6 eV. Helium has an ionization energy of 24.6 eV for the first electron and 54.4 eV for the second electron. Table 8.3 gives the ionization energies of some of the elements, and Figure 8.4 shows the variation of ionization energy with atomic number Z.

3. *Electrical Resistivity.* In bulk materials, an electric current flows when a potential difference (voltage) is applied across the material. The current I and voltage V are related in many materials according to Ohm's law, $V = IR$, where R is the electrical resistance of the material. If the material has a length L and an area A, then the resistance is

$$R = \rho \frac{L}{A}$$

The *resistivity* ρ is characteristic of the kind of material and is measured in units of ohm·cm. A good electrical conductor has a small

Table 8.3 Ionization Energies (in eV) of Neutral Atoms of Some Elements

H	13.60	K	4.34
He	24.59	Cu	7.72
Li	5.39	Kr	14.00
Be	9.32	Rb	4.18
Ne	21.56	Au	9.22
Na	5.14		
Ar	15.76		

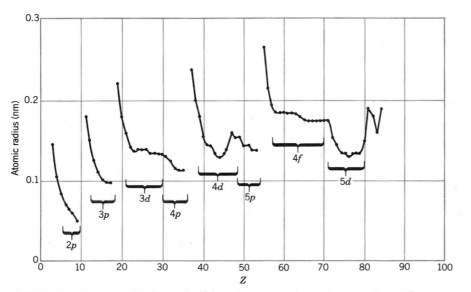

FIGURE 8.3 Atomic radii, determined from ionic crystal atomic separations. These radii are different from the mean radii of the electron cloud for free atoms.

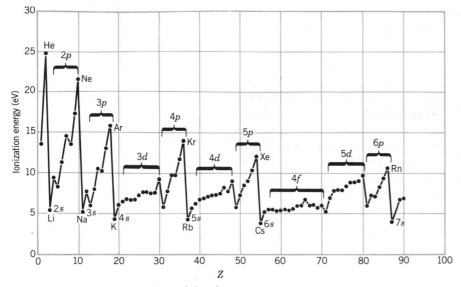

FIGURE 8.4 Ionization energies of the elements.

resistivity ($\rho = 1.7 \times 10^{-6}$ ohm·cm for copper); a poor conductor has a large resistivity ($\rho = 2 \times 10^{17}$ ohm·cm for sulfur). From the atomic point of view, current depends on the movement of relatively loosely bound electrons, which can be removed from their atoms by the applied potential difference, and also on the ability of the electrons to travel from one atom to another. Thus elements with s electrons, which can more often be found further from the nucleus than electrons with larger l-values, would be expected to have small resistivities.

Figure 8.5 shows the variation of electrical resistivity with atomic number.

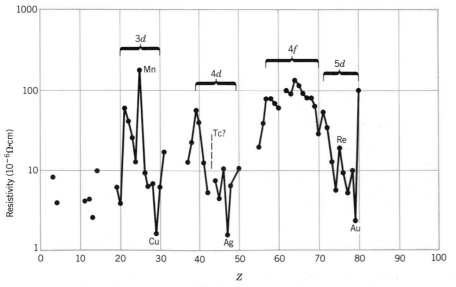

FIGURE 8.5 Electrical resistivities of the elements.

4. **Magnetic Susceptibility.**   When a material is placed in a magnetic field of intensity B, the material becomes "magnetized" and acquires a magnetization M, which is proportional to B:

$$\mu_0 M = \chi B$$

where χ is the *magnetic susceptibility*. (Materials for which $\chi > 0$ are known as *paramagnetic,* and those for which $\chi < 0$ are called *diamagnetic;* materials that remain permanently magnetized even when B is removed are known as *ferromagnetic* and χ is undefined for such materials.)

From the atomic point of view, the magnetism of atoms depends on the **l** and **s** of their outer electrons, since the atomic magnetic moments $\boldsymbol{\mu}_l$ and $\boldsymbol{\mu}_s$ are proportional to **l** and **s** [recall Equations (7.16) and (7.24).] This effect is responsible for paramagnetic susceptibilities and occurs in all atoms. Diamagnetism is caused by the following effect: when a magnetic field is applied to an electrical circuit, an *induced current* flows in the circuit; the induced current sets up a magnetic field which tends to *oppose* the applied field (Lenz's law). In the atomic physics case, the electrical circuit is the circulating electron, and the induced current consists of a slight speeding up or slowing down of the electron in its orbit. This produces a contribution to the magnetization of the material which is opposite to the applied field B, and so the diamagnetic contribution to χ is negative.

Figure 8.6 shows the magnetic susceptibilities of the elements.

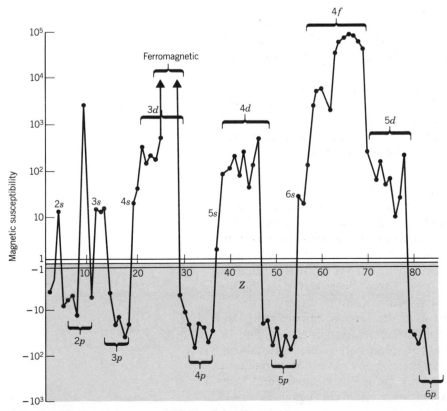

FIGURE 8.6 Magnetic susceptibilities of the elements.

Just by examining Figures 8.3 to 8.6, you can see the remarkable regularities in the properties of the elements. Notice especially how similar the properties of the different sequences of elements are—for example, the electrical resistivity of the d-subshell elements or the magnetic susceptibility of the p-subshell elements. We now look at how the atomic structure is responsible for these properties.

Inert Gases   The inert gases occupy the last column of the periodic table. Since they have only filled subshells, the inert gases do not generally combine with other elements to form compounds; these elements are very reluctant to give up or to accept an electron. At room temperature they are monatomic gases, and since their atoms don't like to join with each other, the boiling points are very low $(-200°C)$. Their ionization energies are much larger than those of neighboring elements, because of the extra energy needed to break open a filled subshell.

p-Subshell Elements   The elements of the column (group) next to the inert gases are the halogens (F, Cl, Br, I, At). These atoms lack one electron from a closed shell and have the configuration np^5. Since a closed p subshell is a very stable configuration, these elements will readily form compounds with other atoms that can provide an extra electron to close the p subshell. The halogens are therefore extremely reactive.

As we move across the series of six elements in which the p subshell is being filled, the atomic radius *decreases*. This "shrinking" occurs because the nuclear charge is increasing and pulling all of the orbits closer to the nucleus. Notice from Figure 8.3 that the halogens have the smallest radii within each p subshell series. (The crystal radii of the inert gases are not known.)

As we increase the nuclear charge, the p electrons also become more tightly bound; from Figure 8.4 we see how the ionization energy increases systematically as the p subshell is filled.

From Figure 8.6 we see that each p subshell series is diamagnetic, with a characteristic negative magnetic susceptibility.

s-Subshell Elements   The elements of the first two columns (groups) are known as the alkalis (configuration ns^1) and alkaline earths (ns^2). The single s electron makes the alkalis quite reactive; the alkaline earths are similarly reactive, in spite of the filled s subshell. This occurs because the s electron wave function can extend rather far from the nucleus, where the electron is screened (by $Z - 1$ other electrons) from the nuclear charge and therefore not tightly bound. (Notice from Figure 8.3 that the ns^1 and ns^2 configurations give the largest atomic radii, and from Figure 8.4 that they have the smallest ionization energies.) For the same reasons, the ns^1 and ns^2 elements are relatively good electrical conductors. From Figure 8.6 we see that these elements are paramagnetic $(\chi > 0)$; in the orbital picture, s orbits are completely flattened ellipses enclosing no area. There is no induced magnetic field (Lenz's law does not operate) and no diamagnetism.

Transition Metals The three rows of elements in which the d subshell is filling (Sc to Zn, Y to Cd, Lu to Hg) are known as the *transition metals*. Many of their chemical properties are determined by the "outer" electrons—those whose wave functions extend furthest from the nucleus. For the transition metals, these are always s electrons, which have a larger mean radius than the d electrons. (Remember that the mean radius depends mostly on n; the s electrons of the transition metals have a larger n than the d electrons.) As the atomic number increases across the transition metal series, we add one d electron and one unit of nuclear charge; the net effect on the s electron is very small, since the additional d electron screens the additional nuclear charge. The chemical properties of the transition metals are therefore very similar, as the small variation in radius and ionization energy shows.

The electrical resistivity of the transition metals shows two interesting features: a sharp rise at the center of the sequence, and a sharp drop near the end (Figure 8.5). The sharp drop near the end of the sequence is responsible for the small resistivity (large conductivity) of copper, silver, and gold. If we filled the d subshell in the expected sequence, copper would have the configuration $4s^2 3d^9$; however the filled d subshell is more stable than a filled s subshell, and so one of the s electrons is transferred to the d subshell, resulting in the configuration $4s^1 3d^{10}$. This relatively free, single s electron makes copper an excellent conductor.

At the center of the sequence of transition metals there is a sharp rise in the resistivity; apparently a half-filled shell is also a stable configuration, and so Mn $(3d^5)$ and Re $(5d^5)$ have larger resistivities than their neighbors. (The element Tc, with configuration $4d^5$, is radioactive and is not found in nature; its resistivity is unknown.) A similar rise in resistivity is seen at the center of the $4f$ sequence.

The transition metals have similar paramagnetic susceptibilities, due to the large orbital angular momentum of the d electrons and also to the large *number* of d-subshell electrons that can couple their *spin* magnetic moments. These two effects are large enough to overcome the diamagnetism of the orbital motion. It is the d electrons, with their large angular momentum, which are also responsible for the ferromagnetism of iron, nickel, and cobalt. As soon as the d subshell is filled, however, the orbital and spin magnetic moments no longer contribute to the magnetic properties (all of the m_l and m_s values, positive as well as negative, are taken); for this reason, copper and zinc are diamagnetic, not paramagnetic like their transition metal neighbors.

Lanthanides (Rare Earths) The rare earths are rather similar to the transition metals in that an "inner" subshell (the $4f$) is being filled after an "outer" subshell (the $6s$). For the same reasons discussed above, the chemical properties of the rare earths should be rather similar; the radii and ionization energies show that this is true.

Because of the larger orbital angular momentum of f-subshell electrons $(l = 3)$ and also because of the larger *number* of f-subshell electrons that can align their spin magnetic moments, the paramagnetic susceptibilities of the rare earths are even larger than those of the transition metals. Even the ferromagnetism of the rare earths is substantially stronger than that of the iron group. Generally, we think of iron as the most magnetic of the elements. If we magnetize a piece of iron, the internal magnetic field (within the piece of iron) is about 28 T. Magnetized holmium metal, a rare earth, has an internal magnetic field of 800 T, roughly 30 times that of iron! Most of the other rare earths have similar magnetic properties. (The rare earth metals do not reveal their "ferromagnetic" properties at room temperature, but must be cooled to lower temperatures. Holmium must be cooled to 20 K to reveal its magnetic properties.)

Actinides   The actinide series of elements should have chemical and physical properties similar to those of the rare earths. Unfortunately, most of the actinide elements (those beyond uranium) are naturally radioactive and do not occur in nature. They are man-made elements and are available only in microscopic quantities. We are thus unable to determine many of their bulk properties.

8.5 X RAYS

X rays, as we discussed in Chapter 3, are electromagnetic radiations with wavelengths approximately from 0.01 to 10 nm (energies approximately from 100 eV to 100 keV). In Chapter 3 we discussed the *continuous* X-ray spectrum emitted by accelerated electrons. In this section we are concerned with the *discrete* X-ray line spectra emitted by atoms.

X rays are emitted in transitions between the normally filled, lower energy levels of an atom. The inner electrons are so tightly bound that the energy spacing of these levels is about right for the emission of photons in the X-ray range of wavelengths. The outer electrons are relatively weakly bound, and the spacing between these outer levels is only a few electron-volts; transitions between these levels give photons in the visible region of the spectrum. These "optical" transitions are discussed in the next section.

Since all of the inner shells of an atom are filled, X-ray transitions will not occur between these levels under normal circumstances. For example a $2p$ electron cannot make a transition to the $1s$ subshell, since all atoms beyond hydrogen have filled $1s$ subshells. In order to observe such a transition, we must remove an electron from the $1s$ subshell. This can be done by bombarding the atom with electrons (or other particles) accelerated to an energy sufficient to knock loose a $1s$ electron following a collision. (This requires accelerating voltages of about 10,000 V.)

Once we have removed an electron from the $1s$ subshell, an electron from a higher subshell will rapidly make a transition to fill that vacancy, emitting an X-ray photon in the process. Of course, the energy of the photon is just equal to the energy difference of the initial and final atomic levels of the electron that makes the transition.

We previously defined a notation in which the $n = 1$ shell is known as the K shell. When we remove a $1s$ electron, we are creating a vacancy in the K shell. The X rays that are emitted in the process of filling this vacancy are known as *K-shell X rays,* or simply K *X rays.* (These X rays are emitted in transitions which come *from* the *L, M, N,* . . . shells, but they are known by the vacancy that they fill, not by the shell in which they originate.) The K X ray that originates with the $n = 2$ shell (L shell) is known as the K_α X ray, and the K X rays originating from higher shells are known as K_β, K_γ, and so forth. Figure 8.7 illustrates these transitions.

It is also possible that the bombarding electrons can knock loose an electron from the L shell, and electrons from higher levels will drop down to fill this vacancy. The photons emitted in these transitions are known as L X rays. The lowest-energy X ray of the L series is known as L_α, and the other L X rays are labeled in order of increasing energy as shown in Figure 8.7.

Of course, it is also possible to have an L X ray emitted directly following the K_α X ray. A vacancy in the K shell can be filled by a transition from the L shell, with the emission of the K_α X ray. However, the electron that made the jump from the L shell left a vacancy there, which will be filled by an electron from a higher shell, with the accompanying emission of an L X ray.

In a similar manner, we label the other X-ray series by M, N, and so forth. Figure 8.8 shows a sample X-ray spectrum emitted by silver.

We have not considered the energy differences of the subshells within the major shells. For example, the L_α X ray could originate from one of the $n = 3$ levels ($3s$, $3p$, $3d$) and end in one of the $n = 2$ levels ($2s$, $2p$). The energies of these different transitions will be slightly different, so that there will be many L_α X rays, but their energy differences are very small compared, for example, with the energy difference between the L_α and L_β X rays. In fact, in many applications, we don't even see this small energy splitting.

Let us consider the K_α X ray in more detail. An electron in the L shell is screened by the two $1s$ electrons, and so it sees an effective nuclear charge of $Z_{eff} \cong Z - 2$. When one of those $1s$ electrons is removed in the creation of a K-shell vacancy, only the remaining $1s$ electron shields the L shell, and so $Z_{eff} \cong Z - 1$. (In this calculation, we neglect the small screening effect of the outer electrons; their probability densities are not zero within the L-shell orbits, but they are sufficiently small that their effect on Z_{eff} can be neglected.) The K_α X ray can thus be analyzed as a transition from the $n = 2$ level to the $n = 1$ level in a one-electron atom with $Z_{eff} = Z - 1$. Using relationships (6.40) and (6.43) for the Bohr atom developed in Chapter 6, the frequency of the K_α

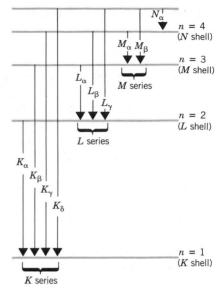

FIGURE 8.7 X-ray series.

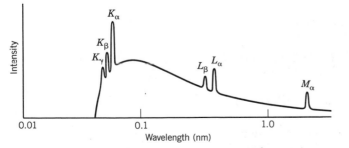

FIGURE 8.8 Characteristic X-ray spectrum of silver, such as might be produced by 30 keV electrons striking a silver target. The continuous distribution is a bremsstrahlung spectrum.

transition in an atom of atomic number Z is found to be

$$\nu = \frac{3cR_\infty}{4} (Z - 1)^2 \tag{8.1}$$

If we were to plot $\sqrt{\nu}$ as a function of Z, we should obtain a straight line with slope $(3cR_\infty/4)^{1/2}$. Figure 8.9 is an example of such a plot. (Incidently, this result is independent of our assumption regarding the exact value of the screening. That is, we could have written $Z_{\text{eff}} = Z - k$, where k is some unknown number, probably close to 1. The only change in our plot would be in the intercept. We would still have a straight line with the same slope.)

This method gives us a powerful and yet simple way to determine the atomic number Z of an atom, as was first demonstrated in 1913 by the young British physicist H. G. J. Moseley, who measured the K_α (and other) X-ray energies of the elements and thus determined their atomic numbers. Moseley was the first to demonstrate the type of linear relationship shown in Figure 8.9; such graphs are now known as Moseley plots. His discovery provided the first direct means of measuring the atomic numbers of the elements. Previously, the elements had been ordered in the periodic table according to increasing mass. Moseley found certain elements listed out of order, in which the element of higher Z happened to have the smaller mass (see, for example, cobalt and nickel or iodine and tellurium). He also found gaps corresponding to yet undiscovered elements; for example, the naturally radioactive element technetium ($Z = 43$) does not exist in nature and was not known at the time of Moseley's work, but Moseley showed the existence of such a gap at $Z = 43$.

Moseley's work was of great importance in the development of atomic physics. Working in the same year as Rutherford and Bohr, Moseley not only provided a direct confirmation of the Rutherford-Bohr model, he also made a strong link between the periodic table, which was previously a rather arbitrary classification scheme of the elements, and atomic theory.

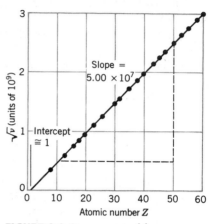

FIGURE 8.9 Square root of frequency versus atomic number for K_α X rays. The slope of the line is 5.00×10^7 s$^{-1/2}$, in good agreement with the expected value $(3cR_\infty/4)^{1/2} = 4.97 \times 10^7$ s$^{-1/2}$. The intercept is close to 1, as expected.

8.6 OPTICAL SPECTRA

As mentioned in the previous section, when we excite or remove one of the *outer* electrons, the resulting transitions fall in the visible range of the spectrum, and are thus known as *optical* transitions. The binding energies of the outer electrons in a typical atom are of the order of several electron-volts, and so it takes a relatively low voltage to remove an outer electron and produce an optical transition. In fact, it is the absorption and reemission of light by these outer electrons that are responsible for the colors of material objects (although in solids the electron energy levels are usually very different from those in isolated atoms). In contrast with X-ray spectra, which vary slowly and smoothly from one element to the next, optical spectra can show large variations between neighboring elements, especially those that correspond to filled subshells.

Beyond hydrogen, the easiest energy-level diagrams to understand are those of the alkali metals (Li, Na, K, Rb, Cs, Fr), which have a single s electron outside an inert core. Many of the excited states then correspond to the excitation of this single electron, and the resulting spectra are very similar to the spectrum of hydrogen, since the nuclear charge of $+Ze$ is shielded by the other $(Z-1)$ electrons. Figure 8.10 shows the energy levels of Li and Na along with some of the emitted transitions, which follow the same $\Delta l = \pm 1$ rule as the transitions in hydrogen.

The ground-state configuration of lithium is $1s^2 2s^1$ and the ground state configuration of sodium is $1s^2 2s^2 2p^6 3s^1$. The excited states in both

Henry G. J. Moseley (1887–1915, England). His work on X-ray spectra provided the first link between the chemical periodic table and atomic physics, but his brilliant career was cut short when he died on a World War I battlefield.

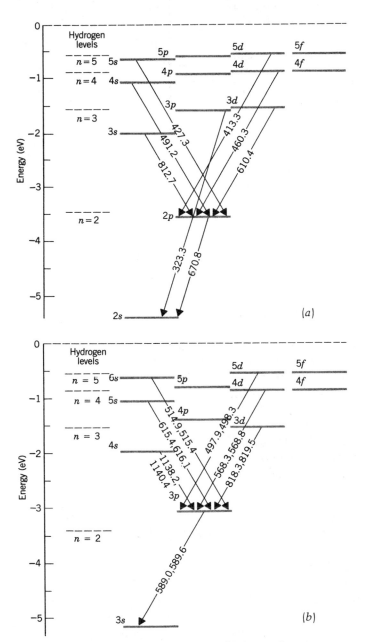

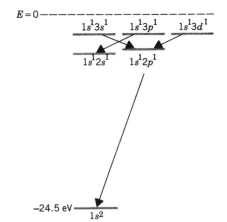

FIGURE 8.10 (*a*) Energy level diagram of lithium, showing some transitions (labeled with wavelength values in nanometers) in the optical region. The corresponding energies of hydrogen are included for comparison. (*b*) Energy level diagram of sodium. The fine-structure splitting of the 3p state makes each transition involving that state into a closely spaced doublet.

FIGURE 8.11 A small portion of the energy level diagram of helium. Note the $\Delta l = \pm 1$ transitions.

cases can be obtained by moving the outer electron to a higher state. For example, the first excited state of Li is $1s^2 2p^1$, with the $2s$ electron moving to the $2p$ level. (The energy necessary to accomplish this can be provided by many means, such as by absorption of a photon or by passage of an electric current through the material as in a gas discharge tube.) The excited electron in the $2p$ state rapidly drops back to the $2s$ state, with the emission of a photon of wavelength 670.8 nm. Since the inert core doesn't participate in this excitation or emission, we can ignore all but the outer electron in studying the levels and transitions in the alkali metals.

The ground-state configuration of helium is $1s^2$. We can produce an excited state by moving one of these electrons up to a higher level, and so some possible excited-state configurations might be $1s^1 2s^1$, $1s^1 2p^1$, $1s^1 3s^1$, and so forth. Photons are emitted when the excited electron drops back to the $1s$ level. The $\Delta l = \pm 1$ selection rule for transitions once again limits those that can occur. Figure 8.11 shows a portion of the energy level diagram for helium.

The phenomenon of *fluorescence* is responsible for the appearance of objects under the so-called "black light," which is a source of ultraviolet radiation. Photons in the ultraviolet region, invisible to the human eye, have higher energies than those in the visible region, and hence if an ultraviolet photon is absorbed by an atom, the outer electron (those responsible for the optical transitions) can be excited to high levels. These electrons make transitions back to their ground state, accompanied by the emission of photons in the visible region. Objects seen in ultraviolet light often show colors in the blue or violet end of the spectrum that are not present when the objects are viewed in sunlight. We can understand this effect by considering the composition of sunlight and the optical excited states of a hypothetical atom shown in Figure 8.12. The intensity of sunlight is concentrated in the center of the visible spectrum, in the yellow region; very little intensity is present in the red or blue ends of the visible spectrum. The "yellow" photons have enough energy to excite the hypothetical atom to levels 1 and 2 shown in Figure 8.12, but not enough to reach level 3. However, the higher-energy ultraviolet photons have sufficient energy to reach all three levels, so the light emitted by the atom has a stronger blue component when that atom is excited by ultraviolet light than when excited by sunlight.

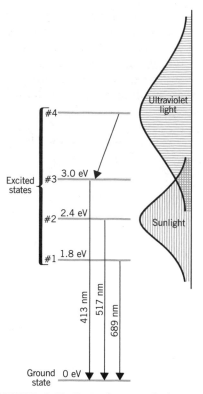

FIGURE 8.12 Excited states of a hypothetical atom. Only excited states 1 and 2 can be reached by exposure to sunlight; exposure to ultraviolet light populates state 4, which in turn populates state 3. Under ultraviolet light, a stronger blue or violet (413 nm) color is revealed than under sunlight.

In an alkali atom such as sodium, the properties of the atom are determined primarily by the properties of the single outer electron; if that electron has quantum numbers (n, l, m_l, m_s) then the entire atom behaves as if it had those same quantum numbers. In a more complicated atom, such as carbon, this is not necessarily true. The electronic configuration of carbon $(Z = 6)$ is $1s^2 2s^2 2p^2$. We will ignore the filled $1s$ and $2s$ subshells for this discussion, since they do not contribute directly to most of the atomic properties of carbon. Suppose the two $2p$ electrons have quantum numbers $(2, l_1, m_{l_1}, m_{s_1})$ and $(2, l_2, m_{l_2}, m_{s_2})$, where $l_1 = l_2 = 1$. The atom then behaves as if it had an angular momentum vector **L**, which is the *vector* sum of the angular momentum *vectors* corresponding to:

$$\mathbf{L} = \mathbf{l}_1 + \mathbf{l}_2 \tag{8.2}$$

*8.7 ADDITION OF ANGULAR MOMENTA

the changes in the atomic configuration giving rise to the optical spectrum are produced by the motion of a single atom

Tota. ang. mom = $\vec{l} + \vec{s}$

orbital   spin
ang.

These vectors are not ordinary vectors, but rather represent quantized angular momenta, so they must be added in a special way:

rules

✻ 1. *Find the maximum value that the magnitude of the z component of the combination could have.* Since the z components add like ordinary numbers, rather than like vectors, the combined z component of **L**, which we call M_L, is found according to

$$M_L = m_{l_1} + m_{l_2} \tag{8.3}$$

The maximum value of M_L for carbon is obviously $+2$, when m_{l_1} and m_{l_2} are each $+1$.   This sum gives the maximum resultant L.

✻ 2. *Find the magnitude of the difference of the z-components when each vector has its maximum z-component.* This gives the minimum value of of L, which is zero when m_{l_1} and m_{l_2} are both $+1$.

✻ 3. *The L quantum number of the combination can take any possible value from the one corresponding to the maximum sum of z components* (2 in this case) *to the one corresponding to the maximum difference* (0 in this case) *in integer steps.* For carbon, the possible values of L would be 0, 1, 2.

✻ 4. *The usual rules for quantized angular momentum vectors then apply to the vector **L**.* That is,

$$|\mathbf{L}| = \sqrt{L(L + 1)}\, \hbar \tag{8.4}$$

and

$$L_z = M_L \hbar \tag{8.5}$$

Each of the two electrons has spin $\frac{1}{2}$, and the spins must also be added to get the total spin S. We apply the same rules as we applied for L: There are a maximum sum and difference in the magnitude of $M_S = m_{s_1} + m_{s_2}$; these are (for carbon) 1 (sum) and 0 (difference).   Values of S are therefore 1 and 0 (from minimum to maximum in integer steps), and

$$|\mathbf{S}| = \sqrt{S(S + 1)}\, \hbar \tag{8.6}$$

$$S_z = M_S \hbar \tag{8.7}$$

The two 2p electrons of carbon can combine to give $L = 0$, 1, or 2 and $S = 0$ or 1. How do we know which of these combinations will be the ground state of carbon? The rules for finding the ground-state quantum numbers are known as *Hund's rules*:

1. First find the maximum value of S consistent with the Pauli principle.
2. Then, for that S, find the maximum value of L consistent with the Pauli principle.

In the case of carbon, the maximum value of S is 1. With only two electrons in the 2p shell (which has a capacity of 6), the Pauli principle places no restrictions on the value of S, since both electrons can have $m_s = +\frac{1}{2}$. (In fact, the Pauli principle allows three electrons in a p subshell to have $m_s = +\frac{1}{2}$.) Our next task is to find the maximum value of L for $S = 1$. The maximum value of m_l for the first electron is $+1$, giving it quantum numbers $(2, 1, +1, +\frac{1}{2})$. The second electron cannot also have $m_l = +1$, because that would violate the Pauli principle, since the

(handwritten margin notes)

Total orbital angular momentum L

Total ang mom = orbit am + spin am.
$$\vec{J} = \vec{l} + \vec{s}$$

$$\vec{S_1} + \vec{S_2} = \vec{S}$$

then $\vec{J} = \vec{L} + \vec{S}$

First method
C: $(2, l_1, m_{l_1}, m_{s_1})$ +
$(2, l_2, m_{l_2}, m_{s_2})$

want to add l_1 & l_2 to obtain $L = l_1 + l_2$

rules (✻)

Hund's rules

for Carbon $2p^2$
L = 0 ① or 2
S = 0 or ①

$m_{s_1} = +\frac{1}{2}$
$m_{s_2} = +\frac{1}{2}$ ⟹ $M_S = +1$ ⟹ S=1

$n=2$ $l=1$ $m_l = \begin{cases} 1 \\ 0 \\ -1 \end{cases}$ $m_s = \begin{cases} +\frac{1}{2} \\ -\frac{1}{2} \end{cases}$

$m_{l_1} = 1$ $m_{l_2} = 0$ $M_L = 1 + 0 = 1$

L=1

second electron also has $m_s = +\frac{1}{2}$. (Remember, we began by *maximizing S*.) Its maximum value of m_l is therefore 0, so $M_L = 1 + 0 = 1$; thus $L = 1$ for the ground state of carbon.

Use Hund's rules to find the ground-state quantum numbers of nitrogen.

EXAMPLE 8.1

Solution

$M_s = \frac{3}{2} \Rightarrow S = \frac{3}{2}$

The electronic configuration of nitrogen is $1s^2 2s^2 2p^3$. We begin by maximizing S for the three $2p$ electrons. Since three electrons in the p subshell are permitted by the Pauli principle to have $m_s = +\frac{1}{2}$, the maximum value of S is $\frac{3}{2}$. Each of the three electrons has quantum numbers $(2, 1, m_l, +\frac{1}{2})$. To maximize L, we assign the first electron the maximum value of m_l, namely $+1$. The maximum value of m_l left for the second electron is 0, and the third electron must therefore have $m_l = -1$. The total M_L is $1 + 0 + (-1) = 0$, so $L = 0$. Thus $L = 0$, $S = \frac{3}{2}$ are the ground-state quantum numbers for nitrogen.

Find the ground-state L and S of oxygen $(Z = 8)$.

EXAMPLE 8.2

Solution

The electronic configuration of oxygen is $1s^2 2s^2 2p^4$. Since only three electrons in the p subshell can have $m_s = +\frac{1}{2}$, the fourth must have $m_s = -\frac{1}{2}$, so $M_S = \frac{1}{2} + \frac{1}{2} + \frac{1}{2} + (-\frac{1}{2}) = 1$, and it follows that $S = 1$. To maximize L, we note that, as for nitrogen, the three electrons with $m_s = +\frac{1}{2}$ will have $m_l = +1$, 0, and -1, and we maximize M_L by giving the fourth electron $m_l = +1$. Thus $M_L = +1$, and $L = 1$.

Let us look now at the energy levels of helium. The ground-state configuration of helium is $1s^2$. Both electrons are s electrons, with $l = 0$, and so the only possible value of L is zero. With two electrons, we might expect a total spin of 0 or 1, as with carbon; however, the Pauli principle requires that the spin of the two electrons be opposite, so that one has $m_s = +\frac{1}{2}$ and the other has $m_s = -\frac{1}{2}$. The *only* possible total M_S is therefore zero, so the ground state of helium has $L = 0$ and $S = 0$. The first excited state has configuration $1s^1 2s^1$. Since both electrons still have $l = 0$, we must again have $L = 0$. However, the total spin S can now be 0 or 1, since the Pauli principle does not apply—the two electrons already have different principal quantum numbers n, and so there is nothing to prevent them from having the same m_s. There are, therefore, *two* "first excited states" of helium, one with $L = 0$ and $S = 0$, and another with $L = 0$ and $S = 1$. (Both of these states have configuration $1s^1 2s^1$.) A state with $S = 0$ is called a *singlet* state (because there is only a single possible M_S value), and a state with $S = 1$ is called a *triplet* state (because there are three possible M_S values: $+1$, 0, -1). The classification of states into singlet and triplet is important when we consider the *selection rules* for transitions between states; these selection rules tell us which transitions are allowed (and therefore likely) to occur and which are not. The selection rules, which involve both L and S, are

$$\Delta L = 0, \pm 1 \qquad (8.8)$$

L and S selection rules

$$\Delta S = 0 \qquad (8.9)$$

(There are no selection rules for n.) Of course, the selection rule $\Delta l = \pm 1$ *for the single electron that makes the transition* still applies. For the

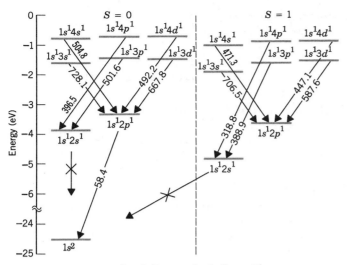

FIGURE 8.13 Energy level diagram for helium. The states are grouped into singlets ($S = 0$) and triplets ($S = 1$). Some of the transitions in the optical and ultraviolet regions are shown. Transitions marked with an X would violate the $\Delta l = \pm 1$ selection rule.

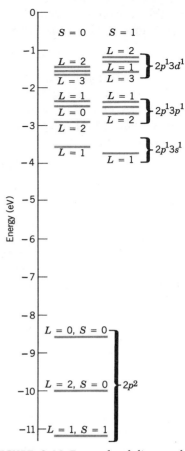

FIGURE 8.14 Energy level diagram for carbon. Each group of levels is labeled with the electron configuration. Each individual level is labeled with the total L and S.

two $1s^12s^1$ states of helium, the Δl rule does not permit either state to make transitions to the $1s^2$ ground state ($2s$ to $1s$ would be $\Delta l = 0$), and in addition, the ΔS rule forbids the triplet ($S = 1$) states from decaying to the $S = 0$ ground state. These transitions can thus only occur by violating these selection rules, and since that is a very unlikely event, the transitions occur with very low probability. Energy levels that have a low probability of decay must "live" for a long time before they decay; such states are known as *metastable* states.

Figure 8.13 shows the energy levels and transitions in helium. The singlet and triplet levels are grouped separately, since transitions between singlet and triplet levels would violate the $\Delta S = 0$ selection rule.

Figure 8.14 shows the energy-level diagram of carbon. Notice the increasing complexity of the diagram, compared with the alkali metals and even with helium. This follows from the coupling of two electrons, both of whose l values may be different from zero. We have already discussed how the $2p^2$ configuration can give $L = 0$, 1, or 2 and $S = 0$ or 1. Only one of these ($L = 1$, $S = 1$) will be the ground state of carbon; the others will be excited states. More excited states can be obtained by promoting one of the $2p$ electrons to a higher level, giving configurations of $2p^13s^1$ ($L = 1$, $S = 0$ or 1), $2p^13p^1$ ($L = 0$, 1, or 2; $S = 0$ or 1), $2p^13d^1$ ($L = 1$, 2, or 3; $S = 0$ or 1) and so forth. Imagine the difficulty of analyzing the energy level diagram of the rare earths or actinides, which have unfilled f subshells ($l = 3$) with as many as 14 electrons!

There are three means by which radiation can interact with the energy levels of atoms (depicted in Figure 8.15). The first two we have already discussed. An atom in an excited state makes a transition to a lower state, with the emission of a photon. (In all the examples we will consider

8.8 LASERS

here, the photon energy will be equal to the energy difference of the two atomic states.) This is *spontaneous emission,* which we represent as

$$\text{atom}^\star \longrightarrow \text{atom} + \text{photon}$$

where the asterisk indicates an excited state.

The second interaction, *induced absorption,* is responsible for absorption spectra and resonance absorption. An atom in the ground state absorbs a photon (of the proper energy) and makes a transition to an excited state. Symbolically:

$$\text{atom} + \text{photon} \longrightarrow \text{atom}^\star$$

The third interaction, which is responsible for the operation of the laser, is *induced (or stimulated) emission.* In this process, an atom is in an excited state. A passing photon of just the right energy (again, equal to the energy difference of the two levels) induces the atom to emit a photon and make a transition to the lower, or ground, state. (Of course, it would eventually have made that transition left on its own, but it makes it *sooner* after being prodded by the passing photon.) Symbolically,

$$\text{atom}^\star + \text{photon} \longrightarrow \text{atom} + 2\text{ photons}$$

The significant detail is that the two photons that emerge are traveling in *exactly the same direction* with *exactly the same energy,* and the associated electromagnetic waves are *perfectly in phase (coherent).*

Suppose we have a collection of atoms, all in the same excited state, as shown in Figure 8.16. A photon passes the first atom, causing induced emission and resulting in two photons. Each of these two photons causes an induced emission process, resulting in four photons. This process continues, doubling the number of photons at each step, until we build up an intense beam of photons, all coherent and moving in the same direction. In its simplest interpretation, this is the basis of operation of the laser. (The word *laser* is an acronym for *L*ight *A*mplification by *S*timulated *E*mission of *R*adiation.)

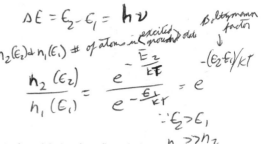

$$\Delta E = E_2 - E_1 = h\nu$$

$$n_2(E_2) \neq n_1(E_1) \; \#\text{ of atoms in } \binom{\text{excited}}{\text{ground}} \text{ state}$$

Boltzmann factor

$$\frac{n_2(E_2)}{n_1(E_1)} = \frac{e^{-\frac{E_2}{kT}}}{e^{-\frac{E_1}{kT}}} = e^{-(E_2-E_1)/kT}$$

$$\therefore E_2 > E_1$$

$$n_1 \gg n_2$$

Induced (stimulated) emission

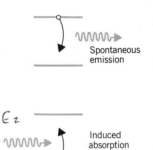

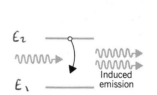

FIGURE 8.15 Interactions of radiation with atomic energy levels.

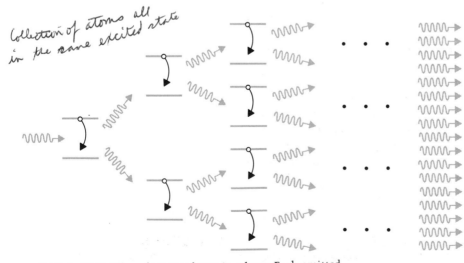

FIGURE 8.16 Buildup of intense beam in a laser. Each emitted photon interacts with an excited atom and produces two photons.

coherent unidirectional beam

This simple model for a laser will not work, for several reasons. The first is simply that it is difficult to keep a collection of atoms in their excited states until they are stimulated to emit the photon (we don't want any *spontaneous* emission). The second reason is that atoms that happen to be in their ground state will undergo absorption and will thus *remove* photons from the beam as it builds up.

A solution to these problems is to choose an atom in which there are three levels, as shown in Figure 8.17. The atoms, originally in the ground state, are "pumped" into the excited state by some external source of energy (an electrical pulse or a flash of light). The excited state decays (by spontaneous emission) very rapidly into a lower excited state, which is a metastable state (the atom remains in that level for a long time.) The transition from the metastable state back to the ground state is the "lasing" transition, induced by a passing photon. This system (a "three-level" laser) solves the first of our problems that arose with the two-level laser—placing the collection of atoms in their excited states—but it doesn't solve the second problem: any atom in the ground state will absorb the lasing transition and remove photons from the beam.

The four-level laser illustrated in Figure 8.18 relieves this remaining difficulty. The ground state is pumped to an excited state that decays rapidly to the metastable state, as with the three-level laser. The lasing transition proceeds from the metastable state to yet another excited state, which in turn decays *rapidly* to the ground state. *The atom in its ground state thus cannot absorb at the energy of the lasing transition*, and we have a workable laser.

The helium-neon laser is an example of such a four-level laser. A mixture of helium and neon gas (about 90 percent helium) is contained in a narrow tube, as shown in Figure 8.19. An electrical current in the gas "pumps" the helium from its ground state to the excited state at an energy of about 20.6 eV. You will recall that this is a metastable state of helium—the atom remains in that state for a relatively long time because a $2s$ electron is not permitted to return to the $1s$ level by photon emission. Occasionally, an excited helium atom will collide with a ground-state neon atom. When this occurs, the 20.6 eV of excitation energy may be transferred to the neon atom, since neon just happens to have an excited state at 20.6 eV, and the helium atom will return to its ground state. Symbolically,

$$\text{helium}^* + \text{neon} \longrightarrow \text{helium} + \text{neon}^*$$

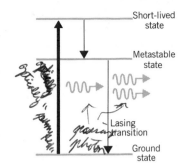

FIGURE 8.17 A three-level atom.

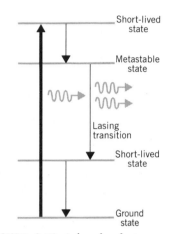

FIGURE 8.18 A four-level atom.

Helium-neon laser

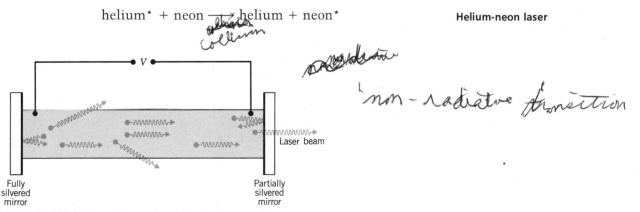

FIGURE 8.19 Schematic diagram of a He-Ne laser.

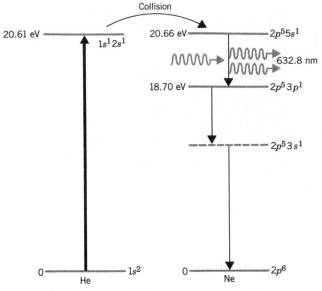

FIGURE 8.20 Sequence of transitions in a He-Ne laser.

where once again the excited state is indicated by the asterisk. The excited state of neon corresponds to removing one electron from the filled $2p$ subshell and promoting it to the $5s$ subshell. From there it decays to the $3p$ level and eventually returns to the $2p$ ground state. Figure 8.20 illustrates this sequence of events and the level schemes. (The level shown with a dashed line, the neon $3s$ level, is not important for the basic operation of the laser, but it is necessary as an intermediate step in the return to the neon ground state, since the $\Delta l = 0$ transition $3p \rightarrow 2p$ is not allowed, but the sequence $3p \rightarrow 3s \rightarrow 2p$ is permitted.)

At any given time, there are more neon atoms in the $5s$ state than in the $3p$ state, since the good energy matchup of the $5s$ state with the helium excited state gives a much higher probability of the $5s$ state in neon being excited. This situation, in which a higher state has a greater population than a lower state, is known as *population inversion*, and is essential to the operation of the laser.

Occasionally a neon atom in the $5s$ state will emit a photon (at a wavelength of 632.8 nm) parallel to the axis of the tube. This photon will cause stimulated emission by other atoms, and a beam of coherent (in-phase) radiation will eventually build up traveling along the tube axis. Mirrors are carefully aligned at the ends of the tube to help in the formation of the coherent wave, as it bounces back and forth between the two ends of the tube, causing additional stimulated emission. One of the mirrors is only partially silvered, allowing a portion of the beam to escape through one end.

The laser is not a particularly efficient device; the small helium-neon lasers you have probably seen used for laboratory or demonstration experiments have a light output of perhaps a few milliwatts; the electric power required to operate such a device may be of the order of 10 to 100 W, and thus the efficiency (power out ÷ power in) of such a device is only about 10^{-4} to 10^{-5}. It is the *coherence* and *directionality* of the laser beam and its *energy density* that make the laser such a useful

Population inversion

[handwritten margin notes] Wish to make $n_2 > n_1$ create a "negative temperature" state

device—its power can be concentrated in a beam only a few millimeters in diameter and thus even a small laser can deliver 100 to 1000 W/m². Larger lasers in the megawatt (10^6 W) range are presently readily available, and research laboratories are developing terawatt (10^{12} W) lasers for special applications. These powerful lasers do not operate continuously, but are instead *pulsed,* producing short (perhaps 10^{-8} s) pulses at rates of order 100 Hz. (Such a pulse is, in fact, an excellent example of a wave packet.)

SUGGESTIONS FOR FURTHER READING

$N_A = 6 \times 10^{23}$ molecules/mole

Some additional basic features of the vector representation of atomic states are in:

K. W. Ford, *Classical and Modern Physics* (Lexington, Xerox College Publishing, 1974), Volume 3, Chapter 24.

Two slightly more advanced works including more detail are:

H. Semat and J. R. Albright, *Introduction to Atomic and Nuclear Physics* (New York, Holt, Rinehart and Winston, 1972).
J. C. Willmott, *Atomic Physics* (Chichester, Wiley, 1975).

A still higher level reference is:

H. G. Kuhn, *Atomic Spectra* (New York, Academic Press, 1969).

Finally, three classic works that include some introductory and advanced material covering almost all aspects of atomic structure:

A. C. Candler, *Atomic Spectra and the Vector Model* (Cambridge, Cambridge University Press, 1937).
G. Herzberg, *Atomic Spectra and Atomic Structure* (New York, Prentice-Hall, 1937).
H. E. White, *Introduction to Atomic Spectra* (New York, McGraw-Hill, 1934).

Many reference works, both popular and technical, are available about lasers. A good introductory work is:

B. A. Lengyel, *Lasers* (New York, Wiley, 1971).

Two popular articles by the 1981 Nobel prize recipient (for his work with lasers):

A. W. Schawlow, "Optical Masers," *Scientific American 204,* 52 (June 1961).
A. W. Schawlow, "Laser Light," *Scientific American 219,* 120 (September 1968). The entire September 1968 issue is devoted to light and includes other articles on lasers.

QUESTIONS

1. Continue Figure 8.1 upward, showing the next two major groups. What will be the atomic number of the next inert gas below Rn? What will be the structure of the eighth row (period) of the periodic table? Where do you expect the first g subshell to begin filling? What properties would you expect the g-subshell elements to have? What will be the atomic number of the second inert gas below radon?

2. Why do the $4s$ and $3d$ subshells appear so close in energy, when they belong to different principal quantum numbers n?

3. Would you expect element 107 to be a good conductor or a poor conductor? How about element 111? Do you expect element 112 to be paramagnetic or diamagnetic?

4. Zirconium frequently is present as an impurity in hafnium metal. Why?

5. Do you expect ytterbium (Yb) to become ferromagnetic at sufficiently low temperatures? What type of magnetic behavior would be expected at ordinary temperatures for polonium (Po)? For francium (Fr)?

6. As we move across the series of transition metal or rare earth elements, we add electrons to the d or f subshells. In chemical compounds, many of these elements show valence states of $+2$ or $+3$, which correspond to removing two s electrons. Explain this apparent paradox.

7. What can you conclude about the electronic configuration of an atom that has both $L = 0$ and $S = 0$ in the ground state?

8. Suppose we do a Stern-Gerlach experiment using an atom that has angular momenta L and S in its ground state. Into how many components will the beam split? Do you expect them to be equally spaced?

9. What is the degeneracy of a state of total orbital angular momentum L that has $S = 0$? What is the degeneracy of a state of total spin angular momentum S that has $L = 0$? What is the *total* degeneracy of a state in which both L and S are nonzero?

10. What L and S values must an atom have in order to show the *normal* Zeeman effect? Does this apply only to the ground state or to excited states also? Can an atom show the normal Zeeman effect in some transitions and the anomalous Zeeman effect in other transitions? Could the same atom even show no Zeeman effect at all in some transitions?

11. Based on the rules for coupling electron l and s values to give the total L and S, explain why filled subshells don't contribute to the magnetic properties of an atom.

12. If an atom in its ground state has $S = 0$, can you infer whether it has an even or an odd number of electrons? What if $L = 0$?

13. The L atomic shell actually contains three distinct levels: a $2s$ level and two $2p$ levels (a fine-structure doublet). If we look carefully at the K_α X ray under high resolution, we see two, not three, different components. Explain this discrepancy.

14. The K_α energies computed using Equation (8.1) are about 0.1 percent low for $Z = 20$, 1 percent low for $Z = 40$, and 10 percent low for $Z = 80$. Why does the simple theory fail for large Z? Could it be because the screening effect has not been handled correctly and that Z_{eff} is not $Z - 1$? If not, can you suggest an alternative reason?

15. The first excited state in sodium is a fine-structure doublet; the wavelengths emitted in the decay of these states are 589.59 nm and 589.00 nm, a difference of 0.59 nm. The excited $4s$ state in sodium (see Figure 8.10) decays to the $3p$ doublet with the emission of radiation at the wavelengths 1138.15 nm and 1140.38 nm, a difference of 2.23 nm. Explain how the $3p$ fine structure can give a wavelength difference of 0.59 nm in one case and 2.23 nm in the other case.

16. Suppose we had a three-level atom, like Figure 8.17, in which the metastable state were the higher excited state; the lasing transition would then be the upper transition. Does this atom solve the problem of absorption of the lasing transition? Would such an atom make a good laser?

PROBLEMS

1. (a) List all elements with a p^3 configuration. (b) List all elements with a d^7 configuration.

2. Give the electronic configuration of (a) P; (b) V; (c) Sb; (d) Pb.

3. Derive Equation (8.1).

4. A certain element emits a K_α X ray of wavelength 0.1940 nm. Identify the element.

5. Compute the K_α X ray energies of calcium $(Z = 20)$, zirconium $(Z = 40)$, and mercury $(Z = 80)$. Compare with the measured values of 3.69 keV, 15.8 keV, and 70.8 keV. (See Question 14.)

6. Draw a Moseley plot, similar to Figure 8.9, for the K_β X rays using the following energies in keV:

Ne 0.858	Mn 6.51	Zr 17.7
P 2.14	Zn 9.57	Rh 22.8
Ca 4.02	Br 13.3	Sn 28.4

Determine the slope and compare with the expected value. [Equation (8.1) only applies to K_α X rays; you will need to derive a similar equation for the K_β X rays.] Determine the Z-axis intercept and give its interpretation.

7. Repeat Problem 6 for the L_α X rays (energies in keV):

Mn 0.721	Rh 2.89
Zn 1.11	Sn 3.71
Br 1.60	Cs 4.65
Zr 2.06	Nd 5.72

Give interpretations of the slope and intercept.

8. In an X-ray tube electrons strike a target after being accelerated through a potential difference V. Estimate the minimum value of V required to observe the K_α X rays of copper.

9. Chromium has the electron configuration $4s^1 3d^5$ beyond the inert argon core. What are the ground-state L and S values?

10. Use Hund's rules to find the ground-state L and S of:
    (a) Ce, configuration [Xe]$6s^2 4f^1 5d^1$
    (b) Gd, configuration [Xe]$6s^2 4f^7 5d^1$
    (c) Pt, configuration [Xe]$6s^1 4f^{14} 5d^9$

11. Using Hund's rules, find the ground-state L and S of (a) fluorine $(Z = 9)$; (b) magnesium $(Z = 12)$; (c) titanium $(Z = 22)$; (d) iron $(Z = 26)$.

12. A certain excited state of an atom has the configuration $4d^1 5d^1$. What are the possible L and S values?

13. Assuming a magnetic moment of one Bohr magneton, find the effective magnetic field that produces the fine-structure splitting of the $3p$ state of sodium. (See Problem 16 or Figure 8.10 for the splittings.)

14. Using the wavelengths given in Figure 8.10, compute the energy difference between the $3d$ and $4d$ states of lithium; do the same for sodium. Compare those values with the corresponding $n = 4$ to $n = 3$ energy difference in hydrogen. Why is the agreement so good, considering the different values of Z?

15. (a) Using the information for lithium given in Figure 8.10, compute the energy difference of the $3p$ and $3d$ states. (b) Compute the energy of the $3s$, $4s$, and $5s$ states above the ground state. (c) The ionization energy of lithium in its ground state is 5.39 eV. What is the ionization energy of the $2p$ state? Of the $3s$ state?

16. Consider the following transitions from different fine-structure doublets in sodium (all wavelengths in nm):

$$3p-3s:\ \ 588.995,\ 589.592$$
$$4p-3s:\ \ 330.303,\ 330.241$$
$$5p-3s:\ \ 285.307,\ 285.286$$
$$6p-3s:\ \ 268.047,\ 268.038$$

Compute the splitting, in eV, of these four p states. Determine the dependence of this splitting on the quantum number n, and compare with the simple theory of Section 7.8.

17. Using the wavelengths given in Figure 8.13, compute the energy difference between the $1s^14p^1$ and $1s^13p^1$ singlet $(S = 0)$ states in helium. Compare this energy difference with the value expected using the Bohr model, assuming that the p electron is screened by the s electron. Repeat the calculation for the $3d$ and $4d$ triplet $(S = 1)$ states.

18. (a) List the six possible sets of quantum numbers (n, l, m_l, m_s) of a $2p$ electron. (b) Suppose we have an atom such as carbon, which has two $2p$ electrons. Ignoring the Pauli principle, how many different possible combinations of quantum numbers of the two electrons are there? (c) How many of the possible combinations of part (b) are eliminated by applying the Pauli principle? (d) Suppose carbon is in an excited state with configuration $2p^13p^1$. Does the Pauli principle apply? How many different sets of quantum numbers for the two electrons will there be?

19. Use the degeneracies of the states with all possible total L and S to find how many different levels the $2p^13p^1$ excited state of carbon includes. (See Figure 8.14.) Compare this result with the result of counting the individual m_l and m_s values from Problem 18(d). (See also Question 9.)

20. (a) What is the longest wavelength of the absorption spectrum of lithium? (b) What is the longest wavelength of the absorption spectrum of helium? In what region of the spectrum does this occur? (c) What are the *shortest* wavelengths in the absorption spectra of helium and lithium? In what region of the electromagnetic spectrum are these?

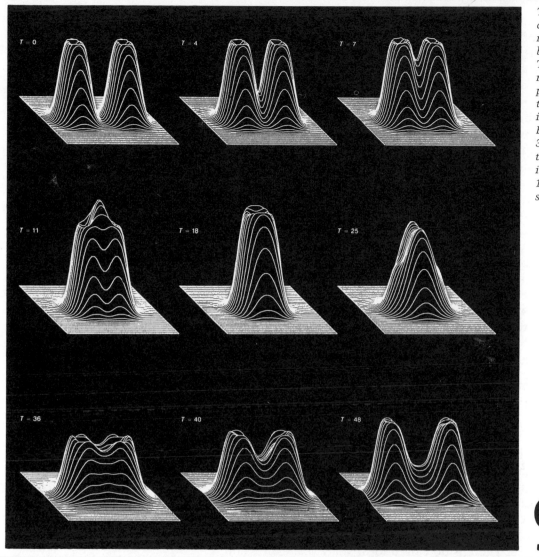

This sequence of computer drawings represents the collision between two ^{12}C nuclei. The vertical axis represents the density of protons and neutrons in the separate nuclei and in their combination. Each unit of time T is 3.3×10^{-24} second, and the collision area shown is a square about 18×10^{-15} meter on each side.

9

NUCLEAR STRUCTURE AND RADIOACTIVITY

When we probe deep inside the atom, we find at its center the nucleus, occupying only 10^{-15} of the volume of the atom, but providing the electrical force that holds the atom together. Were it not for the Coulomb attraction provided by the nucleus, the mutual repulsion of the electrons would cause the atom to fly apart. What keeps the nucleus itself from flying apart, under the repulsive force of the positive charges? A positive charge sitting at the surface of the nucleus experiences an electrostatic repulsive force that gives a potential energy of roughly 100 MeV. In order to keep that positive charge inside the nucleus, the nuclear force must provide a binding energy in excess of 100 MeV—thousands of times larger than typical atomic binding energies!

There are many similarities between atomic structure and nuclear structure, which will make our study of the properties of the nucleus somewhat easier. Nuclei are subject to the laws of quantum physics. They have ground and excited states and emit photons (known as *gamma rays*) in transitions between the excited states. Just like atomic states, nuclear states are labeled by their total angular momentum.

There are, however, two major differences between the study of atomic and nuclear properties. In atomic physics, the electrons experience the force provided by the nucleus; in nuclear physics, there is no such external agent. The constituents of the nucleus move about under the influence of a force that *they themselves* provide. The mutual interactions of the electrons have an influence on the atomic level scheme, but it is a relatively small influence, and we saw how we could understand a great deal of atomic structure based primarily on the study of the interaction of the electron and the nucleus; we treat the effect of the other electrons as a minor perturbation. In nuclear physics, the mutual interaction of the nuclear constituents is just what provides the nuclear force, so we cannot treat this many-body problem as a perturbation. We therefore cannot avoid the mathematical difficulties in the nuclear case, as we did in the atomic case.

The second problem associated with nuclear physics is that we cannot write the nuclear force in a simple form like the Coulomb force or the gravitational force. There is no closed-form analytical expression that can be written to describe the mutual forces of the nuclear constituents.

In spite of these difficulties, we can learn a great deal about the properties of the nucleus by studying the interactions between different nuclei, the radioactive decay of nuclei, and the properties of some nuclear constituents. In this chapter and the next we describe these studies and how we learn about the nucleus from them.

9.1 NUCLEAR CONSTITUENTS

From the work of Rutherford, Bohr, and their contemporaries, it was learned that the positive charge of the atom is confined in a very small nuclear region at the center of the atom, that the nucleus has a charge of $+Ze$, and that the nucleus provides most (99.9 percent) of the atomic mass. It was also known that the masses of the atoms were very nearly integer multiples of the mass of hydrogen, the lightest atom; a glance at Appendix B confirms this observation. We call this integer A, the *mass number*. It is therefore reasonable to suppose that the hydrogen nucleus is composed of a fundamental unit of positive charge. This fundamental unit is the *proton*, with a mass equal to the atomic mass of hydrogen, less the electronic mass and binding energy, and an electric charge of $+e$. If a heavier nucleus contained A of these protons, it would have a nuclear charge of Ae rather than Ze; since $A > Z$ for all atoms heavier than hydrogen, this model gives too much positive charge to the nucleus. This difficulty was removed by the *proton-electron model*, in which it was postulated that the nucleus also contained $(A - Z)$ electrons. Under these assumptions, the nuclear mass would be about A times the mass of the proton (since the mass of the electrons is negligible) and the nuclear electric charge would be $A(+e) + (A - Z)(-e) = Ze$, in agreement with experiment. However, this model leads to several difficulties. First, as we discovered in Chapter 4, the presence of electrons in the nucleus is not consistent with the uncertainty principle, which would require those electrons to have extremely large kinetic energies. A more serious problem concerns the total *intrinsic spin* of the nucleus.

From measurements of the *very* small effect of the nuclear magnetic moment on the atomic transitions (called the *hyperfine splitting*), we know that the proton has an intrinsic spin of $\frac{1}{2}$, just like the electron. Consider an atom of deuterium, sometimes known as "heavy hydrogen." It has a nuclear charge of $+e$, just like ordinary hydrogen, but a mass of two units, twice that of ordinary hydrogen. The proton-electron nuclear model would then require that the deuterium *nucleus* contain two protons and one electron, giving a net mass of two units and a net charge of one. Each of these three particles has a spin of $\frac{1}{2}$, so the rules for adding angular momenta in quantum mechanics would lead to a spin of deuterium of either $\frac{1}{2}$ or $\frac{3}{2}$. However, the measured total spin of deuterium is 1. For these and other reasons, the hypothesis that electrons are a nuclear constituent must be discarded.

The resolution of this dilemma came in 1932 with the discovery of the *neutron*, a particle of roughly the same mass as the proton (actually about 0.1 percent more massive) but having no electric charge. According to the *proton-neutron* model, a nucleus consists of Z protons and $(A - Z)$ neutrons, giving a total charge of Ze and a total mass of roughly A, since the proton and neutron masses are roughly the same.

Since the proton and neutron are, except for their electric charges, so similar to one another, they are classified together as *nucleons*. Some properties of the two nucleons are listed in Table 9.1.

The chemical properties of a certain element depend on the atomic number Z, but not on the mass number A. It is possible to have two different nuclei, with the same Z but with different A (i.e., with different numbers of neutrons). Atoms of these nuclei will be identical in all their chemical properties, differing only in mass and in those properties that depend on mass. Nuclei with the same Z but different A are called *iso-*   **Isotopes**

Table 9.1 Properties of the Nucleons

Name	Charge	Mass Energy	Spin
Proton	$+e$	938.28 MeV	$\frac{1}{2}$
Neutron	0	939.57 MeV	$\frac{1}{2}$

iso topes. Hydrogen, for example, has three isotopes: ordinary hydrogen ($Z = 1$, $A = 1$), deuterium ($Z = 1$, $A = 2$), and the radioactive isotope tritium ($Z = 1$, $A = 3$). All of these are indicated by the chemical symbol H. When we discuss nuclear properties it is important to distinguish among the different isotopes. We do this by indicating, along with the chemical symbol, the atomic number Z, the mass number A, and the *neutron number* $N = A - Z$ in the following format:

$$\underline{\underline{{}^{A}_{Z}X_N}}$$

where X is any chemical symbol. The three isotopes of hydrogen would be indicated as ${}^{1}_{1}H_0$, ${}^{2}_{1}H_1$, and ${}^{3}_{1}H_2$.

EXAMPLE 9.1

Give the proper symbol for the following:
(a) The isotope of helium with mass number 4.
(b) The isotope of tin with 66 neutrons.
(c) An isotope with mass number 235 that contains 143 neutrons.

Solution

(a) From the periodic table, we find that helium has $Z = 2$. Since $A = 4$, $N = A - Z = 2$. Thus the symbol would be ${}^{4}_{2}He_2$.
(b) Again from the periodic table, we know that for tin (Sn), $Z = 50$. Since we are given $N = 66$, then $A = Z + N = 116$. The symbol is ${}^{116}_{50}Sn_{66}$.
(c) Given that $A = 235$ and $N = 143$, we know that $Z = A - N = 92$. From the periodic table, we find that this element is uranium, and so the proper symbol for this isotope is ${}^{235}_{92}U_{143}$.

In Appendix B you will find a list of isotopes and some of their properties.

9.2 NUCLEAR SIZES AND SHAPES

It is as difficult to define precisely the radius of a nucleus as it is the radius of an atom. The probability distribution for atomic electrons makes the atom appear like a "fuzzy ball" of charge with no sharp boundary; the charge distribution does not stop at any clearly defined point. We can, however, take the average or most probable radius of the outer electronic orbit to represent the radius of an atom.

The nucleus must be treated in much the same way, although there are not neutron or proton orbits that we can use for this purpose. Most nuclei are rather spherical (although some are slightly stretched or flattened) and the radial dependence of the nuclear density can be represented as shown in Figure 9.1. From a variety of experiments, we know many remarkable features of the nuclear density. We have already discussed how strong the nuclear force must be to keep the Coulomb repulsion from blowing the nucleus to pieces. We might expect that this force would cause the protons and neutrons to congregate at the center of the nucleus, giving an increasing density in the central region. However, Figure 9.1 suggests that this is not the case—the density remains quite constant. Some mechanism keeps the nucleus from collapsing toward the center. This fact gives one important clue about a property of the nuclear force, as we discuss in Section 9.4. Another interesting feature of Figure 9.1 is that the density of a nucleus seems not to depend on the

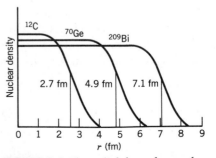

FIGURE 9.1 The radial dependence of the nuclear charge density.

mass number A—very light nuclei have roughly the same density as very heavy nuclei. Stated another way, the number of neutrons and protons per unit volume is approximately constant over the entire range of nuclei:

$$\frac{\text{number of neutrons and protons}}{\text{volume of nucleus}} = \frac{A}{\frac{4}{3}\pi R^3} \cong \text{constant}$$

Thus

$$A \propto R^3$$

and so we have a proportionality between the nuclear radius R and the cube root of the mass number:

$$R \propto A^{1/3}$$

or, defining a constant of proportionality R_0,

Nuclear radii

$$R = R_0 A^{1/3} \qquad (9.1)$$

The constant R_0 must be determined by experiment, and a typical experiment might be to scatter charged particles (alpha particles or electrons, for example) from the nucleus and to infer the radius of the nucleus from the distribution of scattered particles. From such experiments, we know the value of R_0 is approximately 1.2 × 10⁻¹⁵ m. (The exact value depends, as in the case of atomic physics, on exactly how we define the radius, and values of R_0 usually range from 1.0×10^{-15} m to 1.5×10^{-15} m.) The length 10^{-15} m is one femtometer (fm), but physicists often refer to this length as one fermi, in honor of the Italian-American physicist, Enrico Fermi.

(handwritten: 1 fm = 10⁻¹⁵ m)

Compute the approximate nuclear radii of carbon ($A = 12$), germanium ($A = 70$), and bismuth ($A = 209$).

EXAMPLE 9.2

Carbon: $R = (1.2 \text{ fm})A^{1/3} = (1.2 \text{ fm})(12)^{1/3} = 2.7 \text{ fm}$
Germanium: $R = (1.2 \text{ fm})A^{1/3} = (1.2 \text{ fm})(70)^{1/3} = 4.9 \text{ fm}$
Bismuth: $R = (1.2 \text{ fm})A^{1/3} = (1.2 \text{ fm})(209)^{1/3} = 7.1 \text{ fm}$

Solution

Compute the density of a typical nucleus, and find the resultant mass if we could manufacture a nucleus with a radius of 1 cm.

EXAMPLE 9.3

$$\rho = \frac{m}{V} = \frac{A}{\frac{4}{3}\pi R^3} = \frac{A}{\frac{4}{3}\pi R_0^3 A} = \frac{1 \text{ u}}{7 \times 10^{-45} \text{ m}^3} = 2 \times 10^{17} \text{ kg/m}^3$$

Solution

The mass of our hypothetical nucleus would be

$$m = \rho V = \left(2 \times 10^{17} \frac{\text{kg}}{\text{m}^3}\right)\left(\frac{4}{3}\pi\right)(0.01 \text{ m})^3 = 8 \times 10^{11} \text{ kg}$$

about the mass of a 1-km sphere of ordinary matter!

This calculation illustrates the extreme density of what physicists call *nuclear matter*. Although examples of such nuclear matter in bulk are not found on Earth (a sample of nuclear matter the size of a large building would have a mass as great as that of the entire Earth), they are found

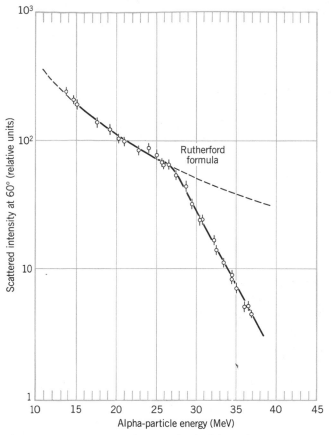

FIGURE 9.2 Deviations from Rutherford formula in scattering from ^{208}Pb at high alpha-particle energies.

Enrico Fermi (1901–1954, Italy–United States). There is hardly a field of modern physics to which he did not make contributions in theory or experiment; perhaps Newton was the only other scientist with such skills in both theoretical and experimental work. He developed the statistical laws for spin-$\frac{1}{2}$ particles and in the 1930s he proposed a theory of beta decay that is still used today. He was the first to demonstrate the transmutation of elements by neutron bombardment (for which he received the 1938 Nobel prize), and he directed the construction of the first nuclear reactor.

in certain extremely hot stars, in which the high temperature "boils off" the electrons from atoms, leaving only the bare nuclei, which then form an aggregate nucleus out of the entire stellar material.

One way of measuring the size of a nucleus is to scatter charged particles, such as alpha particles, as in Rutherford scattering experiments. As long as the alpha particle is outside the nucleus, the Rutherford scattering formula holds, but when the distance of closest approach is less than the nuclear radius, deviations from the Rutherford formula occur. Figure 9.2 shows the results of a Rutherford scattering experiment in which such deviations are observed.

Other scattering experiments can also be used to measure the nuclear radius. Figure 9.3 shows a sort of "diffraction pattern" that results from the scattering of energetic electrons by a nucleus. In each case the first diffraction minimum is clearly visible. (The intensity at the minimum doesn't fall to zero because the nuclear density doesn't have a sharp edge, as illustrated in Figure 9.1.) For diffraction by a circular disc of diameter D, the first minimum should appear at an angle of $\theta = \sin^{-1}(1.22\ \lambda/D)$, where λ is the wavelength of the scattered radiation. [You will recall that the diffraction pattern produced by a slit of width a has a minimum at $\theta = \sin^{-1}(\lambda/a)$.] At an extremely relativistic energy of 420 MeV, the electron deBroglie wavelength is 2.95 fm, and the observed minima of about 44° for ^{16}O and 50° for ^{12}C enable us to calculate

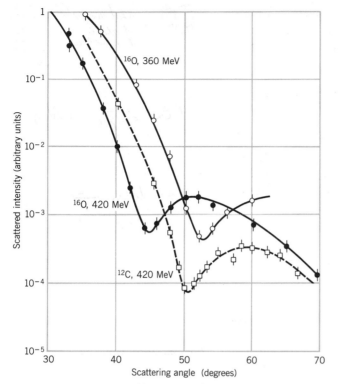

FIGURE 9.3 Electron diffraction by nuclei.

the nuclear diameter and therefore the radius. The results are 2.6 fm for ^{16}O and 2.3 fm for ^{12}C, in good agreement with the values 3.0 fm and 2.7 fm computed from Equation (9.1). The effect of changing the wavelength of the scattered radiation to 3.44 fm (corresponding to an energy of 360 MeV) can also be seen in Figure 9.3; the diffraction minimum shifts to about 52°, corresponding to a nuclear radius of 2.7 fm for ^{16}O, in agreement with the 420-MeV value.

Suppose we have a proton and an electron at rest separated by a large distance. The total energy of this system is just the total rest energy of the two particles, $m_p c^2 + m_e c^2$. Now we bring the two particles together to form a hydrogen atom in its ground state. In the process, several photons are emitted, the *total* energy of which is 13.6 eV. Thus the total energy of this system is the rest energy of the hydrogen atom, $m_H c^2$, plus 13.6 eV. Conservation of energy demands that the total energy of the system of isolated particles must equal the total energy of atom plus photons:

$$m_e c^2 + m_p c^2 = m_H c^2 + 13.6 \text{ eV}$$

or

$$m_e c^2 + m_p c^2 - m_H c^2 = 13.6 \text{ eV}$$

That is, the mass energy of the combined system (the hydrogen atom) is less than the mass energy of its constituents by 13.6 eV. This energy difference is the *binding energy* of the atom. We can regard the binding en-

9.3 NUCLEAR MASSES AND BINDING ENERGIES

infinitely separated proton & electron

$m_p c^2 \quad --- \quad \infty - \underset{m_e c^2}{\bigcirc}$

ergy as either the "extra" energy we obtain when we assemble an atom from its components (the 13.6 eV of photons) or else the energy we must supply to disassemble the atom into its components.

Nuclear binding energies are calculated in much the same way. Consider, for example, the nucleus of deuterium, $^2_1H_1$, which is composed of one proton and one neutron. The binding energy of deuterium would be

$$B = m_n c^2 + m_p c^2 - m_D c^2 \qquad (9.2)$$

where m_D is the mass of the deuterium nucleus. In using mass tables to do these calculations, it is important to remember that the tabulated masses are *atomic masses, not nuclear masses*. To calculate nuclear binding energies, however, we must use nuclear masses. The relationship between atomic and nuclear masses is:

$m(\text{atom}) = m(\text{nucleus}) + Z \cdot m(\text{electron})$

+ *total electron binding energy* (9.3)

Nuclear mass energies are of the order of 10^9 to 10^{11} eV, electron mass energies are of the order of 10^6 to 10^8 eV, and electron binding energies are of the order of 1 to 10^4 eV. Thus the last term of Equation (9.3) is very small compared with the other two terms, and we can safely neglect it to the accuracy we need for these calculations.

The *nuclear* mass of hydrogen (the proton mass) is then just the *atomic* mass of hydrogen (1.007825 u) less the mass of one electron. The *nuclear* mass of deuterium is the *atomic* mass of deuterium (2.014102 u) less the mass of one electron. Substituting for the *nuclear* masses that appear in Equation (9.2), we can find the binding energy in terms of the *atomic* masses:

$$B = m_n c^2 + [m(^1H) - m_e]c^2 - [m(^2H) - m_e]c^2$$

$$= [m_n + m(^1H) - m(^2H)]c^2$$

Notice that the electron mass cancels out in this calculation, as it would in any such calculation, since the constituents will include Z hydrogen atoms (with Z electrons) and the atom of atomic number Z will also include Z electrons. We can therefore generalize this equation to give the total binding energy of any nucleus $^A_ZX_N$:

$$B = [Nm_n + Zm(^1_1H_0) - m(^A_ZX_N)]c^2 \qquad (9.4)$$

Nuclear binding energy *(in terms of atomic masses)*

The masses that appear in Equation (9.4) are *atomic* masses. For deuterium, we would then have

$B = (1.008665\text{ u} + 1.007825\text{ u} - 2.014102\text{ u})931.5\text{ MeV/u}$

$= 2.224\text{ MeV}$

Find the total binding energy B and also the binding energy per nucleon B/A for $^{56}_{26}Fe_{30}$ and $^{238}_{92}U_{146}$.

EXAMPLE 9.4

From Equation (9.4), for $^{56}_{26}Fe_{30}$ with $N = 30$ and $Z = 26$,

Solution

$B = (30 \times 1.008665\text{ u} + 26 \times 1.007825\text{ u} - 55.934939\text{ u})931.5\text{ MeV/u}$

$= 492.3\text{ MeV}$

$$\frac{B}{A} = (492.3 \text{ MeV})/56 = \underline{8.791 \text{ MeV per nucleon}}$$

For $^{238}_{92}\text{U}_{146}$,

$$B = (146 \times 1.008665 \text{ u} + 92 \times 1.007825 \text{ u} - 238.050786 \text{ u})931.5 \text{ MeV/u}$$

$$= 1802 \text{ MeV}$$

$$\frac{B}{A} = (1802 \text{ MeV})/238 = \underline{7.571 \text{ MeV per nucleon}}$$

^{56}Fe is more lightly bound than ^{238}U

Example 9.4 gives us insight into an important aspect of nuclear structure. The values of B/A we calculated show us that the nucleus ^{56}Fe is *relatively* more tightly bound than is the nucleus ^{238}U—the binding energy *per nucleon* is greater for ^{56}Fe than for ^{238}U. Alternatively, this calculation shows that, given a large supply of protons and neutrons, we would release more energy by assembling those nucleons into nuclei of ^{56}Fe than we would by assembling them into nuclei of ^{238}U.

Repeating the previous calculation for the entire range of nuclei, we would obtain the results shown in Figure 9.4. The binding energy per nucleon starts at small values (0 for the proton and neutron, 1.11 MeV for deuterium), rises to a maximum of 8.79 MeV for ^{56}Fe, and then falls to values of 7.5 MeV for the heavy nuclei.

Figure 9.4 suggests that we can liberate energy from the nucleus in two different ways. If we split a heavy nucleus into two lighter nuclei, energy is released, because the binding energy per nucleon is greater for the two lighter fragments than it is for the original nucleus. This process

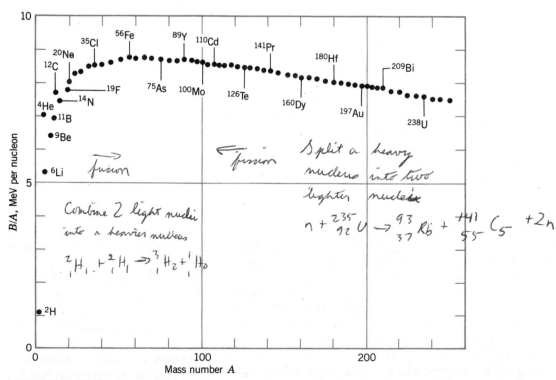

fusion *fission* *Split a heavy nucleus into two lighter nuclei*

Combine 2 light nuclei into a heavier nucleus

$$^{2}_{1}H_{1} + ^{2}_{1}H_{1} \rightarrow ^{3}_{1}H_{2} + ^{1}_{1}H_{0}$$

$$n + ^{235}_{92}U \rightarrow ^{93}_{37}Rb + ^{141}_{55}Cs + 2n$$

FIGURE 9.4 The binding energy per nucleon.

is known as *nuclear fission.* Alternatively, we could combine two light nuclei into a heavier nucleus; again, energy is released when the binding energy per nucleon is greater in the final nucleus than it is in the two original nuclei. This process is known as *nuclear fusion.* We consider fission and fusion in greater detail in Chapter 10.

Our clues to the structure of the hydrogen atom came from studies of transitions between its excited states. The Balmer series and the Ritz combination principle provided information that helped us to understand the forces that hold the atom together. Moreover, understanding atomic structure was helped in hydrogen by the simplicity of its structure—a single electron revolving about the nucleus, with no other electrons to complicate the structure.

With such experience to guide us, it therefore seems advisable to begin our study of the nuclear force by looking at the simplest system in which that force operates—the deuterium nucleus. A study of the excited states of deuterium ought to give us, as in the case of the hydrogen atom, some valuable insight into the nuclear force. Unfortunately, nature conspires against us—the deuterium nucleus has *no* excited states. When we bring a proton and an electron together to form a hydrogen atom, a whole spectrum of photons is emitted as the electron drops into its ground state; from this spectrum we learn the energies of the excited states. When we bring a proton and a neutron together to form a deuterium nucleus, only one photon (of energy 2.224 MeV) is emitted as the system drops directly into its ground state.

Since we can't study this simple system, we must turn to the study of more complicated systems. By doing a variety of different experiments with nuclei, we have learned many features of the nuclear force:

1. It is a very different kind of force from electromagnetism, gravitation, or the other kinds of forces we commonly encounter. It is also the strongest of the known forces, and so it is sometimes known as the *strong* force.

2. The strong nuclear force has a very short range—the region in which the force acts is limited to nuclear dimensions (10^{-15} m or so). There are two major items of evidence for this short range. The first comes from our study of the density of nuclear matter. As we add nucleons to the nucleus, the density remains roughly constant. This suggests to us that each nucleon we add feels a force only from its nearest neighbors, and *not* from all of the other nucleons in the nucleus. The second bit of evidence for the short range comes from the binding energy per nucleon (Figure 9.4). Since the binding energy per nucleon is roughly constant, total nuclear binding energies are roughly proportional to A. A force with a long range (an infinite range, for instance, like the electric or gravitational force) would be expected to have a binding energy proportional to A^2. (For example, the total electrostatic repulsion between the protons in the nucleus is proportional to $Z(Z - 1)$, which is roughly Z^2 for large Z. This comes about because *each* of the Z protons feels the repulsion of the other $Z - 1$.) Figure 9.5 illustrates a representative dependence of the nuclear binding energy on the separation distance between the nucleons.

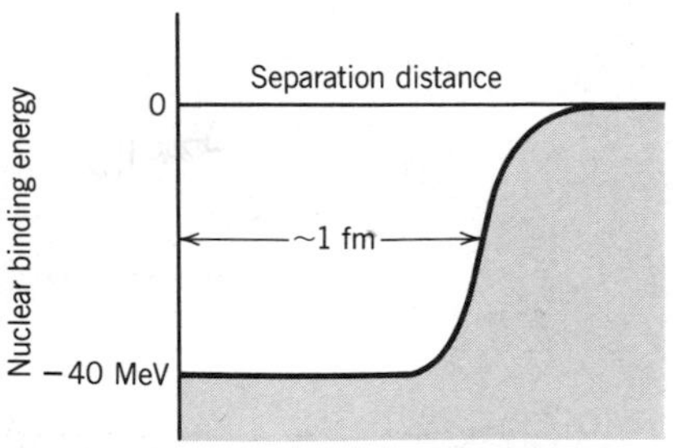

9.4 THE NUCLEAR FORCE

Forces

Strong > Electromagnetic > Weak > Gravitation
  1   2   3   4

decay ↓ Weak

$$\frac{Z(Z-1)e^2}{r^2} \propto Z(Z-1) \sim Z^2$$

Z protons   BE ∝ A

FIGURE 9.5 Dependence of nuclear binding energy on the separation distance of nucleons.

$$\frac{Z(Z-1)}{2} \quad \text{pairs}$$

3. The <u>nuclear force</u> between any two nucleons does not depend on whether the nucleons are protons or neutrons—the *n-p* nuclear force is the same as the *n-n* nuclear force, which is in turn the same as the *p-p* nuclear force.

A successful model for the origin of this short-range force is the *exchange force*. Suppose we have a neutron and a proton in the nucleus. The neutron emits a particle, and it also exerts a strong attractive force on that particle. The nearby proton, if the emitted particle happens to be close to it, can also exert a strong force on the particle, perhaps strong enough to absorb the particle. The proton can then emit a particle that can be absorbed by the neutron. The proton and neutron each exert a strong force on the exchanged particle, and thus they appear to exert a strong force on each other. The situation is similar to that shown in Figure 9.6, in which two people play catch with a ball to which each is attached by a spring. Each player exerts a force on the ball, and the effect is as if each exerted a force on the other.

How can a neutron of rest energy $m_n c^2$ emit a particle of rest energy mc^2 and still remain a neutron, without violating conservation of energy? The answer to this question can be found by remembering the uncertainty principle, $\Delta E \, \Delta t \sim \hbar$. We don't know that energy has been conserved unless we measure it, and we can't measure it more accurately than the uncertainty ΔE in a time interval Δt. We can therefore "violate" energy conservation by an amount ΔE for a short enough time interval $\Delta t = \hbar/\Delta E$. The amount by which we are violating energy conservation in our neutron-proton exchange force model is mc^2, the rest energy of the exchanged particle. This particle can thus only exist for a time interval (in the laboratory frame)

$$\Delta t = \frac{\hbar}{mc^2} \tag{9.5}$$

The longest distance this particle can possibly travel in the time Δt is $x = c \, \Delta t$, since it can't move faster than the speed of light. We therefore have a relationship between the range of the exchange force and the mass of the exchanged particle:

$$x = c \, \Delta t = c \left(\frac{\hbar}{mc^2} \right)$$

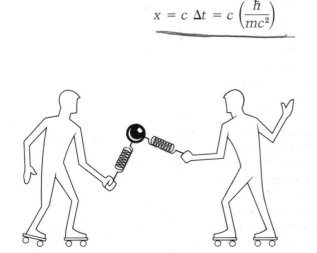

FIGURE 9.6 An attractive exchange force.

or

$$mc^2 = \frac{c\hbar}{x} \qquad (9.6)$$

We know that the range of the nuclear force is about 10^{-15} m, and this allows us to estimate the rest energy of the exchanged particle:

$$mc^2 \cong 200 \text{ MeV}$$

The exchanged particle cannot be a real particle, since the emission of a real particle by the neutron would (by conservation of momentum) cause the neutron to recoil backward, and the absorption of the particle by the proton would also cause the proton to recoil backward. It must thus be a *virtual* particle; if we look carefully at the nucleus, we can see the strong attractive force between the neutron and the proton, but we can't see the virtual particle being exchanged. If we hit the nucleus hard enough in a nuclear reaction, the projectile can perhaps strike a neutron or a proton in such a way as to supply the recoil momentum, allowing the virtual particle to become a real particle and appear in the laboratory. When we carry out this experiment, using for example accelerated protons as projectiles, we find appearing in the laboratory a particle with a rest energy of 140 MeV, remarkably close to the estimate of 200 MeV. This discovery provides strong support for the exchange force model of nuclear forces. We will study the properties of this particle, known as a pi meson, in Chapter 11.

9.5 NUCLEAR STABILITY AND DECAY

One of the remarkable properties of certain nuclei is their ability to transform themselves spontaneously from one value of Z and N to another. Other nuclei are stable and do not decay into different nuclei. Usually, for each A value there are one or two stable nuclei. All other nuclei with that A value would be unstable and would undergo some sort of decay process, until stability was reached.

Figure 9.7 shows a plot of all the known stable nuclei. For light nuclei, the neutron and proton numbers are roughly equal. However, for heavy nuclei, the Coulomb repulsion term $Z(Z - 1)$ begins to grow rapidly, so extra neutrons are needed to supply additional binding energy. Thus heavy stable nuclei all have $N > Z$. There are no stable nuclei with $A = 5$ or 8. The alpha particle ${}^4_2\text{He}_2$ is a particularly stable nucleus $(B/A = 7.07 \text{ MeV})$; a nucleus with $A = 5$, such as ${}^5_2\text{He}_3$ or ${}^5_3\text{Li}_2$, will quickly $(10^{-21}$ s) disintegrate into an alpha particle and a neutron or proton, and a nucleus with $A = 8$ such as ${}^8_4\text{Be}_4$ will quickly break apart into two alpha particles.

The existence of stable nuclei beyond $Z = 92$ is presently the subject of active and intense research. None of these nuclear species exist in nature, and so all must be produced artificially. The man-made elements beyond uranium with $Z = 93$ to 107 are all radioactive, and decay into known stable species. However, there is strong theoretical evidence that leads us to expect the existence of much heavier nuclei, with $Z \cong 114$, which may be stable. These nuclei do not exist in nature, because all naturally occurring nuclear species heavier than ${}^{56}\text{Fe}$ were produced by capturing neutrons and by successive decays. Since all of the nuclei immediately beyond uranium are unstable, there is no way that neutron

[handwritten margin notes:]
for A small, $\boxed{Z=N}$
45° line ${}^4_2\text{He}_2$
${}^8_4\text{Be}_4$

for $A = Z + N$ large,
$\boxed{N > Z}$

${}^{75}_{33}\text{As}_{42}$ ${}^{115}_{49}\text{In}_{66}$

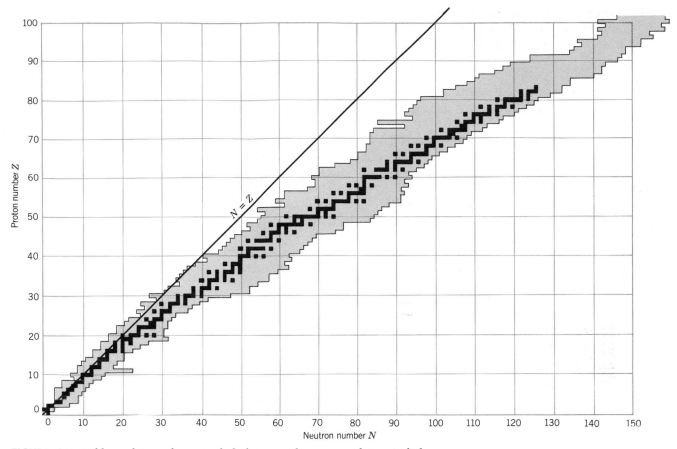

FIGURE 9.7 Stable nuclei are shown in dark; known radioactive nuclei are in light shading.

capture can bridge the gulf from uranium to the next heavier "island of stability."

Unstable nuclei are transformed into other nuclear species by means of two different decay processes that change the Z and N of a nucleus. (Excited states of nuclei can emit photons, nuclear gamma rays, in making transitions that will eventually lead to the ground state. However, no change of Z or N occurs.) These two processes are _alpha decay_ and _beta decay_. The three decay processes (alpha, beta, and gamma decay) are examples of the general subject of _radioactive decay_. In the next section, we establish some of the basic properties of radioactive decay, and in the following sections we treat alpha, beta, and gamma decay separately.

γ decay $[E]^{*} \xrightarrow[photon]{} [G]$

$\alpha = He$

$\beta = e$

9.6 RADIOACTIVE DECAY

Suppose we have a small, laboratory-sized sample (of the order of grams) of radioactive material. The rate at which the radioactive nuclei decay is called the _activity_ of the sample. The greater the activity, the more nuclear decays per second. (The activity has nothing to do with the _kind_ of decays or of radiations emitted by the sample, or with the _energy_ of the emitted radiations. The activity is determined only by the _number_ of decays per second.)

The basic unit for measuring activity is the *curie*.* Originally, the curie was defined as the activity of one gram of radium; that definition has since been replaced by a more convenient one:

$$1 \text{ curie (Ci)} = 3.7 \times 10^{10} \text{ decays/s}$$

One curie is quite a large activity, and so we work more often with units of millicurie (mCi), equal to 10^{-3} Ci, and microcurie (μCi), equal to 10^{-6} Ci.

Our laboratory-sized sample contains the order of 10^{23} atoms. If our sample had the large activity of 1 Ci, about 10^{10} of its nuclei would decay every second. We could also say that any one nucleus has a 10^{-13} probability of decaying during each second. This quantity, the decay probability per nucleus per second, is called the *decay constant* and is represented by λ. We assume that λ is a small number, and that it is a constant—the probability of any one nucleus decaying doesn't depend on the age of the sample. The activity a depends on the number N of radioactive nuclei in the sample and also on the probability λ for each nucleus to decay:

$$a = \lambda N \qquad (9.7)$$

Both a and N are functions of the time t. As our sample decays, N certainly decreases—there are fewer radioactive nuclei left. If N decreases and λ is constant, then a must also decrease with time, so the number of decays per second becomes smaller with increasing time.

We can regard a as the change in the number of radioactive nuclei per unit time—the more nuclei decay per second, the larger is a.

$$a = -\frac{dN}{dt} \qquad (9.8)$$

(We have included a minus sign because dN/dt is negative, since N is decreasing with time, and we want a to be a positive number.) From Equations (9.7) and (9.8) we have

$$\frac{dN}{dt} = -\lambda N \qquad (9.9)$$

or

$$\frac{dN}{N} = -\lambda \, dt \qquad (9.10)$$

This equation can be integrated directly to yield

$$\ln N = -\lambda t + c \qquad (9.11)$$

where c is the constant of integration. We can rewrite this as

$$N = e^{-\lambda t + c} \qquad (9.12)$$

or

$$\boxed{N = N_0 e^{-\lambda t}} \qquad (9.13)$$

---

* The SI unit of activity is the becquerel (Bq), named for the discoverer of radioactivity. One becquerel equals one decay/s, so 1 Ci = 3.7×10^{10} Bq. This unit has not yet been generally accepted, and the curie continues to be the most widely used unit of activity.

(Handwritten margin notes:)

All unstable nuclei follow the radioactive decay law

parent → daughter

$X \longrightarrow X'$

α or β

For any unstable nucleus one instant is like the other ⟹

Probability that an unstable nucleus will decay spontaneous

i) independent of past history of parent
ii) same for all nuclei of same type
iii) independent of external influences (temp, pressure)

Activity

Can't predict the precise time when it will decay

During dt (probability that the nucleus decays in time dt) = $\lambda \, t$

decay / disintegration constant

probability that the nucleus survives time dt = $1 - \lambda \, dt$

$p(t + dt) = (1 - \lambda \, dt) p(t)$

$p(t + dt) - p(t) = -\lambda \, p(t) \, dt$

$\frac{dp}{dt} = -\lambda \, p(t) \qquad \frac{dp}{p} = -\lambda \, dt$

$\ln p(t) = -\lambda t + c_{opt}$

$p(t) = p(0) \, e^{-\lambda t}$

Radioactive decay law

where we have replaced e^c by N_0. At $t = 0$, $N = N_0$, so N_0 is just the original number of radioactive nuclei. Equation (9.13) is the *exponential law of radioactive decay*, which tells us how the number of radioactive nuclei in a sample decreases with time. We can't really measure N, but we can put this equation in a more useful form by multiplying on both sides by λ, which gives

$$a = a_0 e^{-\lambda t} \quad = \lambda N_0 \, e^{-\lambda t} \quad (9.14)$$

where a_0 is the original activity.

Suppose we count the number of decays of our sample in one second (by counting for one second the radiations resulting from the decays). We wait for a while, and then repeat the measurement. Continuing this process, we could then plot the activity a as a function of time, as shown in Figure 9.8. This plot shows the exponential dependence expected on the basis of Equation (9.14).

The half-life, $t_{1/2}$, of the decay is the time that it takes for the activity to be reduced by half, as shown in Figure 9.8. That is, $a = a_0/2$ when $t = t_{1/2}$.

$$\frac{a_0}{2} = a_0 e^{-\lambda t_{1/2}}$$

$$a(t_{1/2}) = a_0 \, e^{\lambda t_{1/2}}$$

from which we find

$$t_{1/2} = \frac{1}{\lambda} \ln 2$$

$$= \frac{0.693}{\lambda}$$

$\quad (9.15) \quad$ **Half-life**

It is often more useful to plot a as a function of t on a semilogarithmic scale, as shown in Figure 9.9. On this kind of plot, Equation (9.14) would be a straight line; fitting a straight line to the data gives the value of λ.

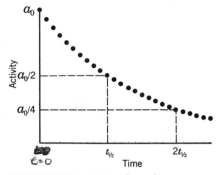

FIGURE 9.8 Activity of a radioactive sample as a function of time.

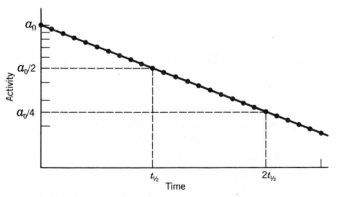

FIGURE 9.9 Semilog plot of activity versus time.

The half-life of ^{198}Au is 2.70 days. (a) What is the decay constant of ^{198}Au? (b) What is the probability that any ^{198}Au nucleus will decay in one second? (c) Suppose we had a 1.00-μg sample of ^{198}Au. What is its activity? (d) How many decays per second occur when the sample is one week old?

EXAMPLE 9.5

(a)

$$\lambda = \frac{0.693}{t_{1/2}} = \frac{0.693}{2.70 \text{ d}} \cdot \frac{1 \text{ d}}{24 \text{ h}} \cdot \frac{1 \text{ h}}{3600 \text{ s}}$$

$$= 2.97 \times 10^{-6} \text{ s}^{-1}$$

(b) The decay probability per second is just the decay constant, so the probability of any ^{198}Au nucleus decaying in one second is 2.97×10^{-6}.

(c) The number of atoms in the sample is:

$$N = 1.00 \times 10^{-6} \text{ g} \cdot \frac{1 \text{ mole}}{198 \text{ g}} \cdot \frac{6.02 \times 10^{23} \text{ atoms}}{\text{mole}}$$

$$= 3.04 \times 10^{15} \text{ atoms}$$

$$a = \lambda N = (2.97 \times 10^{-6} \text{ s}^{-1})(3.04 \times 10^{15})$$

$$= 9.03 \times 10^{9} \text{ decays per second}$$

$$= 0.244 \text{ Ci}$$

(d) The activity decays according to Equation (9.14)

$$a = a_0 e^{-\lambda t}$$

$$= (9.03 \times 10^{9} \text{ decays/s}) e^{-(0.693/2.70 \text{ d})(7 \text{ d})}$$

$$= 1.50 \times 10^{9} \text{ decays/s}$$

The half-life of ^{235}U is 7.04×10^{8} y. A sample of rock, which solidified with the Earth 4.55×10^{9} years ago, contains N atoms of ^{235}U. How many ^{235}U atoms did the same rock have at the time it solidified?

EXAMPLE 9.6

The age of the rock corresponds to

Solution

$$\frac{4.55 \times 10^{9} \text{ y}}{7.04 \times 10^{8} \text{ y}} = 6.46 \text{ half-lives}$$

Since each half-life reduces N by a factor of 2, the overall reduction in N has been

$$2^{6.46} = 88.2$$

The original rock therefore contained $88.2N$ atoms of ^{235}U.

Our study of radioactive decays and nuclear reactions reveals that nature is not arbitrary in selecting the outcome of decays or reactions, but rather that certain laws limit the possible outcomes. We call these laws *conservation laws* and we believe these laws to give us important insight into the fundamental workings of nature. Some of these laws, which we will apply to radioactive decay processes, are discussed as follows.

1. *Conservation of energy.* Perhaps the most important of the conservation laws, it tells us which decays are energetically possible and enables us to calculate rest energies or kinetic energies of decay products. A nucleus X will decay into a lighter nucleus X', with the emission of one or more particles we call collectively x, only if the mass of X is

9.7 CONSERVATION LAWS IN RADIOACTIVE DECAYS

greater than the total mass of $X' + x$. The excess *mass energy* is known as the Q *value* of the decay:

$$m_N(X)c^2 = m_N(X')c^2 + m_N(x)c^2 + Q$$

$$Q = [m_N(X) - m_N(X') - m_N(x)]c^2 \qquad (9.16)$$

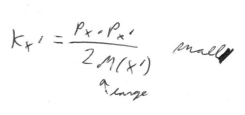

where the m_N's are *nuclear rest masses*. Of course, the decay is only possible if this Q value is positive. The excess energy Q appears as kinetic energy of the decay products (assuming X is initially at rest):

$$Q = K_{X'} + K_x \qquad (9.17)$$

2. *Conservation of linear momentum.* If the initially decaying nucleus is at rest, then the total momentum of all of the decay products must sum to zero

$$\mathbf{p}_{X'} + \mathbf{p}_x = 0 \qquad (9.18)$$

Usually the emitted particle or particles x are much less massive than the residual nucleus X', and the *recoil momentum* $p_{X'}$ yields a very small kinetic energy $K_{X'}$.

If there is only one emitted particle x, Equations (9.17) and (9.18) can be solved simultaneously for $K_{X'}$ and K_x. If x represents two or more particles, we have more unknowns than we have equations, and no unique solution is possible. In this case, a range of values from some minimum to some maximum is permitted for the decay products.

3. *Conservation of angular momentum.* There are two types of angular momentum: the spin angular momentum $\mathbf{s}$ and the motional or orbital angular momentum $\mathbf{l}$. In the rest frame of X, before the decay the total angular momentum is $\mathbf{s}_X$. After the decay, we have the spins of the residual nucleus X' and the decay products x, and also the angular momenta $\mathbf{l} = \mathbf{r} \times \mathbf{p}$ of x and X', which are moving relative to the point in space that was previously occupied by X. Thus

$$\mathbf{s}_X = \mathbf{s}_{X'} + \mathbf{s}_x + \mathbf{l}_{X'} + \mathbf{l}_x \qquad (9.19)$$

The intrinsic spins $\mathbf{s}$ are properties of the particles or nuclei; you have already learned that the electron has $s = \frac{1}{2}$, and nuclei have intrinsic spins also, which originate from the constituent protons and neutrons. The intrinsic spin of a nucleus has values which can be integer or half-integer multiples of $\hbar$, depending on whether A is even or odd. The angular momentum $\mathbf{l}$ is always quantized in integer multiples of $\hbar$.

4. *Conservation of electric charge.* This is such a fundamental part of all decay and reaction processes that it hardly needs elaborating. The total net electric charge before and after the decay must not change.

5. *Conservation of mass number.* In some decay processes, we can create particles (photons or electrons, for example) which did not exist before the decay occurred. (This of course must be done out of the available energy—that is, it takes 0.511 MeV of energy to create an electron.) However, nature does *not* permit us to create or destroy protons and neutrons, although in certain decay processes we can convert neutrons into protons or protons into neutrons. *The total mass number A does not change in decay or reaction processes.* In some decay processes, A remains constant because both Z and N remain unchanged; in other processes Z and N both change in such a way as to keep their sum constant.

9.8 ALPHA DECAY

In alpha decay, an unstable nucleus disintegrates into a lighter nucleus and an alpha particle (a nucleus of ${}^4_2\text{He}$), according to

$$^A_Z X_N \longrightarrow {}^{A-4}_{Z-2} X'_{N-2} + {}^4_2\text{He}_2 \tag{9.20}$$

where X and X' represent different nuclear species.

Decay processes of this sort liberate energy, since the decay products are more tightly bound than the initial nucleus. The liberated energy, which appears as the kinetic energy of the alpha particle and the "daughter" nucleus X', can be found from the masses of the nuclei involved according to Equation (9.16):

$$Q = [m(X) - m(X') - m(\alpha)]c^2 \tag{9.21}$$

(As we did in our calculations of binding energy, we can show that the electron masses cancel in Equation (9.21), and so we can use *atomic* (m) *masses.*) This energy appears as kinetic energy:

$$Q = K_{X'} + K_\alpha \tag{9.22}$$

assuming we choose a reference frame in which X is at rest. Linear momentum is also conserved in the decay process, as shown in Figure 9.10, so that

$$M_X \cdot V_X = M_\alpha \, V_\alpha$$
$$M_{X'} \, K_{X'} = 4 \, K_\alpha$$

$$p_\alpha = p_{X'} \tag{9.23}$$

From Equations (9.22) and (9.23) we eliminate $p_{X'}$ and $K_{X'}$, since we normally don't observe the daughter nucleus in the laboratory (an alpha particle can travel relatively much further through matter than a heavy nucleus). Typical alpha decay energies are a few MeV; thus the kinetic energies of the alpha particle and the nucleus are much smaller than their corresponding rest energies, and so we can use nonrelativistic mechanics to find

$$K_{X'} = \frac{4}{A} Q$$

$$K_\alpha \cong \frac{A - 4}{A} Q \tag{9.24}$$

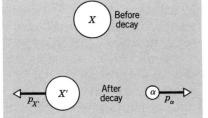

FIGURE 9.10 A nucleus X alpha decays, resulting in a nucleus X' and an alpha particle.

Alpha particle kinetic energy

EXAMPLE 9.7

Find the kinetic energy of the alpha particle emitted in the alpha decay of ${}^{226}\text{Ra}$.

Solution

From the list of isotopes in Appendix B we find the decay process to be:

$$^{226}_{88}\text{Ra}_{138} \longrightarrow {}^{222}_{86}\text{Rn}_{136} + \alpha$$

$$Q = [m({}^{226}\text{Ra}) - m({}^{222}\text{Rn}) - m(\alpha)]c^2$$

$$= [226.025406 \text{ u} - 222.017574 \text{ u} - 4.002603 \text{ u}]931.5 \text{ MeV/u}$$

$$= 4.871 \text{ MeV}$$

$$K_\alpha = \frac{A - 4}{A} Q = \left(\frac{222}{226}\right) 4.871 \text{ MeV}$$

$$= 4.785 \text{ MeV}$$

Alpha decay is an example of barrier penetration, as we discussed in Chapter 5. Let us imagine that two neutrons and two protons happen to come together inside a nucleus to form an alpha particle. (Picture the neutrons and protons swimming around within the nucleus, occasionally clumping together and breaking apart again.) That alpha particle is

$$\alpha = {}^4_2\text{He}_2 = \text{nucleus of He}$$

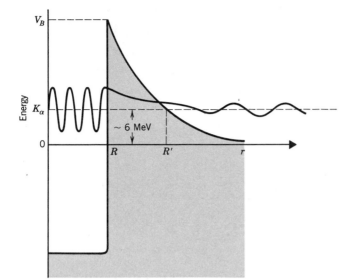

FIGURE 9.11 Barrier penetration by an alpha particle.

bound inside the nucleus by the nuclear force. Once it gets beyond the nuclear radius R, it feels the Coulomb repulsion of the daughter nucleus. The potential energy in such a situation might be represented as in Figure 9.11. The height of the barrier in a typical heavy nucleus is 30 to 40 MeV. Typically, alpha particles have energies of 4 to 8 MeV, and so it is impossible for the alpha particle to surmount the barrier; the only way the alpha particle can escape is to "tunnel" through the barrier.

A: very large

$K_\alpha = Q$

$K_{x'} = 0$

The probability per unit time λ for the alpha particle to appear in the laboratory is just the probability of it penetrating the barrier multiplied by the number of times per second the alpha particle strikes the barrier in its attempt to escape. If the alpha particle is moving at speed v inside a nucleus of radius R, it will strike the barrier as it bounces back and forth inside the nucleus at time intervals of $2R/v$. In a heavy nucleus with $R \sim 6$ fm, the α particle strikes the "wall" of the nucleus about 10^{22} times per second!

The probability for the alpha particle to penetrate the barrier can be found by solving the Schrödinger equation for the potential shown in Figure 9.11. We will not solve this problem, but we will try to account qualitatively for the enormous range of alpha-decay half lives and decay probabilities listed in Table 9.2. One way of interpreting this alpha

Table 9.2 Some Alpha Decay Energies and Half-Lives

Isotope	K_α(MeV)	$t_{1/2}$	$\lambda(s^{-1})$
^{232}Th	4.01	1.4×10^{10} y	1.6×10^{-18}
^{238}U	4.19	4.5×10^{9} y	4.9×10^{-18}
^{230}Th	4.69	8.0×10^{4} y	2.8×10^{-13}
^{238}Pu	5.50	88 y	2.5×10^{-10}
^{230}U	5.89	20.8 d	3.9×10^{-7}
^{220}Rn	6.29	56 s	1.2×10^{-2}
^{222}Ac	7.01	5 s	0.14
^{216}Rn	8.05	45 μs	1.5×10^{4}
^{212}Po	8.78	0.30 μs	2.3×10^{6}

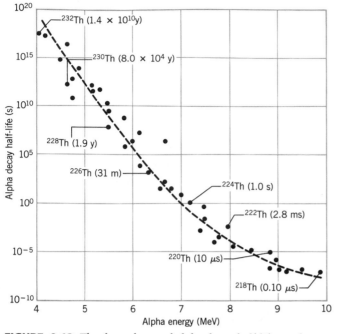

FIGURE 9.12 The dependence of alpha-decay half-life on the kinetic energy of the alpha particle. Values are marked for some isotopes of thorium.

decay data is to plot, on a log scale, the decay probability against the kinetic energy of the alpha particles. (The decay constant λ also depends on Z, but the dependence is much weaker and less dramatic than the dependence on kinetic energy, since Z changes by a relatively small amount over the range of heavy nuclei which alpha decay.) Such a plot is shown in Figure 9.12.

In Chapter 5, we learned that the probability to penetrate a barrier of constant height V_0 was of the form $P \propto e^{-2kL}$, where L is the penetration distance and $k = \sqrt{(2m/\hbar^2)(V_0 - E)}$. Although the "flat" barrier of height V_0 is *not* a good approximation to the barrier shown in Figure 9.11, we can learn something of the effect of the barrier height and thickness by studying transmissions through the "flat" barrier. We expect that the transmission probability will depend on the thickness of the barrier and on the difference between the height of the barrier and the energy of the particle. The maximum barrier height V_B is just the Coulomb energy of the alpha particle at the nuclear surface

$$V_B = \frac{1}{4\pi\varepsilon_0} \frac{2(Z - 2)e^2}{R} \qquad (9.25)$$

where R is the nuclear radius and where the factor "2" in the numerator comes from the electric charge of the alpha particle. (The factor $Z - 2$ occurs because the residual nucleus is responsible for the electrostatic force.) Since the difference between the barrier height and the kinetic energy of the particle varies from $V_B - K_\alpha$ at the nuclear surface to 0 at the radius R', where it "leaves" the barrier, we will take an average value of $\frac{1}{2}(V_B - K_\alpha)$ as a representative value of $(V_0 - E)$, the height of the "flat" barrier above the energy of the particle. For the effective thickness

L of the barrier, we will similarly take $\frac{1}{2}(R' - R)$, where R is the nuclear radius $(= R_0 A^{1/3})$ and R' is the radial coordinate at which the potential energy $V = (1/4\pi\varepsilon_0)[2(Z - 2)e^2/R']$ is just equal to the kinetic energy K_α of the alpha particle when it is far from the nucleus.

$$R' = \frac{e^2}{4\pi\varepsilon_0} \frac{2(Z - 2)}{K_\alpha} \tag{9.26}$$

A rough estimate of the alpha decay probability is then

$$\lambda = \frac{v}{2R} e^{-k(R'-R)} \tag{9.27}$$

where $k = \sqrt{(2m/\hbar^2)\cdot\frac{1}{2}(V_B - K_\alpha)}$. Values computed from Equation (9.27) range from 10^5 s^{-1} to 10^{-21} s^{-1} and correspond roughly to the range of experimentally measured values. This agreement is somewhat accidental since our calculation was based on a very unrealistic set of assumptions. However, you can see even from this rough calculation how the great range of half-lives that are observed for alpha decay originates from a barrier penetration effect, with high-energy alpha particles needing to penetrate a much thinner and lower barrier than low-energy alpha particles.

9.9 BETA DECAY

In beta decay a neutron changes into a proton (or a proton into a neutron); Z and N each change by one unit, but A doesn't change. In the most basic beta decay process, a neutron decays into a proton and an electron: $n \rightarrow p + e$. When this decay process was first studied, the emitted particles were called beta particles; later they were shown to be electrons.

The electron that is emitted in beta decay is *not* an orbital electron. It also is not an electron that was previously present within the nucleus, for as we have seen (Example 4.7) the uncertainty principle forbids electrons of the observed energies to exist inside the nucleus. The electron is "manufactured" by the nucleus out of the available energy. If the rest energy *difference* between the nuclei is at least $m_e c^2$, this will be possible.

Early experiments with beta decay revealed two difficulties. The intrinsic spins of the proton, neutron, and electron are all $\frac{1}{2}$. After the decay of the neutron, the proton and electron spins can be parallel (total spin = 1) or antiparallel (total spin = 0), but in neither case can they combine to give a spin of $\frac{1}{2}$, the spin of the original neutron. The decay process would therefore seem not to conserve angular momentum. An even more serious problem followed from measuring the energy of the emitted electrons—the energy spectrum of the electrons was continuous, from zero up to some maximum value K_{max}, as shown in Figure 9.13. For example, in the neutron decay, the Q value is

$$Q = (m_n - m_p - m_e)c^2 \tag{9.28}$$

In calculating Q values for beta decay, we must exercise a bit of care, since the electron masses may *not* cancel out at they did in our previous calculations using atomic masses. This is because the initial and final nuclei have different Z values, and therefore different numbers of elec-

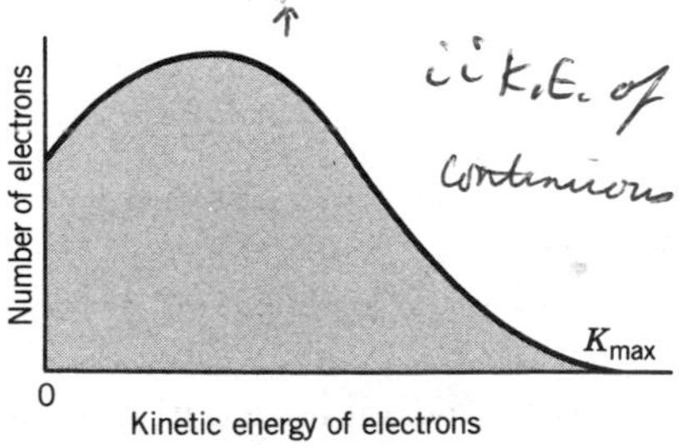

FIGURE 9.13 Spectrum of electrons emitted in beta decay.

trons. For neutron beta decay, we can group m_p and m_e together to give the atomic mass $m(^1_1H_0)$, and so

$$Q = [m_n - m(^1_1H_0)]c^2 \qquad (9.29)$$

which we can calculate to be 0.782 MeV.

Except for a very small correction, which accounts for the recoil energy of the proton, all of this energy should appear as kinetic energy of the electron, and all emitted electrons should have *exactly* this energy. We find by experiment that all the emitted electrons have less than this energy. Experimental physicists attacked the problem of this "missing" energy with considerable vigor and ingenuity before 1930, but without success.

Wolfgang Pauli found the solution to the apparent violations of conservation of angular momentum and energy in 1930; he suggested that there is a *third* particle emitted in beta decay. Since electric charge is already conserved by the proton and electron, this new particle cannot have electric charge. If it has spin $\frac{1}{2}$, it will satisfy conservation of angular momentum, since we can combine the spins of the three decay particles to give $\frac{1}{2}$. The "missing" energy is just the energy carried away by this third particle, and the observed fact that the energy spectrum extends all the way to the value $Q = [m_n - m(^1_1H_0)]c^2$ suggests that this particle has zero rest mass, just like the photon. (We know this new particle can't be a photon, because a photon has a spin of 1.)

This new particle is called the *neutrino* ("little neutral one" in Italian) and has the symbol ν. As we discuss in Chapter 11, every particle has an *antiparticle*, and the antiparticle of the neutrino is the *antineutrino* $\bar{\nu}$. It is, in fact, the antineutrino that is emitted in neutron beta decay. The complete decay process is thus

$$n \longrightarrow p + e^- + \bar{\nu} \qquad (9.30)$$

Since the antineutrino has zero rest mass, the Q value we calculated above is correct. This energy appears as the kinetic energy of the electron, the energy of the antineutrino, and the recoil kinetic energy of the proton (which is very small). Note that the kinetic energy of the electron is *not* small compared with its rest energy, so we *must* use relativistic energy and momentum when discussing beta decay.

Neutron decay can also occur in a nucleus, in which a nucleus with Z protons and N neutrons decays to a nucleus with $Z + 1$ protons and $N - 1$ neutrons:

$$^A_Z X_N \longrightarrow ^{\ A}_{Z+1} X'_{N-1} + e^- + \bar{\nu} \qquad (9.31)$$

The Q value for this decay is

$$Q = [m(^A X) - m(^A X')]c^2 \qquad (9.32) \qquad \textbf{Negative beta decay Q value}$$

It can be shown (Problem 27 at the end of the chapter) that the electron masses cancel in calculating Q, so it is *atomic masses* that appear in Equation (9.32).

The energy released in the decay (the Q value) appears as the energy of the antineutrino, the kinetic energy of the electron, and a small (usually negligible) recoil kinetic energy of the nucleus X'. The electron has its maximum kinetic energy when the antineutrino has an energy of approximately zero. Figure 9.13 shows the energy distribution of electrons emitted in a typical negative beta decay.

Another beta decay process is

β⁺ decay

$$p \longrightarrow n + e^+ + \nu \qquad (9.33)$$

in which a *positive electron, or positron,* is emitted. The positron is the antiparticle of the electron; it has the same mass as the electron but the opposite electric charge. This decay has a negative Q value, and so it is never observed in nature for free protons. (This is indeed fortunate—if the free proton were unstable to beta decay, stable hydrogen atoms, the basic material of the universe, could not exist!) Protons in nuclei can undergo such decay processes:

$$_Z^A X_N \longrightarrow {}_{Z-1}^A X'_{N+1} + e^+ + \nu \qquad (9.34)$$

The Q value for this process is (Problem 27) *electron mass*

$$Q = [m(_Z^A X) - m(_{Z-1}^A X') - 2m_e]c^2 \qquad (9.35)$$

Positive beta decay Q value

in which the masses are *atomic masses*. Figure 9.14 shows the energy distribution of positrons emitted in a typical positive beta decay.

A nuclear decay process that competes with positron emission is *electron capture*; the basic electron capture process is

$$p + e^- \longrightarrow n + \nu \qquad (9.36)$$

in which a proton captures an electron from its orbit and converts into a neutron plus a neutrino. The electron necessary for this process is one of the inner orbital electrons in an atom, and we identify the capture process by the shell from which the electron comes: *K*-shell capture, *L*-shell capture and so forth. (Of course, the electronic orbits that come closest to, or even penetrate, the nucleus have the higher probability to be captured.) The electron capture process does not occur for free protons, but in nuclei the process is

$$_Z^A X_N + e^- \longrightarrow {}_{Z-1}^A X'_{N+1} + \nu \qquad (9.37)$$

and the Q value, using atomic masses, is

$$Q = [m(^A X) - m(^A X')]c^2 \qquad (9.38)$$

Electron capture decay Q value

Table 9.3 gives some typical beta decay processes, along with their Q values and half-lives.

FIGURE 9.14 Spectrum of positrons emitted in beta decay.

Table 9.3 Typical Beta Decay Processes

Decay	Type	Q (MeV)	$t_{1/2}$
$^{19}O \rightarrow {}^{19}F + e^- + \bar{\nu}$	β^-	4.82	27 s
$^{176}Lu \rightarrow {}^{176}Hf + e^- + \bar{\nu}$	β^-	1.19	3.6×10^{10} y
$^{25}Al \rightarrow {}^{25}Mg + e^+ + \nu$	β^+	3.26	7.2 s
$^{124}I \rightarrow {}^{124}Te + e^+ + \nu$	β^+	2.14	4.2 d
$^{15}O + e^- \rightarrow {}^{15}N + \nu$	EC	2.75	122 s
$^{170}Tm + e^- \rightarrow {}^{170}Er + \nu$	EC	0.31	129 d

neutron

K

much less probable (catching an electron)

^{23}Ne decays to ^{23}Na by negative beta emission. What is the maximum kinetic energy of the emitted electrons?

EXAMPLE 9.8

This decay is of the form given by Equation (9.31):

$$^{23}_{10}\text{Ne}_{13} \longrightarrow \,^{23}_{11}\text{Na}_{12} + e^- + \bar{\nu}$$

and the Q value is found from Equation (9.32), using *atomic* masses:

$$Q = [m(^{23}\text{Ne}) - m(^{23}\text{Na})]c^2$$

$$= (22.994466 \text{ u} - 22.989770 \text{ u})931.5 \text{ MeV/u}$$

$$= 4.374 \text{ MeV}$$

Except for a small correction for the kinetic energy of the recoiling nucleus, the maximum kinetic energy of the electrons is just equal to this. (This occurs when the neutrino has a negligible energy. Similarly, the maximum *neutrino* energy occurs when the electron has a negligibly small kinetic energy.)

^{40}K is an unusual isotope, in that it decays by positive beta emission, negative beta emission, and electron capture. Find the Q values for these decays.

EXAMPLE 9.9

The process for negative beta decay is given by Equation (9.31):

$$^{40}_{19}\text{K}_{21} \longrightarrow \,^{40}_{20}\text{Ca}_{20} + e^- + \bar{\nu}$$

and the Q value is found from Equation (9.32) using atomic masses:

$$Q_{\beta^-} = [m(^{40}\text{K}) - m(^{40}\text{Ca})]c^2$$

$$= (39.963999 \text{ u} - 39.962591 \text{ u})931.5 \text{ MeV/u}$$

$$= 1.312 \text{ MeV}$$

Equation (9.34) gives the decay process for positive beta emission:

$$^{40}_{19}\text{K}_{21} \longrightarrow \,^{40}_{18}\text{Ar}_{22} + e^+ + \nu$$

and the Q value is given by Equation (9.35)

$$Q_{\beta^+} = [m(^{40}\text{K}) - m(^{40}\text{Ar}) - 2m_e]c^2$$

$$= (39.963999 \text{ u} - 39.962383 \text{ u} - 2 \times 0.000549 \text{ u})931.5 \text{ MeV/u}$$

$$= 0.483 \text{ MeV}$$

For electron capture,

$$^{40}_{19}\text{K}_{21} + e^- \longrightarrow \,^{40}_{18}\text{Ar}_{22} + \nu$$

and from Equation (9.38)

$$Q_{ec} = [m(^{40}\text{K}) - m(^{40}\text{Ar})]c^2$$

$$= (39.963999 \text{ u} - 39.962383 \text{ u})931.5 \text{ MeV/u}$$

$$= 1.505 \text{ MeV}$$

Following alpha or beta decay, the final nucleus may be left in an excited state. Just as an atom does, the nucleus will reach its ground state after emitting one or more photons, known as *nuclear gamma rays*. The energy of each photon is just the energy difference between the initial and final nuclear states, less a negligibly small correction for the recoil energy of the nucleus. These energies are typically in the range of

9.10 GAMMA DECAY

100 keV to a few MeV. Nuclei can likewise be excited from the ground state to an excited state by absorbing a photon of the appropriate energy, in a process similar to the resonant absorption by atomic states.

Figure 9.15 shows a typical energy level diagram of excited nuclear states and some of the gamma ray transitions that can be emitted. Typical values for the half-lives of the excited states are 10^{-9} to 10^{-12} s; the exact values of the half-lives (and the *selection rules* which allow certain transitions and forbid others) depend on rather sophisticated details of nuclear structure beyond the level of this book. Occasionally it happens that these details result in a half-life that is very long—hours or even days. Such states are known as *isomeric states* or *isomers*.

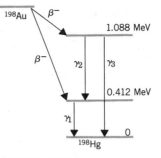

FIGURE 9.15 Some gamma rays emitted following beta decay.

The study of nuclear gamma emission is an important tool of the nuclear physicist; the energies of the gamma rays can be measured with great precision, and they provide a powerful means of deducing the energies of the excited states of nuclei.

In calculating the energies of alpha and beta particles emitted in radioactive decays, we have assumed that no gamma rays are emitted. If there are gamma rays emitted, the available energy (Q value) must be shared between the other particles and the gamma ray.

---

^{12}N beta decays to an excited state of ^{12}C, which subsequently decays to the ground state with the emission of a 4.43-MeV gamma ray. What is the maximum kinetic energy of the emitted beta particle?

EXAMPLE 9.10

To determine the Q value for this decay, we first need to find the mass of the product nucleus ^{12}C *in its excited state*. In the ground state, ^{12}C has a mass of 12.000000 u, so its mass in the excited state is

Solution

$$12.000000 \text{ u} + \frac{4.43 \text{ MeV}}{931.5 \text{ MeV/u}} = 12.004756 \text{ u}$$

The Q value is therefore

$$Q = (12.018613 \text{ u} - 12.004756 \text{ u} - 2 \times 0.000549 \text{ u})931.5 \text{ MeV/u}$$

$$= 11.89 \text{ MeV}$$

(Notice that we could have just as easily found the Q value by first finding the Q value for decay to the *ground state*, 16.32 MeV, and then subtracting the excitation energy of 4.43 MeV, since the decay to the excited state has that much less available energy.)

Neglecting the small correction for the recoil kinetic energy of the ^{12}C nucleus, the maximum electron kinetic energy is 11.89 MeV.

All of the elements beyond the very lightest (hydrogen and helium) were produced by nuclear reactions in the interiors of stars. These reactions produce not only stable elements, but radioactive ones as well. Most of the radioactive elements have half-lives of the order of days or years, much smaller than the age of the Earth (about 4.5×10^9 y). Therefore, most of the radioactive elements that may have been present when the Earth was formed have decayed to stable elements. However, a few of the radioactive elements created long ago have half-lives that are of the same order as the age of the Earth, and so are still present and can still be observed to undergo radioactive decay. These elements are part of the background of *natural radioactivity* in which we live.

9.11 NATURAL RADIOACTIVITY

Radioactive decay processes either change the mass number A of a nucleus by four units (alpha decay) or don't change A at all (beta or gamma decay). A radioactive decay process can be part of a sequence or series of decays if a *radioactive* element of mass number A decays to another *radioactive* element of mass number A or $A - 4$. Such a series of processes will continue until a stable element is reached. A hypothetical such series is illustrated in Figure 9.16. Since gamma decays don't change Z or A, they are not shown; however, most of the alpha and beta decays are accompanied by gamma-ray emissions.

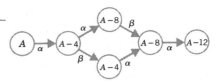

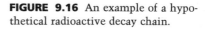

FIGURE 9.16 An example of a hypothetical radioactive decay chain.

The A values of the members of such decay chains differ by a multiple of 4 (including zero as a possible multiple) and so we expect four possible decay chains, with A values that can be expressed as $4n$, $4n + 1$, $4n + 2$, and $4n + 3$, where n is an integer. One of the four naturally occurring radioactive series is illustrated in Figure 9.17. Each series begins with a relatively long-lived member, proceeds through many α and β decays, which may have very short half-lives, and finally ends with a stable isotope. Three of these series begin with isotopes having half-lives comparable to the age of the Earth, and so are still observed today. The neptunium series $(4n + 1)$ begins with ^{237}Np which has a half-life of "only" 2.1×10^6 y, much less than the 4.5×10^9 y since the formation of the Earth. Thus all of the ^{237}Np that was originally present has long since decayed to ^{209}Bi.

$4n \qquad ^{232}\text{Th}$

$4n+1 \qquad ^{237}N_p$

$4n+2 \quad ^{238}\text{U}$

$4n+3 \quad ^{235}\text{U}$

$4n+4 \Rightarrow 5n$

EXAMPLE 9.11

Compute the Q value for the ^{238}U $\rightarrow$ ^{206}Pb decay chain, and find the rate of energy production per gram of uranium.

Solution

The uranium decay chain consists of eight alpha and six beta decays. We recall that for β^- decays, the electron masses combine with the nuclear masses in the computation of the Q value and we can therefore use atomic masses. Thus for the entire decay chain

$$Q = [m(^{238}\text{U}) - m(^{206}\text{Pb}) - 8m(^4\text{He})]c^2$$

where we use *atomic* masses.

$$Q = [238.050786 \text{ u} - 205.974455 \text{ u} - 8 \times 4.002603 \text{ u}]931.5 \text{ MeV/u}$$

$$= 51.7 \text{ MeV}$$

One gram of ^{238}U is $\frac{1}{238}$ mole and therefore contains $\frac{1}{238} \times 6 \times 10^{23}$ atoms. The half-life of the decay is 4.5×10^9 y, so λ, the decay probability per atom, is

$$\lambda = \frac{0.693}{4.5 \times 10^9 \text{ y}} \times \frac{1 \text{ y}}{3.16 \times 10^7 \text{ s}} = 4.9 \times 10^{-18} \text{ s}^{-1}$$

Thus, on the average, the number of decays of the ^{238}U is

$$\left(\frac{1}{238} \times 6 \times 10^{23} \text{ atoms}\right) \times 4.9 \times 10^{-18} \frac{\text{decays}}{\text{atom·s}} = 12,000 \text{ decays/s}$$

Each decay liberates 51.7 MeV, and so the rate of energy liberation is

$$12,000 \frac{\text{decays}}{\text{s}} \times 51.7 \frac{\text{MeV}}{\text{decay}} \times 10^6 \frac{\text{eV}}{\text{MeV}} \times 1.6 \times 10^{-19} \frac{\text{J}}{\text{eV}} = 1.0 \times 10^{-7} \text{ W}$$

This may seem like a very small rate of energy release, but if the energy were to appear as thermal energy and were not dissipated by some

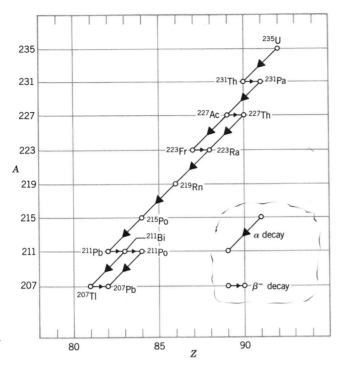

FIGURE 9.17 The ^{235}U decay chain.

means (radiation or conduction to other matter, for example) the 1-g sample of ^{238}U would increase in temperature by 25°C per year and would be melted and vaporized in the order of one century! This calculation suggests that we can perhaps account for some of the internal heat of planets through natural radioactive processes.

If we examine by chemical means a sample of uranium-bearing rock, we can find the ratio of ^{238}U atoms to ^{206}Pb atoms. If we assume that all of the ^{206}Pb was produced by the uranium decay and none was present when the rock was originally formed (an assumption that must be examined with care both theoretically and experimentally), then this ratio can be used to find the age of the sample, as shown in the following example.

Radioactive dating

Three different rock samples have ratios of numbers of ^{238}U atoms to ^{206}Pb atoms of 0.5, 1.0, and 2.0. Compute the ages of the three rocks.

EXAMPLE 9.12

Since all of the other members of the uranium series have half-lives that are much shorter than the half-life of ^{238}U $(4.5 \times 10^9$ y), we ignore the intervening decays and consider only the ^{238}U decay. Let N_0 be the original number of ^{238}U atoms, so that $N_0 e^{-\lambda t}$ would be the number that are still present today, and $N_0 - N_0 e^{-\lambda t}$ would be the number that have decayed and are presently observed as ^{206}Pb. The ratio R of ^{238}U to ^{206}Pb is thus

Solution

$$R = \frac{\text{number of } ^{238}\text{U}}{\text{number of } ^{206}\text{Pb}} = \frac{N_0 e^{-\lambda t}}{N_0 - N_0 e^{-\lambda t}} = \frac{1}{e^{\lambda t} - 1}$$

Solving for t, we find:

$$t = \frac{1}{\lambda} \ln \left(\frac{1}{R} + 1 \right)$$

and recalling that $\lambda = 0.693/t_{1/2}$ we can write this as

$$t = \frac{t_{1/2}}{0.693} \ln \left(\frac{1}{R} + 1 \right)$$

We can then find the values of t corresponding to the three values of R,

$$R = 0.5 \qquad t = 7.1 \times 10^9 \text{ y}$$
$$R = 1.0 \qquad t = 4.5 \times 10^9 \text{ y}$$
$$R = 2.0 \qquad t = 2.6 \times 10^9 \text{ y}$$

The oldest rocks on Earth, dated by this and other means, have ages of about 4.5×10^9 y. The age of the first rock analyzed above, 7.1×10^9 y, suggests either that the rock had an extraterrestrial origin, or else that our assumption of no initial ^{206}Pb was incorrect. The age of the third rock suggests that it solidified only 2.6×10^9 y ago; previous to that time it was molten and the decay product ^{206}Pb may have "boiled away" from the ^{238}U.

There are a number of other naturally occurring radioactive isotopes that are not part of the decay chain of the heavy elements. A partial list is given in Table 9.4; some of these can also be used for radioactive dating.

Other radioactive elements are being produced continuously in the Earth's atmosphere as a result of nuclear reactions between air molecules and the high-energy particles known as "cosmic rays." The most notable and useful of these is ^{14}C, which beta decays with a half-life of 5730 y. When a living plant absorbs CO_2 from the atmosphere, a small fraction (about 1 in 10^{12}) of the carbon atoms will be ^{14}C, and the remainder will be stable ^{12}C (99 percent) and ^{13}C (1 percent). When the plant dies, its intake of ^{14}C stops, and the ^{14}C decays. If we assume that the composition of the Earth's atmosphere and the flux of cosmic rays have not changed significantly in the last few thousand years, we can find the age of specimens of organic material by comparing their ^{14}C/^{12}C ratios to those of living plants. The following example shows how this *radiocarbon dating* technique is used.

Table 9.4 Some Naturally Occurring Radioactive Isotopes

Isotope	$t_{1/2}$
^{40}K	1.28×10^9 y
^{87}Rb	4.8×10^{10} y
^{92}Nb	3.2×10^7 y
^{113}Cd	9×10^{15} y
^{115}In	5.1×10^{14} y
^{138}La	1.1×10^{11} y
^{176}Lu	3.6×10^{10} y
^{187}Re	4×10^{10} y

Radiocarbon dating

(a) A sample of carbon dioxide gas from the atmosphere fills a vessel of volume 200.0 cm³ to a pressure of 2.00×10^4 Pa (1 Pa = 1 N/m², about 10^{-5} atm) at a temperature of 295 K. Assuming that all of the ^{14}C beta decays were counted, how many counts would be accumulated in one week? (b) An old sample of wood is burned, and the resulting carbon dioxide is placed in the same vessel at the same pressure and temperature. After one week, 1420 counts have been accumulated. What is the age of the sample?

EXAMPLE 9.13

(a) We first find the number of moles present in the vessel, using the ideal gas law:

Solution

$$n = \frac{PV}{RT} = \frac{(2.00 \times 10^4 \text{ N/m}^2)(2.00 \times 10^{-4} \text{ m}^3)}{(8.314 \text{ J/mole·K})(295 \text{ K})}$$

$$= 1.63 \times 10^{-3} \text{ mole}$$

Since each mole of CO_2 contains 6.02×10^{23} molecules, the number of molecules N is

$$N = (6.02 \times 10^{23} \text{ molecules/mole})(1.63 \times 10^{-3} \text{ mole})$$

$$= 9.82 \times 10^{20} \text{ molecules}$$

Each molecule has one carbon atom, so N is also the number of carbon atoms in the sample. If the fraction of ^{14}C atoms is 10^{-12}, there are 9.82×10^{8} atoms of ^{14}C present. The activity is therefore

$$a = \lambda N = \frac{0.693}{5730 \text{ y}} \cdot \frac{1 \text{ y}}{3.16 \times 10^{7} \text{ s}} \cdot 9.82 \times 10^{8}$$

$$= 3.76 \times 10^{-3} \text{ decays/s}$$

In one week the number of decays is 2280.

(b) An identical sample which gives only 1420 counts must be old enough for only 1420/2280 of its original activity to remain.

$$\frac{1420}{2280} = e^{-\lambda t}$$

$$t = \frac{1}{\lambda} \ln \frac{2280}{1420}$$

$$= \frac{5730}{0.693} \ln \frac{2280}{1420} = 3920 \text{ y}$$

A typical atomic transition has a lifetime of about 10^{-8} s and an energy of a few electron-volts (for visible light). When an atom emits a photon of energy E_γ and momentum p_γ, the atom must recoil so that momentum is conserved. If we assume that the atom was initially at rest, its recoil momentum p_R is then just equal to the photon momentum (but in the opposite direction) and its recoil kinetic energy is $K = p_R^2/2M$, where M is the mass of the atom. (We assume that $K \ll Mc^2$, and so nonrelativistic kinematics will be applicable.) Since $p_R = p_\gamma = E_\gamma/c$,

$$K = \frac{E_\gamma^2}{2Mc^2} \tag{9.39}$$

With $E_\gamma \sim 1$ eV and $Mc^2 \sim 1000$ MeV $\times A$, even for the lightest atoms K is of the order of 10^{-10} eV. Thus if the energy available from the atomic transition is E (the energy difference of the electronic levels) conservation of energy demands that the photon energy be $E_\gamma = E - K$, where for the atomic case, $K \ll E$.

One productive way of studying atomic systems is to do *resonance* experiments. In such experiments, radiation from a collection of atoms in an excited state is incident on a collection of identical atoms in their ground state. The ground-state atoms can absorb the photons and jump to the corresponding excited state. However, as we have seen, the emitted photon energy is less than the transition energy by the recoil kinetic energy K; moreover, it is less than the photon energy required for resonance by $2K$, since the absorbing atom must recoil also. The absorption experiment is still possible, because the excited states don't have "exact" energies—a state with a mean lifetime τ ($\tau = 1/\lambda$; see Problem 21) has an energy uncertainty ΔE which is given by the uncertainty rela-

*9.12 THE MÖSSBAUER EFFECT

tionship: $\Delta E\ \tau \sim \hbar$. That is, the state lives on the average for a time τ, and during that time we can't determine its energy to an accuracy less than ΔE. For typical atomic states, $\tau \sim 10^{-8}$ s, so $\Delta E \sim 10^{-7}$ eV. Since K, which is of the order of 10^{-10} eV, is much less than the width ΔE, the "shift" caused by the recoil is not large, and the widths of the emitting and absorbing atomic states cause sufficient overlap for the absorption process to occur. Figure 9.18 illustrates this case.

The situation is different for nuclear gamma rays. A typical lifetime might be 10^{-10} s, and so the widths are the order of $\Delta E \sim 10^{-5}$ eV. The photon energies are typically 100 keV $= 10^5$ eV, and so K is of order 1 eV. This situation is depicted in Figure 9.19, and you can immediately see that since K is so much larger than the width ΔE, no overlap of emitter and absorber is possible.

In 1958, Rudolf Mössbauer made a discovery for which he was awarded the 1961 Nobel prize in physics. He was working on resonant absorption experiments using the 129-keV γ ray of ^{191}Ir, and he discovered that he could eliminate most of the recoil by placing the radioactive nuclei and the absorbing atoms in crystals. Since the crystalline binding energies are large compared with K, the individual atoms are held tightly to their positions in the crystal lattice and are not free to recoil; if any recoil is to occur, it must be the whole crystal that recoils. This effect is to make the mass M that appears in Equation (9.39) not the mass of an atom, but the mass of the entire crystal, perhaps 10^{20} times larger than an atomic mass. (As an analogy, imagine the difference between striking a brick with a baseball bat, and striking a brick wall!) Once again the recoil kinetic energy is made small, and resonant absorption can occur (Figure 9.20).

In order to demonstrate the resonant absorption, the source of radiation can be slowly moved relative to the absorber, so that the gamma-ray energies are Doppler shifted in the vicinity of the resonance peak. We can, in fact, estimate the speeds necessary to achieve this effect. The frequency shift due to the motion is given by [see Equation (2.7)]

$$\nu' = \nu \left(1 + \frac{v}{c}\right) \tag{9.40}$$

where we ignore the $\sqrt{1 - v^2/c^2}$ term since $v \ll c$. Since the photon energy E is just $h\nu$, we have

$$E' = E \left(1 + \frac{v}{c}\right) \tag{9.41}$$

If we take the width ΔE as a representative estimate of how far we would like to Doppler shift the photon energy, then $E' \cong E + \Delta E$, and so

$$E + \Delta E \cong E + E\frac{v}{c} \tag{9.42}$$

and, solving for v,

$$v \cong c\frac{\Delta E}{E} \tag{9.43}$$

We have estimated $\Delta E \sim 10^{-5}$ eV (the width of the state) and $E \sim$

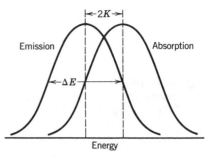

FIGURE 9.18 Representative emission and absorption energies in an atomic system.

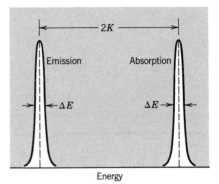

FIGURE 9.19 Representative emission and absorption energies in a nuclear system.

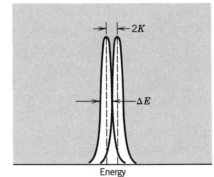

FIGURE 9.20 Emission and absorption energies for nuclei bound in a crystal lattice.

100 keV (the energy of the photon). Thus

$$v \cong (3 \times 10^8 \text{ m/s}) \frac{10^{-5}}{10^5} \cong 3 \text{ cm/s}$$

Such low speeds can easily and accurately be produced in the laboratory.

Figure 9.21 shows a diagram of the apparatus to measure the Möss-bauer effect. The resonant absorption is observed by looking for de-creases in the number of gamma rays that are transmitted through the absorber. At resonance, more gamma rays are absorbed and so the transmitted intensity decreases. Typical results are shown in Figure 9.22.

The Mössbauer effect is an extremely precise method for measuring small changes in the energies of photons. In one particular application, the Zeeman splitting of *nuclear* (not atomic) states can be observed. When a nucleus is placed in a magnetic field, the Zeeman effect causes an energy splitting of the nuclear m states, similar to the atomic case. However, nuclear magnetic moments are about 2000 times smaller than atomic magnetic moments, and a typical energy splitting would be about 10^{-6} eV. To observe such an effect directly we would need to measure photon energies to 1 part in 10^{11} (a photon energy of 10^5 eV is shifted by 10^{-6} eV), but using the Mössbauer effect, this is not difficult.

In Chapter 15 we discuss another application of this extremely pre-cise technique, in which the energy gained when a photon "falls" through several meters of the Earth's gravitational field is measured in order to test Einstein's general theory of relativity.

The following are some intermediate-level, comprehensive nuclear physics texts.

B. L. Cohen, *Concepts of Nuclear Physics* (New York, McGraw-Hill, 1971).

J. G. Cunninghame, *Introduction to the Atomic Nucleus* (Amsterdam, Elsevier, 1964).

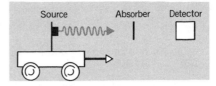

FIGURE 9.21 Mössbauer effect appa-ratus. A source of gamma rays is made movable, in order to Doppler shift the photon energies. The intensity of radia-tions transmitted through the absorber is measured as a function of the speed of the source.

SUGGESTIONS FOR FURTHER READING

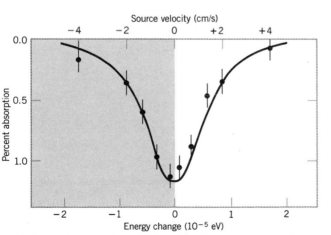

FIGURE 9.22 Typical results in a Mössbauer effect experi-ment. A velocity of 2 cm/s Doppler shifts the gamma rays enough to move the emission and absorption energies off reso-nance.

R. D. Evans, *The Atomic Nucleus* (New York, McGraw-Hill, 1955).
I. Kaplan, *Nuclear Physics* (Reading, Addison-Wesley, 1962).

A complete tabulation of nuclear masses, decay properties, isotopic abundances, excited states is:

C. M. Lederer and V. S. Shirley, editors, *Table of Isotopes, 7th Edition* (New York, Wiley, 1978).

QUESTIONS

1. The magnetic moment of a deuterium nucleus is about $\frac{1}{2000}$ Bohr magneton. What does this imply about the presence of an electron in the nucleus, as the proton-electron model requires?

2. Suppose we have a supply of 20 protons and 20 neutrons. Do we liberate more energy if we assemble them into a single ^{40}Ca nucleus or into two ^{20}Ne nuclei?

3. Atomic rest masses are usually given to a precision of about the sixth decimal place in atomic mass units (u). This is true both for stable and radioactive nuclei, even though the uncertainty principle requires that an atom with a lifetime Δt has a rest energy uncertain by $\hbar/\Delta t$. Based on the typical lifetimes given for nuclear decays, are we justified in expressing atomic masses to such precision? At what lifetimes would such precision not be justified?

4. Only two stable nuclei have $Z > N$. (a) What are these nuclei? (b) Why don't more nuclei have $Z > N$?

5. In a deuterium nucleus, the proton and neutron spins can be either parallel or antiparallel. What are the possible values of the total spin of the deuterium nucleus? (It is not necessary to consider any orbital angular momentum.) The magnetic moment of the deuterium nucleus is measured to be nonzero. Which of the possible spins is eliminated by this measured value?

6. What is the difference between a real particle and a virtual particle? Can real particles sometimes be virtual particles? Could a virtual particle be one that is never real?

7. The electromagnetic interaction can be interpreted as an exchange force, in which photons are the exchanged particle. What does Equation (9.6) imply about the range of such a force? Is this consistent with the conventional interpretation of the electromagnetic force? What would you expect for the rest energy of the exchanged particle that carries the gravitational force?

8. What is meant by assuming that the decay constant λ is a constant, independent of time? Is this a requirement of theory, an axiom, or an experimental conclusion? Under what circumstances might λ change with time?

9. If we focus our attention on a specific nucleus in a radioactive sample, can we know exactly how long that nucleus will live before it decays? Can we predict which half of the nuclei in a sample will decay during one half-life? What part of quantum physics is responsible for this?

10. A certain radioactive sample is observed to undergo 10,000 decays in 10 s. Can we conclude that a = 1000 decays/s if (a) $t_{1/2} \gg$ 10 s; (b) $t_{1/2} \cong$ 10 s; (c) $t_{1/2} \ll$ 10 s?

11. Suppose we wish to do radioactive dating of a sample whose age we guess to be t. Should we choose an isotope whose half-life is (a) $\gg t$; (b) $\sim t$; (c) $\ll t$?

12. Figure 9.12 seems to suggest that $\log t_{1/2} \propto 1/K$. Is Equation (9.27) consistent with this?

13. The alpha particle is a particularly tightly bound nucleus. Based on this fact, explain why heavy nuclei alpha decay and light nuclei don't.

14. Estimate the recoil kinetic energy of the residual nucleus following alpha decay. (This energy is large enough to drive the residual nucleus out of certain radioactive sources; if the residual nucleus is itself radioactive, there is the chance of spread of radioactive material. A thin coating over the source is necessary to prevent this.)

15. Why does the electron energy spectrum (Figure 9.13) look different from the positron energy spectrum (Figure 9.14) at low energies?

16. Will electron capture always be energetically possible when positron beta decay is possible? Will positron beta decay always be energetically possible when electron capture is possible?

17. All three beta decay processes involve the emission of neutrinos (or antineutrinos). In which processes do the neutrinos have a continuous energy spectrum? In which is the neutrino monoenergetic?

18. Neutrinos always accompany electron capture decays. What other kind of radiation always accompanies electron capture? (*Hint.* It is not nuclear radiation.) What other kind of nonnuclear radiation might accompany β^- or β^+ decays in bulk samples?

19. The positron decay of ^{15}O goes directly to the ground state of ^{15}N; no excited states of ^{15}N populated and no γ rays follow the beta decay. Yet a source of ^{15}O is found to emit γ rays of energy 0.51 MeV. Explain the origin of these γ rays.

20. Would ^{92}Nb be a convenient isotope to use for determining the age of the earth by radioactive dating? (See Table 9.4.) What about ^{113}Cd?

21. The natural decay chain $^{238}_{92}$U $\rightarrow$ $^{206}_{82}$Pb consists of several alpha decays, which decrease A by 4 and Z by 2, and negative beta decays, which *increase* Z by 1. (a) How many alpha decays must occur in the chain? (b) For that number of alpha decays, how many beta decays must occur to make the final Z come out correctly? (c) As shown in Figure 9.17, sometimes a decay chain can proceed through different branches. Do your answers to (a) and (b) depend on this branching?

22. It has been observed that there is an increased level of radon gas ($Z = 86$) in the air just before an earthquake. Where does the radon come from? How is it produced? How is it released? How is it detected?

23. In Figure 9.22, only 1 percent of the gamma intensity is absorbed, even at resonance. For complete resonance, we would expect 100 percent absorption. What factors might contribute to this small absorption?

PROBLEMS

1. Give the proper isotopic symbols for:
   (a) The isotope of fluorine with mass number 19.
   (b) An isotope of gold with 120 neutrons.
   (c) An isotope of mass number 107 with 60 neutrons.

2. Tin has more stable isotopes than any other element; they have mass numbers 114, 115, 116, 117, 118, 119, 120, 122, 124. Give the proper symbols for these isotopes.

3. (a) Compute the Coulomb repulsion energy between two nuclei of ^{16}O that just touch at their surfaces. (b) Do the same for two nuclei of ^{238}U.

4. What is the nuclear radius of (a) ^{197}Au; (b) ^{4}He; (c) ^{20}Ne?

5. Figure 9.2 suggests that the Rutherford scattering formula fails for 60° scattering when K is about 28 MeV. Use the results derived in Chapter 6 to find the closest distance between alpha particle and nucleus for this case, and compare with the nuclear radius of ^{208}Pb. Suggest a possible reason for any discrepancy.

6. Find the total binding energy, and the binding energy per nucleon, for (a) ^{208}Pb; (b) ^{133}Cs; (c) ^{90}Zr; (d) ^{59}Co.

7. Find the total binding energy, and the binding energy per nucleon, for (a) ^{4}He; (b) ^{20}Ne; (c) ^{40}Ca; (d) ^{55}Mn.

8. Calculate the total nuclear binding energy of ^{3}He and ^{3}H. Account for any difference by considering the Coulomb interaction of the extra proton of ^{3}He.

9. The *neutron separation energy* S_n is the amount of energy we must supply to remove a neutron from a nucleus.
   (a) Show that S_n can be found from the masses according to:

$$S_n = [m(n) + m(^{A-1}_Z X_{N-1}) - m(^A_Z X_N)]c^2$$

   (b) Find the neutron separation energy of ^{17}O, ^{7}Li, and ^{57}Fe.

10. Find the *proton separation energy* (see Problem 9) of ^{4}He, ^{12}C, and ^{40}Ca.

11. The nuclear attractive force must turn into a repulsion at very small distances to keep the nucleons from crowding too close together. What is the mass of an exchanged particle that will contribute to the repulsion at separations of 0.25 fm?

12. The weak interaction (the force responsible for beta decay) is thought to originate from an exchanged particle with a mass of roughly 75 GeV. What is the range of this force?

13. What fraction of the original number of nuclei present in a sample will remain after (a) two half-lives; (b) four half-lives; (c) 10 half-lives?

14. A certain sample of a radioactive material decays at a rate of 548 per second at $t = 0$. At $t = 48$ minutes, the counting rate has fallen to 213 per second. (a) What is the half-life of the sample? (b) What is its decay constant? (c) What will be the counting rate at $t = 125$ minutes?

15. What is the decay probability per second per nucleus of a substance with a half-life of 5.0 hours?

16. Tritium, the hydrogen isotope of mass 3, has a half-life of 12.3 y. What fraction of the tritium atoms remains in a sample after 50.0 y?

17. What is the activity of a container holding 125 cm^3 of tritium (^{3}H, $t_{1/2}$ = 12.3 y) at a pressure of 5.0×10^5 Pa (about 5 atm) at $T = 300$ K?

18. Suppose we have a sample containing 2.00 mCi of radioactive ^{131}I ($t_{1/2}$ = 8.04 d). (a) How many decays per second occur in the sample? (b) How many decays per second will occur in the sample after four weeks?

19. Ordinary potassium contains 0.012 percent of the naturally occurring radioactive isotope ^{40}K, which has a half-life of 1.3×10^9 y. (a) What is the activity of 1.0 kg of potassium? (b) What would have been the fraction of ^{40}K in natural potassium 4.5×10^9 y ago?

20. A radiation detector is in the form of a circular disc of diameter 3.0 cm. It is held 25 cm from a source of radiation, where it records 1250 counts per second. Assuming that the detector records every radiation incident upon it, find the activity of the sample (in curies).

21. With a radioactive sample originally of N_0 atoms, we could measure the mean, or average, lifetime of a nucleus by measuring the number N_1 that live for a time t_1 and then decay, the number N_2 that decay after t_2 and so on:

$$t_{av} = \frac{1}{N_0} (N_1 t_1 + N_2 t_2 + \cdots)$$

   (a) Show that this is equivalent to $t_{av} = \lambda \int_0^\infty e^{-\lambda t} t \, dt$
   (b) Show that $t_{av} = 1/\lambda$.

(c) Is t_{av} longer or shorter than $t_{1/2}$?

The time t_{av} is sometimes called the mean lifetime τ, and is the time necessary for $1/e$ of the nuclei to decay.

22. Complete the following decays:
    (a) $^{27}Si \rightarrow {}^{27}Al +$
    (b) $^{74}As \rightarrow {}^{74}Se +$
    (c) $^{228}U \rightarrow \alpha +$
    (d) $^{93}Mo + e^- \rightarrow$
    (e) $^{131}I \rightarrow {}^{131}Xe +$

23. Derive Equation (9.24) from Equations (9.22) and (9.23).

24. Find the kinetic energy of the alpha particle emitted in the decay of ^{234}U.

25. ^{239}Pu alpha decays with a half-life of 2.41×10^4 y. Compute the power output, in watts, which could be obtained from 1.00 gram of ^{239}Pu.

26. ^{228}Th alpha decays to an excited state of ^{224}Ra, which in turn decays to the ground state with the emission of a 217-keV photon. Find the kinetic energy of the alpha particle. The mass of ^{228}Th is 228.028726 u.

27. Derive Equations (9.32), (9.35), and (9.38).

28. Compute the recoil proton kinetic energy in neutron beta decay (a) when the electron has its maximum energy; (b) when the neutrino has its maximum energy.

29. Find the maximum kinetic energy of the electrons emitted in the negative beta decay of ^{11}Be.

30. ^{15}O decays to ^{15}N by positron beta decay. (a) What is the Q value for this decay? (b) What is the maximum kinetic energy of the positrons?

31. ^{75}Se decays by electron capture to ^{75}As. Find the energy of the emitted neutrino.

32. In the beta decay of ^{24}Na, an electron is observed with a kinetic energy of 2.15 MeV. What is the energy of the accompanying neutrino?

33. The $4n$ radioactive decay series begins with $^{232}_{90}Th$ and ends with $^{208}_{82}Pb$. (a) How many alpha decays are in the chain? (See Question 21.) (b) How many beta decays? (c) How much energy is released in the complete chain? (d) What is the radioactive power produced by 1.00 kg of $^{232}Th(t_{1/2} = 1.40 \times 10^{10}$ y)?

34. A piece of wood from a recently cut tree shows 12.4 ^{14}C decays per minute. A sample of the same size from a tree cut thousands of years ago shows 3.5 decays per minute. What is the age of this sample?

35. The first excited state of ^{57}Fe decays to the ground state with the emission of a 14.4-keV photon in a mean lifetime of 141 ns. (a) What is the width ΔE of the state? (b) What is the recoil kinetic energy of an atom of ^{57}Fe that emits a 14.4-keV photon? (c) If the kinetic energy of recoil is made negligible by placing the atoms in a solid lattice, resonant absorptions will occur. What velocity is required to Doppler shift the emitted photon so that resonance does not occur?

A nuclear physics reaction in which one cubic centimeter of matter is converted into energy.

10

NUCLEAR REACTIONS

The knowledge of the nucleus that we can obtain from studying radioactive decays is limited, because only certain radioactive processes occur in nature, only certain isotopes are made in those processes, and only certain excited states of nuclei (those that happen to follow radioactive decays) can be studied. Nuclear reactions, however, give us a controllable way to study *any* nuclear species, and to select any excited states of that species.

In this chapter we discuss some of the different nuclear reactions that can occur, and we study the properties of those reactions. Two nuclear reactions are of particular importance: fission and fusion; we pay special attention to those processes and we discuss how they are useful as sources of energy (or, more correctly, as *converters* of nuclear energy into thermal or electrical energy).

We conclude our study of nuclear physics with an introduction to some of the ways that methods of nuclear physics can be applied to problems in a variety of different areas.

In a typical nuclear reaction laboratory experiment, a beam of particles of type x is incident on a target containing nuclei of type X. After the reaction, an outgoing particle y is observed in the laboratory, leaving a residual nucleus Y. Symbolically, we write the reaction as

$$x + X \longrightarrow y + Y$$

For example,

$$_1^2\text{H}_1 + _{29}^{63}\text{Cu}_{34} \longrightarrow n + _{30}^{64}\text{Zn}_{34}$$

Like a chemical reaction, a nuclear reaction must be balanced—the number of protons and neutrons must be the same on both sides of the equation. (The forces responsible for nuclear beta decay can change neutrons into protons, but these forces act on a time scale of at least 10^{-10} s. The projectile and target nuclei are only within the range of one another's nuclear forces for an interval of at most 10^{-20} s, so there is not enough time for the proton-neutron conversion to take place.)

Since a nuclear reaction takes place under the influence only of internal forces between the projectile and target, we expect the reaction to conserve energy, linear momentum, and angular momentum.

In most experiments, we observe only the outgoing light particle y; the heavy residual nucleus Y usually loses all its kinetic energy (by collisions with other atoms) and therefore stops within the target.

We assume that we do the reaction by bombarding target nuclei X, initially at rest, with projectiles x of kinetic energy K_x. The product particles then share this kinetic energy, plus or minus any additional energy from the mass energy difference of the initial and final nuclei. (We will consider the energetics of nuclear reactions in detail in the next section.)

The bombarding particles x can be either charged particles, supplied by a suitable nuclear accelerator, or neutrons, whose source may be a nuclear reactor. Accelerators for charged particles, illustrated in Figures 10.1 and 10.2, are of two basic types. In a cyclotron, a particle is held in a circular orbit by a magnetic field and receives a small "kick" by an electric field twice each time it travels around the circle; a particle may

10.1 TYPES OF NUCLEAR REACTIONS

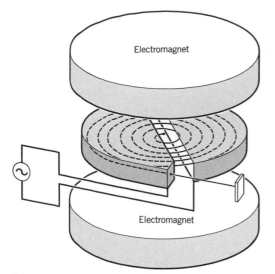

(a)

(b)

FIGURE 10.1 (a) Schematic diagram of cyclotron accelerator. Charged particles are bent in a circular path by the magnetic field and are accelerated by an electric field each time they cross the gap. (b) A cyclotron accelerator. The magnets are in the large cylinders at top and bottom. The beam is visible as it collides with air molecules after leaving the cyclotron (courtesy Argonne National Laboratory).

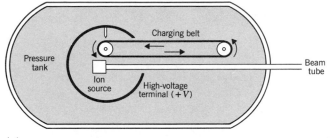

(a)

(b)

FIGURE 10.2 (a) Diagram of a Van de Graaff accelerator. Particles are accelerated from $+V$ to ground. (b) A typical Van de Graaff accelerator laboratory. The beam line and high-voltage terminal are inside the large pressure tank (courtesy Purdue University).

make perhaps 100 orbits before finally emerging with an energy of the order of 10 to 20 MeV per unit of electric charge. In the Van de Graaff accelerator, a particle is accelerated only once from a single high-voltage terminal, which may be at a potential of 6 million volts; the energy of the particle is then about 6 MeV per unit of charge.

In nuclear reaction experiments, we usually measure two basic properties of the particle y: its energy, and its probability to emerge at a certain angle with a certain energy. We look briefly at these two types of measurements.

If neither the residual nucleus Y nor the outgoing particle y had excited states, then by using conservation of energy and momentum, we should be able to calculate exactly the energy of y when measured at a certain angle. If the nucleus Y is left in an excited state, then the kinetic energy of y is reduced by (approximately) the energy of the excited state above the ground state, since the two particles Y and y must still share the same amount of total energy. Each higher excited state of the nucleus Y corresponds to a certain reduced energy of the particle y, and a

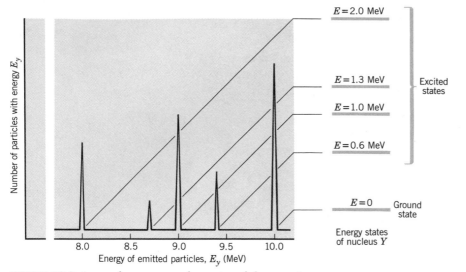

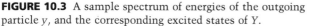

FIGURE 10.3 A sample spectrum of energies of the outgoing particle *y*, and the corresponding excited states of *Y*.

measurement of the different energies that the particle *y* can have tells us about the excited states of the nucleus *Y*. Figure 10.3 shows an example of a typical set of experimental results and the corresponding deduced excited states of the residual nucleus. Each peak in Figure 10.3 corresponds to a specific energy of *y*, and therefore to a specific excited state of *Y*; that is, when particles with energy 9.0 MeV are observed, the nucleus *Y* is left in the excited state with energy 1.0 MeV.

Notice that the different peaks in Figure 10.3 have different heights. This feature of the results of our experiment tells us that it is more probable for the reaction to lead to one excited state than to another. This is an example of the reaction probability, the second of the properties of *y* that we can determine. For example, Figure 10.3 shows that the probability of leaving *Y* in its second excited state (1.0 MeV) is about twice the probability of leaving *Y* in its first excited state. If it were possible to use the Schrödinger equation with the nuclear potential (if we knew what it was), we could calculate these reaction probabilities and compare them with experiment. Since we can't do that, we must work backward by measuring the reaction probabilities and then trying to infer some properties of the nuclear force.

Reaction probabilities are usually expressed in terms of the *cross section*, which is a sort of effective area presented by the target nucleus to that projectile for a specific reaction, for all possible energies and directions of travel of the outgoing particle *y*. In general, the cross section depends on the energy of the incident particle, K_x.

Cross section of nuclear reactions

The cross section σ is expressed in units of area, but the area is a very small one, of the order of 10^{-28} m². Nuclear physicists use this as a convenient unit of measure for cross section measurements, and it is known as one *barn* (b): 1 barn = 10^{-28} m². Notice that the area of the disc of a single nucleus of medium weight is about 1 barn; however, reaction cross sections can often be very much greater or less than one barn. For example, consider the cross section for these reactions involving certain isotopes of the neighboring elements iodine and xenon:

$$I + n \longrightarrow I + n \text{ (inelastic scattering)} \qquad \sigma = 4 \text{ b}$$

$$Xe + n \longrightarrow Xe + n \text{ (inelastic scattering)} \qquad \sigma = 4 \text{ b}$$

$$I + n \longrightarrow I + \gamma \text{ (neutron capture)} \qquad \sigma = 7 \text{ b}$$

$$Xe + n \longrightarrow Xe + \gamma \text{ (neutron capture)} \qquad \sigma = 10^6 \text{ b}$$

You can see how, although the neutron inelastic scattering cross sections of I and Xe are similar, the neutron capture cross sections are very different. These measurements are therefore telling us something interesting and unusual about the properties of the nucleus Xe.

Suppose a beam of particles is incident on a thin target of area A, which contains a total of N nuclei. The effective area of each nucleus is just the cross section σ, and so the total effective area of all the nuclei in the target is (ignoring shadowing effects) σN. The fraction of the target area which this represents is just $\sigma N/A$, and as long as this ratio is small, shadowing effects will be negligible. This fraction is the probability for the reaction to occur.

Suppose the incident particles strike the target at a rate of I_0 particles per second, and suppose the outgoing particles y are emitted at a rate of I per second. (This is also the rate at which the product nucleus Y is formed.) Then the reaction probability can also be expressed as the probability to find y per incident particle x, or I/I_0. Combining the two expressions for the reaction probability,

$$\frac{I}{I_0} = \frac{\sigma N}{A} \qquad (10.1)$$

we have a relationship between the reaction cross section and the rate of emission of y.

EXAMPLE 10.1

For a certain incident proton energy the reaction

$$p + {}^{56}\text{Fe} \longrightarrow n + {}^{56}\text{Co}$$

has a cross section of 0.60 b. If we bombard a target in the form of a 1.0-cm square, 1.0-μm thick iron foil, with a beam of protons equivalent to a current of 3.0 μA, and if the beam is spread uniformly over the entire surface of the target, at what rate are the neutrons produced?

Solution

We first calculate the number of nuclei in the target. The volume of the target is

$$V = 1.0 \text{ cm} \times 1.0 \text{ cm} \times 1.0 \text{ } \mu\text{m} = 1.0 \times 10^{-4} \text{ cm}^3$$

and since the density of iron is 7.9 g/cm³, the mass of the target is 7.9×10^{-4} g. One mole of ${}^{56}\text{Fe}$ has a mass of 56 grams, and so we have 1.4×10^{-5} mole. Since each mole contains Avogadro's number of atoms, it follows that the number of atoms in the target is

$$N = 1.4 \times 10^{-5} \times 6.02 \times 10^{23}$$

$$= 8.5 \times 10^{18}$$

Next we need to find the number of particles per second in the incident beam. We are given that the current is 3.0×10^{-6} A $= 3.0 \times 10^{-6}$ C/s and since each proton has a charge of 1.6×10^{-19} C, the beam intensity is

$$I_0 = \frac{3.0 \times 10^{-6} \text{ C/s}}{1.6 \times 10^{-19} \text{ C/particle}}$$

$$= 1.9 \times 10^{13} \text{ particle/s}$$

From Equation (10.1) we can now find I

$$I = \frac{N\sigma I_0}{A}$$

$$= \frac{(8.5 \times 10^{18} \text{ nuclei})(0.60 \text{ b/nucleus})(10^{-24} \text{ cm}^2/\text{b})(1.9 \times 10^{13} \text{ particle/s})}{1 \text{ cm}^2}$$

$$= 9.7 \times 10^7 \text{ particles/s}$$

About 10^8 neutrons/s are emitted from the target.

In Figure 10.4, the cross section is shown for reactions of the sort

$$^{A}_{Z}X_N + {}^{4}_{2}\text{He}_2 \longrightarrow {}^{A+3}_{Z+2}X'_{N+1} + n$$

$$^{A}_{Z}X_N + {}^{4}_{2}\text{He}_2 \longrightarrow {}^{A+2}_{Z+2}X'_{N} + 2n$$

$$^{A}_{Z}X_N + {}^{4}_{2}\text{He}_2 \longrightarrow {}^{A+1}_{Z+2}X'_{N-1} + 3n$$

10.2 THE COMPOUND NUCLEUS

and so forth. You can see that as we increase the energy of the incident alpha particle, each reaction probability reaches a peak and then begins to decrease; the more energy we put into the nucleus, the more neutrons we are likely to find leaving the nucleus. It becomes *less probable* to emit a single neutron than several neutrons as the incident energy increases, and this suggests that the reaction mechanism is *not* the alpha particle bumping into and knocking loose one or more neutrons, but rather that the incident energy is shared throughout the entire nucleus, and that one or more neutrons simply "evaporate" from the nucleus much like water molecules evaporate from a pot of water when we heat it. The more energy we put in, the more particles evaporate.

An alpha particle with an energy of 10 MeV (typical for nuclear reaction studies) has a deBroglie wavelength of about 5 fm. Just as we can't use a light wave to see an object that is smaller than the wavelength of the light, the deBroglie wave which represents the alpha particle will not "see" anything smaller than about 5 fm. The alpha particle therefore "sees" the whole target nucleus (since 5 fm is a typical

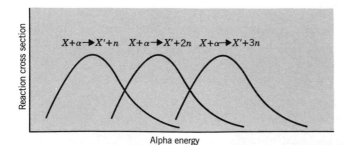

FIGURE 10.4 Cross sections for typical reactions in which different numbers of neutrons are produced.

nuclear radius) and does not see the individual protons and neutrons, whose size is about 1 fm. It is therefore *wrong* to think of such nuclear reactions as a sort of "billiard-ball" collision between the incident particle and a single nucleon in the nucleus; rather, the alpha particle shares its incident energy with all of the nucleons in the nucleus. There is an intermediate stage in a nuclear reaction, after the incident particle has been absorbed and has distributed its energy, but before the outgoing particle y is emitted; this is known as the *compound nucleus.* We don't observe this state directly, and in fact it only lives for a very short time (perhaps 10^{-15} s), but indirect evidence leads us to believe in its existence.

Compound nucleus reaction model

An important assumption of the compound nucleus reaction model is that once the incident energy has been distributed, the compound nucleus "forgets" how it was formed and chooses one of several possible decay modes based purely on statistical considerations. For example the reactions $p + {}^{63}$Cu and $\alpha + {}^{60}$Ni lead to the same compound nucleus, ^{64}Zn*. (The asterisk reminds us that the ^{64}Zn is a compound nuclear state and not a final reaction product.) The ^{64}Zn* can then break up in any number of ways, such as ^{63}Zn $+ n$ or ^{62}Cu $+ d$ or ^{61}Ni $+ {}^{3}$He. The compound nuclear model then requires that if, for example, the probability (and therefore the cross section) for $p + {}^{63}$Cu $\rightarrow n + {}^{63}$Zn was twice that for $p + {}^{63}$Cu $\rightarrow d + {}^{62}$Cu, then the cross section for $\alpha + {}^{60}$Ni $\rightarrow n + {}^{63}$Zn would similarly be twice that for $\alpha + {}^{60}$Ni $\rightarrow d + {}^{62}$Cu.

Symbolically,

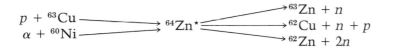

The relative probability for any of the decay products to be formed is independent of the means of formation of the compound nucleus.

This model has been tested in many experiments and found to be an accurate description of the way that certain reactions occur, especially in medium and heavy nuclei. Figure 10.5 shows the result of a typical experimental test; you can see that the cross sections for the three different final products of the compound nucleus ^{64}Zn* are essentially identical, even though the initial particles are different.

10.3 RADIOISOTOPE PRODUCTION IN NUCLEAR REACTIONS

Often we use nuclear reactions to produce radioactive isotopes. In this procedure, a stable (nonradioactive) isotope X is irradiated with the particle x to form the radioactive isotope Y; the outgoing particle y is of no interest and is not observed.

We would like now to calculate the activity of the isotope Y which is produced from a given exposure to a certain quantity of the particle x for a certain time t. Let R represent the constant rate at which Y is produced; this quantity is related to the cross section and to the intensity of the beam of x, as given in Equation (10.1). (In fact, you should convince yourself that the rate R at which Y is formed is identical to the rate I at which y is emitted.) In a time interval dt, the number of Y nuclei produced is just $R\,dt$. Since the isotope Y is radioactive, the number

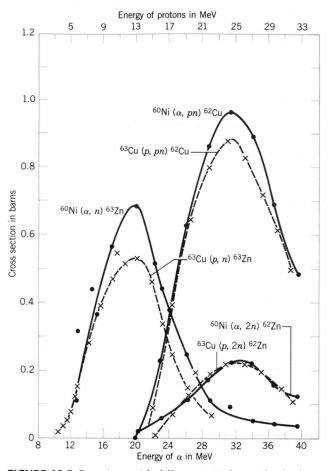

FIGURE 10.5 Reactions with different initial particles leading to the same compound nucleus and the same final particles have similar cross sections, demonstrating that the decay of a compound nucleus is independent of its means of formation. The reactions indicated as $X(x,y)Y$ mean $x + X \rightarrow y + Y$.

of nuclei of Y that decay in the interval dt is $\lambda N\,dt$, where λ is the decay constant $(\lambda = 0.693/t_{1/2})$ and N is the number of Y nuclei present. The net change dN in the number of Y nuclei is

$$dN = R\,dt - \lambda N\,dt \qquad (10.2)$$

or

$$\frac{dN}{dt} = R - \lambda N \qquad (10.3)$$

The solution to this differential equation is

$$N(t) = \frac{R}{\lambda}(1 - e^{-\lambda t}) \qquad (10.4)$$

and the activity is

$$\mathcal{a}(t) = \lambda N = R(1 - e^{-\lambda t}) \qquad (10.5)$$

Notice that, as expected, at $t = 0$, $\mathcal{a} = 0$ (there are no nuclei of type Y present at the start). For large irradiation times $t \gg t_{1/2}$, this expression

Activity produced in a nuclear reaction

approaches the constant value R. When t is small compared with the half-life $t_{1/2}$, the activity increases linearly with time:

$$\mathcal{A}(t) = R[1 - (1 - \lambda t + \cdots)]$$

$$\mathcal{A}(t) \cong R\lambda t \qquad (t \ll t_{1/2}) \tag{10.6}$$

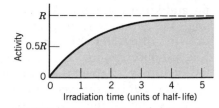

FIGURE 10.6 Formation of activity in a nuclear reaction.

Figure 10.6 shows the relationship between $\mathcal{A}(t)$ and t. As you can see, not much activity is gained by irradiating for more than about two half-lives.

In a reactor, the intensity of neutrons is usually expressed in terms of the neutron rate per unit area, or *neutron flux* ϕ (neutrons/cm²/s). The neutron cross section is σ (square centimeter per nucleus per incident neutron). The rate R also depends on the number of target nuclei. Suppose the mass of the target is m; the number of target nuclei is just $(m/M)N_A$, where M is the molecular weight, which can be the mass number A if the target is a pure atomic substance, and N_A is Avogadro's number (6.02×10^{23} atoms per mole). Thus, for neutron-induced reactions, using Equation (10.1)

$$R = \phi\sigma\frac{m}{M}N_A \tag{10.7}$$

EXAMPLE 10.2

Thirty milligrams of gold are exposed to a neutron flux of 3.0×10^{12} neutron/cm²/s for one minute. The neutron capture cross section of gold is 99 b. Find the resultant activity of ^{198}Au.

Solution

From the table of isotopes in the appendix we find that the stable isotope of gold has a mass number of $A = 197$, and that radioactive ^{198}Au has a half-life of 2.7 d. Thus:

$$R = \left(3.0 \times 10^{12}\,\frac{\text{neutrons}}{\text{cm}^2\cdot\text{s}}\right)\left(99 \times 10^{-24}\,\frac{\text{cm}^2}{\text{neutron}\cdot\text{nucleus}}\right)\left(\frac{0.030\,\text{g}}{197\,\text{g/mole}}\right)$$

$$\times\,(6.02 \times 10^{23}\,\text{atoms/mole})$$

$$= 2.7 \times 10^{10}\,\text{s}^{-1}$$

Since $t \ll t_{1/2}$ in this case, we can use Equation (10.6):

$$\mathcal{A} = (2.7 \times 10^{10}\,\text{s}^{-1})\left(\frac{0.693}{2.7\,\text{d}}\right)\left(1\,\text{min} \times \frac{1\,\text{h}}{60\,\text{min}} \times \frac{1\,\text{d}}{24\,\text{h}}\right)$$

$$= 4.8 \times 10^{6}\,\text{s}^{-1}$$

$$= 130\,\mu\text{Ci}$$

We can also use charged particle beams to produce radioactive isotopes. The calculation of the rate R would be carried out in a way identical to Example 10.1, and the activity can then be calculated directly as in Example 10.2. (See Problem 10 at the end of the chapter.)

10.4 LOW-ENERGY REACTION KINEMATICS

We assume for this discussion that the velocities of the nuclear particles are sufficiently small that we can use nonrelativistic kinematics. We consider a projectile x moving with momentum $\mathbf{p}_x$ and kinetic energy K_x. The target is at rest, and the reaction products have momenta $\mathbf{p}_y$ and

$\mathbf{p}_Y$, and kinetic energies K_y and K_Y. The particles y and Y are emitted at angles θ_y and θ_Y with respect to the direction of the incident beam. Figure 10.7 illustrates this reaction. We assume that the resultant nucleus Y is not observed in the laboratory (if it is a heavy nucleus, moving relatively slowly, it generally stops within the target).

As we did in the case of radioactive decay, we use energy conservation to compute the Q value for this reaction (assuming X is initially at rest):

$$\text{initial energy} = \text{final energy}$$

$$m_xc^2 + K_x + m_Xc^2 = m_yc^2 + K_y + m_Yc^2 + K_Y \qquad (10.8)$$

The m's in Equation (10.8) represent the actual *nuclear* masses of the reacting particles. However, as we have discussed, the number of protons and neutrons must *each* be balanced in a nuclear reaction:

since
$$Z_x + Z_X = Z_y + Z_Y \quad \therefore \quad \text{nuclear} \qquad (10.9)$$
$$\text{masses} \rightarrow \text{atomic masses}$$

We can therefore add equal numbers of electron masses to each side of Equation (10.8) and, neglecting as usual the electron binding energy, the nuclear masses become atomic masses with no additional corrections needed. Rewriting Equation (10.8)

$(m_x + m_X - m_y - m_Y)c^2 =$

$$m_xc^2 + m_Xc^2 - m_yc^2 - m_Yc^2 = K_y + K_Y - K_x \qquad (10.10)$$

The mass difference between the initial particles and final particles is defined to be the Q *value* of the reaction

$$Q = (m_x + m_X - m_y - m_Y)c^2 \qquad (10.11)$$

and, combining (10.10) and (10.11), we see that the Q value is just equal to the difference in kinetic energy between the final particles and initial particle:

$$Q = K_y + K_Y - K_x \qquad (10.12)$$

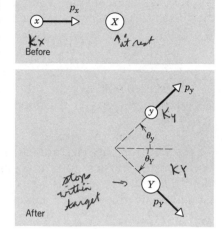

FIGURE 10.7 Momenta of reaction particles before (x, X) and after (y, Y) reaction.

Reaction Q value

$Q = $ L.H.S by definition (initial energy)

EXAMPLE 10.3

(a) Compute the Q value for the reaction

$$^2\text{H} + {}^{63}\text{Cu} \longrightarrow n + {}^{64}\text{Zn}$$

(b) Deuterons of energy 12.00 MeV are incident on a ^{63}Cu target, and neutrons are observed with 16.85 MeV of kinetic energy. Find the kinetic energy of the ^{64}Zn.

Solution

(a) The atomic masses can be found in Appendix B: *provided on test*

^{2}H: 2.014102 u
^{63}Cu: 62.929599 u
n: 1.008665 u
^{64}Zn: 63.929145 u

The Q value can be found using Equation (10.11):

$Q = (2.014102\text{ u} + 62.929599\text{ u} - 1.008665\text{ u} - 63.929145\text{ u})931.5\text{ MeV/u}$

$\quad = 5.487\text{ MeV}$

(b) From Equation (10.12)

$$K_Y = Q + K_x - K_y$$
$$= (5.487 + 12.00 - 16.85)\text{MeV}$$
$$= 0.64\text{ MeV}$$

Reactions for which $Q > 0$ convert nuclear energy to kinetic energy of y and Y and are called *exothermic* or *exoergic* reactions. Reactions with $Q < 0$ require energy input, in the form of the kinetic energy of x, to be converted into nuclear binding energy. These are known as *endothermic* or *endoergic* reactions.

In an endoergic reaction, we must supply at least enough kinetic energy to provide the additional rest mass of the reaction products. There is thus some minimum, or *threshold*, kinetic energy of x, below which the reaction will not take place. This threshold kinetic energy not only must supply the mass excess of the products, but also must supply some kinetic energy of the products; even at the minimum energy, the products cannot be at rest, for that would violate conservation of linear momentum—the momentum p_x before the collision would not be equal to the momentum of the final products after the collision.

This problem is most easily analyzed in the center-of-mass reference frame. In the lab frame, before the reaction, the center of mass moves with speed $v = m_x v_x/(m_x + m_X)$. If we travel with that speed and observe the reaction, we would see x moving with speed $v_x - v$ and X moving with speed $-v$, as shown in Figure 10.8. If x is moving with the threshold kinetic energy, in this reference frame the reaction products y and Y will appear to be at rest.

We must conserve total relativistic energy $K + m_0c^2$ in the reaction, and we restrict our discussion to small velocities $v \ll c$ so that the classical expression for the kinetic energy can be used. Energy conservation in the center-of-mass frame gives:

$$\tfrac{1}{2}m_x(v_x - v)^2 + \tfrac{1}{2}m_X(-v)^2 + m_xc^2 + m_Xc^2 = m_yc^2 + m_Yc^2 \qquad (10.13)$$

where v_x represents the threshold speed in the lab frame. Substituting the value of v and doing a bit of algebra, we can find the threshold kinetic energy (in the laboratory reference frame):

$$K_{th} = -Q\left(1 + \frac{m_x}{m_X}\right) \qquad (10.14)$$

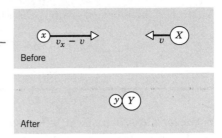

Before

After

FIGURE 10.8 Reaction at threshold in center-of-mass reference frame.

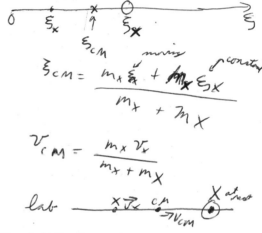

Threshold kinetic energy

Calculate the threshold kinetic energy for the reaction

$$p + {}_1^3H_2 \longrightarrow {}_1^2H_1 + {}_1^2H_1$$

(a) If protons are incident on ^3H at rest.
(b) If ^3H (tritons) are incident on protons at rest.

The atomic masses are:

$$p: \quad 1.007825 \text{ u}$$
$$^3H: \quad 3.016049 \text{ u}$$
$$^2H: \quad 2.014102 \text{ u}$$

The Q value is thus

$$Q = (1.007825 \text{ u} + 3.016049 \text{ u} - 2 \times 2.014102 \text{ u})931.5 \text{ MeV/u}$$

$$= -4.033 \text{ MeV}$$

When protons are incident on ^3H, the identification is $x = {}^1H$, $X = {}^3H$, so

$$K_{th} = (4.033 \text{ MeV})\left(1 + \frac{1.007825}{3.016049}\right) \quad \text{incoming mass / target}$$

$$= 5.381 \text{ MeV}$$

EXAMPLE 10.4

Solution

When ^{3}H is incident on protons the identification of x and X is reversed, so

$$K_{th} = (4.033 \text{ MeV}) \left(1 + \frac{3.016049}{1.007825} \right)$$

$$= 16.10 \text{ MeV}$$

10.5 FISSION

In the process of fission, a heavy nucleus such as uranium splits into two lighter nuclei. Since the lighter nuclei are about 1 MeV per nucleon more tightly bound than the heavy nucleus (see Figure 9.4), there is an energy conversion of about 200 MeV (200 nucleons × 1 MeV per nucleon) in each fission process. Compare this with typical electronic processes in atoms that involve a few electron-volts of energy—the energy converted per atom is roughly 10^8 times greater in nuclear reactions than in chemical reactions!

In a nucleus, there is a competition between the nuclear force, which holds the nucleus together, and the electrostatic repulsion of the nuclear protons, which tries to tear the nucleus apart. For most nuclei, the nuclear force dominates this competition, but for heavy nuclei there is a delicate balance between the nuclear and electric forces, a balance that is easily upset.

We can imagine a stable heavy nucleus as a sort of liquid drop, with a slightly elongated equilibrium shape, as shown in Figure 10.9. When that nucleus is disturbed, such as by absorbing a neutron or a high-energy photon, the drop begins to vibrate and quiver. The shape of the nucleus changes rapidly back and forth from more elongated to rather spherical. When the nucleus is stretched to highly elongated shapes, the Coulomb repulsion energy is not changed very much, but the nuclear force is reduced, because of the increased surface area (nucleons on the surface are less tightly bound). With sufficient stretching the center becomes "pinched off" somewhat, and the nucleus readily splits into two pieces, with the Coulomb repulsion driving the two pieces apart, a process illustrated in Figure 10.10. The energy necessary to cause fission is about 5 to 6 MeV; that is, if we consider an intermediate compound nucleus, about 5 or 6 MeV of excitation will cause fission.

The sizes of the two fragments may vary; Figure 10.11 shows the mass distribution of the fragments from the fission of ^{235}U, and we can see that it is most likely that one fragment will have a mass number of about 90 and the other about 140.

Of the 200 MeV or so released in fission, most of the energy goes into kinetic energy of the two fission fragments. We can understand that statement with a rough calculation of the Coulomb potential energy of two electric charges of $Z_1 \cong 40$ and $Z_2 \cong 50$ (as expected for $A_1 \cong 90$ and $A_2 \cong 140$) separated by a distance of $R = R_1 + R_2$, where R_1 and R_2 are the radii of the two fragments (which are assumed to be just touching at their surfaces). The potential energy $Z_1 Z_2 e^2/4\pi\varepsilon_0 R$ is easily calculated to be about 200 MeV. The two fragments separate rapidly, with the poten-

FIGURE 10.9 The elongated shape of a heavy nucleus.

FIGURE 10.10 Sequence of nuclear shapes in fission.

tial energy converted into about 200 MeV of kinetic energy. Although this is a very rough calculation, it does show that we might expect most of the fission energy to go to the fragments; in fact about 80 percent of the energy released in fission does appear as the kinetic energy of the fragments, and the remaining 20 percent appears as decay products (beta and gamma decays) and kinetic energy of neutrons emitted in the fission process. The neutrons typically have energies of one to several MeV.

A typical fission nuclear reaction might be

$$^{235}_{92}U_{143} + n \longrightarrow {}^{93}_{37}Rb_{56} + {}^{141}_{55}Cs_{86} + 2n$$

Of course, many different fission reactions are possible, with many different final products; the mass distribution of fragments was illustrated in Figure 10.11. The number of neutrons produced in the fission process can likewise vary, but the average is about 2.5. Each neutron can then initiate another fission process, resulting in the emission of still more neutrons, followed by more fissions, and so forth. This avalanche or *chain reaction* of fission events, each with the release of about 200 MeV of energy, can either occur under very rapid and uncontrolled conditions, as in a nuclear weapon, or else under slower and carefully controlled conditions, as in a nuclear reactor.

A schematic diagram of the processes that can occur in fission reactions is shown in Figure 10.12.

There are three features of the fission reaction that make it useful as a means to generate electrical energy. First of all, as we have already discussed, most of the energy is released as kinetic energy of the fission fragments. These relatively heavy fragments do not travel very far through the reactor fuel element before they dissipate most of their kinetic energy as heat, following collisions with the atoms of the fuel element. This heat can be used to boil water, and the resulting steam can then be used in a conventional way to drive a turbine to generate electricity.

The second feature of the fission reaction that makes it useful is

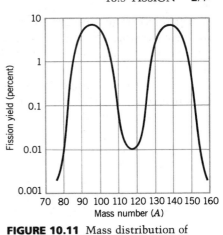

FIGURE 10.11 Mass distribution of fragments from fission of ^{235}U.

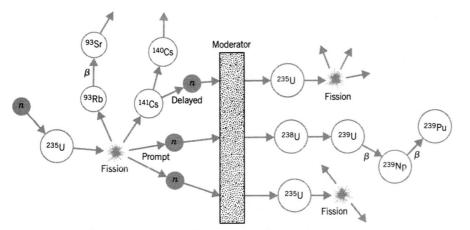

FIGURE 10.12 A typical sequence of processes in fission. A ^{235}U nucleus absorbs a neutron and fissions; two prompt neutrons and one delayed neutron are emitted. Following moderation, two neutrons cause new fissions and the third is captured by ^{238}U resulting finally in ^{239}Pu.

that the average number of neutrons produced is greater than one, making possible the chain reaction. How much greater than one it must be, in order to achieve a chain reaction, depends on the construction of the reactor.

The third advantage of the fission process is the one that enables an operator or mechanical system to control the reaction and keep it from proceeding too rapidly. The two neutrons emitted in the fission process

$$^{235}_{92}U_{143} + n \longrightarrow {}^{93}_{37}Rb_{56} + {}^{141}_{55}Cs_{86} + 2n$$

are *prompt neutrons*—they are emitted essentially at the instant of fission. About 1 percent of the neutrons in the fission process are *delayed neutrons* emitted following the decays of the heavy fragments. For example, the fission products in the above reaction are unstable, and decay according to the following sequences:

Delayed neutrons

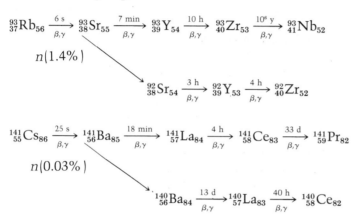

As you can see, ^{93}Rb decays with a half-life of 6 s and emits a neutron in 1.4 percent of the decays; ^{141}Cs decays with a half-life of 25 s and emits a neutron in 0.03 percent of its decays. Were it not for these delayed neutrons, mechanical control of the reaction rate would not be possible. In practice this control is accomplished by inserting into the core of the reactor a rod of material, such as cadmium, which has a high absorption cross section for neutrons. With the control rod fully inserted, enough neutrons are absorbed so that the average number of neutrons available to cause new fissions is less than one per fission reaction; as the rod is slowly withdrawn, the average number of available neutrons climbs until it is just equal to one per reaction, at which time the reactor is said to be *critical*. During operation, the position of the control rod can be continually adjusted, so that the chain reaction rate and the power level can be held constant. No mechanical system can respond rapidly enough to control the fluctuations in the reaction rate caused by prompt neutrons, but if the reactor is carefully designed to be less than critical for prompt neutrons, and critical for prompt plus delayed neutrons, mechanical control is possible.

There are, however, several technological problems that required solutions before the nuclear reactor became a useful power generator. First, the only naturally occurring material with a reasonably large fission cross section is the isotope ^{235}U. Naturally occurring uranium is only 0.7 percent ^{235}U; the other 99.3 percent is ^{238}U, which is for all practical purposes not fissionable. In order to build a fission reactor or a fission weapon, the concentration of ^{235}U must be substantially increased. This process is known as *enrichment*. Since ^{235}U and ^{238}U are chemically identical, the only means of enrichment is to take advantage of their small mass difference. This is a relatively difficult process, but one that is today accomplished with large quantities of uranium. For example, the gaseous diffusion plant at Oak Ridge, Tennessee, is based on the less massive ^{235}U diffusing through materials more easily than ^{238}U. Another easily fissionable material is ^{239}Pu. This substance does not occur in nature but can be produced through neutron capture by the nonfissionable ^{238}U; the resulting ^{239}U beta decays to ^{239}Np, which in turn decays to ^{239}Pu:

$$^{238}_{92}U_{146} + n \longrightarrow {}^{239}_{92}U_{147} \longrightarrow {}^{239}_{93}Np_{146} + e^- + \bar{\nu}$$

$$^{239}_{93}Np_{146} \longrightarrow {}^{239}_{94}Pu_{145} + e^- + \bar{\nu}$$

The plutonium can then be separated by chemical means from the uranium. This process of plutonium fuel production from uranium is known as *breeding* and a reactor that is designed to produce plutonium fuel is known as a *breeder*.

A second difficulty in trying to produce a chain reaction is the energy of the neutrons emitted in the fission process. Typically, the kinetic energies of these neutrons are a few MeV; such energetic neutrons have a relatively low probability of inducing new fissions, since the fission cross section generally decreases rapidly with increasing neutron energy. We therefore must slow down, or *moderate*, these neutrons in order to increase their chances of initiating fission events. The fissionable material is surrounded by a *moderator*, and the neutrons lose energy in collisions with the atoms of the moderator. When a neutron is scattered from a heavy nucleus like uranium, the energy of the neutron is changed hardly at all, but in a collision with a very light nucleus, the neutron can lose substantial energy. The most effective moderator would be one whose atoms are just as massive as a neutron; hydrogen is therefore the first choice. Ordinary water is frequently used as a moderator, since collisions with the protons are very effective in slowing the neutrons; however, neutrons have a relatively high probability of being absorbed by the water according to the reaction $p + n \rightarrow {}^2_1H_1 + \gamma$. So-called "heavy water," in which the hydrogen is replaced by deuterium, is more useful as a moderator, since it has virtually no neutron absorption cross section. A heavy-water reactor, since it has more available neutrons, can use ordinary (nonenriched) uranium as fuel; a reactor using ordinary water as moderator has fewer neutrons available to produce fission, and must therefore have more ^{235}U in its core.

Carbon is a light material that is solid, stable, and abundant, and that has a relatively small neutron absorption cross section. The first nuclear reactor was constructed by Enrico Fermi and his co-workers in 1942 at the University of Chicago; this reactor used carbon, in the form of graphite blocks, as moderator.

There is also a problem in reactor design associated with neutrons that do not produce fission reactions. If every neutron produced a fission reaction, then a self-sustaining chain reaction could occur if the average number of neutrons produced per fission were exactly 1. There are, however, many ways that neutrons can get "lost" and be unable to pro-

Neutron moderators

duce fission reactions: (1) escape through the surface; (2) absorption in the moderator; (3) absorption by ^{238}U. Escape through the surface is minimized by making the core of the reactor large enough so that the surface-to-volume ratio is small, and absorption by the moderator can be eliminated by use of a heavy-water moderator.

A final problem in reactor design is the extraction of the fission energy to produce usable power in the form of electricity. Most of the energy released in fission goes to the fission fragments, and those fragments, which are rather heavy, rapidly lose their kinetic energies in collisions with the atoms of the fuel element. This energy appears as heat in the fuel element and must be extracted to serve as a source of power, such as to turn an electrical generator. (It must also be removed for reasons of safety, since enough heat is produced to melt the core and cause a serious accident; for this reason, much effort has gone into the design of emergency cooling systems that prevent the core from melting if the heat were not properly removed.)

There are in use at present at least three systems for extracting the fission energy from the core of the reactor.

1. *Boiling-water reactors.* As indicated in Figure 10.13, a stream of water circulates through the core. The heat turns the water to steam, which is then used to generate electricity. The disadvantage of this system is that the water can become radioactive, and a rupture of the pipes near the turbines could result in a serious accident, with the spread of radioactive materials.

2. *Pressurized-water reactors.* In this case, as shown in Figure 10.14, the heat is extracted in a two-step process. Water circulates through the core under great pressure, to prevent its turning to steam. This hot water then in turn heats a second water system, which actually delivers steam to the turbine. Since the steam never enters the reactor core, it is not radioactive, and so there is no radioactive material in the vicinity of the turbine.

3. *Liquid-metal reactors.* The disadvantage of using water is that it has a rather small heat capacity and thus is not an efficient medium for extracting the heat from the core. A metallic medium would be much better for heat transfer. Liquid sodium, for example, could replace the

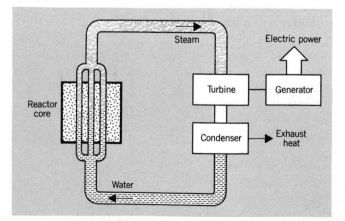

FIGURE 10.13 A boiling-water reactor.

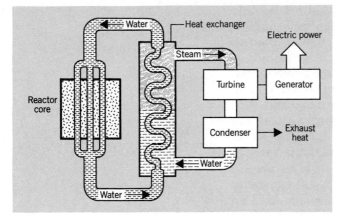

FIGURE 10.14 A pressurized-water reactor.

pressurized water of Figure 10.14; since the boiling point of sodium is well above the operating temperature, the high pressures are not necessary to keep the coolant a liquid.

There are yet other technological problems associated with nuclear power that are the subjects of active debate and investigation. Some of the radioactive isotopes among the fission fragments have very long half-lives, of the order of many years. The radioactive waste from reactors must be stored in a manner that prevents leakage of radioactive material into the biological environment. Many people are concerned about the safety of nuclear reactors, not only regarding proper design and operation, but about their resistance to such external forces as earthquakes and acts of terrorism or sabotage. Finally, as in all heat engines, the removal and disposal of the exhaust or waste heat (as the steam is recondensed to water, for example) generates considerable thermal pollution, which may have an adverse effect on the environment. Nuclear power plants do not necessarily produce more thermal pollution per unit of power generated than conventional coal- or oil-burning plants, but the economics of nuclear power favors the building of large power plants, in the 1000-megawatt range. One 1000-MW nuclear plant may produce just as much thermal pollution as 10 100-MW conventional plants, but 10 conventional plants spread over a wide geographical region do not release their thermal pollution in a single location as the nuclear plant does.

We conclude this section with a fascinating example of nature at work—the first sustained nuclear fission reactor on Earth was *not* the one constructed by Fermi in Chicago in 1942, but a *natural* fission reactor in Africa, which is believed to have operated two billion years ago for a period of perhaps several hundred thousand years. This reactor of course used naturally occurring uranium as a fuel, and naturally occurring water as a moderator. As we have discussed, such a reactor would not be possible to build today, because the capture of neutrons by the protons in water results in too few neutrons remaining to sustain a chain reaction in uranium with only 0.7 percent of ^{235}U. However, two billion years ago, naturally occurring uranium contained a much larger fraction of ^{235}U than does present-day uranium. The reason for this is

Natural fission reactor

that both ^{235}U and ^{238}U are radioactive, but the half-life of ^{235}U is only about one-sixth as great as the half-life of ^{238}U. If we go back in time about 2×10^9 y, which is half of one half-life of ^{238}U, there was about 40 percent more ^{238}U than there is today, but there was $2^3 = 8$ times as much ^{235}U. Naturally occurring uranium was then about 3 percent ^{235}U, and, at such enrichments, ordinary water can serve as an effective moderator. A deposit of such uranium, in a large enough mass and with ground water present to act as moderator, could have "gone critical" and begun to react.The reaction could have been controlled by the boiling of the water—when enough heat had been generated to evaporate some of the water, the reaction would slow down and perhaps stop, because of the lack of a moderator. When the uranium had cooled sufficiently to allow more liquid water to collect, the reactor would have started up again. This cycle could in principle have continued indefinitely, until enough ^{235}U were used up or until geological changes resulted in the removal of the water.

The discovery of this reactor followed the observation of a French researcher that the uranium that was being mined from that region in Africa contained too little ^{235}U. The discrepancy was a very small one—the samples contained 0.7171 percent ^{235}U, compared with the usual 0.7202 percent—but it was enough to stimulate the curiosity of the French workers. They guessed that the only mechanism that could result in the consumption of ^{235}U was the nuclear fission process, and this guess was tested by searching in the ore for stable isotopes that result from the radioactive decay of fission products. When such isotopes were found, and in particular when they were found in abundances very different from what would be expected from "natural" mineral deposits, the existence of the natural reactor was confirmed. An interesting account of this reactor is given in the reference listed at the end of this chapter.

10.6 FUSION

Energy may also be released in nuclear reactions in the process of fusion, in which two light nuclei combine to form a heavier nucleus. The energy released in this process is just the excess binding energy of the heavy nucleus compared with the lighter nuclei; from Figure 9.4, we see that this process will release energy as long as the final nucleus is below ^{56}Fe.

For example, consider the reaction

$$^2_1H_1 + {}^2_1H_1 \longrightarrow {}^3_1H_2 + {}^1_1H_0$$

The Q value is 4.0 MeV, and so this nuclear reaction liberates about 1 MeV per nucleon, roughly the same as the fission reaction. This reaction can be performed in the laboratory, by accelerating a beam of deuterons on to a deuterium target. In order to observe the reaction, we must get the incident and target deuterons close enough that the nuclear force can produce the reaction; that is, we must overcome the mutual Coulomb repulsion of the two particles. We can estimate this Coulomb repulsion by calculating the electrostatic repulsion of two deuterons when they are just touching. The radius of a deuteron is about 1.5 fm, and the electrostatic potential energy of the two charges separated by about 3 fm is about 0.5 MeV. A deuteron with 0.5 MeV of kinetic energy can overcome the Coulomb repulsion and initiate a reaction in which 4.5 MeV of energy (0.5 MeV of incident kinetic energy plus the 4-MeV Q value) is released.

We can produce such a beam of deuterons in many of the accelerators available in nuclear physics laboratories. The beam currents of such accelerators are typically of the order of microamperes. If every particle in the beam produced a reaction (hardly a reasonable assumption!), the total power produced would be 4 W. An output of 4 W (assuming we could extract every bit of the energy liberated in the reaction, which appears as the kinetic energies of the products ^{1}H and ^{3}H) hardly makes this device a useful power source!

A more promising approach would consist of heating deuterium gas to a high enough temperature that each atom of deuterium has about 0.25 MeV of thermal kinetic energy (hence the name *thermo*nuclear fusion). Then in a collision between two deuterium atoms, the total of 0.5 MeV of kinetic energy would be sufficient to overcome the Coulomb repulsion. If we could extract the fusion energy from the deuterium in a cup of "heavy water" (D_2O), we would have an energy of about 5×10^{12} J available; even if the conversion were done over the time of one day, the power output would be about 50 MW! Ordinary water contains about 0.015 percent D_2O; the fusion energy from the deuterium in a liter of ordinary water is equivalent to the chemical energy obtained from burning about 300 liters of gasoline.

The difficulty with this approach is in heating the deuterium gas to a sufficient temperature; from the expression $\frac{3}{2}kT$ for the thermal kinetic energy of a gas molecule, we can calculate that an energy of 0.25 MeV corresponds to a temperature of the order of 10^9 K. Even assuming that barrier penetration (Section 5.7) would allow a reasonable probability to penetrate the Coulomb barrier at lower kinetic energies (perhaps corresponding to one-tenth of the calculated temperature), it is hard to imagine any conditions under which such temperatures can be obtained. Nevertheless, fusion processes not only support all life on Earth, since they provide the energy by which the sun shines, but also are believed by many scientists and engineers to hold a promising future for providing essentially unlimited quantities of electrical power.

Let us begin with a brief look at the fusion processes inside the sun. In the basic fusion process, which can occur through several different paths, four protons combine to make one ^{4}He. Since the sun is composed of ordinary hydrogen, rather than deuterium, it is first necessary to convert the hydrogen to deuterium. This is done according to the reaction

Proton-proton cycle

$$^1_1H_0 + {}^1_1H_0 \longrightarrow {}^2_1H_1 + e^+ + \nu$$

This process involves converting a proton to a neutron and is analogous to the beta-decay processes we have studied previously. Once we have obtained ^{2}H (deuterium), the next reaction that can occur is

$$^2_1H_1 + {}^1_1H_0 \longrightarrow {}^3_2He_1 + \gamma$$

followed by

$$^3_2He_1 + {}^3_2He_1 \longrightarrow {}^4_2He_2 + 2{}^1_1H_0$$

Note that the first two reactions must occur *twice* in order to produce the two ^{3}He we need for the third reaction; see the schematic dia-

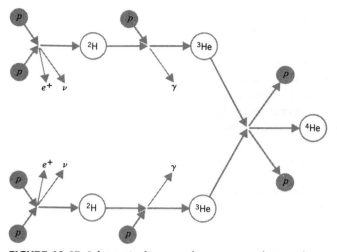

FIGURE 10.15 Schematic diagram of processes in fusion of protons to form helium.

gram of Figure 10.15. We can write the net process as

$$4{}_1^1H_0 \longrightarrow {}_2^4He_2 + 2e^+ + 2\nu + 2\gamma$$

For the calculation of the Q value in terms of *atomic* masses, four electrons must be added to the left side to make four neutral hydrogen atoms. To balance the reaction we must also add four electrons to the right side; two of these are associated with the ${}^4He$ atom, and the other two will combine with the two positrons according to the reaction $e^+ + e^- \rightarrow 2\gamma$, so that the additional gamma rays are available energy from the reaction. Since the two positrons disappear in this process, the only masses remaining are four hydrogen *atoms* and the one helium *atom*, and so

$$Q = (m_i - m_f)c^2$$

$$= (4 \times 1.007825 \text{ u} - 4.002603 \text{ u})931.5 \text{ MeV/u}$$

$$= 26.7 \text{ MeV}$$

Each fusion reaction liberates about 26.7 MeV of energy. Let us try now to calculate the rate at which these reactions occur in the sun. The solar power reaching the Earth is typically about 1.4×10^3 W/m^2. Since we are about 1.5×10^{11} m from the sun, the sun's energy is spread over a sphere of area $4\pi r^2 = 28 \times 10^{22}$ m^2, and thus the energy output from the sun is about 4×10^{26} W, which corresponds to about 2×10^{39} MeV/s. Each fusion reaction liberates about 26 MeV, and thus there must be about 10^{38} fusion reactions per second, consuming about 4×10^{38} protons per second. (Don't worry about running out of protons—the sun's mass is about 2×10^{30} kg, which corresponds to about 10^{57} protons, enough to burn for the next few billion years.)

The sequence of reactions described above is called the *proton-proton* cycle and probably represents the source of the sun's energy. However, it is probably *not* the primary source of fusion energy in many stars, because the first reaction (in which two protons combine to form a deuteron), which is a sort of beta decay, takes place only on a very long time scale (as we discuss in the next chapter), and is therefore very

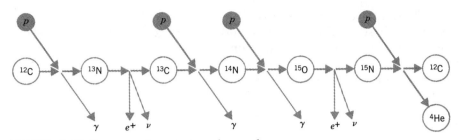

FIGURE 10.16 Sequence of events in carbon cycle.

unlikely to occur. A more likely sequence of reactions is the *carbon cycle:*

$$^{12}C + {}^1H \longrightarrow {}^{13}N + \gamma$$

$$^{13}N \longrightarrow {}^{13}C + e^+ + \nu$$

$$^{13}C + {}^1H \longrightarrow {}^{14}N + \gamma$$

$$^{14}N + {}^1H \longrightarrow {}^{15}O + \gamma$$

$$^{15}O \longrightarrow {}^{15}N + e^+ + \nu$$

$$^{15}N + {}^1H \longrightarrow {}^{12}C + {}^4He$$

Carbon cycle

A symbolic diagram of the process is shown in Figure 10.16. Notice that the ^{12}C plays the role of catalyst; we neither produce nor consume any ^{12}C in these reactions, but the presence of the carbon permits this sequence of reactions to take place at a much greater rate than the previously discussed proton-proton cycle. The net process is still described by $4{}^1H \rightarrow {}^4He$, and of course the Q value is the same. Since the Coulomb repulsion between H and C is larger than the Coulomb repulsion between two H nuclei, more thermal energy and a correspondingly higher temperature are needed for the carbon cycle. The carbon cycle probably becomes important at a temperature of about 20×10^6 K, while the sun's interior temperature is "only" 15×10^6 K.

When all of the hydrogen has been converted to helium, the sun will contract and heat up until the temperature rises by a factor sufficient to allow helium burning to occur, by processes such as

$$3{}^4He \longrightarrow {}^{12}C$$

Two He nuclei have a larger mutual Coulomb repulsion than two H nuclei, so helium fusion needs more thermal energy than hydrogen fusion.

When the helium is used up, a still higher temperature will allow carbon fusion to make even heavier elements like ^{24}Mg. Such processes will continue until ^{56}Fe is reached; beyond this point (see Figure 9.4) no further energy is gained by fusion. The production of elements in fusion processes is discussed in greater detail in Chapter 15.

For a controlled thermonuclear reactor, several reactions could be used, such as

$$^2H + {}^2H \longrightarrow {}^3H + {}^1H \qquad Q = 4.0 \text{ MeV}$$

$$^2H + {}^2H \longrightarrow {}^3He + n \qquad Q = 3.3 \text{ MeV}$$

$$^2H + {}^3H \longrightarrow {}^4He + n \qquad Q = 17.6 \text{ MeV}$$

The third reaction, known as the D-T (deuterium-tritium) reaction, has the largest energy release and is perhaps the best candidate for a fusion reactor. The most difficult technological problem facing the development of fusion reactors is to obtain a sufficiently high temperature (perhaps 10^8 K) so that the Coulomb repulsion is overcome, and simultaneously to maintain a high enough density so that the probability of two particles colliding is reasonably high. There are presently two methods of solution to these problems under active investigation: *magnetic confinement* and *inertial confinement*.

In magnetic confinement, the deuterium gas is confined in a large chamber at a high temperature. When the gas reaches a high enough temperature, collisions of the gas atoms result in the removal of electrons, and what remains is a gas of hot, ionized particles which is known as a *plasma*. Mutual repulsion would tend to force the atoms to the walls, where they would immediately lose energy in collisions with the cooler walls of the chamber. Thus the problem of confinement means the plasma must be kept toward the center of the chamber, and its density and temperature must be sufficient to allow fusion to occur. The geometries of fusion reactors can be *linear* or *toroidal*. A typical linear device, shown in Figure 10.17, operates on the *magnetic mirror* principle. The charged gas atoms spiral around the magnetic field lines in the center of the device and are thus kept away from the walls. As the particles enter the region of high magnetic field, the spirals become tighter and tighter until the particle is reflected back. An example of a device that operates in the toroidal geometry is shown in Figure 10.18. An electric current passed through the plasma helps to heat it and at the same time creates a magnetic field that tends to confine the plasma. Another magnetic field is produced along the axis of the toroid by current-carrying conductors. The combination of these two magnetic fields helps to confine the plasma. This system is known as the *Tokamak* design.

In the inertial confinement method, a small pellet (diameter about 0.1 to 1 mm) containing deuterium and tritium is struck simultaneously from many directions by intense laser beams that first vaporize the pellet and convert it to a plasma, and then heat and compress the pellet

Magnetic confinement

Inertial confinement

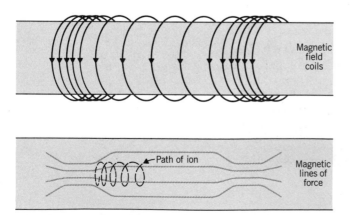

FIGURE 10.17 The magnetic mirror principle of plasma confinement. The ionized atoms are "reflected" from the regions of high magnetic field.

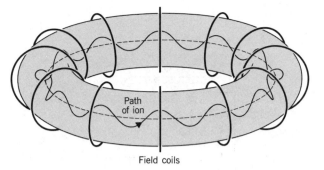

FIGURE 10.18 The toroidal geometry of plasma confinement. The ionized atoms circulate around the ring, spiraling around the magnetic field lines.

to a point at which fusion can occur. (See Figure 10.19.) The absorption of the laser light causes an "implosion" that compresses the pellet until its density is about 10^3 to 10^4 times ordinary densities. At such densities, the alpha particles that result from the D-T reaction cannot escape from the fuel pellet; they lose all their energy in collisions and contribute to the heating of the pellet.

The laser input energy must be in the range of 100,000 J, and the energy must be delivered in a pulse which lasts only 10^{-9} s. Such power levels, 10^{14} W, sound incredibly large, but that power need be delivered only for a short time. A typical power plant might "implode" 100 pellets per second, and might produce 1000 MW of power. The "average" power needed for the laser is only 10^7 W, since it needs its power of 10^{14} W for only 10^{-7} (10^{-9} second per pulse × 100 pulses per second) of the time. The power gain is therefore a factor of 100, although this is reduced somewhat in allowing for inefficiency in powering the lasers (probably 10 percent efficiency is the best that can be expected) and in recovering the energy produced by the fusions.

In the D-T fusion reaction, most of the energy is carried by the neutrons (recall in the fission reaction only a small fraction of the energy went to the neutrons). This presents some difficult problems for the recovery of the energy and its conversion into electrical power. One possibility for a fusion reactor design is shown in Figure 10.20. The reaction area is surrounded by lithium, which captures neutrons by the reaction

$$^6_3Li_3 + n \longrightarrow {}^4_2He_2 + {}^3_1H_2$$

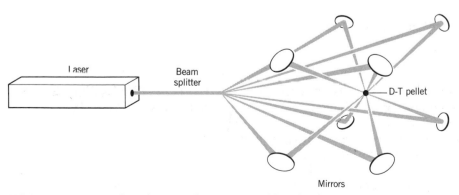

FIGURE 10.19 Inertial confinement fusion initiated by a laser.

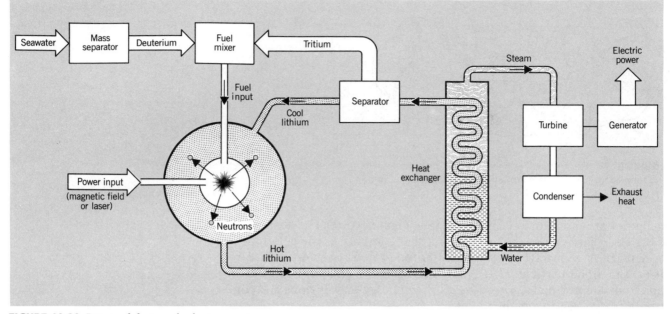

FIGURE 10.20 Proposed design of a fusion reactor.

The kinetic energies of the reaction products are rapidly dissipated as heat, and the thermal energy of the liquid lithium can be used to convert water to steam in order to generate electricity. This reaction has the added advantage of producing tritium (^{3}H), which is needed as a fuel for the fusion reactor.

One difficulty with the D-T fusion reaction is the large quantity of neutrons released in the reaction. Although fusion reactors will not produce the radioactive wastes that fission reactors do, the neutrons are sure to make radioactive the immediate area surrounding the reactor, and the structural damage to materials resulting from exposure to large fluxes of neutrons may weaken critical parts of the reactor vessel. Here once again the lithium is helpful, since a 1-m thickness of lithium should be sufficient to stop essentially all of the neutrons.

Fusion energy is the subject of vigorous research in many laboratories in the United States and around the world; the technological problems are being attacked with a variety of methods, and researchers are hopeful that solutions can be found during the next 20 years so that fusion can help to supply our electrical power needs.

In this chapter, we have discussed how fission and fusion reactions can be used to generate electrical power, and in the last chapter we discussed how the radioactive decay of various isotopes can be used to date the historical origin of material containing those isotopes. These are but a few of the many ways that nuclear decays and reactions can be applied to the solution of practical problems. In this section we discuss briefly some other applications of the techniques of nuclear physics.

10.7 APPLICATIONS OF NUCLEAR PHYSICS

Medical Radiation Physics   One of the most important applications of nuclear physics has been in the field of medicine, both for diagnostic and therapeutic purposes. Radioisotopes can be introduced into the body in chemical forms that have an affinity for a certain biological system, such as bone or thyroid gland. With sensitive detectors of the gamma rays from the decay of these isotopes, it is possible to reconstruct, using computers, a three-dimensional view of the system that can reveal abnormalities, tumors, or other difficulties. One treatment method presently used for tumors is to expose them to large quantities of radiation. This treatment is based on the fact that often tumors are chemically different from the surrounding tissue, and so may have a larger cross section for certain kinds of radiation. Properly aimed and focused, the radiation can produce substantially more damage to the tumor than to the surrounding tissue. Among the radiations that have been used for such experiments are gamma rays and pi mesons; the latter have the advantage of being able to be focused (since they are charged particles) and causing nuclear reactions at the site of the tumor, which do more local damage and are more effective in destroying the tumor.

Neutron Activation Analysis   Nearly every radioactive isotope emits characteristic gamma rays, and so many chemical elements can be identified by their gamma ray spectra. For example, when ^{59}Co (the only stable isotope of cobalt) is placed in a flux of neutrons (such as is found near the core of a reactor), neutron absorption results in the production of the radioactive isotope ^{60}Co, which beta decays with a half-life of 5.27 years. Following the beta decay, ^{60}Ni emits two gamma rays of energies 1.17 MeV and 1.33 MeV and of equal intensity. If we place in a flux of neutrons a material of unknown composition, and if we observe, following the neutron bombardment, two gamma rays of equal intensity and energies 1.17 MeV and 1.33 MeV, it is a safe bet that the unknown sample contained cobalt. In fact, from the rate of gamma emission we could deduce exactly how much cobalt the material contains, assuming that we know the neutron flux and the neutron capture cross section of ^{59}Co. This technique is known as *neutron activation analysis,* and has been used in many applications in which the elements are present in such small quantities that chemical identification is not practical. Typically, neutron activation analysis can be used to identify elements in quantities of the order of 10^{-9} g, and sensitivity down to 10^{-12} g is often possible.

Such a sensitive and precise technique finds application in a variety of areas, in which the chemical composition must be determined for samples which are only available in microscopic quantities or which must be analyzed in a nondestructive manner. For example, the chemical composition of various types of pottery can help us trace the geographical origin of the clay from which they were made; such analyses of pottery shards can trace the trading routes of prehistoric people. Art forgeries can be detected by a knowledge of the chemical composition of paints, since techniques for producing pigments have changed over the last four centuries with corresponding changes in the level of impurities in paints. The chemical analysis of minute quantities of material such as paint, gunshot residues, soil, or hair can provide important evidence in criminal investigations. As an extreme example, neutron activation analysis of samples of the hair of such historical figures as Napoleon or Newton has revealed the chemicals to which they were exposed centuries ago.

Synthetic Elements   The elements up through Fe were made by means of fusion reactions in stars. As we have seen, it is not energetically favorable for fusion reactions to continue beyond ^{56}Fe. The remaining elements were made by neutron capture, producing isotopes with an excess of neutrons, followed by beta decay in which a neutron converts to a proton. Thus,

$$^{56}_{26}\text{Fe}_{30} + n \longrightarrow {}^{57}_{26}\text{Fe}_{31} \quad \text{(stable)}$$

$$^{57}_{26}\text{Fe}_{31} + n \longrightarrow {}^{58}_{26}\text{Fe}_{32} \quad \text{(stable)}$$

$$^{58}_{26}\text{Fe}_{32} + n \longrightarrow {}^{59}_{26}\text{Fe}_{33} \xrightarrow[\beta]{} {}^{59}_{27}\text{Co}_{32} \quad \text{(stable)}$$

The stable isotope ^{59}Co is produced following the radioactive decay of ^{59}Fe, which is produced from ^{56}Fe following the capture of three neutrons. This process continues as follows:

$$^{59}_{27}\text{Co}_{32} + n \longrightarrow {}^{60}_{27}\text{Co}_{33} \xrightarrow[\beta]{} {}^{60}_{28}\text{Ni}_{32} \quad \text{(stable)}$$

Notice that each beta decay *increases* the atomic number by one unit. In principle this process of n capture and β decay could go on indefinitely, producing atoms of larger and larger Z. However, the atoms beyond uranium ($Z = 92$) are *all* radioactive, with half-lives short compared with the age of the Earth. They are therefore not present in terrestrial matter, but they can be produced in the laboratory. The production process for the series of elements beginning with neptunium ($Z = 93$), called *trans-uranic* elements, follows the same process outlined above: neutron capture followed by beta decay. In this way elements all the way up to $Z = 107$ have been produced. Many of the elements in this series have half-lives of only minutes or seconds, and thus the production and identification of these elements requires painstaking experimental

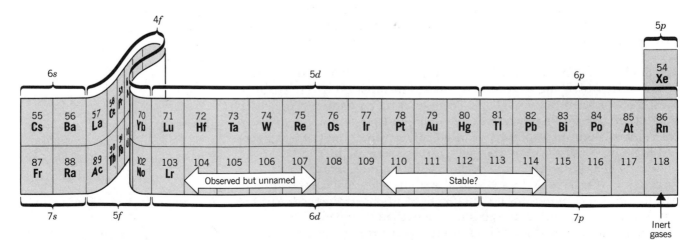

FIGURE 10.21 The location of possible new elements in the periodic table.

efforts—the isotopes are often produced in quantities of a few atoms! Although most of these elements have not been produced in sufficient quantity to study their chemical properties, it is expected that their place in the periodic table will be as shown in Figure 10.21, up to the inert gas $Z = 118$.

The extreme instability of these transuranic elements results from the increased Coulomb repulsion of the nuclear protons as Z increases; these elements decay by alpha decay or by spontaneous fission (without first having to absorb a neutron). However, there is strong theoretical evidence which suggests that, for complicated reasons based on the nuclear structure, elements around $Z = 114$, $N = 184$ should be stable against alpha decay, beta decay, and spontaneous fission. Owing to the short half-lives and microscopic quantities of the elements between $Z = 100$ and $Z = 114$, it is unlikely that the neutron-capture, beta-decay process could be used to produce and observe element 114. Other nuclear reactions, with a light or medium mass projectile bombarding a heavy target (such as $^{40}_{18}Ar + ^{248}_{96}Cm$), are not feasible, because not enough neutrons are available to supply the 184 necessary for stability. It is necessary to use neutron-rich nuclei for projectile *and* target, such as $^{238}U + {}^{238}U$, in the hope that an intermediate state (a sort of compound nucleus) would be formed that would then decay (fission, perhaps) to the stable isotope at $Z = 114$.

Although the production and observation of such *superheavy* nuclei are not likely to have any immediate applications, their study would be of great interest as a test of our understanding of the ordering of the periodic table. Many transuranic elements, however, have already found applications in research and technology. These applications result primarily from the alpha decay of these elements. For example, the alpha decay of ^{238}Pu has found use as a compact and totally reliable power source for cardiac pacemakers as well as for the Voyager spacecraft, which photographed Jupiter and Saturn. Other alpha-emitting isotopes have been used in the Surveyor spacecraft on the moon and the Viking spacecraft on Mars to study the chemical composition of their surfaces by alpha-scattering techniques. The isotope ^{241}Am is used in commerical smoke detectors; when the emitted alpha particles are scattered by particles of smoke, the alarm is triggered.

SUGGESTIONS FOR FURTHER READING

See the references listed in Chapter 9 for more detail on nuclear reactions. Other references at about the same level as this chapter are:

H. Semat and J. R. Albright, *Introduction to Atomic and Nuclear Physics* (New York, Holt, Rinehart and Winston, 1972).

M. R. Wehr, J. A. Richards, and T. W. Adair, *Physics of the Atom* (Reading, Addison-Wesley, 1978). Particularly good discussion of fission reactors.

The natural fission reactor is discussed in:

G. A. Cowan, "A Natural Fission Reactor," *Scientific American* 235, 36 (July 1976).

Fusion power has frequently been treated in general articles, including the following:

B. Coppi and J. Rem, "The Tokamak Approach in Fusion Research," *Scientific American* 227, 65 (July 1972).

9. Neutron capture in sodium occurs with a cross section of 0.53 b and leads to radioactive ^{24}Na $(t_{1/2} = 15$ h$)$. What is the activity which results when 1.0 μg of Na is placed in a neutron flux of 2.5×10^{13} neutrons/cm^2/s for 4.0 h?

10. The radioactive isotope ^{61}Cu $(t_{1/2} = 3.41$ h$)$ is to be produced by alpha particle reactions on ^{59}Co. A target of cobalt, 2.5 μm thick, is placed in a 12.0-μA beam of alpha particles; the beam uniformly covers an area of the target 1.5 by 1.5 cm. For the particular alpha particle energy used, the reaction has a cross section of 0.640 b. (a) At what rate is the ^{61}Cu produced? (b) What is the resulting activity of ^{61}Cu after 2.0 h of irradiation? (c) If the irradiation stops after 2.0 h, what is the activity 3.0 h later?

11. A 2.0-mg sample of copper (69% ^{63}Cu, 31% ^{65}Cu) is placed in a reactor where it is exposed to a neutron flux of 5.0×10^{12} neutrons/cm^2/s. After 10.0 min the resulting activities are 72 μCi of ^{64}Cu $(t_{1/2} = 12.7$ h$)$ and 1.30 mCi of ^{66}Cu $(t_{1/2} = 5.1$ min$)$. Find the cross sections of ^{63}Cu and ^{65}Cu.

12. A small sample of paint is placed in a neutron flux of 3.0×10^{12} neutrons/cm^2/s for a period of 2.5 min. At the end of that period the activity of the sample is found to include 105 decays/s of ^{51}Ti $(t_{1/2} = 5.8$ min$)$ and 12 decays/s of ^{60}Co $(t_{1/2} = 5.27$ y$)$. Find the amount, in grams, of titanium and cobalt in the original sample. Use the following information: Cobalt is pure ^{59}Co, which has a cross section of 19 b; titanium is 5.25 percent ^{50}Ti, which has a cross section of 0.14 b.

13. A beam of neutrons of intensity I is incident on a thin slab of material of area A, thickness dx, density ρ, and atomic weight M. The neutron absorption cross section is σ. (a) What is the loss in intensity dI of this beam in passing through the material? (b) A beam of original intensity I_0 passes through a thickness x of the material. Show that the intensity of the emerging beam is $I = I_0 e^{-n\sigma x}$, where n is the number of absorber nuclei per unit volume.

14. Assume that the total cross section for neutrons incident on copper is 5.0 b. What fraction of the intensity of a neutron beam is lost after traveling through copper of thickness (a) 1.0 mm; (b) 1.0 cm; (c) 1.0 m? (See Problem 13.)

15. Derive Equation (10.14) from Equation (10.13).

16. Find the Q values of the reactions:
(a) ^{6}Li $+ n \rightarrow {}^3$H $+ {}^4$He
(b) $p + {}^2$H $\rightarrow 2p + n$
(c) ^{7}Li $+ d \rightarrow {}^8$Be $+ n$

17. Find the Q value of the reactions:
(a) $p + {}^{55}$Mn $\rightarrow {}^{54}$Fe $+ 2n$
(b) ^{3}He $+ {}^{40}$Ar $\rightarrow {}^{41}$K $+ {}^2$H

18. (a) What is the Q value of the reaction $p + {}^4$He $\rightarrow {}^2$H $+ {}^3$He? (b) What is the threshold energy for protons incident on ^{4}He at rest? (c) What is the threshold energy if ^{4}He are incident on protons at rest?

19. In the reaction ^{2}H $+ {}^3$He $\rightarrow p + {}^4$He deuterons of energy 5.000 MeV are incident on ^{3}He at rest. Both the proton and the alpha particle are observed to travel along the same direction as the incident deuteron. Find the kinetic energies of the proton and the alpha particle.

20. The nucleus ^{113}Cd captures a thermal neutron $(K = 0.025$ eV$)$, producing ^{114}Cd in an excited state; the excited state of ^{114}Cd decays to the ground state by emitting a photon. Find the energy of the photon.

21. A reaction in which two particles join to form a single excited nucleus, which then decays to its ground state by photon emission, is known as *radiative capture*. Find the energy of the gamma ray emitted in the radiative

capture of an alpha particle by ^{7}Li. Assume alpha particles of very small kinetic energy are incident on ^{7}Li at rest.

22. How much energy is required (in the form of gamma-ray photons) to break up ^{7}Li into ^{3}H + ^{4}He? This reaction is known as *photodisintegration*.

23. We can understand why ^{235}U is readily fissionable, and ^{238}U is not, with the following calculation. (a) Find the energy difference between ^{235}U + n and ^{236}U. We can regard this as the "excitation energy" of a sort of compound nuclear state of ^{236}U. (b) Repeat for ^{238}U + n and ^{239}U. (c) Comparing your results for (a) and (b) explain why ^{235}U will fission with very low energy neutrons, while ^{238}U requires fast neutrons of 1 to 2 MeV of energy to fission. (d) From a similar calculation, predict whether ^{239}Pu requires low-energy or higher-energy neutrons to fission.

24. Find the energy released in the fission of 1.00 kg of uranium that has been enriched to 3.0 percent in the isotope ^{235}U.

25. Find the Q value (and therefore the energy released) in the fission reaction ^{235}U + $n \rightarrow$ ^{93}Rb + ^{141}Cs + 2n. Use $m(^{93}$Rb$) = 92.92172$ u and $m(^{141}$Cs$) = 140.91949$ u.

26. A neutron of kinetic energy K makes a head-on elastic collision with a ^{12}C nucleus in the moderator of a nuclear reactor. (a) Find an expression for the energy lost by the neutron in the collision. (b) How much energy is lost by a 3.00-MeV neutron? (c) How many such collisions must a 3.00-MeV neutron make before its energy is reduced to 0.03 eV (thermal energies)?

27. To what temperature must helium gas be heated before the Coulomb barrier is overcome and fusion reactions begin?

28. Show that the D-T fusion reaction releases 17.6 MeV of energy.

29. In the D-T fusion reaction, the kinetic energies of ^{2}H and ^{3}H are small, compared with typical nuclear binding energies. (Why?) Find the kinetic energy of the emitted neutron.

30. Suppose we have 100.0 cm^3 of water, which is 0.015 percent D$_2$O. (a) Compute the energy that could be obtained if all the deuterium were consumed in the ^{2}H + ^{2}H $\rightarrow$ ^{3}H + p reaction. (b) As an alternative, compute the energy released if two-thirds of the deuterium were fused to form ^{3}H, which is then combined with the remaining one-third in the D-T reaction.

31. In the reactions of the carbon cycle of fusion, calculate (a) the energies of the three gamma rays; (b) the maximum energies of the positrons and neutrinos; (c) the kinetic energy of the final alpha particle. (Assume the reacting particles have kinetic energies in the keV range, which are negligible compared with the Q values.)

32. Find the energy released when three alpha particles combine to form ^{12}C.

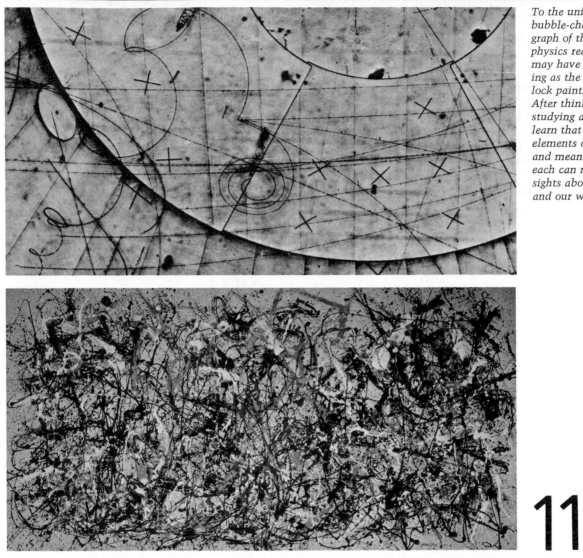

To the uninitiated, the bubble-chamber photograph of the particle physics reaction (top) may have as little meaning as the Jackson Pollock painting (bottom). After thinking and studying about them, we learn that each contains elements of art, beauty, and meaning, and that each can reveal to us insights about ourselves and our world.

11

ELEMENTARY PARTICLES

The search for the basic building blocks of nature has occupied the thoughts of scientific investigators since the Greeks introduced the idea of atomism 2500 years ago. As we look carefully at complex structures, we find underlying symmetries and regularities, which help us to understand the laws that determine how they are put together. The regularities of crystal structure, for example, suggest to us that the atoms of which the crystal is composed must follow certain rules for arranging themselves and joining together. As we look more deeply, we find that although nature has constructed all material objects out of roughly 100 different kinds of atoms, we can understand these atoms in terms of only three fundamental particles: the electron, proton, and neutron. Our attempts to look further within the electron meet with failure—the electron seems indeed to be a fundamental particle, with no internal structure. However, when we do reactions with nucleons at high energy, we seem to find more complexity rather than simplicity; hundreds of new particles can emerge as products of these reactions. If there are hundreds of basic building blocks, it seems unlikely that we could ever uncover any fundamental dynamic laws of their behavior. However, in recent years, experiments have suggested a new, underlying regularity that seems to be consistent with an analysis in terms of a small number of fundamental particles called *quarks.*

In this chapter, we examine the properties of many of the particles of physics, the laws that govern their behavior, and the classifications of these particles. We also show how the quark model helps us to understand the properties of the particles.

In the early days of the study of atomic physics, before the insights of the Bohr model helped us to understand atomic structure, all we had available were such guidelines as the Balmer formula, the Ritz combination principle, and so forth. These were rules based not on a fundamental understanding of atomic structure, but rather on the regularities which were revealed in experimental results. The study of particle physics was for many years in much the same state as atomic physics before Bohr. Just as the regularities and classifications of atomic properties led to the Bohr model and to the new mechanics of quantum physics, the regularities and classifications of particle properties have led to the quark model and to a new mechanical system called *quantum chromodynamics.* The mathematical level of this new mechanics is beyond the level of this book, but the fundamental details of the model and the properties of the quarks can be appreciated without the mathematics.

11.1 PARTICLES AND FIELDS

On the most fundamental level, the particles of physics are subject to four basic forces: strong, electromagnetic, weak, and gravitational. The strong force is the one responsible for nuclear binding, and is the strongest of the known interactions. Next in strength,* about $\frac{1}{100}$ of the strong force, comes the electromagnetic interaction. The weak force,

---

* We must take care how we define the relative strengths of the four forces. What is really meant is that two protons, separated by a distance of about 1 fm, exert forces on one another whose strengths are as given. For separations much larger than 1 fm, the strong and weak forces have no strength at all; for particles other than protons, the electromagnetic or strong force might have no strength.

which is responsible for such processes as nuclear beta decay, is only about 10^{-13} the strength of the strong interaction. The gravitational interaction, with a relative strength of only 10^{-38}, has a negligible effect at the level of particle physics.

We can classify particles in quite a variety of ways. One classification is based on the types of interactions which the particles undergo in their reactions and decays. Of the particles we will study, all interact through the weak interaction; a subset of those interacts through the electromagnetic interaction, and a still smaller subset interacts through the strong interaction. A *strongly interacting* particle is one that can interact with other particles by means of the strong interaction. (It will not do so in every instance, however. Two protons interact by means of the strong interaction if they are sufficiently close together, but a proton and an electron will *never* interact through the strong force. The electron is able to ignore the strong force field of the proton and respond only to the electromagnetic field of force.) Even though two protons in a nucleus interact by means of the strong, electromagnetic, and weak interactions, the strong force dominates and in many (but certainly not all) cases, the others can be ignored. This is similar to classical physics, in which we consider only the more important forces (friction and the Earth's gravity, in the case of a block sliding down an inclined plane) and neglect those whose influence is expected to be very small (air resistance or the sun's gravity in the same case).

This classification scheme might be as follows, for a few of the elementary particles:

Weak Interaction	Electromagnetic Interaction	Strong Interaction
e	e	π
μ	μ	K
ν	π	p
π	K	n
K	p	Λ
p	n	
n	Λ	
Λ		

Another classification scheme might be according to the masses of the particles. In the early days of particle physics, it was observed that the very light particles seemed to have a similar sort of behavior, the heaviest group of particles showed in common quite a different sort of behavior, and a middle group showed yet a different set of properties. Recent research in high-energy physics has shown that these classifications are not strictly valid. We have been able to create particles which, by their masses, would fall into the heaviest group, but whose properties are more similar to those of the light or middle group. Even though these classifications are obsolete, we retain the names of the groups, which now describe a set of particles with common properties. These names are *lepton* for the lightest group, *meson* for the middle group, and *baryon* for the heaviest group.

This classification might look like this (approximate rest energies are shown in parentheses):

Leptons	Mesons	Baryons
e (0.5 MeV)	π (140 MeV)	p (938 MeV)
μ (100 MeV)	K (500 MeV)	n (940 MeV)
ν ($\cong 0$ MeV)		Λ (1116 MeV)

Comparing this classification scheme with the previous one, we find that the leptons can interact through the weak or electromagnetic interactions, but *never* through the strong interaction. The mesons and baryons can interact through the strong, electromagnetic, or weak interactions.

Another classification scheme might be based on the *intrinsic spins* of the particles. Every particle has such an intrinsic spin; you will recall from Chapter 7 that the electron has an intrinsic spin of $\frac{1}{2}$. When we classify the particles in this way, we find that to some extent our previous classification scheme is duplicated. The leptons all have spin $\frac{1}{2}$. The mesons all have integral spins of 0 or 1. The baryons all have half-integral spins of $\frac{1}{2}$ (or sometimes $\frac{3}{2}$). The particle spins might be classified as follows:

Spin $\frac{1}{2}$	Spin 0
e	π
μ	K
ν	
p	
n	
Λ	

Another classification scheme might be based on the lifetimes of those particles that are unstable with respect to decay. In this scheme, we find that decays that take place according to the strong interaction have the smallest lifetimes, typically in the range of 10^{-20} to 10^{-23} s. Decays that are caused by the electromagnetic interaction are typically much slower, with lifetimes of 10^{-15} to 10^{-18} s, and decays that are caused by the weak interaction have lifetimes that range from about 10^{-10} s to as long as 15 min (for the neutron).

One final property that is used to classify a particle is the nature of its *antiparticle*. Every particle has an antiparticle, which is identical to the particle in mass and in many of its other properties, but which has an electric charge of the opposite sign.* (We will soon learn that there are other properties of the particle that differ in sign from the corresponding properties of its antiparticle.) We have already encountered the

Antiparticles

---

* We will use two systems to indicate antiparticles. Sometimes the symbol for the particle will be written along with the electric charge to indicate particle or antiparticle, as, for example, e^+ and e^-, or μ^+ and μ^-. Other times the antiparticle will be written with a bar over the symbol, for example, ν and $\bar{\nu}$ or p and $\bar{p}$.

antiparticle to the electron—the *positron.* The antiparticle of the proton, the antiproton has (like the proton) a mass of 938 MeV and a spin of $\frac{1}{2}$, but it has a charge of $-e$. An atom of antihydrogen would consist of a positron and an antiproton and would have all of the same properties as the ordinary hydrogen we have previously studied. A neutron has no electric charge, and so a neutron and an antineutron are not so easy to distinguish. One convenient way is by the magnetic dipole moment—a neutron's magnetic moment is opposite to its spin, but an antineutron's magnetic moment is parallel to its spin. A beam of neutrons in a magnetic field will align their spins opposite to the field direction, but a beam of antineutrons *in the same magnetic field* (*not* in a magnetic field produced by antielectrons flowing through anticopper wires!) will align in the opposite direction.

When a particle meets its antiparticle the *annihilation reaction* occurs—both particle and antiparticle vanish, and instead two (or more) photons or gamma rays are produced. Conservation of energy requires that, neglecting the kinetic energies of the particles, when two photons are emitted each must have an energy equal to the rest energy of the particle. Examples of annihilation reactions are

$$e^- + e^+ \longrightarrow \gamma_1 + \gamma_2 \qquad (E_{\gamma_1} = E_{\gamma_2} = 0.511 \text{ MeV})$$

$$p + \bar{p} \longrightarrow \gamma_1 + \gamma_2 \qquad (E_{\gamma_1} = E_{\gamma_2} = 938 \text{ MeV})$$

We call the kind of stuff of which we are made *matter* and the other kind of stuff *antimatter.* There may indeed be galaxies composed of antimatter, but we cannot tell by the ordinary techniques of astronomy, because *light and antilight are identical!* To put it another way, the photon and antiphoton are the same particle. We cannot tell by looking at the light (or other electromagnetic radiation) which reaches us from distant galaxies whether they are made from matter or antimatter. The only way to tell is by sending a chunk of our matter to the distant galaxy and seeing whether or not it is destroyed with the corresponding emission of a burst of photons. (It is indeed possible, but *highly unlikely*, that the first astronaut to travel to another galaxy may suffer such a fate! The first intergalactic handshake would indeed be quite an event!)

In our classification scheme it is usually easy to distinguish particles from antiparticles. We begin by defining *particles* to be the stuff of which ordinary matter is made—electrons, protons, and neutrons. Ordinary matter is not composed of neutrinos, so we have no basis for distinguishing a neutrino from an antineutrino, but the conservation laws in the beta decay process can be understood most easily if we define the *antineutrino* to be the particle that accompanies negative beta decay and the *neutrino* to be the particle that accompanies positron decay and electron capture. For a heavy baryon, such as the Λ, we take advantage of its radioactive decay, which leads eventually to ordinary protons and neutrons; that is, the Λ is the particle that decays to n, and the $\bar{\Lambda}$ ("antilambda") therefore decays to $\bar{n}$. Similarly, in the case of the leptons, the μ^- and the μ^+ are antiparticles of of one another; since μ^- decays to ordinary e^- (and has many properties in common with the electron) it is the *particle*, while μ^+ is the antiparticle. The case of the particles known as π mesons is not so easy to resolve. The π's come with three different electric charges: π^+, π^0, π^-. The π^0 seems to be, like the photon, its own antiparticle and the π^+ and π^- seem to be antiparticles of one another, but which (π^+ or π^-) is the particle and which the antiparticle? Ordinary matter is not composed of π mesons, so we get no clue there. The decays of the π^+ and π^- always give one lepton and one antilepton so again we get no clue there either. There seems in fact to be no way (or *no need*) to distinguish particles from antiparticles for the π mesons, and so we regard the set of π's as three *particles* π^+, π^0, π^- whose antiparticles are π^-, π^0, π^+.

In a similar way the entire list of particles and antiparticles can be constructed.

As you can see, even the very few particles that we have listed in this section have many ways of being classified. In Section 11.3 we discuss how the properties of these particles are used in assigning them to the different groups or families.

11.2 CONSERVATION LAWS

We have frequently used the conservation of energy, of linear momentum, and of angular momentum in our analysis of physical phenomena. These conservation laws are, as we discussed in Chapter 2, closely connected with the fundamental properties of space and time; we believe those laws to be absolute and inviolable.

We also use other conservation laws in analyzing various types of decay and reaction processes. For example, when we combine two elements in a chemical reaction, such as hydrogen + oxygen $\rightarrow$ water, we must *balance* the reaction in the following way:

$$2H_2 + O_2 \longrightarrow 2H_2O$$

The process of balancing such an equation, by which the amounts of matter on each side are made the same, can also be regarded as a way of *conserving electrons*—the number of electrons on each side of the equation must be the same. This is an example of another type of conservation law, in which the number of one type of particle, electrons in this case, must be the same on both sides. We can also interpret this as a statement of the conservation of electric charge, which requires that the net electric charge must be the same on both sides of the equation. In nuclear processes, we are concerned not with electrons, but with protons and neutrons. In the alpha decay of a nucleus, such as

$$^{235}_{92}U_{143} \longrightarrow {}^{231}_{90}Th_{141} + {}^4_2He_2$$

or in a reaction such as

$$p + {}^{63}_{29}Cu_{34} \longrightarrow {}^{63}_{30}Zn_{33} + n$$

we balance the number of protons and also the number of neutrons. We might therefore be led to guess at a conservation law for nuclear reactions and decays, in which both the number of protons and the number of neutrons must be conserved. (Conservation of proton number can also be interpreted as an alternative form of conservation of electric charge.)

These two conservation laws for nuclear processes work quite well until we try to apply them to beta decay. For instance, the decay

$$n \longrightarrow p + e^- + \bar{\nu}$$

does not conserve either proton number or neutron number. However, it does conserve the total neutron number *plus* proton number, which is equal to one on both sides of the equation. (Neutron number plus proton number is automatically conserved if we conserve *both* neutron number *and* proton number, so this new conservation law includes the previous law as a special case.)

We can thus say that all of our nuclear decays and reactions are consistent with conserving both *electric charge* and *nucleon number*.

The existence of the electron following beta decay is a problem from the standpoint of conservation laws. Our analysis of chemical reactions was based on the conservation of electron number, and here we seem to have a violation of that conservation law. We can remedy this violation by considering the electron and neutrino to be members of the same family, just like the neutron and proton were considered as two members of a common family, the nucleons. This family is the *leptons*, and we keep track of the number of leptons by assigning each particle a *lepton number L*. The electron and neutrino have lepton numbers of $+1$,   **Lepton number** and the positron and antineutrino have lepton numbers of -1. The proton and neutron have lepton numbers of zero. Thus our beta decay process has total lepton number zero on the left (because there are no leptons) and also zero on the right (because there is one lepton, with $L = +1$, and one antilepton with $L = -1$). We see now why we have an *antineutrino* in the case of *electron* beta decay and a *neutrino* in the case of *positron* beta decay—the lepton number remains zero in both decays.

According to the lepton conservation law, these processes would be *impossible:*

$$e^- + p \longrightarrow n + \bar{\nu}$$
$$L = 1 + 0 \longrightarrow 0 + (-1)$$

$$p \longrightarrow e^+ + \gamma$$
$$L = 0 \longrightarrow -1 + 0$$

Here is another example of the way conservation laws work. When a proton accelerated to sufficiently high energy collides with another proton, the following reactions are possible (among many others):

$$p + p \longrightarrow p + n + \pi^+$$
$$p + p \longrightarrow p + p + \pi^0$$
$$p + p \longrightarrow p + n + \pi^+ + \pi^0$$
$$p + p \longrightarrow p + p + \pi^+ + \pi^-$$

(The π meson, which we introduced previously, can be positively charged, negatively charged, or uncharged.) There is no conservation law for mesons, so any number of mesons may appear in these reactions. These reactions conserve electric charge ($+2$ units of charge on both sides) also nucleon number (also 2 on both sides—the π's are mesons, not nucleons). The following reactions are *not* possible:

$$p + p \longrightarrow p + \pi^+$$
$$p + p \longrightarrow p + p + n$$

These reactions conserve electric charge, but do not conserve nucleon number ($2 \rightarrow 1$ and $2 \rightarrow 3$, respectively). However, consider the following reaction, which was responsible for the discovery in 1956 of the antiproton:

$$p + p \longrightarrow p + p + p + \bar{p}$$

On the left side, we have two nucleons; on the right side, there are three nucleons and one antinucleon. If we count $A = 1$ for the nucleons and $A = -1$ for the antinucleon we can understand why this reaction conserves nucleon number and the reaction $p + p \rightarrow p + p + \bar{n}$ does not and is therefore forbidden to occur. Similarly the decay $p \rightarrow \pi^+ + \pi^0$ is forbidden by conservation of nucleon number; otherwise all the protons in the universe would have decayed away.

In our discussion of elementary particles we encounter other conservation laws besides nucleon number and lepton number. When we apply these conservation laws, as we did previously, we assign a certain number to the particle and a number of the opposite sign to the antiparticle. If the total of these numbers doesn't balance on the two sides of the decay or reaction, the process is either forbidden or else very improbable.

11.3 FAMILIES OF PARTICLES

In Table 11.1 are shown some of the members of the three families of elementary particles that we have discussed—the leptons, mesons, and baryons. (The photon is a unique particle in a class by itself.)

Table 11.1 Families of Elementary Particles

Family	Particle	Rest Energy (MeV)	Electric Charge	Spin	Strange-ness	Anti-particle	Lifetime (s)	Principal Decay
Leptons	e^-	0.511	-1	$\frac{1}{2}$	0	e^+	∞	Stable
	μ^-	105.7	-1	$\frac{1}{2}$	0	μ^+	2.2×10^{-6}	$e^- + \bar{\nu}_e + \nu_\mu$
	ν_e	0	0	$\frac{1}{2}$	0	$\bar{\nu}_e$	∞	Stable
	ν_μ	0	0	$\frac{1}{2}$	0	$\bar{\nu}_\mu$	∞	Stable
Mesons	$\pi^\pm$	139.6	± 1	0	0	$\pi^\mp$	2.6×10^{-8}	$\mu + \nu_\mu$
	π^0	135.0	0	0	0	π^0	8.3×10^{-17}	$\gamma + \gamma$
	$K^\pm$	493.7	± 1	0	± 1	$K^\mp$	1.2×10^{-8}	$\mu + \nu_\mu$
	K^0	497.7	0	0	$+1$	$\bar{K}^0$	0.9×10^{-10}	$\pi^+ + \pi^-$
	η	548.8	0	0	0	η	7.7×10^{-19}	$\gamma + \gamma$
	η'	957.6	0	0	0	η'	2.4×10^{-21}	$\eta + \pi + \pi$
Baryons	p	938.3	$+1$	$\frac{1}{2}$	0	$\bar{p}$	∞	Stable
	n	939.6	0	$\frac{1}{2}$	0	$\bar{n}$	920	$p + e^- + \bar{\nu}_e$
	Λ^0	1115.6	0	$\frac{1}{2}$	-1	$\bar{\Lambda}^0$	2.6×10^{-10}	$p + \pi^-$
	Σ^+	1189.4	$+1$	$\frac{1}{2}$	-1	$\bar{\Sigma}^+$	0.8×10^{-10}	$p + \pi^0$
	Σ^0	1192.5	0	$\frac{1}{2}$	-1	$\bar{\Sigma}^0$	5.8×10^{-20}	$\Lambda^0 + \gamma$
	Σ^-	1197.3	-1	$\frac{1}{2}$	-1	$\bar{\Sigma}^-$	1.5×10^{-10}	$n + \pi^-$
	Ξ^0	1314.9	0	$\frac{1}{2}$	-2	$\bar{\Xi}^0$	2.9×10^{-10}	$\Lambda^0 + \pi^0$
	Ξ^-	1321.3	-1	$\frac{1}{2}$	-2	$\bar{\Xi}^-$	1.6×10^{-10}	$\Lambda^0 + \pi^-$
	Δ^*	1232	$+2, +1, 0, -1$	$\frac{3}{2}$	0	$\bar{\Delta}^*$	6×10^{-24}	$p + \pi, n + \pi$
	Σ^*	1385	$+1, 0, -1$	$\frac{3}{2}$	-1	$\bar{\Sigma}^*$	2×10^{-23}	$\Lambda^0 + \pi$
	Ξ^*	1530	$-1, 0$	$\frac{3}{2}$	-2	$\bar{\Xi}^*$	6×10^{-23}	$\Xi + \pi$
	Ω^-	1672.2	-1	$\frac{3}{2}$	-3	$\bar{\Omega}^-$	8.2×10^{-11}	$\Lambda^0 + K^-$

Leptons

The lepton family is composed of non-strongly interacting particles with spin $\frac{1}{2}$. In spite of numerous experimental efforts, no evidence for internal structure of the leptons has been discovered. The leptons appear to be fundamental particles—they cannot be further split into smaller particles. Table 11.1 shows four members of the lepton family—the electron, the muon, and their associated neutrinos (plus their four anti-particles). The electron neutrino we have already discussed in connection with nuclear beta decay. The muon was first encountered in studies of cosmic rays; we discussed the decay of cosmic-ray muons in Chapter 2 as providing evidence for the time dilation effect in special relativity. The muon has an associated neutrino that has the same basic properties as the electron neutrino—zero rest mass, spin $\frac{1}{2}$, and a similar antiparticle. Since the electron neutrino and the muon neutrino are both bits of nothing, how do we know they are different? The electron neutrino can be captured by a neutron to give the reaction

$$\nu_e + n \longrightarrow p + e^-$$

while the muon neutrino gives the similar reaction

$$\nu_\mu + n \longrightarrow p + \mu^-$$

Apparently, there are two different kinds of lepton numbers, one for electrons (and electron neutrinos) and a different lepton number for muons and muon neutrinos. We call these numbers L_e and L_μ. The following examples illustrate the conservation of these lepton numbers:

$$\bar{\nu}_e + p \longrightarrow e^+ + n$$
$$L_e = -1 + 0 \longrightarrow -1 + 0$$

$$\nu_\mu + n \longrightarrow \mu^- + p$$
$$L_\mu = 1 + 0 \longrightarrow 1 + 0$$

$$\mu^- \longrightarrow e^- + \bar{\nu}_e + \nu_\mu$$
$$L_e = 0 \longrightarrow 1 + (-1) + 0$$
$$L_\mu = 1 \longrightarrow 0 + 0 + 1$$

$$\pi^- \longrightarrow \mu^- + \bar{\nu}_\mu$$
$$L_\mu = 0 \longrightarrow 1 + (-1)$$

Studying these examples, we can understand why sometimes neutrinos appear and sometimes antineutrinos.

Mesons

Most of the frequently encountered members of the *meson* family have masses between those of the leptons and the baryons (although mesons heavier than the lighter baryons have recently been discovered). The mesons are all unstable; they decay into lighter mesons or into leptons. The mesons have spins of 0 or 1. Since there is no conservation law for mesons, there is no "meson number" (as there is a lepton number). Also, the mesons do not have distinct antiparticles in the same way that leptons or baryons do. It is clear, for instance, that electrons are particles and positrons are antiparticles; the π^- and π^+ are antiparticles of each other, but there is no way to decide which is the particle and which is the antiparticle.

We have already encountered the π mesons in our discussion of the nuclear force. The next heaviest meson, the K meson, has some unusual properties that distinguish it from other mesons. For example, the uncharged η and π^0 mesons decay very rapidly ($10^{-16} - 10^{-18}$ s) into two photons; on the basis of the systematic behavior of mesons, we would expect the K^0 to decay similarly to two photons in a comparable time. The observed decay of the K^0 takes place much more slowly (10^{-10} s); moreover, the decay products are not photons, but π mesons and leptons. On the basic premise that anything that is not observed must be forbidden by some rule of nature, we suspect that there is a reason for the observed decay modes of the K^0. (Similarly, the decay $\mu^- \rightarrow e^- + \gamma$ is not forbidden by any previously known law of physics, but the decay is not observed. The failure to observe this decay leads us to the new law of lepton conservation, and we explain the failure of the decay to occur by its violation of both electron and muon lepton number conservation.) We therefore guess that there is some "number" associated with the K^0 meson that forbids its decay into two photons. As another example, the heavy charged mesons are all strongly interacting particles, and we expect such particles to decay into lighter strongly interacting particles (π mesons, for example) in times of the order of 10^{-23} s. But the decay $K^+ \rightarrow \pi^+ + \pi^0$ occurs very slowly, in a time of the order of 10^{-8} s, and in fact the different decay mode $K^+ \rightarrow \mu^+ + \nu_\mu$ is more probable. This unusual behavior is also explained by the introduction of a new quantum number. This number is called the *strangeness S*, and we can use it to explain the properties of the K-meson decays. The K^0 and K^+ are assigned strangeness of $S = +1$; the π mesons and leptons are nonstrange particles ($S = 0$). The decay $K^0 \rightarrow \gamma + \gamma$, which is an electromagnetic decay (as indicated by the photons), is forbidden because the electromagnetic interaction conserves strangeness ($S = +1$ on the left, $S = 0$ on the right). The decay $K^+ \rightarrow \pi^+ + \pi^0$ does not occur in the typical strong interaction time of 10^{-23} s because the strong interaction cannot change strangeness. It occurs in 10^{-8} s (and the corresponding decay $K^+ \rightarrow \mu^+ + \nu_\mu$ occurs) because the weak interaction *does not* conserve strangeness; decays that are caused by the weak interaction can change the strangeness by one unit. (Recall our discussion in the first section of this chapter that a decay time of the order of 10^{-10} s is typical of the weak interaction.)

Strangeness

Strangeness conservation also helps to explain another curious aspect of the behavior of K mesons. Pi mesons are produced in nuclear collisions; for example $p + p \rightarrow p + p + \pi$ or $p + p \rightarrow p + p + \pi + \pi$. Since there is no meson conservation law, any number of π mesons can be produced. But K mesons and all other "strange" particles are always produced in pairs: $p + p \rightarrow p + p + K + K$ or $p + p \rightarrow p + \Lambda^0 + K^0$. If one of the K mesons in the pair is a K^+ or K^0 with $S = +1$ and the other is a K^- or $\bar{K}^0$ with $S = -1$, conservation of strangeness explains this phenomenon of *associated production*. Similarly if the K^0 has strangeness $+1$ and the Λ^0 has strangeness -1, strangeness is conserved in the second reaction.

The *baryons* are spin $\frac{1}{2}$ or $\frac{3}{2}$ particles, of which the nucleons are the lightest and most familiar. They all have distinct antiparticles like the leptons. Also like the leptons, there is a conservation law for baryons. Each baryon is assigned a *baryon number* $B = +1$; the antibaryons are

Baryons and baryon number

assigned $B = -1$. *All reactions and decays conserve baryon number.* (The conservation of *nucleon number A*, which we discussed previously, is just a special case of baryon number conservation; from now on we will use B instead of A for all baryons, including the nucleons.)

The baryons also come in strange and nonstrange varieties. Looking at the lifetimes, we see that the Λ^0 decays into $p + \pi^-$ with a lifetime of about 10^{-10} s, while we would expect a strongly interacting particle to decay to other strongly interacting particles in 10^{-23} s. With the strangeness of the Λ^0 assigned as -1, these decays change S and are forbidden to go by the strong interaction, and so must be due to the weak interaction, with the expected 10^{-10}-s lifetime. The strangeness violation also tells us why the electromagnetic decay $\Lambda^0 \rightarrow n + \gamma$ does not occur (while the decay $\Sigma^0 \rightarrow \Lambda^0 + \gamma$ does occur, with a typical electromagnetic lifetime of 10^{-19} s). Also, the weak decay can change the strangeness by only *one* unit, and so $\Xi^0 \rightarrow n + \pi^0$ ($S = -2 \rightarrow S = 0$) is forbidden.

The lepton number, baryon number, and strangeness are useful concepts for describing the occurrence and nonoccurrence of various decays and reactions. We don't understand why lepton number or baryon number are conserved, or what strangeness really represents, but the hope is that one day a complete theory of the structure of particles and their interactions will provide that understanding.

11.4 PARTICLE INTERACTIONS AND DECAYS

In this section we briefly summarize the properties of the elementary particles and how they are measured.

Atoms and molecules can be taken apart relatively easily and nonviolently, enabling us to study their structure. However, the elementary particles, most of which are unstable and do not exist in nature, must be created in violent collisions. (The particle theorist Richard Feynman once compared this process with studying fine Swiss watches by smashing them together and looking at the pieces that emerge from the collision.) For this purpose we will need a high-energy beam of particles and a suitable target of elementary particles. The only strongly interacting, stable elementary particle is the proton, and thus a hydrogen target is a logical choice. To get a reasonable density of target atoms, it is necessary to use liquid, rather than gaseous, hydrogen.

For a suitable beam, we must be able to accelerate a particle to very high energies (so high that the energy of the particle may be hundreds of times its rest energy m_0c^2). A stable charged particle is the logical choice for the beam; stability is required because of the relatively long time necessary to accelerate the particle to such an energy, and a charged particle is required so that electromagnetic fields may be used to accelerate the particle. Once again the proton is the logical choice, and so many of our particle physics reactions will be initiated by accelerating protons on to a proton target, which gives

$$p + p \longrightarrow \text{product particles}$$

Among the product particles may be a variety of mesons or even heavier particles of the baryon family, of which the nucleons are the lightest members. The study of the nature and properties of these particles is the goal of particle physics.

In many cases, conservation laws restrict the nature of the product particles, and it would be desirable to have other types of beams available. One possibility is indicated in Figure 11.1. A proton beam is incident on a target—the nature of the target is not important. Like Feynman's Swiss watch parts, many different particles emerge. By suitable focusing and selection of the momentum, we can extract a beam of the *secondary* particles created in the reactions. Once again, the secondary beam particles should be charged, so that we can select them from the many particles produced in the first collision. Also, the particle must live long enough to be delivered to a second target. This second target might be tens of meters away, and so even if the particle were traveling at the speed of light, it would need about 10^{-7} s to make its journey. Although this is a very short time interval by ordinary standards, on the time scale of elementary particles, it is a very long time— in fact, none of the mesons or baryons (except the neutron) lives that long. Although our efforts to make a secondary beam would seem to be in vain, we have forgotten one very important detail. The lifetime of the particle is measured in its rest frame, while we are observing its flight in the laboratory frame, in which the particle is moving at speeds extremely close to the speed of light. The *time dilation* factor results in a lifetime, observed in our frame of reference, which might be hundreds of times longer than the *proper lifetime.* This factor extends the range of available secondary beams to those particles with lifetimes as short as 10^{-10} s, and makes it possible to obtain secondary beams to study such reactions as

$$\pi + p \rightarrow \text{particles}$$

and

$$K + p \rightarrow \text{particles}$$

where the lifetime of the π is about 10^{-8} s and the lifetime of the K meson is about the same value.

Once we have selected a beam and target, we must select a detection system to observe the products of the reaction. We would like to measure the masses, linear momenta, and kinetic energies of the product particles, and the most convenient way of doing this is to find a way to photograph the tracks left by the particles as they emerge after the reaction. One particularly clever way of doing this is by means of the *bubble chamber,* which is simply a large container filled with liquid hydrogen. As the charged particles travel through the liquid hydrogen, they ionize the hydrogen atoms along the paths, and the ionization energy causes bubbles in the supercooled liquid. When light shines on these tiny bubbles, a visible track is produced, which can be photographed. Figure 11.2 is an example of such a photograph. The curved paths of the particles result from the application of a magnetic field; from the radius of curvature of a charged particle in a magnetic field we can deduce the linear momentum of the particle.

From a careful analysis of the paths of particles, such as revealed in bubble chamber photographs, we can deduce the desired quantities of mass, linear momentum, and energy. The other important property we would like to know is the lifetime of the decay of the product particles, since many of the products are often unstable. If we know the speed of a

Richard P. Feynman (1918–   , United States). Seldom is one person known for both exceptional insights into theoretical physics and exceptional methods of teaching first-year physics. He received the Nobel prize for his work on the theory that couples quantum mechanics to electromagnetism, and his text and film *Lectures on Physics* give unusual perspectives to many areas of basic physics and are enjoyed by undergraduate students, graduate students, and instructors.

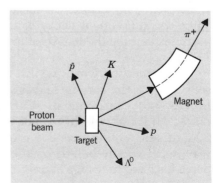

FIGURE 11.1 The production of secondary particle beams. The magnet helps to select the mass and momentum of the desired particle.

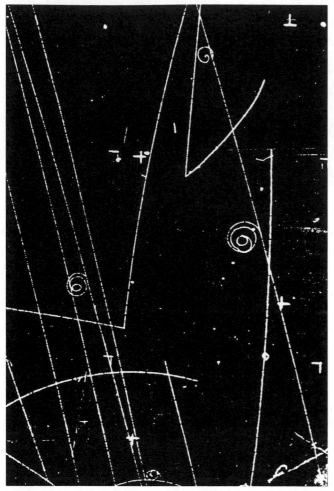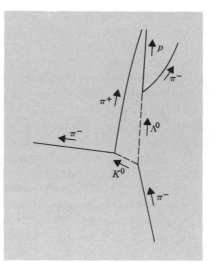

FIGURE 11.2 A bubble chamber photograph of a reaction between particles. At right is shown a diagram indicating the particles that participate in the reaction (courtesy Lawrence Radiation Laboratory).

particle, we can find its lifetime by simply observing the length of its track in a bubble chamber photograph. (Even for uncharged particles, which leave no tracks, we can use this method to deduce the lifetime, since the subsequent decay of the uncharged particle into two charged particles defines the length of its path rather clearly, as shown in Figure 11.2.)

This method works well if the lifetime is of the order of 10^{-10} s or so, such that the particle leaves a track long enough to be measured (millimeters to centimeters). With careful experimental technique and clever data analysis, this can be extended to track lengths of the order of 10^{-6} m, and so lifetimes down to about 10^{-16} s can be measured in this way (with a little help from the time dilation factor). But many of our particles have lifetimes of only 10^{-23} s, and a particle moving at even the speed of light travels only the diameter of a nucleus in that time! How can we measure such a lifetime? Furthermore, how do we even know such a particle exists at all? Consider the reaction

$$\pi + p \longrightarrow \pi + p + x$$

where x is an unknown particle with a lifetime of about 10^{-23} s, which decays into two π mesons according to $x \rightarrow \pi + \pi$. How do we distinguish the above reaction from the reaction

$$\pi + p \longrightarrow \pi + p + \pi + \pi$$

which leads to the same particles as actually observed in the laboratory?

Experimental evidence suggests that the two π mesons combine for an instant (10^{-23} s) to form an entity with all of the usual properties of a particle—a definite mass, charge, spin, lifetime, etc. Such states are known as *resonance particles*, and we now look at the indirect evidence from which we infer their existence.

Suppose you receive a package in the mail from a friend. When you open it, you find it contains many small, irregular pieces of broken glass. How do you learn whether your friend sent you a beautiful glass vase that was broken in shipment or a package of broken glass as a practical joke? You try to put the pieces together! If the pieces fit together, it is a good assumption that the vase was once whole, although the mere fact that they fit together doesn't *prove* that it was once whole. It's just the simplest possible assumption *consistent with our experience.* (An alternative assumption that the pieces were manufactured separately and just happen by chance to fit together, is highly improbable.)

How then do we detect a "particle" which lives for only 10^{-23} s? We look at its decay products (which live long enough to be seen in the laboratory), and putting the pieces back together, we infer that they once may have been a whole particle.

For example, suppose in the laboratory we observe two π mesons emitted as shown in Figure 11.3. We measure the direction of travel and the linear momentum of the π mesons as shown. A second and a third event each produces two π mesons as also shown in the figure. Are these three events consistent with the existence of the same resonance particle?

Let us assume that in each case, a particle moving at an unknown speed decayed into the two particles as shown. Since each decay must conserve energy and momentum, we can use the decay information to work backward and find the energy and momentum of the decaying particle, and therefore we can find its rest energy according to $mc^2 = \sqrt{(E_1 + E_2)^2 - c^2(\mathbf{p}_1 + \mathbf{p}_2)^2}$. Carrying out the calculation, we find that, for the decay shown in part (a) of Figure 11.3, $mc^2 = 764$ MeV, while for part (b), $mc^2 = 775$ MeV. It is therefore possible that these two events result from the decays of identical particles. Part (c) of the figure gives $mc^2 = 498$ MeV, which differs considerably from parts (a) and (b).

Of course, these two events are not sufficient to identify conclusively the existence of the resonance particle. It could be a mere accident, just like the chance fitting together of two pieces of broken glass. What is needed is a large (statistically significant) number of events, in which we can combine the momenta of the two emitted π mesons in such a way that the deduced mass of the resonance particle is always the same. Figure 11.4 is an example of such a result. There is a background of events with a continuous distribution of energies, like beta decay electrons; these come from events like part (c) of Figure 11.3. There is also present a very prominent peak at 770 MeV. We identify this energy as the rest energy of the resonance particle, which is known

Resonance particles

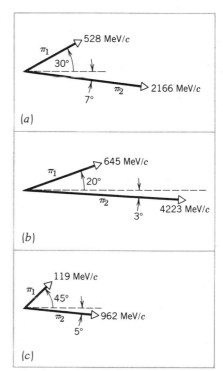

FIGURE 11.3 Three possible decays of an unknown particle into two pi mesons. The direction and momentum of each pi meson are indicated.

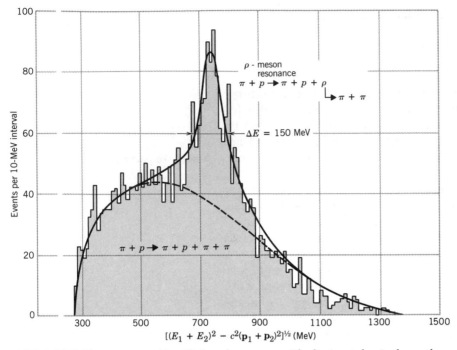

FIGURE 11.4 The resonance identified as the ρ meson. The horizontal axis shows the energy and momenta of the two decay pi mesons, combined to be equivalent to the mass of the resonance particle.

as the ρ (rho) meson. (How do we know it is a meson? It must be a strongly interacting particle, since it decays so rapidly. The only possibilities are therefore mesons, with integral spin, or baryons, with half-integral spin. Since π mesons have integral spin, and since two integral spins can combine to give only another integral spin, it must be a meson.)

We can also infer the lifetime of the particle from Figure 11.4. The particle lives only for about 10^{-23} s, and so if we are to measure its rest energy we have only 10^{-23} s in which to do it. But the uncertainty principle requires that an energy measurement made in a time interval Δt be uncertain by an amount roughly $\Delta E \cong \hbar/\Delta t$. This energy uncertainty ΔE is observed as the *width* of the peak in Figure 11.4. We don't always deduce the same value 770 MeV for the rest energy of the ρ meson; sometimes our value is a bit larger and sometimes a bit smaller. *The width of the resonance peak tells us the lifetime of the particle.* (The width is not really precisely defined, but physicists usually take as the width the interval between the two points where the height of the resonance is one-half its maximum value, as shown in Figure 11.4.) The width of 150 MeV leads to a value of 4.4×10^{-24} s for the lifetime of the ρ meson.

11.5 ENERGETICS OF PARTICLE DECAYS

In analyzing the decays and reactions of elementary particles, we apply many of the same laws that we used for nuclear decays and reactions: energy, linear momentum, and total angular momentum must be conserved, and the total value of the quantum numbers associated with electric charge, lepton number, and baryon number (which we previously called nucleon number) must be the same before and after the decay or reaction. In reactions of elementary particles, we are most often concerned with the production of new varieties of particles. The energy necessary to manufacture these particles comes from the kinetic energy of the reaction constituents (usually the incident particle), and since this energy is usually quite large (hence the name *high-energy physics* for this type of research), *relativistic equations* must be used for energy and momentum.

The decays of elementary particles can be analyzed in a way similar to the decays of nuclei, following the same two basic rules:

1. The energy available for the decay (assuming the decaying particle is at rest) is just the difference in mass energy between the initial decaying particle and the particles which are produced in the decay. By analogy with our study of nuclear decays, we will call this the Q value:

Q value of particle decay

$$Q = (m_i - m_f)c^2 \tag{11.1}$$

where $m_i c^2$ is the mass energy of the initial particle and $m_f c^2$ is the total mass energy of all the product particles. (Of course, the decay will occur only if Q is positive.)

2. The available energy Q is shared as kinetic energy of the decay products in such a way as to conserve linear momentum. As in the case of nuclear decays, for a decay into two final particles, the particles have equal and opposite momenta, and we can find unique values for the energies of the two final particles. For decays into three or more particles, each particle has a spectrum or distribution of energies from zero up to some maximum value (as was the case with nuclear beta decay).

Compute the energies of the proton and π meson that result from the decay of the Λ^0.

EXAMPLE 11.1

The decay process is:

Solution

$$\Lambda^0 \longrightarrow p + \pi^-$$

Using the rest energies from Table 11.1, we have:

$$Q = (m_{\Lambda^0} - m_p - m_{\pi^-})c^2$$
$$= (1115.6 - 938.3 - 139.6) \text{ MeV}$$
$$= 37.7 \text{ MeV}$$

and so the total kinetic energy of the decay products must be:

$$K_p + K_\pi = 37.7 \text{ MeV}$$

Using the relativistic formula for kinetic energy this can be written as:

$$(\sqrt{c^2 p_p^2 + m_p^2 c^4} - m_p c^2) + (\sqrt{c^2 p_\pi^2 + m_\pi^2 c^4} - m_\pi c^2) = 37.7 \text{ MeV}$$

Conservation of momentum requires $p_p = p_\pi$, and solving for the momentum we find

$$p_\pi = p_p = 100.4 \text{ MeV}/c$$

and the kinetic energies can be found to be

$$K_\pi = 32.3 \text{ MeV}$$

$$K_p = 5.4 \text{ MeV}$$

What is the maximum kinetic energy of the electron emitted in the decay $\mu^- \rightarrow e^- + \bar{\nu}_e + \nu_\mu$?

EXAMPLE 11.2

The Q value for this decay is $Q = m_\mu c^2 - m_e c^2 = 105.2$ MeV, since the neutrinos have no rest mass. If the μ^- is at rest, this energy is shared by the electron and the neutrinos: $Q = K_e + E_{\bar{\nu}_e} + E_{\nu_\mu}$. When the electron has its maximum kinetic energy, the two neutrinos carry away the minimum energy. This minimum cannot be zero, because that would violate momentum conservation: the electron would be carrying momentum that would not be balanced by the neutrino momenta to give a net of zero (since we assumed the μ^- to be at rest $p_{\text{initial}} = p_{\text{final}} = 0$). We assume that the electron has its maximum energy when the neutrinos are emitted in exactly the opposite direction to the electron; otherwise some of the decay energy is "wasted" by providing transverse momentum components for the neutrinos, and not as much energy will be available for the electron. Since it does not matter which of the neutrinos carry the energy and momentum (they may even share it in any proportion), we let E_ν and p_ν be the total neutrino energy and momentum; these are of course related by $E_\nu = cp_\nu$, since neutrinos are massless and travel at the speed of light. If we let E_e and p_e represent the energy and momentum of the electron, then linear momentum conservation gives

Solution

$$p_e - p_\nu = 0$$

For the electron, $E_e = \sqrt{c^2 p_e^2 + m_e^2 c^4}$. Together, these equations give:

$$Q = E_e - m_e c^2 + cp_\nu$$

$$= E_e - m_e c^2 + cp_e$$

$$= E_e - m_e c^2 + \sqrt{E_e^2 - m_e^2 c^4}$$

Solving, we find:

$$E_e = \frac{Q^2}{2(Q + m_e c^2)} + m_e c^2$$

$$K_e = E_e - m_e c^2 = Q^2/2m_\mu c^2 = 52.3 \text{ MeV}$$

The mass energy of the μ^- is shared essentially equally by the electron and the two neutrinos in this case ($K_e \cong Q/2$). Note how different this is from the case of the beta decay of the neutron, where the heavy proton resulting from the decay could absorb considerable recoil momentum at a cost of very little energy, so nearly all of the available decay energy could be given to the electron ($K_e \cong Q$).

Find the maximum energy of the positrons and of the π mesons produced in the decay $K^+ \rightarrow \pi^0 + e^+ + \nu_e$

EXAMPLE 11.3

The Q value for this decay is

Solution

$$Q = (m_K - m_\pi - m_e)c^2$$

$$= 493.7 \text{ MeV} - 135.0 \text{ MeV} - 0.5 \text{ MeV}$$

$$= 358.2 \text{ MeV}$$

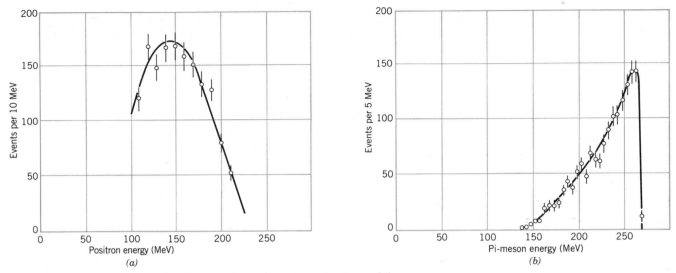

FIGURE 11.5 The spectrum of positrons and pi mesons from the decay of the K^+ meson.

This energy must be shared among the three products:

$$Q = K_\pi + K_e + E_\nu$$

The electron and π meson will have their maximum energies when the neutrino has negligible energy:

$$Q = K_\pi + K_e$$

and conservation of momentum in this case (if the neutrino has negligible momentum) requires $p_\pi = p_e$. Using the relativistic equation for kinetic energy

$$Q = \sqrt{(pc)^2 + (m_\pi c^2)^2} - m_\pi c^2 + \sqrt{(pc)^2 + (m_e c^2)^2} - m_e c^2$$

where $p = p_e = p_\pi$. Inserting the numbers, we have:

$$493.7 \text{ MeV} = \sqrt{(pc)^2 + (135.0 \text{ MeV})^2} + \sqrt{(pc)^2 + (0.5 \text{ MeV})^2}$$

Clearing the two radicals will involve quite a bit of algebra, but we can simplify the problem if we inspect this expression and notice that the solution must have a large value of pc, certainly greater than 100 MeV. (Otherwise the two terms could not sum to nearly 500.) Thus $(pc)^2 \gg (0.5 \text{ MeV})^2$, and we can neglect the electron mass energy term in the second radical:

$$493.7 = \sqrt{(pc)^2 + (135.0)^2} + pc$$

Solving, we find $pc = 228.4$ MeV, which gives $(E_e)_{max} = 228.4$ MeV and $(E_\pi)_{max} = 265.3$ MeV. Figure 11.5 shows the observed energy spectra of e^+ and π^0 from the K^+ decay, and the energy maxima are in agreement with the calculated values. (The shapes of the energy distributions are determined by statistical factors, like nuclear beta decay. The statistical factors are different for e^+ and π^0, since the π^0 also has its maximum energy when the e^+ appears at rest and the ν carries the recoil momentum.)

You should repeat the above calculation and convince yourself that (1) the π^0 has its maximum energy also when $K_e = 0$ ($E_e = m_e c^2$) and (2) the e^+ does *not* have its maximum energy when $K_\pi = 0$.

The basic experimental technique of particle physics reactions consists of studying the product particles that result from a collision between an incident particle (accelerated to high energies) and a target particle (usually at rest). The kinematics of the reaction process must be analyzed using relativistic formulas, since the kinetic energies of the particles are usually comparable to or greater than their rest energies. In this section we derive some of the relationships that are needed to analyze these reactions, using the formulas for relativistic kinematics we derived in Chapter 2. Since an important purpose of these reactions is the production of new varieties of particles, we concentrate on calculating the threshold energy needed to produce these particles. (You might find it helpful to review the discussion in Chapter 10 on *nonrelativistic* reaction thresholds.)

Consider the following reaction:

$$m_1 + m_2 \longrightarrow m_3 + m_4 + m_5 + \cdots$$

Any number of particles can be produced in the final state. We will let m_1 represent the incident particle (all m's are *rest masses*), which has total energy E_1, kinetic energy $K_1 = E_1 - m_1c^2$, and momentum $p_1 = \sqrt{E_1^2 - m_1^2c^4}$ in the *laboratory* frame of reference. The target particle m_2 is at rest in the laboratory. Figure 11.6 illustrates this reaction in the laboratory frame of reference.

Just as we did for nuclear reactions, we define the Q value to be

$$Q = [m_1 + m_2 - (m_3 + m_4 + m_5 + \ldots)]c^2 \qquad (11.2)$$

If Q is positive, energy is "liberated" (actually, rest energy is turned into kinetic energy, so that the product particles $m_3, m_4, m_5, \ldots$ have more combined kinetic energy than the initial particles m_1 and m_2). If Q is negative, some of the initial kinetic energy of m_1 is turned into mass energy.

11.6 ENERGETICS OF PARTICLE REACTIONS

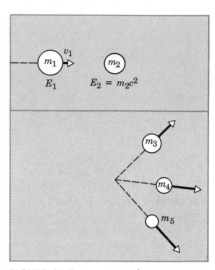

FIGURE 11.6 A reaction between particles in the laboratory reference frame.

Q value of particle reaction

Compute the Q values for the reactions

$$\pi^- + p \longrightarrow K^0 + \Lambda^0$$

$$K^- + p \longrightarrow \Lambda^0 + \pi^0$$

From Table 11.1 the rest energies are as follows:

π^0	135.0 MeV
π^-	139.6 MeV
K^-	493.7 MeV
K^0	497.7 MeV
p	938.3 MeV
Λ^0	1115.6 MeV

For the first reaction we have:

$$Q = [m_{\pi^-} + m_p - (m_{K^0} + m_{\Lambda^0})]c^2$$

$$= [139.6 + 938.3 - (497.7 + 1115.6)]\text{MeV}$$

$$= -535.4 \text{ MeV}$$

This reaction has a negative Q value, and energy must be supplied in the form of initial kinetic energy to produce the additional mass energy of the products. For

EXAMPLE 11.4

Solution

the second reaction we have:

$$Q = [m_{K^-} + m_p - (m_{\Lambda^0} + m_{\pi^0})]c^2$$

$$= [493.7 + 938.3 - (1115.6 + 135.0)]\text{MeV}$$

$$= 181.4 \text{ MeV}$$

A positive Q value indicates that there is enough mass energy in the initial particles to produce the final particles; in fact there is 181.4 MeV of mass energy left over for kinetic energy of the Λ^0 and π^0 (in addition to any kinetic energy of the incident particle).

Only for negative Q values do we have a *threshold energy* K_{th}, a minimum kinetic energy that m_1 must have in order to initiate the reaction. As in the nuclear physics case, the threshold kinetic energy K_{th} is larger than the magnitude of Q. We must not only create the additional mass, but the product particles must be given sufficient kinetic energy so that linear momentum is conserved in the reaction.

This calculation, like the nuclear physics case, is easiest to do if we switch to the center-of-mass (or center-of-momentum) reference frame, in which the total momentum is zero. Above threshold, the reaction viewed in this frame (which we call the CM frame) would look like the illustration in Figure 11.7; the total momentum is zero, both before and after the collision. If we supply only the threshold energy required to create the product particles, the reaction in the CM frame would look like Figure 11.8; this same reaction viewed in the laboratory would be as illustrated in Figure 11.9. In the CM frame, the product particles are at rest (this is possible because the total momentum is zero in this frame). In the laboratory frame, they move together with a common speed which is just the transformation speed between the laboratory and CM frames.

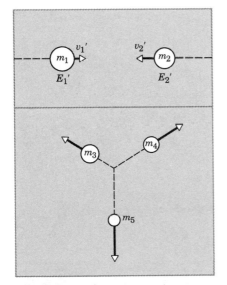

FIGURE 11.7 The same reaction as Figure 11.6, but viewed in the CM reference frame.

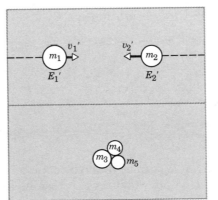

FIGURE 11.8 A reaction in the CM reference frame when m_1 has the threshold kinetic energy.

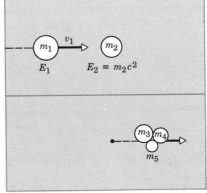

FIGURE 11.9 The reaction of Figure 11.8 in the laboratory reference frame.

We begin by deriving a *relativistic* expression for this transformation speed. In the laboratory frame, m_1 moves with speed v_1 and therefore has momentum $p_1 = m_1 v_1 / \sqrt{1 - v_1^2/c^2}$. In the CM frame, m_1 moves with speed v_1' and momentum p_1', and m_2 moves with speed v_2' and momentum p_2'. Let v be the transformation speed; we wish to find v in terms of m_1, m_2, and v_1.

From Chapter 2 we take the relativistic expression for velocity transformation:

$$v_x' = \frac{v_x - v}{1 - v_x v/c^2} \tag{11.3}$$

and so

$$v_1' = \frac{v_1 - v}{1 - v_1 v/c^2} \tag{11.4}$$

and

$$v_2' = -v \tag{11.5}$$

The last result follows directly from $v_2 = 0$. Since $p_1' = p_2'$ in this frame, we must have:

$$\frac{m_1 v_1'}{\sqrt{1 - v_1'^2/c^2}} = \frac{m_2 v_2'}{\sqrt{1 - v_2'^2/c^2}} \tag{11.6}$$

Substituting our results (11.4) and (11.5) for v_1' and v_2', and doing considerable algebra, we can derive the result

$$v = \frac{m_1 v_1}{m_1 + m_2 \sqrt{1 - v_1^2/c^2}} \tag{11.7}$$

That is, if we were to travel at speed v (in the direction of v_1), then the reaction shown in Figure 11.6 would appear to us as the illustration in Figure 11.7.

When m_1 is given just the threshold kinetic energy, the final products remain at rest in the CM frame. That is, the total energy of the reaction products m_3, m_4, m_5, . . . is just their total rest energy $m_3 c^2 + m_4 c^2 + m_5 c^2 + \ldots$. Since energy is conserved in all frames of reference, the total energy of m_1 and m_2 before the collision must be equal to the total energy after the collision. Letting E_1' and E_2' be the total energies of m_1 and m_2 in the CM frame, the *threshold condition* is

$$E_1' + E_2' = m_3 c^2 + m_4 c^2 + m_5 c^2 + \ldots \tag{11.8}$$

where

$$E_1' = \frac{m_1 c^2}{\sqrt{1 - v_1'^2/c^2}} \tag{11.9}$$

$$E_2' = \frac{m_2 c^2}{\sqrt{1 - v_2'^2/c^2}} \tag{11.10}$$

The total energy in the CM frame, $E_1' + E_2'$, can be found by adding Equations (11.9) and (11.10) and substituting our previous expressions for v_1' and v_2', Equations (11.4) and (11.5). Using our deduced value for v, after considerable algebraic manipulation we find:

$$E_1' + E_2' = \sqrt{m_1^2 c^4 + m_2^2 c^4 + 2E_1 m_2 c^2} \tag{11.11}$$

This is a general expression for the total CM energy when m_1 has *laboratory* total energy E_1; this expression is always valid, not only at threshold.

We now apply the threshold condition (11.8):

$$\sqrt{m_1^2 c^4 + m_2^2 c^4 + 2E_1 m_2 c^2} = m_3 c^2 + m_4 c^2 + m_5 c^2 + \dots \quad (11.12)$$

This expression can be solved for E_1, and the threshold *kinetic energy* K_{th} is just $E_1 - m_1 c^2$; after a bit of algebra the result is as follows:

$$K_{th} = (-Q) \frac{m_1 + m_2 + m_3 + m_4 + m_5 + \dots}{2m_2} \quad (11.13)$$

Reaction threshold kinetic energy

This can also be written as

$$K_{th} = (-Q) \frac{\text{total mass of all particles involved in reaction}}{2 \times \text{mass of target particle}} \quad (11.14)$$

You will be asked to show in Problem 15 at the end of this chapter that the relativistic threshold formula reduces to the nonrelativistic formula for nuclear reactions derived in Chapter 10.

Calculate the threshold kinetic energy to produce π mesons from the reaction $p + p \longrightarrow p + p + \pi^0$.

EXAMPLE 11.5

The rest energies of the constituents are:

Solution

$$m_p c^2 = 938.3 \text{ MeV}$$

$$m_\pi c^2 = 135.0 \text{ MeV}$$

The Q value is thus:

$$Q = m_p c^2 + m_p c^2 - (m_p c^2 + m_p c^2 + m_\pi c^2)$$

$$= -m_\pi c^2 = -135.0 \text{ MeV}$$

Using Equation (11.13) we can find the threshold kinetic energy

$$K_{th} = (-Q) \frac{4m_p + m_\pi}{2m_p}$$

$$= (135.0 \text{ MeV}) \frac{4 \times 938.3 \text{ MeV} + 135.0 \text{ MeV}}{2 \times 938.3 \text{ MeV}}$$

$$= 279.7 \text{ MeV}$$

Such energetic protons are produced at many accelerators throughout the world, and as a result the properties of the π mesons can be carefully investigated.

In 1956 an experiment was performed at Berkeley to search for the antiproton, which could be produced in the reaction

EXAMPLE 11.6

$$p + p \longrightarrow p + p + p + \bar{p}$$

What is the threshold energy for this reaction?

Since the rest energy of the antiproton is identical to the rest energy of the proton (938.3 MeV), the Q value is

Solution

$$Q = m_p c^2 + m_p c^2 - (4 \times m_p c^2)$$

$$= -2m_p c^2$$

Thus

$$K_{\text{th}} = (2m_p c^2) \frac{6m_p c^2}{2m_p c^2}$$

$$= 6m_p c^2 = 5630 \text{ MeV} = 5.630 \text{ GeV}$$

For the discovery of the antiproton produced in this reaction, Owen Chamberlain and Emilio Segrè were awarded the Nobel prize in physics in 1959.

(a)

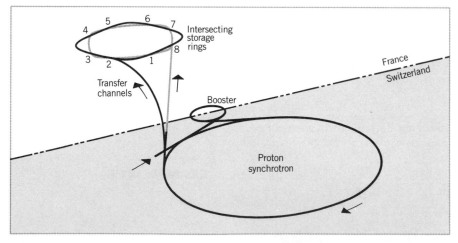

(b)

FIGURE 11.10 A colliding beam accelerator at CERN, the European Organization for Nuclear Research. An aerial view is shown in (a); the large circle in the foreground shows the location of the 28-GeV proton accelerator (synchrotron), and the circular area in the background shows the 300-m rings in which beams of protons circulate in opposite directions. A schematic diagram is shown in (b); the protons collide at eight points around the rings. One of the collision regions is shown in (c). (Photos courtesy CERN)

(c)

It is interesting to compute the "efficiency" of these reactions; that is, how much of the initial kinetic energy we supply actually goes into producing the final particles, and how much is "wasted" in the laboratory kinetic energies of the reaction products. In the first example, we supply 279.7 MeV of kinetic energy to produce 135.0 MeV of rest energy, for an efficiency of about 50 percent. In the second example, $6m_pc^2$ of kinetic energy produces only $2m_pc^2$ of rest energy, for an efficiency of only 33 percent. As the rest masses of the product particles become larger, the efficiency decreases, and relatively more energy must be supplied. For example, there is presently much interest in trying to produce particles with rest energies in the range of 50 to 100 GeV. To produce a particle with a 50-GeV rest energy in a proton-proton collision, we need to supply about 1250 GeV of initial kinetic energy. Only 4 percent of the energy supplied actually goes into producing the new particles; the remaining 96 percent must go to kinetic energy of the products in order to balance the large initial momentum of the incident particle. To produce a 100-GeV particle requires not twice as much energy, but four times as much.

This is obviously not a pleasant situation for particle physicists, who must build increasingly more powerful accelerators to accomplish their goals of producing more massive particles. One way out of this difficulty would be to do an experiment in the CM frame, where at threshold the production of new particles is 100 percent efficient—*none* of the initial kinetic energy goes into kinetic energy of the products, which are at rest in the CM frame. Thus a 50-GeV particle could be produced by a head-on collision between two protons with as little as 25 GeV kinetic energy. Of course, this great gain in efficiency is at a cost of the technological difficulty of making such collisions occur. There are now accelerators, under development and in operation, in which beams of particles (electrons or protons) travel in opposite directions around a circular path and are occasionally made to collide. Such a *colliding beam* accelerator is illustrated in Figure 11.10.

Although the classes and properties of the elementary particles seem like a complicated and disordered collection, there is an underlying order that suggests that a scheme of remarkable simplicity is at work. We can illustrate this order if we plot a diagram that has strangeness along the y axis and electric charge along the x axis. Putting the families of particles in their proper locations on the graphs, regular geometrical patterns begin to emerge. Figures 11.11 to 11.13 show such plots for the spin-0 mesons, the spin-$\frac{1}{2}$ baryons, and the spin-$\frac{3}{2}$ baryons. It was recognized independently and simultaneously in 1964 by Murray Gell-Mann and George Zweig that such regular patterns are evidence of an underlying structure in the particles, and it was shown that one could duplicate these patterns if it was assumed that all of the mesons and baryons were composed of three fundamental particles, which soon became known as *quarks*. The three quarks, known as up (u), down (d), and sideways or strange (s), are assumed to have the properties listed in Table 11.2.

Let us see how the quark model works in the case of the mesons. Since the quarks have spin $\frac{1}{2}$ and the mesons have spin zero, the simplest scheme would be to combine two quarks, with their spins directed oppositely, to make a meson. However the mesons have baryon number $B = 0$, while a combination of two quarks would have $B = \frac{1}{3} + \frac{1}{3} = \frac{2}{3}$. A combination of a quark and an antiquark, on the other hand, would have $B = 0$, since the antiquark has $B = -\frac{1}{3}$. For example, suppose we combine a u quark with a $\bar{d}$ ("antidown") quark, obtaining the combination $u\bar{d}$. This combination has spin zero and electric charge $\frac{2}{3} + \frac{1}{3} = +1$. (A d quark has charge $-\frac{1}{3}$, so $\bar{d}$ has charge $+\frac{1}{3}$.) The properties of this combination are identical with the π^+ meson, and so we assume that the π^+ consists of the combination $u\bar{d}$. Continuing in this way, we find nine possible combinations of a quark and an antiquark, as listed in Table 11.3, and plotting those nine combinations on a graph of strangeness against electric charge, we obtain Figure 11.14, which looks identical to Figure 11.11.

The baryons have $B = 1$ and spin $\frac{1}{2}$ or $\frac{3}{2}$, which suggests immediately that three quarks make a baryon. The 10 possible combinations of three quarks are listed in Table 11.4 and we can arrange them into two patterns as shown in Figures 11.15 and 11.16, which are identical to those for the spin $\frac{1}{2}$ and spin $\frac{3}{2}$ baryons.

The remarkable success of the quark model goes beyond the drawing of simple geometrical patterns. The mathematical complexity of the quantum theory of quarks is beyond the scope of this text; however, the model has had outstanding success in predicting such properties of particles as rest masses, decay modes, lifetimes, magnetic moments, and so forth. Although much work, both experimental and theoretical,

11.7 THE QUARK MODEL

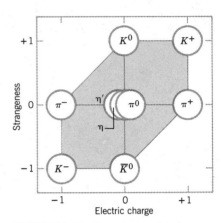

FIGURE 11.11 The relationship between electric charge and strangeness for the spin-0 mesons.

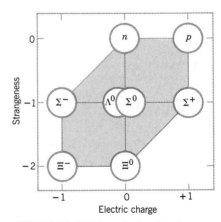

FIGURE 11.12 The relationship between electric charge and strangeness for the spin-$\frac{1}{2}$ baryons.

Table 11.2 The Three Fundamental Quarks

Name	Symbol	Charge	Spin	Baryon Number	Strangeness	Antiquark
Up	u	$+\frac{2}{3}$	$\frac{1}{2}$	$\frac{1}{3}$	0	$\bar{u}$
Down	d	$-\frac{1}{3}$	$\frac{1}{2}$	$\frac{1}{3}$	0	$\bar{d}$
Strange	s	$-\frac{1}{3}$	$\frac{1}{2}$	$\frac{1}{3}$	-1	$\bar{s}$

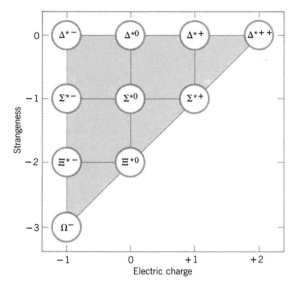

FIGURE 11.13 The relationship between electric charge and strangeness for the spin-$\frac{3}{2}$ baryons.

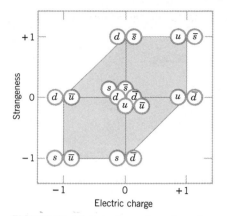

FIGURE 11.14 Spin-0 quark-antiquark combinations; compare with Figure 11.11.

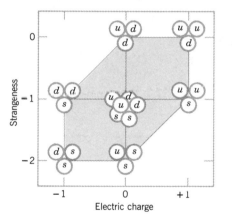

FIGURE 11.15 Spin-$\frac{1}{2}$ three-quark combinations; compare with Figure 11.12.

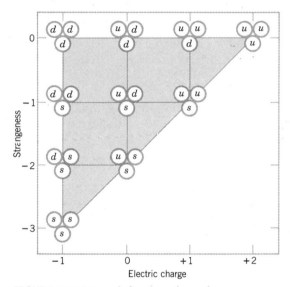

FIGURE 11.16 Spin-$\frac{3}{2}$ three-quark combinations; compare with Figure 11.13.

Table 11.3 Possible Quark-Antiquark Combinations

Combination	Charge	Spin	Baryon Number	Strangeness
$u\bar{u}$	0	0, 1	0	0
$u\bar{d}$	+1	0, 1	0	0
$u\bar{s}$	+1	0, 1	0	+1
$d\bar{u}$	−1	0, 1	0	0
$d\bar{d}$	0	0, 1	0	0
$d\bar{s}$	0	0, 1	0	+1
$s\bar{u}$	−1	0, 1	0	−1
$s\bar{d}$	0	0, 1	0	−1
$s\bar{s}$	0	0, 1	0	0

Table 11.4 Possible Three-Quark Combinations

Combination	Charge	Spin	Baryon Number	Strangeness
uuu	+2	$\frac{3}{2}$	1	0
uud	+1	$\frac{1}{2}, \frac{3}{2}$	1	0
udd	0	$\frac{1}{2}, \frac{3}{2}$	1	0
uus	+1	$\frac{1}{2}, \frac{3}{2}$	1	−1
uss	0	$\frac{1}{2}, \frac{3}{2}$	1	−2
uds	0	$\frac{1}{2}, \frac{3}{2}$	1	−1
ddd	−1	$\frac{3}{2}$	1	0
dds	−1	$\frac{1}{2}, \frac{3}{2}$	1	−1
dss	−1	$\frac{1}{2}, \frac{3}{2}$	1	−2
sss	−1	$\frac{3}{2}$	1	−3

remains to be done in particle physics, the quark model seems to have provided a firm basis upon which to build.

Using the quark model, we can analyze the decays and reactions of the elementary particles, based on two rules:

1. Quark-antiquark pairs can be created from energy quanta, and conversely can annihilate into energy quanta. For example,

$$d + \overline{d} \longrightarrow \text{energy} \qquad \text{or} \qquad \text{energy} \longrightarrow u + \overline{u}$$

This energy can be in the form of gamma rays (as in electron-positron annihilation), or else it can be transferred to or from other particles in the decay or reaction.

2. The weak interaction can change one type of quark into another; the strong and electromagnetic interactions cannot.

EXAMPLE 11.7

Analyze the reaction $\pi^- + p \rightarrow \Lambda^0 + K^0$ and the decay $\Lambda^0 \rightarrow p + \pi^-$ in terms of the constituent quarks.

Solution

The reaction can be rewritten as follows:

$$d\overline{u} + uud \longrightarrow uds + d\overline{s}$$

Each side contains one *u* quark and two *d* quarks, which don't change in the reaction. The remaining transformation is:

$$\overline{u} + u \longrightarrow s + \overline{s}$$

The *u* and $\overline{u}$ annihilate, and from the resulting energy *s* and $\overline{s}$ are created.

The Λ^0 decay process can be rewritten as

$$uds \longrightarrow uud + d\overline{u}$$

The basic process is the decay $s \rightarrow d$ by the weak interaction, along with the creation of the $u\overline{u}$ pair from the decay energy.

A significant, but unanswered, question is: What is a quark? Is there really such a particle, or is it just a mathematical device (which is nevertheless of great use in constructing a theory)? In recent years, there have been many searches for quarks, primarily in high-energy particle collisions. Despite the best efforts of particle physicists, no free quark has yet been seen. The explanations for these failures are varied. Perhaps the quarks are so massive *and* so tightly bound that our accelerators are not yet powerful enough to liberate one. Perhaps the force between quarks increases with separation (like a spring) rather than decreases (like electromagnetism or gravity), and becomes extremely large at separations larger than the size of a nucleon. Perhaps there is a rule of nature that forbids an isolated quark from existing. Or perhaps the quarks themselves don't exist. The ultimate answer to this question is one of the goals of particle physics.

The last few years have been exciting ones for particle physics. We summarize here a few recent discoveries; for additional information more extensive articles are listed in the references at the end of this chapter.

11.8 RECENT DEVELOPMENTS

1. *The charmed quark.* In 1974, in simultaneous experiments at the Brookhaven National Laboratory in New York and at the Stanford Linear Accelerator Center in California, a new meson was discovered. This meson, with a rest energy about three times the nucleon rest energy, should have decayed very rapidly into lighter mesons, but instead its decay was slowed significantly. This occurrence was explained by introducing a fourth quark, called the charmed quark (c), with a charge of $+\frac{2}{3}$. This new meson is interpreted as the combination of the charmed quark and antiquark, $c\bar{c}$; the cause of its "slow" decay is similar to the cause of the slow decay of the strange particles—decays that change charmed quarks into one of the three uncharmed quarks are slowed, just like decays that change strangeness. For this discovery, the leaders of the two research teams, S. C. C. Ting at Brookhaven and B. Richter at Stanford, received the 1976 Nobel prize in physics. In the following years, other mesons were discovered that were interpreted as combinations of a charmed quark and an uncharmed antiquark.

The hypothesis of a fourth quark complicates our simple geometrical figures from two dimensions to three dimensions, as shown in Figure 11.17, and a new quantum number, C, is introduced. The c quark has $C = 1$, and all other quarks have $C = 0$. The charm quantum number is now plotted along the z axis. Six new spin-0 mesons, called F and D, have been found with masses about 2000 MeV. These F and D mesons are interpreted as combinations of a charmed quark and an uncharmed antiquark. Charmed baryons, with quark combinations such as uuc and udc, have also recently been found. Not all of the possible combinations have yet been discovered, but the charmed mesons and baryons that have been found provide additional support for the quark model, and the search for the predicted but as yet undiscovered combinations will provide challenges for particle physicists in the future.

The addition of a fourth quark is satisfying in another respect—we have now four fundamental leptons (e, μ, ν_e, ν_μ) and four fundamental quarks (u, d, s, c). This symmetry between the fundamental strongly interacting particles and the fundamental weakly interacting particles is very striking, but we will find that it may be a temporary state of affairs.

2. *Heavier leptons.* In 1975, experiments at Stanford showed the existence of a new lepton, called τ, with a rest energy of 1.784 GeV (about twice as massive as a nucleon). Other experiments have suggested that this lepton, like e and μ, has its own neutrino, and so we now have six leptons, rather than four, and the symmetry between leptons and quarks has been lost. It is an open question, requiring exper-

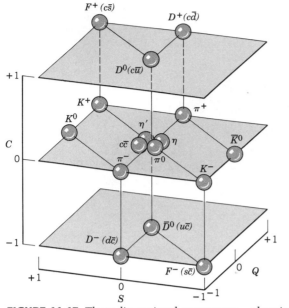

FIGURE 11.17 Three-dimensional strangeness—electric charge
—charm diagram for spin-0 mesons. Figure 11.11 is the central
plane of this diagram, and the new charmed mesons are in the
upper and lower planes.

iments with more powerful accelerators, whether yet heavier leptons
exist.

3. *New quarks.* In 1977, researchers at the Fermi National Laboratory near Chicago announced the discovery of a new particle that was
interpreted as being composed of yet another quark (called *b*, for
"bottom"). This quark, together with its expected but yet undiscovered
partner *t* ("top") may restore the symmetry between the number of
quarks and number of leptons at six.

The question of where it all ends now probably occurs to you. Will
we one day find that the present 100 or so "elementary" particles have
been replaced with 100 "elementary" leptons and quarks? Or will it end
with a small number, and a few basic laws governing their behavior? No
one knows the answer to this question, but calculations suggest that the
total number will remain small; an upper limit of about 16 has been
suggested, and the actual number may prove to be even less. Whatever
the number may prove to be, it is clear that such studies of the most elementary and fundamental building blocks of matter will continue to be
an exciting field of discovery and to occupy the very forefront of research
in physics.

4. *Proton decay and neutrino mass.* A recent theoretical development, which has been responsible for showing that the weak and electromagnetic interactions can be viewed as two different parts of the
same basic force, has led to two predictions that are contrary to two
basic assumptions we have repeatedly made in this text. The first prediction of the new theory is that the proton is not stable but may in fact
decay into lighter particles, with the accompanying breakdown of
matter. The predicted lifetime is of the order of 10^{30} y, so there is no
need to panic, but if the decay is actually seen, it will be a great triumph
for this unified theory (for which Steven Weinberg, Abdus Salam, and

Sheldon Glashow were awarded the 1979 Nobel prize in physics). The second prediction is that the neutrinos are not massless but have small rest masses, perhaps a few eV/c^2. This prediction, which is under active investigation as of this writing, could help to explain such paradoxes as why the measured number of neutrinos reaching us from the sun is only about one-third of what we would predict based on our knowledge of fusion processes, and also where the matter is, which could supply enough gravitational attraction to halt the expansion of the universe and begin a contraction. By the time you read this, these questions may have been resolved, but that is once again evidence for the rapid development and far-reaching significance of discoveries in particle physics.

5. *The force between quarks.* One final question: What holds the quarks together? In field theory, we regard the force between objects as arising from the exchange of particles. The strong interaction between nucleons arises from the interchange of π mesons, and the electromagnetic field is carried by photons. It has been proposed that a particle called the *graviton* is exchanged between objects that have a gravitational attraction and that a particle called the *vector boson* carries the weak interaction. The vector boson was discovered in 1983 by experimenters at the CERN $p\bar{p}$ collider; the graviton has not yet been observed, but most physicists are fairly certain of its existence. What then is the particle which is exchanged by quarks to provide the apparently enormous forces that bind the quarks together? This particle has been somewhat whimsically named the *gluon*. Since experiments to observe a quark directly have not been successful, the direct observation of a gluon seems hopeless, but indirect evidence for its existence comes from a number of recent studies.

SUGGESTIONS FOR FURTHER READING

Advanced books on particle physics tend to be mathematically difficult, full of field theory and relativistic quantum mechanics. Fortunately there are many popular-level books and articles that can be read for background material. These are generally descriptive and nonmathematical. The two books by Capra and Zukav listed in Chapter 1 are of this type; another excellent recent book is:

G. Feinberg, *What is the World Made Of?* (Garden City, Anchor Press, 1977).

A fascinating and easily readable account of recent developments in particle physics is:

J. C. Polkinghorne, *The Particle Play* (Oxford, W. H. Freeman, 1981).

Other general books that are now somewhat out of date but are still interesting for their background material are:

D. H. Frisch and A. M. Thorndike, *Elementary Particles* (Princeton, Van Nostrand, 1964).
R. Gourian, *Particles and Accelerators* (New York, McGraw-Hill, 1967).
C. N. Yang, *Elementary Particles* (Princeton, Princeton University Press, 1961).

Developments occur so rapidly in particle physics that textbooks are often two years outdated when they are published. A better source of current, popular-level information on particle physics is the magazine *Scientific American*, which usually includes a major article on current developments in particle physics every other month or so. A good background survey article is:

V. F. Weisskopf, "The Three Spectroscopies," *Scientific American 218*, 15 (May 1968).

Two excellent summaries of the quark model are:

S. L. Glashow, "Quarks with Color and Flavor," *Scientific American 233*, 38 (October 1975).

Y. Nambu, "The Confinement of Quarks," *Scientific American 235*, 48 (November 1976).

QUESTIONS

1. Some conservation laws are based on fundamental properties of nature, while others are based on systematics of decays and reactions and have as yet no fundamental basis. Give the basis for the following conservation laws: energy, linear momentum, angular momentum, electric charge, baryon number, lepton number, strangeness.

2. Does the presence of neutrinos among the decay products of a particle always indicate that the weak interaction is responsible for the decay? Do all weak interaction decays have neutrinos among the decay products? Which decay product indicates an electromagnetic decay?

3. Do all strongly interacting particles also feel the weak interaction?

4. In what ways would physics be different if there were another member of the lepton family less massive than the electron? What if there were another lepton more massive than the tau?

5. Suppose a proton is moving with high speed, so that $E \gg m_0 c^2$. Is it possible for the proton to decay, such as into $n + \pi^+$ or $p + \pi^0$?

6. On planet anti-Earth, antineutrons beta decay into antiprotons. Is a neutrino or an antineutrino emitted in this decay?

7. The Σ^0 can decay to Λ^0 without changing strangeness, so it goes by the electromagnetic interaction; the charged $\Sigma^\pm$ decay to p or n by the weak interaction in characteristic lifetimes of 10^{-10} s. Why can't $\Sigma^\pm$ decay to Λ^0 by the strong interaction in a much shorter time?

8. The Ω^- particle decays to $\Lambda^0 + K^-$. Why doesn't it also decay to $\Lambda^0 + \pi^-$?

9. Can antibaryons be produced in reactions between baryons and mesons?

10. Is it reasonable to describe a resonance as a definite particle, when its mass is uncertain (and therefore variable) by 20 percent?

11. Why are most particle physics reactions endothermic ($Q < 0$)?

12. Although doubly charged baryons have been found, no doubly charged mesons have yet been found. What would be the effect on the quark model if a meson with charge $+2e$ were found? How could such a meson be interpreted within the quark model?

13. The quark transformation $s \rightarrow d$ involves no change of electric charge, so no additional particles must be involved. How can the transformation $d \rightarrow u$ be accomplished? (Remember, only the weak interaction can change one kind of quark to another.)

14. The newly discovered charmed D mesons decay to π and K mesons with a lifetime of 10^{-13} s. (a) Why is the lifetime so much slower than a typical strong interaction lifetime? Is a quantum number not conserved in the decay? (b) What interaction is responsible for the decay?

15. The Δ^* baryons are found with electric charges $+2, +1, 0$, and -1. Based on the quark model, why do we expect no Δ^* with charge -2?

16. Although we cannot observe quarks directly, indirect evidence for quarks in nucleons comes from the scattering of high energy particles, such as electrons. When the deBroglie wavelength of the electrons is small compared with the size of a nucleon (~ 1 fm), the electrons appear to be scattered from massive, compact objects much smaller than a nucleon. To which phenom-

enon discussed previously in this text is this similar? Can the scattering be used to deduce the mass of the struck object? How does the scattering depend on the electric charge of the struck object? What would be the difference between scattering from a particle of charge e and one of charge $2e/3$?

1. Identify the interaction responsible for the following decays:

   (a) $\Delta^{\star} \rightarrow p + \pi$ (c) $K^+ \rightarrow \mu^+ + \nu_\mu$ (e) $\eta' \rightarrow \eta + 2\pi$

   (b) $\eta \rightarrow \gamma + \gamma$ (d) $\Lambda^0 \rightarrow p + \pi^-$ (f) $K^0 \rightarrow \pi^+ + \pi^-$

2. Name the conservation law that would be violated in each of the following decays:

   (a) $\pi^+ \rightarrow e^+ + \gamma$ (d) $\Lambda^0 \rightarrow \pi^- + \pi^+$ (g) $\Xi^0 \rightarrow \Sigma^0 + \pi^0$

   (b) $\Lambda^0 \rightarrow p + K^-$ (e) $\Lambda^0 \rightarrow n + \gamma$ (h) $\mu^- \rightarrow e^- + \gamma$

   (c) $\Omega^- \rightarrow \Sigma^- + \pi^0$ (f) $\Omega^- \rightarrow \Xi^0 + K^-$

3. Each of the following reactions violates one (or more) of the conservation laws. Name the conservation law violated in each case:

   (a) $\nu_e + p \rightarrow n + e^+$ (d) $\pi^- + n \rightarrow K^- + \Lambda^0$

   (b) $p + p \rightarrow p + n + K^+$ (e) $K^- + p \rightarrow n + \Lambda^0$

   (c) $p + p \rightarrow p + p + \Lambda^0 + K^0$

4. Table 11.1 lists the most likely decay mode of the K^+ meson; Example 11.3 gives another possible decay. List four other possible decays that are allowed by the conservation laws.

5. Supply the missing particle in each of the following decays:

   (a) $K^- \rightarrow \pi^0 + e^- +$ (c) $\eta \rightarrow \pi^+ + \pi^- +$

   (b) $K^0 \rightarrow \pi^0 + \pi^0 +$

6. List one possible decay mode of the following antiparticles: (a) $\overline{\Lambda^0}$; (b) $\overline{\Omega^-}$; (c) $\overline{K^0}$; (d) $\bar{n}$.

7. Carry out the calculations of mc^2 for the three decays of Figure 11.3.

8. Repeat the calculation of Example 11.3 for the case in which the π meson has zero kinetic energy, and show that the electron energy in this case is less than the maximum value.

9. Determine the energy uncertainty or width of (a) η; (b) η'; (c) Σ^0; (d) $\Delta^{\star}$.

10. Find the kinetic energies of each of the two product particles in the following decays (assume the decaying particle is at rest):

    (a) $\Omega^- \rightarrow \Lambda^0 + K^-$ (b) $\pi^+ \rightarrow \mu^+ + \nu_\mu$ (c) $K^0 \rightarrow \pi^+ + \pi^-$

11. Find the Q values of the following decays:

    (a) $\pi^- \rightarrow \mu^- + \nu_\mu$ (c) $K^0 \rightarrow \pi^+ + \pi^-$ (e) $\Sigma^0 \rightarrow \Lambda^0 + \gamma$

    (b) $\pi^0 \rightarrow \gamma + \gamma$ (d) $\Sigma^+ \rightarrow p + \pi^0$

12. List an alternative decay mode for (a) Ω^-; (b) Λ^0; (c) Σ^+, that satisfies the applicable conservation laws.

13. Each of the reactions below is missing a single particle. Supply the missing particle in each case.

    (a) $p + p \rightarrow p + \Lambda^0 +$ (c) $\pi^- + \acute{p} \rightarrow \Xi^0 + K^0 +$ (e) $\bar{\nu}_\mu + p \rightarrow n +$

    (b) $p + \bar{p} \rightarrow n +$ (d) $K^- + n \rightarrow \Lambda^0 +$ (f) $K^- + p \rightarrow K^+ +$

14. In this problem you will derive Equation (11.13) using a slightly different procedure. See Figure 11.9, and let M be the total mass of the product particles, which move together and can therefore be considered as a single particle. Let p_1 be the momentum of m_1. (a) What is the momentum of M? (b) Write an expression for conservation of total relativistic energy. (c) Combine (a) and (b) and solve for the kinetic energy.

15. Show Equation (11.13) reduces to Equation (10.14) in the nonrelativistic limit.

16. Determine the Q values of the following reactions:
    (a) $K^- + p \rightarrow \Lambda^0 + \pi^0$
    (b) $\pi^+ + p \rightarrow \Sigma^+ + K^+$
    (c) $K^- + p \rightarrow \Omega^- + K^+ + K^0$
    (d) $p + p \rightarrow p + \pi^+ + \Lambda^0 + K^0$
    (e) $\gamma + n \rightarrow \pi^- + p$

17. Find the threshold kinetic energy for the following reactions. In each case the first particle is in motion and the second is at rest.
    (a) $p + p \rightarrow n + \Sigma^+ + K^0 + \pi^+$
    (b) $\pi^- + p \rightarrow \Sigma^0 + K^0$
    (c) $p + n \rightarrow p + \Sigma^- + K^+$
    (d) $\pi^+ + p \rightarrow p + p + \bar{n}$

18. A K^0 with a kinetic energy of 276 MeV decays in flight into π^+ and π^-, which move off at equal angles with the original direction of the K^0. Find the energies and directions of motion of the π^+ and π^-.

19. A Σ^- with a kinetic energy of 0.250 GeV decays into $\pi^- + n$. The π^- moves at 90° to the original direction of travel of the Σ^-. Find the kinetic energies of π^- and n and the direction of travel of n.

20. Analyze the following reactions in terms of their constituent quarks:
    (a) $K^- + p \rightarrow \Lambda^0 + \pi^0$
    (b) $\pi^+ + p \rightarrow \Sigma^+ + K^+$
    (c) $K^- + p \rightarrow \Omega^- + K^+ + K^0$
    (d) $p + p \rightarrow p + \pi^+ + \Lambda^0 + K^0$
    (e) $\gamma + n \rightarrow \pi^- + p$

21. Analyze the following decays in terms of the constituent quarks:
    (a) $\Omega^- \rightarrow \Lambda^0 + K^-$
    (b) $n \rightarrow p + e^- + \bar{\nu}_e$
    (c) $\pi^0 \rightarrow \gamma + \gamma$
    (d) $K^0 \rightarrow \pi^+ + \pi^-$
    (e) $\Delta^{*++} \rightarrow p + \pi^+$
    (f) $\Sigma^- \rightarrow n + \pi^-$

An applied statistics laboratory.

12

STATISTICAL PHYSICS

Many experiments are done in physics research that are analyzed as if the interactions take place in single, isolated events. For example, in nuclear reactions, we detect the outgoing particles, y, one at a time. We analyze the reaction as if a single incoming particle, x, reacted and produced a single outgoing particle, y; the rest of the x particles in the beam are assumed to have no influence on the fate of the single x particle on which we focus our attention. We calculate the energy of a *single* outgoing particle y by taking the energy of a *single* incoming particle x and adding or subtracting the energy gained or lost in a *single* nuclear reaction. Even though we observe these y particles at a high rate (perhaps 10^6 per second), each single event is assumed to be independent of all others, and every y particle should emerge with a definite, fixed, and predictable energy. Since experiment shows this to be true, these assumptions are justified.

Nuclear decay processes, Compton scattering, and the photoelectric effect are all similarly treated as isolated events, and experiment once again proves these assumptions to be correct.

On the other hand, there are experiments we can do in which these assumptions are not justified. For example, suppose we have a gas of 10^{20} atoms. If we heat the gas and add 10^{20} eV of energy, the *average* energy per atom will increase by 1 eV, but we should be very surprised to find that the energy of *each* atom increased by *exactly* 1 eV. Even if we added the energy in 1 eV quanta to a gas with atomic excited states 1 eV apart, we still would not expect to find every atom with 1 eV of energy—some unfortunate atoms might acquire no energy at all, while other greedy atoms might absorb 10 eV or even 100 eV.

This sharing of energy among the many parts of a system cannot be simply analyzed in terms of single isolated events. The analysis of such *cooperative* phenomena requires the techniques of *statistical physics*, in which we are concerned not with calculating the *exact* outcome of single, isolated events, but with predicting the *average* outcome of many cooperative events, based on the *statistical distribution* of the possible outcomes.

In this chapter we discuss the laws of statistical physics, and we illustrate some systems that are governed by *classical* statistics and some others that require *quantum* statistics. You should particularly note how these statistical concepts are necessary to understand the bulk properties of matter, which are discussed briefly in this chapter and more extensively in Chapter 14.

12.1 THE NEED FOR STATISTICAL PHYSICS

Consider what happens when we pass an electric current through a tube containing a gas of very low density, such as a mercury vapor tube. The current may excite an atom from the ground state to one of its many excited states, and the atom will then return ("make a transition") to the ground state with the emission of one or more photons. In the case of mercury vapor, we see individual transitions corresponding to green light, blue light, orange light, etc.; each of these individual wavelengths is "sharp"—that is, the wavelength has a certain, definite, reproducible value. Figure 12.1 shows the spectrum of light from a mercury source as it might be seen with a diffraction grating, which permits good separation and therefore accurate measurement of the wavelengths. Of course, as we have learned, the energy of each photon is just equal to the energy difference between two states of the atom. For example, the transition corresponding to green light has a wavelength of 546.0 nm, which corresponds to a photon energy of 2.271 eV. If we were to study the scheme of the excited states of the mercury atom, somewhere we would expect to find two states differing in energy by 2.271 eV, and the green wavelength would be expected to originate as the electron jumps from the higher state to the state 2.271 eV lower. As in the case of nuclear reactions, here also we treat the events as if they occurred singly and in isolation—the density of the gas is so low that a single atom emits its photons before it can encounter a neighbor atom. (Otherwise the neighbor atom might pass nearby and change slightly the electron energy levels and therefore the wavelengths of the emitted transitions. At low density this is unlikely to happen.)

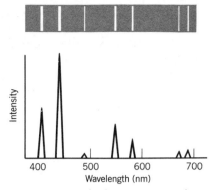

FIGURE 12.1 The line spectrum of mercury in the visible region.

Consider now the contrasting case of the emission of light by the tungsten filament of an ordinary incandescent light bulb. Figure 12.2 shows the spectrum of the light in this case; we see that *all* wavelengths are emitted, from the blue to the red end of the spectrum, and the total effect of all those colors is what we call "white" light. If we were to look carefully at the emitted light, we would find not only "green" photons of wavelength 546.0 nm (and energy 2.271 eV), but also of wavelength 547.0 nm (2.267 eV), 548.0 nm (2.263 eV) and also everything in between and beyond. Surely we don't expect to find, among the energy levels of tungsten, states with energy differences of 2.271 eV, 2.267 eV, 2.263 eV, and so forth, including all possible energy differences corresponding to every possible wavelength emitted from our light bulb! What is happening in this case is a *cooperative* phenomenon; even though a single, isolated atom of tungsten has well-defined energy levels and emits radiation, like mercury, in discrete wavelengths, the presence of the other nearby atoms in the solid tungsten so influences the emission of light that the spectrum becomes continuous. The blackbody radiation discussed in Chapter 3 is another example of a continuous spectrum resulting from this kind of cooperative phenomenon.

Is it possible to understand the emission of light from such a cooperative system as we understand the emission of light from an isolated system? It is indeed, and there are two ways to approach the problem. In the first way, we consider the tungsten filament as being composed of perhaps 10^{15} atoms. If we understood the properties of each of those 10^{15} atoms, as influenced by its $(10^{15} - 1)$ neighbors, we could surely understand the emission of light by our filament. Perhaps 10^{15} scientists working for 10^{15} years could perform enough experiments and write enough equations to solve this problem, but by then no one would be interested!

The second and more reasonable approach is to admit that, not only is it *hopeless* to try to understand the behavior of each atom, as long as there is a large number of atoms it is also *unnecessary* to understand the behavior of each one. Such a problem is solved according to the methods of *statistical physics,* in which we treat a system such as the filament of the light bulb as if it had a very large number of ways in which to distribute some available energy (supplied by the electric current, in this case) among its parts. The properties of the system are then completely described simply by the statistical distribution of the energy and by the

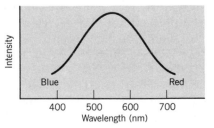

FIGURE 12.2 The continuous spectrum of an incandescent source in the visible region.

number of different ways in which it can be shared by the parts of the system. We will then find that, instead of requiring 10^{15} different properties, the system is well enough described by only a few *statistical* properties, of which the temperature is perhaps the most familiar. It is rather like stepping back from a painting and seeing the individual brush strokes merge and blend to form a unified image; just imagine trying to describe the *Mona Lisa* as being composed of so many brush strokes applied in a certain manner, rather than as the haunting image of a young woman.

Our goal in studying *statistical physics* is to discuss the kinds of systems that can be best understood using its techniques, from those with a relatively small number of particles (such as nuclei, with the order of 10^2 particles) in which we do not understand the dynamical laws, to those with a relatively large number of particles (such as bulk solids, with the order of 10^{23} particles) in which we may well understand the dynamical laws, but can't apply those laws to such large numbers of particles. Indeed, it may be that statistical fluctuations in the density of primordial gas clouds are responsible for the condensation of stars and planets, and so the size of the system to which statistical laws are applicable is virtually unlimited.

12.2 CLASSICAL VERSUS QUANTUM STATISTICS

A system, such as a collection of atoms, has *macroscopic* properties we can measure in the laboratory, such as temperature or spectrum of radiant energy, and *microscopic* properties we can't easily measure, such as the motion of individual atoms or the way in which each atom absorbs or emits energy. The goal of statistical physics is to bridge the gap between the macroscopic and microscopic properties. The great stride toward that goal was taken by the Austrian physicist Ludwig Boltzmann, who introduced the relationship between the number of different arrangements of a system, W, and the *entropy*, S, of a system.

Entropy

The number of arrangements W is usually a measure of how many ways the available energy can be distributed to the parts of the system. For example, 2 quanta of energy might be added to a "gas" of 4 particles in 12 ways, if each particle is restricted to having at most one unit of energy; there are 4 ways to choose the first particle to be given one unit of energy, and then 3 ways to choose the second.

The entropy is a macroscopic property which governs the behavior of interacting systems or interacting parts of a system; natural processes are governed by the law that the entropy must increase. For example, when two systems are placed in contact so that they may exchange energy (heat, for example), the entropy of one system may increase while that of the other may decrease, but this always occurs in such a way that the overall change in entropy must be positive. Boltzmann's relationship is $S = k \ln W$, where k is the Boltzmann constant with a value of 1.38066×10^{-23} J/K.

Let us illustrate the use of statistical principles by considering a simple system: a box containing a "gas" of 10 particles, which we assume not to be atoms but rather hard spheres, which interact only through elastic collisions with one another. We assume that the particles can only absorb energy in the form of translational kinetic energy, and for simplicity let us assume that they can only accept energy in certain units. (This has nothing to do with the energy quanta of quantum physics, but serves only to simplify our example.) In how many different ways can this system accept five units of energy? This number of arrangements is just the W of Boltzmann's equation. One possible way would be to give one particle all five units of energy. Since there are 10 particles, there are obviously 10 different ways we can choose the one particle to which we give the five units, and so $W = 10$ for this particular distribution of energy. Alternatively, we could give one particle four units and another particle one unit; there are 10 ways to choose the particle to which we give the four units, and for each of those choices there are 9 ways to choose the particle to give the remaining unit of energy, so that $W = 90$ for this distribution of energy. In Table 12.1 are shown the number of arrangements associated with the remaining possible distributions of energy.

As you can see, the distribution associated with one particle having all the energy (distribution A) has fewer arrangements, and therefore less entropy, than distribution G with 5 particles each having one unit of energy. If we were to prepare our system with distribution A and then allow the particles to interact, we would expect to find after a time that the energy was distributed according to distribution F since, according to Boltzmann, the entropy tends to increase toward a maximum value. Conversely, if we prepared the system according to distribution F, we should be very surprised to find after a time that the energy had taken on distribution A!

(It is of course not *impossible* to go from F to A, just *improbable*; the larger the number of particles, the more improbable such transformations become. For example, when we slide a block along a table, friction eventually brings the block to rest and its kinetic energy is transformed by the frictional force into thermal energy or heat, which is really just the random motions of the molecules of the block and table. We expect that the heat energy is conducted or radiated away from the area of contact of block and table, eventually becoming indistinguishable from the "background" thermal energy of those objects. You should keep in mind that it is entirely possible for that energy to "concentrate" itself, to return to the point of contact, and spontaneously to set the block in motion again. This process would not violate any law of physics, such as conservation of energy; what keeps it from being observed is that it is, rather than impossible, highly improbable.)

The method we have described for our simple system is not gener-

Table 12.1 Energy Distributions in a Simple System

Distribution Designation	\multicolumn{10}{c}{Energy Units Given to Particle Number:}	Number of Arrangements
	1	2	3	4	5	6	7	8	9	10	
A	5	0	0	0	0	0	0	0	0	0	10
B	4	1	0	0	0	0	0	0	0	0	90
C	3	2	0	0	0	0	0	0	0	0	90
D	3	1	1	0	0	0	0	0	0	0	360
E	2	2	1	0	0	0	0	0	0	0	360
F	2	1	1	1	0	0	0	0	0	0	840
G	1	1	1	1	1	0	0	0	0	0	252
	\multicolumn{10}{c}{Total}	2002

ally useful in statistical physics, because seldom do we have a system containing such a small number of particles that we can enumerate and list all of the possible arrangements, as we did in Table 12.1. Besides, we are often more concerned with the solution to a slightly different problem: What is the *probability* of a particle of our system having a certain energy? To answer this question we must make one of the most important, and often misunderstood, assumptions of statistical physics:

> Any individual arrangement is as likely as any other individual arrangement.

Fundamental assumption of statistical physics

This does *not* mean that distribution A is as likely as distribution G; it means that any one of the individual 10 arrangements of distribution A is as likely as any one of the 252 individual arrangements of distribution G. Thus the reason that it is more likely to find G than A is *not* that G is favored over A by some law of nature, but rather that there are so many more possible arrangements of G than of A, with each arrangement having the same probability to occur.

As another example, in the card game of poker, a "royal flush" consisting of the 10, J, Q, K, A of hearts is just as likely as the *particular* worthless hand consisting of 2♠, 4♥, 5♦, 8♣, 10♠. What makes a royal flush so special and rare is that there are only 4 possible royal flushes of the 2,598,960 possible poker hands, most of which are worthless hands; even though a royal flush is just as likely as a *particular* worthless hand, it is much less likely than *any* worthless hand.

Using the fundamental assumption of statistical physics, we can proceed to find the answer to the question of the probability of finding a particle with a particular value of its energy. For example, let us compute the probability to find a particle with an energy E of one unit; that is, we reach into the box, grab a single particle, and measure its energy. In arrangement A, we never find a particle with $E = 1$. In arrangement B (which occurs with a probability of 90/2002) we find an energy of one unit only if we happen to grab the one particle of the 10 that has one unit of energy, so we find $E = 1$ for that arrangement with a probability of $(1/10) \times 90/2002$. Continuing, we can compute the total probability p to find $E = 1$ as follows

$$p(E = 1) = 0 \times \frac{10}{2002} + \frac{1}{10} \times \frac{90}{2002} + 0 \times \frac{90}{2002}$$

$$+ \frac{2}{10} \times \frac{360}{2002} + \frac{1}{10} \times \frac{360}{2002} + \frac{3}{10} \times \frac{840}{2002}$$

$$+ \frac{5}{10} \times \frac{252}{2002} = 0.247$$

Given 10 particles in a box, sharing five units of energy, the probability to find *any* particle with one unit of energy is 24.7 percent. The other probabilities can be computed similarly, and the result is

Measured Energy Units	0	1	2	3	4	5
Probability	64.3%	24.7%	8.24%	2.25%	0.45%	0.05%

As you can see, the probability decreases rapidly as the measured energy increases; *it is less probable to find the energy "concentrated" in a single element of the system than it is to find it more or less uniformly distributed among many elements of the system.* As we will find in later discussions of this chapter, this is a general property of statistical distribution functions; for classical situations, like we have considered in our simple example, the probability to find a certain value of E in a system with a large number of particles is proportional to e^{-E/E_0}, where E_0 is a fundamental energy of the system. (In our case, we do not expect such a relationship to be valid, since 10 is hardly a large number of particles!) As E/E_0 increases, the probability decreases rapidly, so that large individual values of E become less likely.

The simple calculation we have considered in this section is an example of *classical* statistics; as you might expect, the laws of quantum physics introduce some important changes to classical statistics. In the next section, we discuss the classical distribution function and give some examples of its applications; one of the most familiar is the properties of an ideal gas. However, the classical laws are not sufficient to explain many statistical phenomena, among which are the electrical conductivity and specific heat of metals, the behavior of liquid helium, and blackbody radiation. Later in this chapter we discuss the laws of quantum statistics and their applications to these phenomena.

For the present, let us try to understand why the laws of quantum statistics should differ from the laws of classical statistics. One important distinction between classical physics and quantum physics is the inclusion of the observation process; the laws of classical physics do not include the observer, but the processes of observation and measurement are a fundamental part of the laws of quantum physics. Classical physics might ask "What is the value of x?" but quantum physics can only ask questions like "What value of x is observed in a certain kind of experiment?" The uncertainty principle, for example, tells us *not* that nature has some sort of built-in uncertainty, but rather that our *knowledge* of nature, as derived from *measurement*, is uncertain. In the realm of quantum physics the 10 particles we have treated in our example in this chapter are identical and therefore *indistinguishable*—we can't tell one particle from another. All atoms of a given type are alike, all electrons are alike (except that they may move in different orbits), and so forth. This has important consequences, even for our simple example. There is no possible way to tell any single one of the 252 possible arrangements of distribution G from any of the 251 others. If they cannot be *observed* as separate arrangements, they cannot be *counted* as separate arrangements.

Indistinguishable particles

A second important effect of quantum physics modifies the chance for one particle to have a certain energy when there is already another particle with that same energy. (To carry over this simple example to real calculations in quantum physics, when we are speaking of "energy" you should keep in mind that what we really mean is the complete set of *quantum numbers*; a quantum statistical distribution means assigning different sets of quantum numbers to particles.) The most extreme example of this effect comes from applying the *Pauli principle*, which requires that no two electrons in an atom have the same set of quantum numbers. If the 10 particles in our box obeyed the Pauli principle, and if there were no other label to assign to them, then (ignoring the states with zero energy) distributions D, E, F, and G would all violate the Pauli principle and be forbidden. If the particle had a second label, such as a spin quantum number with two possible values, then only F and G would be forbidden.

All particles that have half-integral spins, like the electron, proton, and neutron, obey the Pauli principle and therefore a special type of quantum statistics. The distribution function that governs the statistical behavior of such particles is known as the *Fermi-Dirac* distribution and the particles are called collectively *fermions*. Particles with integral spins, like the pi meson or alpha particle, do not obey the Pauli principle and therefore must be described by a different sort of quantum statistics; the distribution function for particles with integral spin is known as the *Bose-Einstein* distribution, and the particles are known as *bosons*. The properties of bosons and fermions, and their distribution functions, will be considered in the last half of this chapter.

Bosons and fermions

12.3 THE DISTRIBUTION OF MOLECULAR SPEEDS

Let us imagine a large container filled with a certain type of gas at a reasonably low pressure. As we have discussed in the previous section, the available energy, which we assume to be only in the form of translational kinetic energy, will be distributed over the N molecules in such a way that there are many more molecules with small velocities than with large velocities. We can investigate this assertion in detail by measuring the velocity distribution of the molecules. As we have argued frequently in this text, it makes no sense to ask "How many molecules (per unit volume) have velocity components v_x, v_y, and v_z?", because we are not sure what it means for a molecule to have *exactly* those velocity components. (In other words, how exact is exact?) It makes more sense, and in fact is a better description of what we actually measure, to ask, "How many molecules (per unit volume) have velocities between (v_x, v_y, v_z) and $(v_x + dv_x, v_y + dv_y, v_z + dv_z)$?"

We expect that, if the box is at rest, the distribution of velocities will be symmetric around zero (just as many molecules should be moving to the right as to the left) and we also expect the distribution to be such that finding a molecule with a large velocity is unlikely. (Remember the example of the previous section—it was very unlikely to find one particle with all the energy.) If we were to perform an experiment in which we measured the component v_x in the interval dv_x, we might obtain a result of the form shown in Figure 12.3. We may assume that this is a representation of a smooth curve, and a familiar curve that has the proper form (and that is also generally associated with random statistical processes) is the *normal* or *Gaussian* distribution e^{-x^2}. We therefore assume the proper distribution to be represented by

$$f(v_x)\, dv_x = \frac{1}{A_x} e^{-bv_x^2}\, dv_x \qquad (12.1)$$

where $f(v_x)dv_x$ gives the number of molecules per unit volume with velocity components between v_x and $v_x + dv_x$. Equation (12.1) is the *Maxwell velocity distribution*. The constants A_x and b are as yet not determined.

Maxwell velocity distribution

More often we are interested not in the *velocity* distribution but rather in the *speed* distribution. In order to obtain the speed distribution

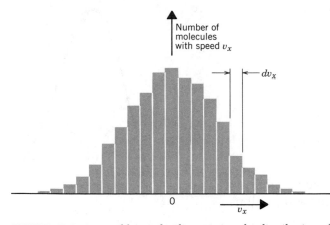

FIGURE 12.3 A possible result of measuring the distribution of one component of the velocity of molecules of a gas.

we must first generalize Equation (12.1) to three dimensions:

$$f(v_x, v_y, v_z)\, dv_x\, dv_y\, dv_z = \frac{1}{A_x A_y A_z}\, e^{-bv_x^2} e^{-bv_y^2} e^{-bv_z^2}\, dv_x\, dv_y\, dv_z$$

$$(12.2)$$

$$= \frac{1}{A'}\, e^{-b(v_x^2 + v_y^2 + v_z^2)}\, dv_x\, dv_y\, dv_z$$

A' is a new constant to be determined. The speed v is just $(v_x^2 + v_y^2 + v_z^2)^{1/2}$, and we wish to know the distribution of speeds $f(v)\, dv$. In order to find this distribution, we must know how dv is related to dv_x, dv_y, and dv_z, or equivalently how many different combinations of v_x, v_y, and v_z give the same value of v. The easiest way to see how this is done is to imagine a coordinate system in which the axes are labeled v_x, v_y, v_z; the points of constant v then form the surface of a sphere of radius v, and allowing for a range dv gives a spherical shell of thickness dv, as shown in Figure 12.4. The volume of the spherical shell is just $4\pi v^2\, dv$, and so replacing the "volume" element $dv_x\, dv_y\, dv_z$ with $4\pi v^2\, dv$, we obtain the *Maxwell speed distribution:*

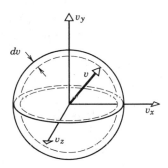

FIGURE 12.4 A spherical shell of radius v and thickness dv in the $v_x v_y v_z$ coordinate system.

$$f(v)\, dv = \frac{4\pi}{A'}\, e^{-bv^2} v^2\, dv \qquad (12.3)$$

Maxwell speed distribution

The function $f(v)$ is plotted in Figure 12.5.

An example of an experiment to measure the distribution of molecular speeds is shown in Figure 12.6. A small hole in the side of an "oven" allows a stream of molecules to escape; we assume the hole to be small enough so that the distribution of speeds inside the oven is not changed. The beam of molecules is made to pass through a slot in a disc attached to an axle rotating at angular velocity ω. At the other end of the axle is a second slotted disc, but the slot is displaced from the first by an angle θ. In order for a molecule to pass through both slots and strike the detector, it must travel the length L of the axle in the same time that it takes the axle to rotate by the angle θ, and thus $L/v = \theta/\omega$. Keeping L and θ fixed, we can vary ω; measuring the number of molecules striking the detector for each different value of ω enables us to measure the Maxwell speed distribution. A set of such experimental results is shown in Figure 12.7,

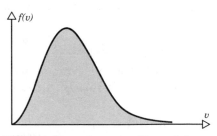

FIGURE 12.5 The Maxwell speed distribution $f(v)$.

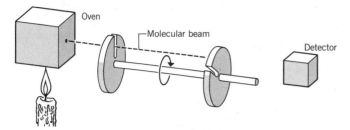

FIGURE 12.6 Apparatus to measure the distribution of molecular speeds.

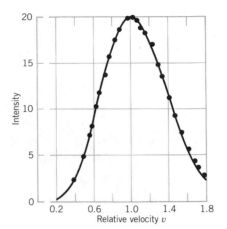

FIGURE 12.7 Result of a measurement of the distribution of atomic speeds of thallium vapor. The solid line is the Maxwell speed distribution corresponding to the oven temperature of 870 K.

and the remarkable agreement between the measured speed distribution and that predicted according to Equation (12.3) justifies the assumptions we have made in its derivation.

(From this example you can also see the importance of the interval dv. What we measure is always the product $f(v)\, dv$, and in this case the range of velocities is determined primarily by the width of the slots in the discs. To make dv very small, and thereby measure v "exactly," we would need to make the slots very small, so that very few molecules could get through. For the "perfect" experiment, we make $dv \to 0$ by making the slot width equal to 0, and no molecules get through the apparatus!)

The final step in the derivation of the Maxwell distribution is the evaluation of the constants A' and b in Equation (12.3). The constant A' is a *normalization* constant, similar to the normalization constant of quantum physics. It is chosen so that the integral over all speeds gives the total number of molecules (per unit volume) of the box:

$$\int_0^\infty f(v)\, dv = n \tag{12.4}$$

where n is the number of molecules per unit volume. Since we have two constants to determine, we need a second equation, and here we turn to a result obtained from kinetic theory: the average kinetic energy per molecule of a gas in thermal equilibrium at absolute temperature T is $\frac{3}{2}kT$, where k is the Boltzmann constant. Computing the average kinetic energy per molecule is equivalent to finding the average value of $\frac{1}{2}mv^2$ over the distribution of speeds:

$$K_{av} = \frac{1}{n}\int_0^\infty (\tfrac{1}{2}mv^2)f(v)\, dv = \tfrac{3}{2}kT \tag{12.5}$$

Definite integrals of the form $\int_0^\infty x^{2j}e^{-ax^2}\, dx$ are not particularly easy to evaluate, but fortunately their values can be found in many handbooks:

$$\int_0^\infty x^{2j}e^{-ax^2}\, dx = \frac{1\cdot3\cdot5\cdot\;\cdot\;\cdot(2j-1)}{2^{j+1}a^j}\sqrt{\frac{\pi}{a}} \tag{12.6}$$

It is left as an exercise to show that Equations (12.4), (12.5), and (12.6) give the values of the constants:

$$b = \frac{m}{2kT}$$

$$\frac{1}{A'} = n\left(\frac{m}{2\pi kT}\right)^{3/2}$$

so that

$$f(v) = 4\pi n \left(\frac{m}{2\pi kT}\right)^{3/2} v^2 e^{-mv^2/2kT} \tag{12.7}$$

Find the mean value of the speed v using the Maxwell speed distribution.

EXAMPLE 12.1

Since the speeds are distributed according to $f(v)$, the mean speed is

Solution

$$v_m = \frac{1}{n}\int_0^\infty v f(v)\, dv$$

$$= 4\pi\left(\frac{m}{2\pi kT}\right)^{3/2}\int_0^\infty v^3 e^{-\frac{mv^2}{2kT}}\,dv$$

$$= \frac{4\pi}{nA'}\int_0^\infty v^3 e^{-bv^2}\, dv$$

let $x = \frac{mv^2}{2k'T}$ $v = \sqrt{\frac{2kT}{m}}\,x$

The integral can be shown to have the value $1/2b^2$, and so

$$v_m = \frac{4\pi}{2b^2 nA'} = 2\pi\left(\frac{m}{2\pi kT}\right)^{3/2}\left(\frac{2kT}{m}\right)^2$$

$$= \left(\frac{8kT}{\pi m}\right)^{1/2}$$

$dv = \left(\frac{2kT}{m}\right)^{1/2}\frac{1}{2}x^{-1/2}dy$

$V_m = 4\pi\left[\left(\frac{m}{2\pi kT}\right)^{3/2}\left(\frac{2kT}{m}\right)^{3/2}\left(\frac{2kT}{m}\right)^{1/2}\left(\frac{1}{2}\right)\right]$

$\int_0^\infty x^{(3/2-1/2)} e^{-x}\,dx$

$\Gamma(2) = 1$

Since one of the conditions used to evaluate A' and b was that the mean kinetic energy per molecule was $\tfrac{3}{2}kT$, it follows that the mean value of v^2 is

see notes 4/13

$$(v^2)_m = \frac{3kT}{m}$$

(note that this is *not* equal to v_m^2), and thus the *root-mean-square* speed v_{rms} is

$$v_{rms} = \sqrt{(v^2)_m}$$

$$= \left(\frac{3kT}{m}\right)^{1/2}$$

The *most probable* speed v_p can be found from $f(v)$ using standard calculus techniques to find maxima. Its value is $(2kT/m)^{1/2}$. The relationship of v_p, v_m, and v_{rms} is shown in Figure 12.8.

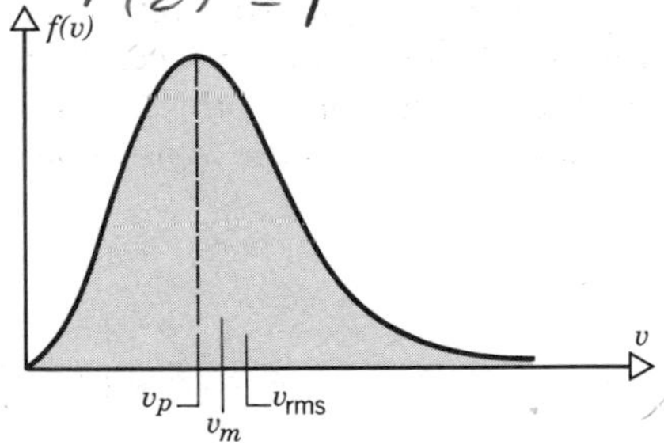

$\frac{d f(v)}{d(v)} = c\left[2v - \frac{mv^3}{kT}\right]e^{-\frac{mv^2}{2kT}} = 0$

FIGURE 12.8 The relationship of the most probable speed v_p, the mean speed v_m, and the root-mean-square speed v_{rms}.

EXAMPLE 12.2

let $v = v_p$

$v_p = \left(2 - \frac{mv_p^2}{kT}\right) = 0$

$v_p = \sqrt{\frac{2kT}{m}}$

Solution

The light that reaches us from distant stars is Doppler shifted not only because the stars are moving relative to us, but also because the atoms of the stars are in rapid thermal motion. Consider a star that happens to be at rest relative to the Earth, and calculate the frequency distribution of the emitted light, assuming that only a single frequency ν_0 is emitted in the rest frame of each atom. (Assume also that the speed of thermal motion is small compared with c.)

The formula for the relativistic Doppler shift is (from Chapter 2):

$$\nu = \nu_0\sqrt{\frac{1 - u/c}{1 + u/c}}$$

Let us rewrite this as

$$\nu = \frac{\nu_0(1 - u/c)}{\sqrt{1 - u^2/c^2}}$$

and assuming $u \ll c$, the factor in the denominator does not differ appreciably from unity. The velocity u, upon which the Doppler shift depends, is the *compo-*

nent v_x of the velocity along the line joining the Earth with the star, and so we must use the Maxwell *velocity* distribution, rather than the *speed* distribution

$$f(v_x)\, dv_x = \frac{1}{A_x}\, e^{-bv_x^2}\, dv_x$$

With

$$\nu = \nu_0\left(1 - \frac{v_x}{c}\right)$$

we have

$$v_x = c\left(1 - \frac{\nu}{\nu_0}\right)$$

and

$$|dv_x| = \frac{c}{\nu_0}\, d\nu$$

so the distribution of observed frequencies is

$$f(\nu)\, d\nu = \frac{1}{A_x}\left(\frac{c}{\nu_0}\right) e^{-bc^2(1-\nu/\nu_0)^2}\, d\nu$$

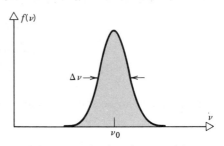

FIGURE 12.9 The distribution of frequencies in a Doppler-broadened spectral line.

This distribution is shown in Figure 12.9. Notice that, since the average value of v_x is zero (recall Figure 12.3) the average value of ν must be ν_0. The effect of the thermal motion of the atoms of the star is not to shift the spectral line, but rather to *broaden* it, and the broadening gives a direct measurement of the temperature. Let us define the broadening to be the range of frequencies $\Delta\nu$ that corresponds to the intensity of the spectral line having half its maximum value; this is called the "full width at half maximum," or FWHM. The only feature of $f(\nu)$ that depends on ν is the exponential, which is unity at $\nu = \nu_0$; $f(\nu)$ thus drops to half of its maximum value when the exponential factor has the value $\frac{1}{2}$:

$$e^{-bc^2(1-\nu/\nu_0)^2} = \frac{1}{2}$$

$$bc^2\left(1 - \frac{\nu}{\nu_0}\right)^2 = \ln 2$$

$$1 - \frac{\nu}{\nu_0} = \pm\sqrt{\frac{\ln 2}{bc^2}}$$

Since ν represents the point at which $f(\nu)$ drops by half, the range $\Delta\nu$ is just $\nu_+ - \nu_-$, where ν_+ and ν_- represent the solutions to the above expression for the positive and negative roots

$$\nu_\pm = \nu_0\left(1 \pm \sqrt{\frac{\ln 2}{bc^2}}\right)$$

$$\Delta\nu = 2\nu_0\sqrt{\frac{\ln 2}{bc^2}}$$

$$= (2\sqrt{\ln 2})\, \nu_0\sqrt{\frac{2kT}{mc^2}}$$

For a star composed of hydrogen atoms, $mc^2 \cong 938$ MeV and we can find

$$\Delta\nu = 7.14 \times 10^{-7}\nu_0\sqrt{T}$$

For a star like the sun, with $T \cong 6000$ K, $\Delta\nu \cong 5.5 \times 10^{-5}\nu_0$. This small width is nevertheless relatively easy to measure with modern spectroscopic equipment, and so the *Doppler broadening* gives us a way to measure the surface temperature of stars.

Study the above example to see how we accomplished a *change of variable* to go from the velocity distribution to a frequency distribution. Notice that there were three steps involved:

1. Find an expression that relates one variable to another.
2. Differentiate the expression to find relationship between the differentials of the two variables.
3. Substitute for the old variable and its differential.

12.4 THE MAXWELL-BOLTZMANN DISTRIBUTION

The Maxwell speed and velocity distributions give us a detailed answer to the problem we posed in our example of the 10 particles in the box. When we add energy (in the form of heat) to a container filled with gas molecules, the distribution of kinetic energies among the individual molecules is given by Equation (12.7), which gives, in effect, the probability of finding any particular value of the kinetic energy as being proportional to $e^{-K/kT}$. Since the kinetic energy was the only form of energy permitted in our previous example, it is of interest to convert Equation (12.7) to an energy distribution; using the methods described at the end of the previous section

$$E = \tfrac{1}{2}mv^2$$

$$dE = mv\,dv$$

$$dv = \frac{dE}{mv} = \frac{dE}{\sqrt{2mE}} \tag{12.8}$$

$$f(E)\,dE = 4\pi n \left(\frac{m}{2\pi kT}\right)^{3/2} \frac{2E}{m} e^{-E/kT} \frac{dE}{\sqrt{2mE}}$$

$$= \frac{2n}{\pi^{1/2}(kT)^{3/2}} E^{1/2} e^{-E/kT}\,dE \tag{12.9}$$

It is now reasonable to ask how this result, which gives the energy distribution when the energy takes the form of translational kinetic energy, can be carried over to other cases in which the energy may take different forms. The normalization factor, for example, was derived from a special assumption regarding the relationship in a gas between temperature and mean molecular kinetic energy—surely this will not be correct for other types of systems. Similarly, the factor $E^{1/2}$ in Equation (12.9) came from the calculation of the relationship between dE and the "volume" element $dv_x\,dv_y\,dv_z$ of the original velocity distribution; this will likewise not be correct in other situations. The remaining exponential factor $e^{-E/kT}$ turns out to be the only characteristic of the velocity and speed distributions that is true in general, no matter what form the energy takes. The *Maxwell-Boltzmann distribution function* is just the exponential factor with its normalization constant:

$$f_{MB}(E) = A^{-1} e^{-E/kT} \tag{12.10}$$

Maxwell-Boltzmann distribution

According to the Maxwell-Boltzmann distribution, the relative number of particles having a particular value of the energy E is:

$$p(E) = g(E) f_{MB}(E) \tag{12.11}$$

The normalization constant A in general can depend on many factors, such as on density and temperature in the case of the molecular gas described by Equation (12.9). The function $g(E)$, known as the *density of states* factor, describes the number of possible ways the system can have a certain value of E. In the case of the molecular gas, the density of states factor comes from the "volume" element $dv_x\, dv_y\, dv_z$, which we modified to the "volume" of the spherical shell $4\pi v^2\, dv$, which is proportional to $E^{1/2}\, dE$, giving $g(E) \propto E^{1/2}$ for the previous case. When we discuss discrete energy levels, such as those of an atomic system, $g(E)$ will be proportional to the *degeneracy* of the atomic levels; recall that the degeneracy gives the number of different ways an atomic system can have the same value of E, which is exactly how we defined $g(E)$.

EXAMPLE 12.3

(a) In a gas of atomic hydrogen at room temperature, what is the relative population of the first excited state at $E = 10.2$ eV? How much hydrogen would be required in order to have a reasonable probability of finding one atom in the first excited state?

(b) At what temperature would we expect to find 10.0 percent of the atoms in the first excited state?

Solution

(a) We recall from our study of atomic hydrogen that the degeneracy of the atomic levels is $2n^2$. Thus $g = 2$ for the ground state $(n = 1)$ and $g = 8$ for the first excited state $(n = 2)$. The constant A is the same for the two states, and so

$$\frac{p(E_1)}{p(E_2)} = \frac{g(E_1)}{g(E_2)} e^{-(E_1 - E_2)/kT}$$

At room temperature $(T = 293$ K$)$, $kT = 0.0252$ eV; thus

$$\frac{p(E_1)}{p(E_2)} = \frac{8}{2} e^{-10.2\,\text{eV}/0.0252\,\text{eV}}$$

$$= 4e^{-405}$$

$$= 0.6 \times 10^{-175}$$

In order to have one atom in the excited state, we therefore require about 1.7×10^{175} atoms of hydrogen, about 3×10^{148} kg, a quantity greater than the mass of the universe!

(b) We now require $p(E_1)/p(E_2) = 0.100/0.900$ and solve for T:

$$0.111 = 4e^{-10.2\,\text{eV}/kT}$$

$$kT = 2.85\text{ eV}$$

$$T = 3.30 \times 10^4\text{ K}$$

EXAMPLE 12.4

A certain atom with total atomic spin $\frac{1}{2}$ has a magnetic moment μ. A collection of such atoms is placed in a magnetic field of strength B. (a) What is the ratio, at temperature T, of the number of atoms with their spins aligned along the field to those with their spins aligned opposite to the field? (b) By making reasonable estimates for μ and B, estimate the temperature at which this ratio becomes 1.1, that is, at which the difference between the numbers parallel and antiparallel to the field becomes 10 percent.

(a) The energy of interaction with the magnetic field is $E = -\boldsymbol{\mu}\cdot\mathbf{B}$ and the degeneracies of the states with $m_s = +\frac{1}{2}$ and $m_s = -\frac{1}{2}$ are identical, so $g(E_1) = g(E_2)$. Therefore we have:

Solution

$$\frac{p(E_1)}{p(E_2)} = e^{-(E_1-E_2)/kT}$$

The energy of those aligned parallel to the field is $E_1 = -\mu B$, while those aligned antiparallel have energy $E_2 = +\mu B$, and thus

$$\frac{p(E_1)}{p(E_2)} = e^{2\mu B/kT}$$

(b) A typical atom has $\mu \cong \mu_B$ (one Bohr magneton) and the largest magnetic fields that can be produced in the laboratory are of order 10 T, and so

$$\mu B \cong \mu_B B = 9.27 \times 10^{-24} \text{ J/T} \times 10 \text{ T}$$
$$= 9.27 \times 10^{-23} \text{ J}$$
$$= 5.79 \times 10^{-4} \text{ eV}$$

When $p(E_1)/p(E_2) = 1.1$,

$$\frac{2\mu B}{kT} = \ln 1.1 = 9.53 \times 10^{-2}$$
$$kT = \frac{2(5.79 \times 10^{-4} \text{ eV})}{9.53 \times 10^{-2}} = 1.22 \times 10^{-2} \text{ eV}$$
$$T = 141 \text{ K}$$

This temperature is only a factor of 2 below room temperature, which suggests that atomic magnetic effects become important at temperatures of the order of room temperature and that the source of ordinary magnetic effects is atomic magnetism. Since nuclear magnetic moments are about 1000 times smaller than atomic magnetic moments, they should become important at temperatures 1000 times smaller.

12.5 QUANTUM STATISTICS

As we discussed in Section 12.2, the distribution functions for the indistinguishable particles of quantum physics are different from those of classical physics. Because of the unusual behavior of quantum systems, we must have separate distribution functions for those particles, like the electron, that obey the Pauli exclusion principle, and those particles that do not. We will not derive these distribution functions, but merely state them and discuss some of their properties.

Particles that do not obey the Pauli principle are those with integral spins (0, 1, 2, . . . , in units of $\hbar$) and are known collectively as *bosons*. The distribution function for bosons is known as the *Bose-Einstein distribution* and has the form

$$f_{BE}(E) = \frac{1}{Ae^{E/kT} - 1} \qquad (12.12)$$

Bose-Einstein distribution

Note the similarity to the Maxwell-Boltzmann distribution function. In particular, for energies not too different from kT, when A is a large number, the Bose-Einstein distribution reduces directly to the Maxwell-Boltzmann distribution. Since A depends *inversely* on the density, this means that the Maxwell-Boltzmann distribution will be a good approximation at low densities, such as for gases, while for liquids and solids of higher densities we expect the quantum statistical distributions to be necessary.

Particles of half-integral spin ($\frac{1}{2}$, $\frac{3}{2}$, . . .) which obey the Pauli principle, like electrons or nucleons, are known as *fermions*, and the correct distribution function is the *Fermi-Dirac distribution:*

$$f_{FD}(E) = \frac{1}{Ae^{E/kT} + 1} \tag{12.13}$$

Fermi-Dirac distribution

How the minor change in sign in the denominator between f_{BE} and f_{FD} gives such a radical change in the form of the distribution function is not immediately obvious, and to show the differences we need to know more about the normalization coefficient A, which is not a constant but depends on T. For the Bose-Einstein distribution, in most cases of practical interest A is either independent of T or depends so weakly on T that the exponential term $e^{E/kT}$ dominates. However, for the Fermi-Dirac distribution, A is strongly dependent on T, and the dependence is usually approximately exponential, so A is written as

$$A = e^{-E_F/kT} \tag{12.14}$$

and the Fermi-Dirac distribution becomes

$$f_{FD}(E) = \frac{1}{e^{(E-E_F)/kT} + 1} \tag{12.15}$$

where E_F is called the *Fermi energy*. (Although the Fermi energy itself may depend on temperature, that dependence is quite weak, and we can regard E_F as constant.)

Let us look qualitatively at the differences between f_{BE} and f_{FD} at low temperatures. For the Bose-Einstein distribution, assuming for the moment $A = 1$, in the limit of small T the exponential factor becomes large for large E, and so $f_{BE} \to 0$ for states with large energies. The only energy levels that have any real chance of being populated are those with $E \cong 0$, for which the exponential factor approaches 1, the denominator becomes very small, and $f_{BE} \to \infty$. Thus when T is small, all of the particles in the system try to occupy the lowest energy state. This effect is known as "condensation" and we will see how it has some rather unexpected consequences.

Such a situation is not possible for fermions, such as electrons. We know that the electrons in an atom, for example, do not all occupy the lowest energy state, no matter what the temperature. Let us see how the Fermi-Dirac distribution function prevents this. The exponential factor in the denominator of f_{FD} is $e^{(E-E_F)/kT}$. For values of $E > E_F$, when T is small the exponential factor becomes large and f_{FD} goes to zero, just like f_{BE}. When $E < E_F$, however, the story is very different, for then $E - E_F$ is negative, and $e^{(E-E_F)/kT}$ goes to zero for small T, and $f_{FD} \cong 1$. *The occupation probability is therefore only one per quantum state,* just as required by the Pauli principle. Even at very low temperatures, fermions do not "condense" into the lowest energy level.

In Figures 12.10 to 12.12, the three distributions f_{MB}, f_{FD}, and f_{BE} are plotted as functions of the energy E. You can see, by comparing these figures, that all of the distribution functions fall to zero at large values of E; when $E \gg kT$, the occupation probability is very small, as we calcu-

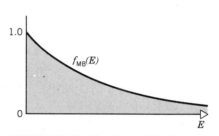

FIGURE 12.10 The Maxwell-Boltzmann distribution function.

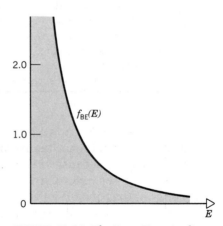

FIGURE 12.11 The Bose-Einstein distribution function.

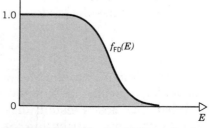

FIGURE 12.12 The Fermi-Dirac distribution function

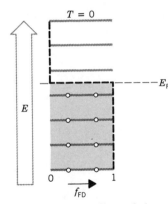

FIGURE 12.13 Filling of electronic levels according to the Fermi-Dirac distribution at $T = 0$.

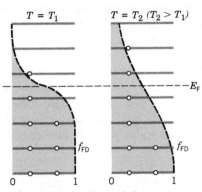

FIGURE 12.14 Filling of electronic levels according to the Fermi-Dirac distribution at $T > 0$.

lated for f_{MB} in the case of the first excited state of the hydrogen atom in Example 12.3. Notice also that, even though f_{MB} becomes large for small E, it remains finite. The Bose-Einstein distribution, f_{BE}, on the other hand, becomes infinite as $E \to 0$; this is the "condensation" effect referred to earlier, in which all of the particles try to occupy the lowest quantum state. You can see that f_{FD} never becomes larger than 1.0, just as we expect for particles that obey the Pauli principle. At $T = 0$, all energy levels up to E_F are occupied ($f_{FD} = 1.0$) and all energy levels above E_F are empty ($f_{FD} = 0$). Figure 12.13 shows a hypothetical set of energy levels and how they would be populated at $T = 0$ by a particle such as an electron, for which $g(E) = 2$. As T increases, some levels above E_F are partially occupied ($f_{FD} > 0$), while some levels below E_F are partially empty ($f_{FD} < 1$). The higher the temperature, the more "spread out" f_{FD} becomes. Figure 12.14 shows how the energy levels of a system might be populated at two different temperatures, T_1 and T_2.

From Equation (12.15) you can see that, when $E = E_F$, f_{FD} is $\frac{1}{2}$, independent of T. Thus an alternative definition of E_F is that point at which the occupation probability is exactly 0.5. The Fermi energy varies only slightly with temperature for most materials, and we can usually regard it as constant for many applications. As we will see in Section 12.7, for electrons in a metal E_F depends on the electron density of the material, which doesn't change much with temperature. For some materials, notably semiconductors, the density of *conduction* electrons can change significantly with temperature, and thus E_F in these materials is temperature dependent.

Blackbody Radiation As we did in our discussion of cavity radiation in Chapter 3, we treat the cavity as a box filled with electromagnetic radiation, but now we consider the radiation to be in the form of photons. For this calculation, we assume the box to be filled with a "gas" of photons, and since photons have spin 1, they are bosons and obey Bose-Einstein statistics. The parameter A of the Bose-Einstein distribution is set equal to unity so that it does not enter into the calculation; the justification for this step is that the purpose of the parameter A is to specify the total number of particles or the density of particles, but

12.6 APPLICATIONS OF BOSE-EINSTEIN STATISTICS

since photons are being absorbed by the walls or emitted by the walls at all times, the number of photons is not fixed. Setting the parameter A equal to 1 is the mathematical way of asserting that the total number of particles is not fixed.

The difficult part of this calculation is the density of states function $g(E)$. In the case of the ordinary molecular gas, you will recall we calculated $g(v)$ by considering the three-dimensional coordinate system v_x, v_y, v_z, and the "volume" of the spherical shell for this case was $4\pi v^2 \, dv$. Since all photons move with the speed of light, this process makes no sense for photons.

Instead, we recall a similar problem we discussed in Chapter 5, where we considered the allowed energies of a particle confined to a region of space in one, two, or three dimensions. In that discussion we learned that requiring $\psi = 0$ at the boundaries resulted in the wave numbers k_x, k_y, and k_z taking a certain set of discrete values $\pi n_x/L$, $\pi n_y/L$, and $\pi n_z/L$, respectively. The case for electromagnetic waves in a box is similar; if the walls are made of a conducting medium, then $E = 0$ is the boundary condition for the electric field and the same set of discrete wave numbers results for the electromagnetic waves. The quantization of wave number is equivalent to the quantization of momentum components, since $p_x = \hbar k_x$, $p_y = \hbar k_y$, and $p_z = \hbar k_z$. Thus

$$
\begin{aligned}
p &= \sqrt{p_x^2 + p_y^2 + p_z^2} \\
&= \hbar \sqrt{k_x^2 + k_y^2 + k_z^2} \\
&= \frac{\hbar \pi}{L} \sqrt{n_x^2 + n_y^2 + n_z^2}
\end{aligned}
$$

(12.16)

For photons, $E = pc$, so

$$
E = \frac{\hbar c \pi}{L} \sqrt{n_x^2 + n_y^2 + n_z^2}
$$

(12.17)

This expression gives the discrete values of E that are permitted; in order to find the density of states $g(E)$ we need to know how many of these discrete values occur between E and $E + dE$. As we did in the case of velocities in the Maxwell distribution, this calculation is easiest if we imagine a coordinate system, in which the axes are labelled n_x, n_y, n_z. Of course, these are not really continuous variables, since they may take only integer values; furthermore, only positive integers are allowed, since n_x, n_y, and n_z count the nodes in the electric field. A spherical shell in this coordinate system has a "radius" $n = \sqrt{n_x^2 + n_y^2 + n_z^2}$ and a "volume" $4\pi n^2 \, dn$. The number of allowed energy values is only one-eight of this, because the positive values of n_x, n_y, and n_z occupy only one-eighth of the coordinate space. Finally, for each value of n, there are two distinct waves, corresponding to the two possible values of the polarization of the wave. The allowed number of states is therefore

$$
g(n) \, dn = 2 \times \tfrac{1}{8} \times 4\pi n^2 \, dn
$$

(12.18)

and since

$$
E = \frac{\hbar c \pi}{L} n
$$

(12.19)

we find

$$g(E)\,dE = \pi\left(\frac{L}{\hbar c\pi}\right)^3 E^2\,dE \tag{12.20}$$

The number of photons having energy in the range E to $E + dE$ is, according to the Bose-Einstein distribution,

$$p(E)\,dE = g(E)f_{BE}\,dE \tag{12.21}$$

$$p(E)\,dE = \pi\left(\frac{L}{\hbar c\pi}\right)^3 E^2 \frac{1}{e^{E/kT} - 1}\,dE \tag{12.22}$$

The total radiant energy carried by photons of energy E is just $E\,p(E)$ and the energy density (energy per unit volume) of photons in the range E to $E + dE$ is

$$u(E)\,dE = \frac{E\,p(E)\,dE}{L^3} \tag{12.23}$$

Substituting, and converting from photon energy E to wavelength, we find

$$u(\lambda)\,d\lambda = \frac{8\pi hc}{\lambda^5}\frac{1}{e^{hc/\lambda kT} - 1}\,d\lambda \tag{12.24}$$ **Planck blackbody formula**

Multiplying by $c/4$, as is necessary when we go from the energy density of the radiation to the spectral radiancy, we find the result that was given as Equation (3.26).

Thus the Planck theory of blackbody radiation, which was so successful in accounting for experimental results, follows directly from the Bose-Einstein distribution for photons.

Einstein Theory of Specific Heats   In an ordinary solid, most of the physical properties originate either with the valence electrons or with the latticework of atoms. Electrical conductivity, for example, originates with the valence electrons, while the propagation of mechanical waves is due to the lattice of atoms. The specific heats of solids have contributions from both lattice and electrons; at all but the lowest temperatures, the lattice contribution is dominant and the electronic contribution can therefore be neglected.

When we add heat to a solid, we increase the thermal motion of the atoms; the difference between a solid and a gas is that in the solid the atoms are fixed to equilibrium positions, and their motion consists of oscillations about the equilibrium position. For a one-dimensional oscillator, we count two degrees of freedom (one for its potential energy and one for kinetic energy); according to the equipartition theorem there should be an energy of $\frac{1}{2}kT$ per degree of freedom, and thus the total energy of the *three-dimensional* solid should be $3kT$ ($\frac{1}{2}kT$ per degree of freedom $\times$ 2 degrees of freedom per dimension $\times$ 3 dimensions) for each atom. Since a mole has Avogadro's number of atoms, the total energy per mole should be

$$\begin{aligned} E &= 3N_A kT \\ &= 3RT \end{aligned} \tag{12.25}$$

Table 12.2 Specific Heats of Common Metals

	Specific Heat			
	Room Temp (300 K)		Low Temp (25 K)	
Metal	J/g · K	J/mole · K	J/g · K	J/mole · K
Al	0.904	24.4	0.0175	0.473
Ag	0.235	25.3	0.0287	3.10
Au	0.129	25.4	0.0263	5.18
Cu	0.387	24.6	0.016	1.0
Fe	0.450	25.1	0.0075	0.42
Pb	0.128	26.5	0.0681	14.1
Sn	0.222	26.3	0.058	6.9

where $N_A k = R$, the universal gas constant. The specific heat C is the energy change (heat energy) per unit change in temperature $\Delta E / \Delta T$, and thus we expect

$$C = 3R \tag{12.26}$$

This classical result is known as the *law of Dulong and Petit* and predicts that the molar specific heat should take the constant value $3R$ (24.9 J/mole·K), independent of the type of material or the temperature.

The values in Table 12.2 show that this simple result gives quite a remarkable agreement with the experimental values at ordinary temperatures. However, at low temperatures, the specific heat begins to deviate considerably from the classical value and tends to zero as T goes to zero.

The solution to this failure of classical physics was first given by Einstein, who assumed that the *oscillations* (not the atoms) of the solid obeyed Bose-Einstein statistics. Just as electromagnetic waves are analyzed as "particles" (quanta of electromagnetic energy, or photons) that obey Bose-Einstein statistics, so are mechanical or acoustic waves analyzed as "particles" (quanta of vibrational energy, called *phonons*) that also obey Bose-Einstein statistics. Einstein made the simplifying assumption that all of the phonons (oscillations) had the same frequency, and so there is no density of states factor to worry about. We have seen in Chapter 5 that a quantized oscillator has an energy of $\hbar\omega(n + \frac{1}{2})$. Each additional value of n represents an additional phonon; to go from a vibrational energy of $\frac{5}{2}\hbar\omega$ to $\frac{7}{2}\hbar\omega$ we must "create" a phonon of energy $\hbar\omega$. One mole of the solid will contain N_A atoms with $3N_A$ phonons of oscillation, each with energy $\hbar\omega$; the probability distribution for the phonons is just f_{BE}, and so the energy per mole will be

$$E = 3N_A \hbar\omega \frac{1}{e^{\hbar\omega/kT} - 1} \tag{12.27}$$

The specific heat can be found from dE/dT:

$$C = \frac{dE}{dT} = 3N_A\hbar\omega \frac{(e^{\hbar\omega/kT})(\hbar\omega/kT^2)}{(e^{\hbar\omega/kT} - 1)^2} \tag{12.28}$$

$$C = 3R \left(\frac{\hbar\omega}{kT}\right)^2 \frac{e^{\hbar\omega/kT}}{(e^{\hbar\omega/kT} - 1)^2} \tag{12.29}$$

Einstein formula for specific heat

Where T is small, the exponential term dominates, and $C \propto e^{-\hbar\omega/kT}$, so indeed C approaches 0 for small T, in agreement with experiment. Figure

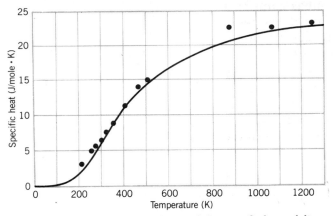

FIGURE 12.15 The Einstein theory of the specific heat of diamond. The dots are the data points and the solid curve is Equation (12.29) with $\hbar\omega = 0.114$ eV.

12.15 compares the experimental values of C with the behavior predicted by Equation (12.29). As you can see, the agreement is reasonably good. (The vibrational energy $\hbar\omega$ is an adjustable parameter of the theory and will have different values for different materials.)

In this calculation we have oversimplified by assuming all of the phonons to have the same frequency. The proper calculation, which was first done by Debye, assumes a distribution of phonon frequencies with a density of states given by an expression of the same form as that for the "photon gas" of blackbody radiation; the predicted low temperature behavior is then $C \propto T^3$, in much better agreement with experiment than the result of the Einstein calculation. Nevertheless, even our simple calculation shows that the correct behavior of C can be found merely by applying the proper sort of statistics to the quantized vibrations of a solid.

Liquid Helium   One of the most remarkable substances we can study in the laboratory is liquid helium. Here are some of its properties:

1. Helium gas is the most inert of the inert gases. It forms absolutely no compounds and has the lowest boiling point, 4.18 K, of any material.

2. Below 4.18 K, helium behaves much like an ordinary liquid. As the helium boils, the escaping gas forms bubbles, like a boiling pot of water. As the liquid is cooled further, a sudden transition occurs at a temperature of 2.18 K, the violent boiling stops, and the liquid becomes absolutely still. (Evaporation continues, but only from the surface.)

3. As the liquid is cooled below 2.18 K, the specific heat and the thermal conductivity both increase suddenly and discontinuously. Figure 12.16 shows the specific heat as a function of temperature. The form of the figure looks rather like the Greek letter λ, and so the transition point at 2.18 K has become known as the *lambda point*. The thermal conductivity rises at the λ point by a factor of perhaps 10^6.

4. When a liquid flows through a narrow capillary tube, its *viscosity* causes some resistance to the flow. When liquid helium flows

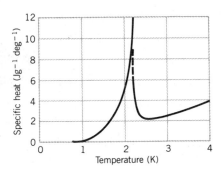

FIGURE 12.16 The specific heat of liquid helium. The discontinuity at 2.18 K is the lambda point.

through a capillary tube, its viscosity drops at the lambda point by a factor of perhaps 10^6. Below the lambda point helium flows easily through narrow capillary tubes (or other narrow constrictions), through which it would not flow above the lambda point.

5. Below the lambda point, liquid helium has the power to seemingly defy gravity, flowing up and over the walls of its container. The helium forms a thin film which lines the walls of the container, the remaining liquid is then drawn up by the film like a siphon, and the helium can be seen dripping from the bottom of the container, as in the photograph of Figure 12.17.

All of these strange properties occur because liquid helium follows Bose-Einstein statistics. Ordinary helium has two electrons filling the $1s$ shell, so the total angular momentum of the electrons is zero. It happens also that the two protons and two neutrons are paired off in the nucleus in a similar way, so that the nucleus also has a spin of zero. Therefore the total spin of the atom (electron spin + nuclear spin) is zero, and a helium atom behaves like a boson. At 2.18 K, a *change of phase* occurs in the helium liquid. Above the lambda point, helium behaves like an ordinary liquid; this phase is called He I. The phase that occurs below the lambda point is called He II, and its special properties are due to its *superfluid* component. As the temperature is decreased from the lambda point toward absolute zero, the relative concentration

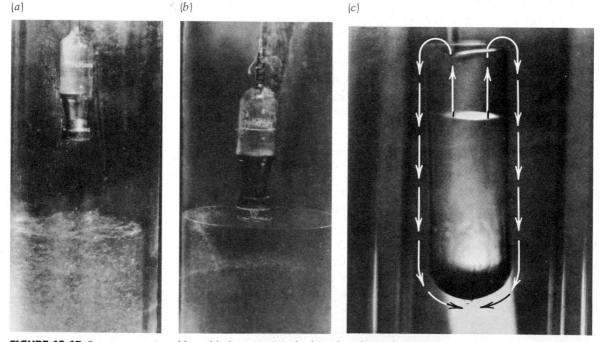

(a)   (b)   (c)

FIGURE 12.17 Some properties of liquid helium. In (a), the liquid is shown boiling violently at a temperature above 2.18 K. Liquid helium is also present in the small container, held in by a porous plug at the bottom. In (b), the helium has been cooled below 2.18 K. The boiling stops, and the viscosity essentially vanishes, allowing the helium to leak out of the small container through the very fine pores of the plug. In (c) the liquid helium escapes from the container by flowing up and over the walls and can be seen dripping from the bottom. [*Source:* Mendelssohn, *The Quest for Absolute Zero.*]

of the normal fluid decreases and that of the superfluid increases. The unusual properties of liquid helium are all caused by the superfluid component, which is also known as a *quantum liquid*. Because the helium atoms obey Bose-Einstein statistics, the Pauli principle does not prevent all of the atoms from being in the same quantum state. This begins to happen at the lambda point, and is an example of the effect referred to as Bose condensation. We can think of the superfluid as being a single quantum state made up of a very large number of atoms; the atoms behave in a cooperative way, giving the superfluid its unusual properties.

By way of comparison, if we try the same kinds of experiments with the rarer isotope of helium, ^{3}He, the behavior is very different. Although ^{3}He has zero atomic spin, just like ^{4}He, it has only three particles rather than four in its nucleus, and its *nuclear* spin is $\frac{1}{2}$. The total atomic (electronic + nuclear) spin is therefore $\frac{1}{2}$, and ^{3}He behaves like a fermion and obeys the Fermi-Dirac distribution. Since the Pauli principle prevents more than one fermion from occupying any quantum state, no Bose condensation is expected for ^{3}He, and indeed none is observed until ^{3}He is cooled to about 0.002 K. At this point the weak coupling of two ^{3}He to form a boson occurs, and the ^{3}He pairs can experience Bose condensation. (A similar effect is responsible for superconductivity; see Section 14.6).

12.7 APPLICATION OF FERMI-DIRAC STATISTICS

In a metal, the valence electrons are not very strongly bound to individual atoms, and consequently travel rather freely throughout the volume of the metal. We can treat these electrons as a "gas" that obeys the Fermi-Dirac distribution. The density of states function can be derived in a way similar to that for the photon gas. (Even the factor of 2, to account for the two different spin states, is the same.) The only difference is that, rather than using $E = pc$, we use the appropriate relationship for nonrelativistic electrons, $E = p^2/2m$. After some manipulations, we find

$$g(E)\,dE = \frac{8\sqrt{2}\pi L^3 m^{3/2}}{h^3}\,E^{1/2}\,dE \qquad (12.30)$$

and therefore the number of electrons with energy between E and $E + dE$ is

$$p(E)\,dE = \frac{8\sqrt{2}\pi L^3 m^{3/2}}{h^3}\,\frac{E^{1/2}\,dE}{e^{(E-E_F)/kT}+1} \qquad (12.31)$$

This function is plotted in Figure 12.18. You can see immediately why we neglected the influence of the electrons when we computed the specific heat of a metal and considered only the lattice contribution. When we add heat to a metal, only a very few of the electrons near E_F absorb the energy and change their state of motion; the vast majority of the electrons are not affected, and thus the influence of the electrons on the specific heat is indeed negligible.

We can find a numerical value for E_F by assuming that the sample

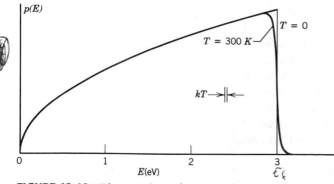

FIGURE 12.18 The number of occupied electronic levels at $T = 0$ and $T = 300$ K, according to the Fermi-Dirac distribution. The Fermi energy E_F is chosen to be 3.0 eV.

contains a total number N of these free electrons:

$$N = \int_0^\infty p(E)\, dE \qquad (12.32)$$

At $T = 0$, the Fermi-Dirac distribution function has the value 1 for $E < E_F$ and 0 for $E > E_F$, so the integral reduces to

$$N = \frac{8\sqrt{2}\pi L^3 m^{3/2}}{h^3} \int_0^{E_F} E^{1/2}\, dE \qquad (12.33)$$

and after some manipulation we find

$$E_F = \frac{h^2}{2m}\left(\frac{3N}{8\pi V}\right)^{2/3} \qquad (12.34) \qquad \textbf{Fermi energy at } \boldsymbol{T = 0}$$

where $V = L^3$, the volume of the metal.

We can also find the average energy of the electrons

$$E_{av} = \frac{1}{N}\int_0^\infty E p(E)\, dE \qquad (12.35)$$

and it is left as an exercise for you to show that

$$E_{av} = \tfrac{3}{5}E_F \qquad (12.36)$$

Compute the Fermi energy E_F for sodium.   **EXAMPLE 12.5**

Each sodium atom contributes one valence electron to the metal, and so the   ***Solution***
number of electrons per unit volume, N/V, is equal to the number of sodium
atoms per unit volume. This can in turn be found from the density of sodium
and the mass of a sodium atom:

$$\frac{N}{V} = \text{atoms per unit volume} = \frac{\rho N_A}{M}$$

$$= 0.971 \text{ g/cm}^3 \cdot \frac{6.02 \times 10^{23} \text{ atoms/mole}}{23.0 \text{ g/mole}}$$

$$= 2.54 \times 10^{22} \text{ cm}^{-3}$$

$$= 2.54 \times 10^{28} \text{ m}^{-3}$$

$$E_F = \frac{h^2}{2m}\left(\frac{3}{8\pi}\frac{N}{V}\right)^{2/3}$$

$$= \frac{(hc)^2}{2mc^2}\left(\frac{3}{8\pi}\cdot 2.54 \times 10^{28} \text{ m}^{-3}\right)^{2/3}$$

$$= \frac{(1240 \text{ eV·nm})^2}{2(0.511 \times 10^6 \text{ eV})}\,(2.09 \times 10^{18} \text{ m}^{-2})\left(\frac{10^{-9}\text{m}}{\text{nm}}\right)^2$$

$$= 3.15 \text{ eV}$$

The average energy of the valence electrons is $\frac{3}{5}E_F$ or 1.89 eV. *Even at the absolute zero of temperature,* the electrons still have quite a large average energy.

From Figure 12.18 we see that the change in $p(E)$ between $T = 0$ and $T = 300$ K (room temperature) is relatively small, and so these values for E_F and E_{av} are approximately correct at room temperature.

The meaning of these numbers is as follows. Instead of isolated atoms with individual energy levels, we consider the metal to be a single system with a very large number of energy levels (at least as far as the valence electrons are concerned). Electrons fill these energy levels, in accordance with the Pauli principle, beginning at $E = 0$. By the time we add 2.54×10^{22} valence electrons to 1 cm³ of sodium, we have filled energy levels up to $E_F = 3.15$ eV; all levels below E_F are filled and all levels above E_F are empty. Electrons have an almost continuous energy distribution (the levels are discrete, but they are very close together) from $E = 0$ to $E = E_F$, with an average energy of 1.89 eV. At $T = 300$ K, a relatively small number of electrons is excited from below E_F to above E_F; the range over which electrons are excited is of order $kT \cong 0.025$ eV, so that only electrons within 0.025 eV of E_F are affected by the change from $T = 0$ to $T = 300$ K.

The change in the number of occupied electron states from $T = 0$ to $T = 300$ K is represented in greater detail by the shaded area of Figure 12.19. If we approximate that area by triangle ABC, the number of electrons n_{ex} excited from below E_F to above E_F is approximately $\frac{1}{2}(AB)(BC)$. The height AB is just $\frac{1}{2}p(E_F)$; since $f_{FD}(E_F) = \frac{1}{2}$, it follows that $p(E_F) = \frac{1}{2}g(E_F)$. On the scale of Figure 12.19, the base BC of the triangle is about $3kT$ so

$$n_{ex} \cong \frac{1}{2} \times \frac{g(E_F)}{4} \times 3kT \qquad (12.37)$$

Combining Equations (12.30) and (12.34), we find

$$g(E_F) = \tfrac{3}{2}N\frac{1}{E_F} \qquad (12.38)$$

and therefore the fraction of electrons that changes state when the metal is heated from $T = 0$ to a temperature T is about

$$\frac{n_{ex}}{N} \cong \frac{9}{16}\frac{kT}{E_F} \qquad (12.39)$$

At room temperature $kT \cong 0.025$ eV, and since E_F has values of several eV, the fraction of electrons that are excited is less than 1 percent. Assuming those electrons to contribute to the specific heat like an ideal

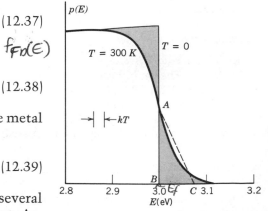

$f_{FD}(E)$

FIGURE 12.19 Detail of Figure 12.18 near E_F.

gas $(C_v = \frac{3}{2}R)$, the net electronic contribution to the specific heat is of order $0.01R$, and is therefore negligible at ordinary temperatures compared with the lattice contribution of $3R$.

At what temperatures would we expect the electronic contribution to the specific heat to become important? Since the fraction of electrons that contribute to C depends on T, by raising T we should be able to increase this fraction. If we increase the temperature until $kT \cong E_F$, we should notice the electrons contributing to the specific heat, but this requires temperatures of the order of 10^4 K, well beyond the boiling point of metals! If we go down to very low temperatures, we find that the electronic contribution to the specific heat is still proportional to T, but the lattice contribution, as discussed in the previous section, is proportional to T^3. At sufficiently low temperatures, the lattice contribution can become smaller than the electronic contribution. At very low temperatures, we do in fact observe $C \propto T$, rather than $C \propto T^3$, thus verifying the contribution of the electrons to the specific heat.

Although the Fermi-Dirac distribution is an accurate and realistic description of the behavior of fermions, we have oversimplified by our use of the free-electron theory. Electrons are not really free, even in conductors. Even though the valence electrons move relatively freely, they see a periodic potential caused by the atoms of the lattice, and this makes significant changes in the density of states function $g(E)$. These changes, and their effect on the behavior of metals, insulators, and semiconductors, will be discussed in Chapter 14.

SUGGESTIONS FOR FURTHER READING

An excellent introduction to statistical physics, including examples of arranging and sorting problems similar to the one in Section 12.2:

D. McLachlan, Jr., *Statistical Mechanical Analogies* (Englewood Cliffs, Prentice-Hall, 1968).

More extensive treatments of statistical physics, at about the same mathematical level as this chapter:

N. Ashby and S. C. Miller, *Principles of Modern Physics* (San Francisco, Holden-Day, 1970). Chapter 2 discusses probability, and Chapters 10 and 11 cover classical and quantum statistics.
F. Reif, *Statistical Physics, Volume 5 of the Berkeley Physics Series* (New York, McGraw-Hill, 1967). Covers only classical statistics.

Advanced texts, with many applications of classical and quantum statistics:

C. Kittel, *Thermal Physics* (New York, Wiley, 1969).
P. M. Morse, *Thermal Physics* (New York, Benjamin, 1965).
R. D. Reed and R. R. Roy, *Statistical Physics for Students of Science and Engineering* (Scranton, Intext, 1971).

The properties of liquid helium are discussed in a very general and highly readable book:

K. Mendelssohn, *The Quest for Absolute Zero* (New York, McGraw-Hill, 1966).

For more recent work on the properties of ^{3}He, see:

N. D. Mermin and D. M. Lee, "Superfluid Helium 3," *Scientific American 235*, 56 (December 1976).

1. Suppose a container filled with a gas moves with speed v. How is the Maxwell velocity distribution different for such a gas, compared with the same container of gas at rest?

2. The population inversion necessary for the operation of a laser is sometimes called a "negative temperature." What is the meaning of a negative temperature? Does it have a physical interpretation?

3. The following figures show two different experimental arrangements used to measure the distribution of molecular speeds. Based on the figures, explain how each apparatus might operate, and try to guess at how the observed distribution of molecules might appear. Where do the fastest molecules land? The slowest?

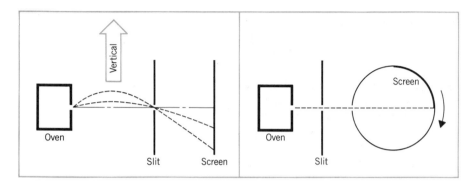

4. It is generally more convenient, wherever possible, to use Maxwell-Boltzmann statistics rather than quantum statistics. Under what circumstances can a quantum system be described by Maxwell-Boltzmann statistics?

5. Suppose we had a gas of hydrogen *atoms* at relatively high density. Do the atoms behave as fermions or as bosons? Would a gas of deuterium atoms behave any differently?

6. The early universe contained a large density of neutrinos. Which statistical distribution would be needed to describe the properties of the neutrinos?

7. Nuclei are often described in terms of a "sea" of pi mesons, which are continuously exchanged between the nucleons. What type of statistics do the pi mesons obey? How would nuclei be different if, instead of pi mesons, the nucleons exchanged pairs of particles with spin $\frac{1}{2}$?

8. If we were to regard a nucleus as a large collection of quarks, which statistical distribution function would describe the quarks?

9. Estimate the mean kinetic energy of the "free" electrons in a metal if they obeyed Maxwell-Boltzmann statistics. How does this compare with the result of applying Fermi-Dirac statistics? Why is there such a difference?

1. Carry through the calculation of the constants b and A' for the Maxwell speed distribution, using Equations (12.4), (12.5), and (12.6).

2. Show that the most probable speed v_p of the Maxwell speed distribution is $(2kT/m)^{1/2}$.

3. For a gas that obeys the Maxwell velocity distribution, find $\Delta v = \sqrt{(v^2)_{av} - (v_{av})^2}$. What is the meaning of this quantity?

4. Equation (12.9) gives the distribution of kinetic energies in a gas. Find:
   (a) The most probable kinetic energy K_p.
   (b) The mean kinetic energy, $K_m = n^{-1}\int_0^\infty Kf(K)dK$.

(c) The rms kinetic energy, $K_{rms} = [n^{-1}\int_0^\infty K^2 f(K)dK]^{1/2}$.

(d) The variation in the kinetic energy, $\Delta K = [(K^2)_{av} - (K_{av})^2]^{1/2}$. Interpret each of these quantities.

5. Consider a collection of N noninteracting atoms with a single excited state at energy E. Assume the atoms obey Maxwell-Boltzmann statistics, and take both the ground state and the excited state to be nondegenerate. (a) At temperature T, what is the ratio of the number of atoms in the excited state to the number in the ground state? (b) What is the average energy of an atom in this system? (c) What is the total energy of the system? (d) What is the specific heat of this system?

6. Suppose we have a gas in thermal equilibrium at temperature T. Each molecule of the gas has mass m. (a) What is the ratio of the number of molecules at the Earth's surface to the number at height h (with potential energy mgh)? (b) What is the ratio of the density of the gas at height h to the density at the surface? (c) Would you expect this simple model to give an adequate description of the Earth's atmosphere?

7. A collection of noninteracting hydrogen atoms is maintained in the $2p$ state in a magnetic field of strength 5.0 T. (a) At room temperature, find the fraction of the atoms in the $m_l = +1, 0$, and -1 states. (b) If the $2p$ state made a transition to the $1s$ state, what would be the relative intensities of the three normal Zeeman components? Ignore any effects of electron spin.

8. The following method is used to measure the molecular weight of very heavy molecules. A liquid containing the molecules is spun rapidly in a centrifuge, which establishes a variation in the density of the liquid. The density is measured, such as by absorption of light, to determine the molecular weight. Assign a fictitious "centrifugal" force to act on the molecules and show that the density varies as $\rho = \rho_0 e^{m\omega^2 x^2/2kT}$ where ω is the angular velocity of the centrifuge and x measures the distance along the centrifuge tube.

9. Find the blackbody energy spectrum $u(E)$, and convert to the wavelength spectrum $u(\lambda)$, Equation (12.24).

10. (a) From the blackbody energy density $u(E)$, the energy per unit volume per unit energy interval at the energy E, find the number of photons of energy E per unit volume per unit energy interval. (b) Show that the total number of photons per unit volume at temperature T is $N = 8\pi(kT/hc)^3\int_0^\infty x^2 dx/(e^x - 1)$. (c) The value of the integral is about 2.404. How many photons per cubic centimeter are there in a cavity filled with radiation at $T = 300$ K? At $T = 3$ K?

11. From the blackbody energy spectrum $u(E)$, find the *energy* at which the maximum radiation is emitted. Compare this result with Wien's displacement law (see Chapter 3) and account for any differences.

12. A blackbody is radiating at a temperature of 2.50×10^3 K.

(a) What is the total energy density of the radiation? (b) What fraction of the energy is emitted in the interval between 1.00 and 1.05 eV? (c) What fraction is emitted between 10.00 and 10.05 eV?

13. (a) Derive Equation (12.34) from Equation (12.33).

(b) Derive Equation (12.36) from Equation (12.35).

14. Compute the Fermi energy E_F and the average electron energy for copper.

15. Calculate the Fermi energy for magnesium, assuming two free electrons per atom.

16. In sodium metal at room temperature, compute the energy difference between the points at which the Fermi-Dirac occupation probability has the values 0.1 and 0.9. What do you conclude about the "sharpness" of the distribution?

Although we can explain and classify the physical processes that determine the structure of simple molecules, our understanding of the physics of complex molecules (such as the DNA molecule represented by the model) is not yet complete.

13

MOLECULAR STRUCTURE

17. A certain metal has a Fermi energy of 3.00 eV. Find the probability to find an electron with energy between 5.00 eV and 5.10 eV for (a) $T = 295$ K; (b) $T = 2500$ K.

18. Assume that the nucleons in a nucleus obey Fermi-Dirac statistics. Find the Fermi energy and the average energy of the protons or neutrons in ^{40}Ca. Are these values reasonable?

19. After a star like our sun stops producing energy by hydrogen fusion, it collapses to a white dwarf star, with a radius about the same as the Earth's. Take the mass of the sun to be 2.00×10^{30} kg and its radius after collapse to be 6.40×10^3 km, and calculate the Fermi energy of the electrons in such a white dwarf. Assume one electron for every two nucleons.

20. When a star is somewhat more massive than the sun, it will collapse to a neutron star, which can be regarded as a giant atomic nucleus, composed of neutrons. For a star about twice the sun's mass, the radius of the neutron star will be about 10 km. Find the Fermi energy of the neutrons in such a neutron star.

In this chapter we consider the combination of atoms into molecules, the excited states of molecules, and the ways that molecules can absorb and emit radiation. From a variety of experiments we learn that the spacing of atoms in molecules is of the order of 0.1 nm, and that the binding energy of an atom in a molecule is of the order of electron-volts. This spacing and binding energy are about what we expect for electronic orbits, and we therefore guess that the forces that bind molecules together originate with the electrons. Our expectations are reinforced by the knowledge that, were it not for the electrons, the Coulomb repulsion of the positively charged nuclei of the atoms would not allow stable molecules to form.

However, we also recognize at once that there is also a Coulomb repulsion of the *electrons*, and it is not apparent why stable molecules form at all. The key to this problem is the existence of the spatial probability densities of atomic orbits, such as we calculated for hydrogen and illustrated in Chapter 7. These probability densities are frequently not spherically symmetric, and very often may show overwhelming preferences for one spatial direction over another. *It is these probability densities that control the nature of molecular bonds and therefore also dictate the nature and properties of molecules.*

As we discuss molecular orbitals in this chapter, you must keep one very important principle in mind: The probability densities of the atomic wave functions "exist" whether or not an electron is actually occupying that particular quantum state. For example, when we bring two hydrogen atoms together to make a molecule of H_2, both atoms are in the $1s$ state. Each $1s$ state can, of course, accommodate two electrons, and so each atom has a "filled" $1s$ orbital and an "empty" $1s$ orbital. The molecule forms because the probability distribution of the filled $1s$ orbital of one atom overlaps with the probability distribution of the empty $1s$ orbital of the other atom, and so the electron can be "shared" between the two atoms. When we try to draw a picture representing this process, the empty and filled $1s$ orbitals are represented by identical probability distributions, even though this seems to contradict our notion that the probability distribution represents the actual probability to locate an electron.

Just as we began to study atomic physics by looking at the simplest atom, we begin our study of molecular physics with an example of the properties of the simplest molecule, H_2^+, the singly ionized hydrogen molecule. We next turn to other simple molecules, such as H_2 and NaCl, and finally we look at how our previous knowledge of atomic orbitals can help us to understand the molecular orbitals which form the basis of organic chemistry.

We conclude the chapter with a brief look at the rotational and vibrational excited states of molecules and at the study of the electromagnetic transitions between those excited states, which give a distinctive signature of the molecule and thus permit the identification of types of molecules. We will see how this type of *molecular spectroscopy* finds application in such diverse areas as identification of atmospheric pollutants and the search for life in outer space.

What complicates molecular structure is also what complicates atomic structure—there are too many electrons present for us to be able to write down and solve the dynamical equations that govern the structure of the atom or molecule. We therefore use the same tactic to study molecular structure that we used for atomic structure: we begin with a molecule that has only one electron. Such a molecule is H_2^+, *the hydrogen molecule ion,* which results when we remove an electron from a molecule of ordinary hydrogen, H_2.

Before we turn to the wave mechanical properties of H_2^+, let us try to guess what it is that holds this simple molecule together. We first realize that it is *not* correct to think of H_2^+ as an atom of hydrogen (proton plus electron) joined to a second proton. Since the atom of hydrogen in such a combination is electrically neutral, there is no electrostatic Coulomb force to hold the two pieces together. In this kind of molecule, at least, it is apparently not correct to identify the electron as belonging exclusively to one or the other of the components. *The electron must somehow be shared between the two parts.* Figure 13.1 shows an example of how this could occur. We regard the two protons as fixed in space, and the single electron "orbits" about both of them. (This will be our only reference in this chapter to the Bohr model of definite, fixed orbits. Recall that the Bohr model did not give the correct properties of atoms but instead provided a sort of rough mental image of how an electron, not an electron wave, might move in an atom. It is similarly true for molecules that the Bohr picture is not a useful one for understanding molecular properties.)

There are apparently two distinct orbits that the electron could take. In the first, it travels an elliptical path about both protons, and in the second it travels a sort of "figure eight" pattern between the two protons. The first type of orbit is not sufficient to bind the molecule together, because the electron cannot provide enough attractive force to overcome the repulsion of the two protons. In the second orbit, the electron spends much of its time in the region between the two protons. Thus, since each proton is attracted to the electron, the molecule becomes bound together. (You can easily see that when the electron is exactly between the two protons the total attractive force $2(1/4\pi\varepsilon_0)[e^2/(R/2)^2]$ is greater than the repulsive force $(1/4\pi\varepsilon_0)(e^2/R^2)$.) *The stability of such a molecule depends on the electron having a high probability of being between the two protons.*

This last statement gives us a valuable clue about the result we expect from a calculation of the properties of H_2^+ using wave mechanics—the bound state of the molecule should correspond to the electron's probability distribution having a large value in the region between the two protons.

As we learned in Chapter 7, an electron in the ground state of hydrogen has an energy of -13.6 eV, a wave function $\psi = (\pi a_0^3)^{-1/2} e^{-r/a_0}$, where a_0 is the Bohr radius, and a probability density proportional to ψ^2. The wave function ψ for one atom is shown in Figure 13.2 for the case when the second proton is so far away that it doesn't affect the neutral H atom. As we bring the second proton closer, we encounter regions of ever increasing ψ^2, and so the electron has a small but nonvanishing chance of being found in the vicinity of the second proton. If we were to

13.1 THE HYDROGEN MOLECULE ION

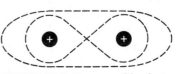

FIGURE 13.1 Two possible electron orbits about the two protons in H_2^+.

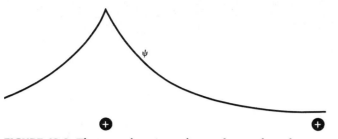

FIGURE 13.2 The wave function ψ for an electron bound to one proton, when the second proton is far away.

move the second proton away again, back to where it began, there is a chance that the electron may now be attached to the second proton!

Another way of viewing this problem is to consider the two protons as forming a sort of double potential well, as shown in Figure 13.3. The electron may thus be imagined as "tunneling" from one potential well to the other, and as we have seen previously, the narrower the barrier between the two wells (the closer the two protons are brought) the higher is the probability of tunneling. Eventually the protons may be brought close enough together that the electron may be found with equal probability in either well and thus "attached" to either proton.

The mathematics of wave mechanics tells us exactly how to deal with such a case—we merely add (superimpose) the wave function representing the electron in one well to the wave function representing the electron in the other well. However, in quantum mechanics we always encounter a problem when we try to add wave functions, because the sign of a wave function can never be determined. (Remember that when we calculate a normalization constant A, such as we did in Equation (5.26), we really calculate the value of A^2. For convenience we have always chosen the positive root, because until now it has not made any difference, but when we add different wave functions with different normalization constants, we must keep in mind that the signs may be different.) We therefore don't know if the two wave functions add with the same sign or with opposite signs.

We can also think of this process as a *coherent* superposition of the two electron waves representing the two different atoms. In making a coherent superposition, we first add the wave displacements and *then* square to find the resultant intensity. Such a process gives constructive interference at some points and destructive interference at other points.

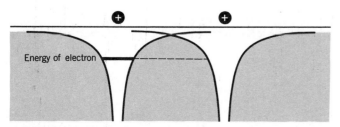

FIGURE 13.3 The double potential well of two protons in H_2^+.

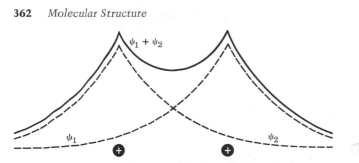

FIGURE 13.4 The overlap of the electronic wave functions in H_2^+. The two wave functions have the same sign, giving a large probability to find the electron between the two protons.

Consider the wave function as represented in Figures 13.4 and 13.5. In one case the two wave functions have the same sign and in the other case they have opposite signs. This difference in sign has a substantial effect on the probability distribution, which is given by ψ^2. In one case, the electron has a relatively large probability to be in the region between the two protons, while in the other case it has a relatively small probability to be there. Only in the case in which the probability is large near the center of the molecule do we expect to find the formation of a stable molecular bond.

We could justify our discussion with an exact calculation of the wave functions and probability density for H_2^+, but since the calculation itself is quite complicated, we merely state its result for the energy of the molecule as a function of the separation R of the two protons. When the two protons are very far apart, the energy of the electron is $-13.6\,\text{eV}$, and the Coulomb repulsion energy of the protons, V_p, is very small and can be neglected. As we bring the two protons closer together, V_p increases and the electron energy decreases, since the electron feels a greater net force and is therefore more tightly bound to the system. For the moment let us ignore the repulsion of the protons and concentrate on the electron binding energy. As we bring the two protons toward $R = 0$, the electron wave function approaches the wave function for an

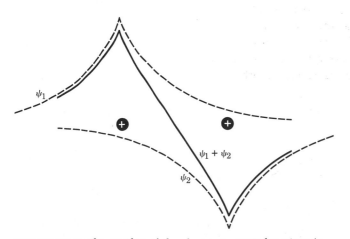

FIGURE 13.5 The overlap of the electronic wave functions in H_2^+. The two wave functions have opposite signs, giving a small probability to find the electron between the two protons.

atom with $Z = 2$. For the case in which the two wave functions have the same sign, ψ has a maximum at $R = 0$ (see Figure 13.4), and this must therefore be the $1s$ state of an atom with $Z = 2$. From Chapter 6 we recall the formula $E_n = (-13.6)Z^2/n^2$ for the energy levels of hydrogenlike atoms, and so the energy must be -54.4 eV. The energy, which we call E_+ (the $+$ indicating the *addition* of the two wave functions), has the value -13.6 eV for $R \to \infty$ and -54.4 eV for $R \to 0$ and has the exact form shown in Figure 13.6. When the two electron wave functions have opposite signs, the combined wave function approaches 0 as $R \to 0$ (see Figure 13.5). From the wave functions tabulated in Chapter 7, we see that the lowest energy level with a wave function that vanishes at $r = 0$ is the $2p$ level, which has an energy, $E_n = -13.6\, Z^2/n^2$, equal to -13.6 eV for our hypothetical $Z = 2$ atom. Thus E_-, the electron energy when the two wave functions have opposite signs, has the value -13.6 eV for $R \to \infty$ and also -13.6 eV for $R \to 0$ (see Figure 13.6).

The proton repulsion V_p has the form $V_p = e^2/4\pi\varepsilon_0 R$, and the total energy of the molecule is $V_p + E_+$ or $V_p + E_-$, depending on the relative signs of the electronic wave functions. From Figure 13.6, you can see that $V_p + E_+$ has a minimum at $R = 0.106$ nm, with $E_{\text{total}} = V_p + E_+ =$

Binding energy of H_2^+

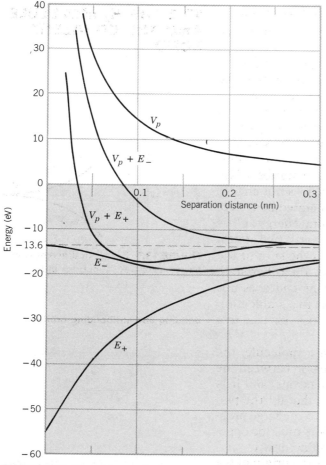

FIGURE 13.6 Dependence of energy on separation distance for H_2^+.

-16.3 eV. This minimum shows us the conditions for a stable H_2^+ molecular ion—a separation of 0.106 nm and a binding energy of $E_{total}(R = \infty) - E_{total}(R = 0.106 \text{ nm}) = 2.7$ eV. Notice also that there is no minimum for $V_p + E_-$—the case in which the two wave functions have opposite signs does not lead to a stable molecule, just as we expected.

It is interesting to note that the stability is achieved at $R \cong 2a_0$. You will recall from Chapter 7 that the probability density for the 1s state of hydrogen has its maximum value at $r = a_0$. Thus the stable configuration of the H_2^+ ion is such that the maximum in the probability density for a single H atom would fall exactly in the middle of the molecule! This is once again consistent with our expectations for the structure of H_2^+—the electron must spend most of its time between the two protons.

In summary, from our study of this simple molecule we have learned that an important feature of molecular bonding concerns the *sharing* of a single electron by two atoms of the molecule. The electron lives first with one atom, then with the other, and this sharing is responsible for the stability of the molecule. With this in mind we can now add a second electron and consider the H_2 molecule.

Suppose we have two hydrogen atoms separated by a very large distance. Associated with each atom there is a 1s electronic orbital, at an energy of -13.6 eV, since the atoms are so far apart that there is no interaction between the electrons. As we bring the atoms closer together to form a H_2 molecule, the electron wave functions begin to overlap, so that the electrons are "shared" between the two atoms. As we have seen in the previous section, this can occur in such a way that the two electron wave functions *add* in the region between the two protons, giving a stable molecule, or *subtract*, leading to no stable molecule. The separate, individual electronic states of the atoms now become *molecular* states, with molecular electronic orbitals and associated quantum numbers. Notice that, as shown in Figure 13.7, the number of states does not change as the separation R is reduced. When the atoms are separated by a large distance, there are two states, each at -13.6 eV, so the total energy at $R = \infty$ is -27.2 eV. When the separation is reduced, there are still two states, but now at different energies. One state corresponds to the sum of the two wave functions and leads to a stable H_2 molecule; the other state corresponds to the difference of the two wave functions and does not give a stable molecule. The molecular orbital that leads to a stable molecule is known as a *bonding* orbital, and the one that does not lead to a stable molecule is an *antibonding* orbital.

As we found previously for H_2^+, in order to form a molecule, the electron probability distribution must be large in the region between the two protons. In the case of H_2, this is true for both electrons, and it is certainly our expectation, based on the Pauli principle, that for the two electrons *both* to occupy the *molecular* orbital leading to the large probability in the central region, their spin quantum numbers must be opposite; that is, one must have spin "up" and the other spin "down."

The sharing of electrons in a molecule such as H_2 is the origin of the *covalent* bond; this type of bonding occurs most frequently in molecules

13.2 THE H_2 MOLECULE AND THE COVALENT BOND

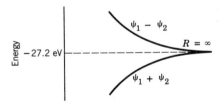

FIGURE 13.7 Energy of different combinations of wave functions in H_2.

Bonding and antibonding orbitals

Covalent bond

containing two identical atoms, and it is therefore also called *homopolar* or *homonuclear* bonding.

Let us summarize the essential features of covalent bonding:

1. As two atoms are brought together, the electrons interact and the separate atomic orbitals and energy levels are transformed into molecular orbitals; two separate, but identical, atomic orbitals become two different molecular orbitals.

2. In one of the molecular orbitals, the electron wave functions overlap in such a way as to give a *lower* energy than the separated atoms had; this is the bonding orbital that leads to the formation of stable molecules.

3. The other molecular orbital (the antibonding orbital) has an increased energy relative to the separated atoms and does not lead to the formation of stable molecules.

4. The restrictions of the Pauli principle apply to molecular orbitals just as they do to atomic orbitals; each molecular orbital has a maximum occupancy of two electrons, corresponding to the two different orientations of electron spin.

The energy of the bonding orbital for H_2 is shown in Figure 13.8; as you can see, there is a minimum with $E = -31.7$ eV at $R = 0.074$ nm. When we separate the H_2 molecule into two neutral H atoms, the total energy is 2×-13.6 eV $= -27.2$ eV, and so the *molecular binding energy* of H_2 is $-27.2 - (-31.7) = 4.5$ eV.

Comparing Figures 13.6 and 13.8 you can immediately see the effect of adding an additional electron to H_2^+: the binding energy is greater (the molecule is more tightly bound), and the nuclei are drawn closer together. Both of these effects are due to the presence of the increased electron density in the region between the two protons.

We can also understand why He does not form the molecule He_2—as two He atoms are brought together, the bonding and antibonding orbitals are formed in much the same way as with H_2. The He_2

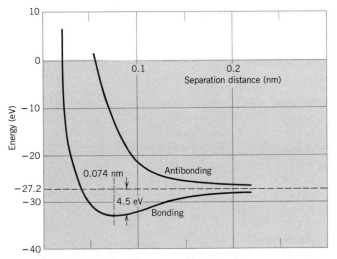

FIGURE 13.8 Bonding and antibonding in H_2.

molecule would have four electrons, two in the bonding orbital and two in the antibonding orbital, and the net effect is that no stable molecule forms. (However, He_2^+ is stable, with two bonding and only *one* antibonding electrons. The binding energy of He_2^+ is 3.1 eV and the separation is 0.108 nm, remarkably close to the corresponding values of H_2^+.)

Other hydrogenlike systems with a single s electron also form stable molecules through covalent bonding; for example, see the values for Li_2 and Na_2 given in Table 13.1. Atoms with valence electrons in p orbitals can also form diatomic molecules through covalent bonds—oxygen and nitrogen, for example. Since there are three atomic p orbitals, there will be six molecular orbitals, and the classification of orbitals can become quite tedious, but we can understand the structure of molecules composed of atoms with p electrons based on the geometry of atomic p orbitals. We discuss three applications of this type of covalent bonding: p-p homopolar molecules, s-p directed bonds and s-p hybrid orbitals. First we review some of the properties of atomic p orbitals.

In Chapter 7 we solved the Schrödinger equation for the H atom and showed the spatial probability distributions for the various possible electronic wave functions. Of course, these solutions for hydrogen will *not* be correct for other atoms, but the essential features of the geometry of the atomic orbitals remains correct. Let us concentrate our attention on the p orbitals, of which there are three, corresponding to $m_l = -1, 0$, and $+1$. The probability distributions corresponding to these m_l values were shown in Figure 7.9. We can imagine these distributions to have a sort of "figure eight" shape with two distinct lobes of large probability. In the $m_l = 0$ case, the "figure eight" has its long axis along the z axis, and the two lobes of maximum probability occur in the $+$ and $-z$ directions. In the $m_l = \pm 1$ cases, the "figure eight" lies in the xy plane, and we can think of it as rotating rapidly about the z axis, counterclockwise for $m_l = +1$ and clockwise for $m_l = -1$, giving again two lobes of large probability, which now fall along a line that has any orientation in the xy plane. For our purposes here, it is not as convenient to use the m_l notation as it is to use a different notation in which we assign each of the three possible p orbitals a label that gives the direction in space corresponding to the lobes of maximum probability. Thus p_z is the orbital

13.3 OTHER COVALENT BONDING MOLECULES

p_x, p_y, p_z Orbitals

Table 13.1 Properties of s-Bonded Molecules

Molecule	Dissociation Energy (eV)	Equilibrium Separation (nm)
H_2	4.52	0.074
Li_2	1.10	0.267
Na_2	0.80	0.308
K_2	0.59	0.392
LiNa	0.91	0.281
KNa	0.66	0.347
LiH	2.43	0.160
Rb_2	0.47	0.422
NaRb	0.61	0.359
Cs_2	0.43	0.450
NaH	2.09	0.189

with regions of large electron probability along the z axis, and similarly for p_x and p_y. Figure 13.9 shows a *schematic* representation of these probability distributions. (Obviously, the p_z orbital corresponds exactly to the $m_l = 0$ orbital; p_x and p_y correspond to *mixtures* of $m_l = +1$ and $m_l = -1$.) We consider the structure of molecules containing p electrons based on this simple model of the three mutually perpendicular p orbitals p_x, p_y, and p_z.

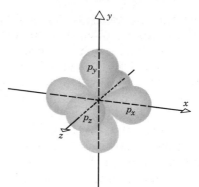

FIGURE 13.9 Probability distributions of three different p electrons.

p-p *Covalent Bonds* Consider what happens when we bring two p-shell atoms together starting from a large separation. (We will consider $2p$ atoms in order to simplify the number of atomic shells to be discussed.) The $1s$ *atomic* orbitals are transformed into the two $1s$ *molecular* orbitals, exactly as we found in the case of H_2. One of the orbitals is bonding, and the other is antibonding. Each atom has a filled $1s$ shell, and so there is a total of four $1s$ electrons; since each *molecular* orbital can hold two electrons, the $1s$ bonding and antibonding orbitals are filled to capacity. The same is true of the $2s$ states. The *atomic* $2s$ levels also form bonding and antibonding orbitals, and since each atom has a filled $2s$ shell, the four $2s$ electrons completely fill the two molecular $2s$ orbitals. Since the atoms we are considering have partially full p shells, the bonding depends critically on the molecular p orbitals. These again behave in a way similar to the s orbitals: For each *atomic* p orbital (p_x, p_y, and p_z) there are corresponding antibonding and bonding *molecular* orbitals. However, the p orbitals that happen to lie along the direction of approach of the two atoms will overlap most, and are therefore most affected. We arbitrarily assume this to be the x direction. At the closest approach of the atoms, the p_x orbitals overlap considerably more than the p_y or p_z orbitals, and so the bonding p_x *molecular* orbital is more stable, and hence lower in energy, than the p_y or p_z bonding orbitals. The same effect occurs for the antibonding molecular orbitals; the overlap of the p_x orbitals is greater, and therefore the antibonding orbital is more successful in *reducing* the electron probability density between the atoms. The antibonding p_x orbital is therefore even less stable and so it rises in energy.

The p_y and p_z orbitals do not overlap as much as the p_x orbital and so are much less affected. Since there is no difference in the way the approach of the atoms affects the p_y and p_z orbitals, they have exactly the same energies.

Figure 13.10 shows the molecular $1s$, $2s$, and $2p$ orbitals as a function of the separation of the nuclei of the two atoms.

In Figure 13.11 we see an illustration that shows why the p_x bonding orbital is more stable than the p_y or p_z bonding orbitals. The p_x molecular bonding orbital has its maximum electron probability directly between the nuclei, and so the electron is most successful in holding the two parts together. In the case of p_y or p_z (only p_y is shown), there are maxima of electron probability above and below the x axis that, to some extent, oppose one another; only the component of the resultant attractive force along the x axis produces bonding. Since the resultant force is smaller than it is in the case of the p_x orbital, the molecular bonding is less stable.

Let us consider three diatomic molecules composed of atoms with

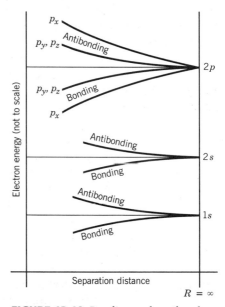

FIGURE 13.10 Bonding and antibonding $2p$ orbitals.

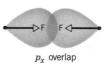

p_x overlap

p_y overlap

FIGURE 13.11 Bonding force from overlap of p_x and p_y orbitals. Because of the off-axis components, the net force is weaker for the p_y orbital.

2*p* valence electrons: N_2, O_2, and F_2. Nitrogen has seven electrons; the 1*s* and 2*s* shells are filled, and three electrons occupy the 2*p* shell. In the N_2 molecule, there are therefore six 2*p* electrons to fill the 2*p* molecular orbitals, each of which can hold two electrons. The three lowest 2*p* molecular orbitals are filled. Since all three of these are bonding orbitals, N_2 forms a very stable diatomic molecule (binding energy = 9.8 eV, separation = 0.11 nm). Because of this stability, nitrogen in the form of N_2 is rather unreactive under most circumstances.

Oxygen has four 2*p* electrons, and so the O_2 molecule must have eight electrons in the 2*p* molecular orbitals. These not only fill up the three bonding orbitals, as with N_2, but also the lowest ($2p_y$ or $2p_z$) *antibonding* orbital. We therefore expect O_2 to be *less* stable than N_2, since it has three bonding and one antibonding molecular orbitals, and experience is consistent with this expectation. The binding energy of O_2 (5.1 eV) is less than that of N_2, and the nuclear separation is greater (0.12 nm). The O_2 molecule is less stable than the N_2 molecule, and the O_2 molecular bonds can be broken by relatively modest chemical reactions, as, for example, the oxidation of metals exposed to air.

Fluorine has five 2*p* electrons, and the F_2 molecule thus has ten electrons in the 2*p* orbitals, filling three bonding and now *two* antibonding orbitals. This makes F_2 even less stable than O_2; the energy necessary to dissociate F_2 into two F atoms is only 1.6 eV, and fluorine gas reacts quite violently with many substances. In fact, since the photons of visible light have energies in excess of 1.6 eV (typically 2 to 4 eV), F_2 is normally unstable and is broken down into F atoms by exposure to light. This process is called *photodissociation*.

Although their molecular orbital diagrams may be more complicated, we expect all molecules based on *p*-shell atoms similar to N_2, O_2, and F_2 to behave in similar ways. Unfortunately, the other np^3 atoms (P, As, Sb, Bi) and np^4 atoms (S, Se, Te, Po) are solids at ordinary temperatures, and their molecular properties can normally not be observed, since the interaction of the atoms in a solid is usually more important than molecular bonding effects (as we will see in the next chapter). However, the remaining halogens (Cl, Br, I, At) have properties similar to F_2—relatively weakly bound, diatomic molecules, which are quite reactive.

s-p Molecular Bonds   It is often the case that a stable molecule is formed from two atoms, one with an *s*-state valence electron and the other with one or more *p*-state valence electrons. Consider, for example, the HF molecule. The electron wave function of H we have described previously, and the F atom has five electrons in the *p* shell, so of the three 2*p* atomic orbitals, two will each have their capacity of two electrons, and the third will have a single electron. We will ignore the four paired electrons, which do not significantly affect the molecular bonding, and concentrate instead on the single unpaired *p* electron. The two-lobed probability distribution corresponds to a two-lobed *p*-state wave function, in which the signs of ψ are opposite for the two lobes. The 1*s* wave function of H has only one sign (Figure 13.12). If, as the H and F atoms approach from a large distance, the H wave function happens to have the same sign as the nearer lobe of the F wave function, an in-

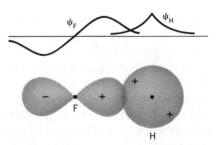

FIGURE 13.12 Overlap of *s* and *p* wave functions.

Table 13.2 Properties of *sp*-Bonded Molecules

Molecule	Dissociation Energy (eV)	Equilibrium Separation (nm)
HF	5.90	0.092
HCl	4.48	0.128
HBr	3.79	0.141
HI	3.10	0.160
LiF	5.98	0.156
LiCl	4.86	0.202
NaF	4.99	0.193
NaCl	4.26	0.236
KF	5.15	0.217
KCl	4.43	0.267

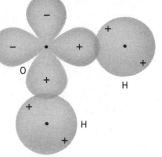

FIGURE 13.13 Overlap of electronic wave functions in H_2O.

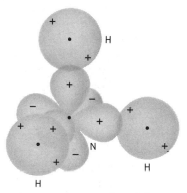

FIGURE 13.14 Overlap of electronic wave functions in NH_3.

Directed bonds

creased electron probability results, and hence a *bonding sp* orbital is formed. It is also possible to have *antibonding sp* orbitals, which result from the H and F wave functions having opposite signs.

Table 13.2 gives dissociation energies and nuclear separation distances for some *sp*-bonded diatomic molecules.

Let us now consider the water molecule, H_2O. Oxygen has eight electrons, four of which occupy the $2p$ shell. When we place these electrons in the $2p$ *atomic* orbitals, we begin with one electron each in the $p_x, p_y,$ and p_z orbitals, and then the fourth $2p$ electron must pair with one of the first three. Let us arbitrarily assume that it fills the p_x orbital. An oxygen atom therefore has two unpaired $2p$ electrons, each of which can form a bond with the $1s$ electron of H to form a molecule of H_2O. Figure 13.13 shows a representation of the electron probability distributions we might expect for an oxygen atom and for a molecule of H_2O. Such a molecule has *directed* bonds, which have a fixed, measurable relative direction in space. The expected angle between the two bonds is 90°; this angle can be measured experimentally by, for example, measuring the electric dipole moment of the atom, and the result, 104.5°, is somewhat larger than we expect. This discrepancy can be interpreted as arising from the Coulomb repulsion of the two H atoms, which tends to spread the bond angle somewhat.

As another example, consider the NH_3 (ammonia) molecule. With $Z = 7$, the nitrogen atom has three unpaired p electrons, one each in the $p_x, p_y,$ and p_z atomic orbitals. Each of these can form a bond with a H atom to form the NH_3 molecule, and we expect to find three mutually perpendicular sp bonds (Figure 13.14). The measured bond angle is 107.3°, again indicating some repulsion between the H atoms.

In Table 13.3 are listed some bond angles measured for other molecules that have *sp*-directed bonds. As you can see, the bond angle does indeed approach 90° in many cases. Based on the discussion given above, you should be able to explain why this happens as the Z of the central atom increases.

s-p *Hybrid Orbitals*
One example of a $2p$ atom we have so far not considered is carbon, and for a special reason: carbon forms a great variety of different types of molecular bonds, with a resulting diversity in the type and complexity of molecules containing carbon. It is this diversity that is the basis for the many kinds of *organic molecules* that can form, based

Table 13.3 Bond Angles of *sp*-Directed Bonds

Molecule	Bond Angle
H_2O	104.5°
H_2S	93.3°
H_2Se	91.0°
H_2Te	89.5°
NH_3	107.3°
PH_3	93.3°
AsH_3	91.8°
SbH_3	91.3°

on various kinds of carbon molecular bonds, and so an understanding of the physics of carbon molecular bonds is essential to the understanding of many fundamental questions of structure and processes in molecular biology.

The properties of many molecules containing carbon are explained by the concept of *sp hybridization*. Carbon, with six electrons, has the configuration $1s^2 2s^2 2p^2$, so that we expect carbon under ordinary circumstances to show a valence of 2, with the two $2p$ electrons contributing to the structure, and we might therefore expect to form stable molecules such as CH_2, with directed *sp* bonding (similar to H_2O) and a bond angle of roughly 90°. Instead, what forms is CH_4 (methane) in a tetrahedral structure (Figure 13.15), with four equivalent bonds. For another example, the elements of the third column of the periodic table (boron, aluminum, gallium . . .) have the outer configuration $ns^2 np$ ($n = 2$ for boron, $n = 3$ for aluminum, etc.), and we expect these elements to form compounds as if they had a single valence electron. We therefore expect halides such as BCl or GaF, oxides such as B_2O or Al_2O, nitrides such as B_3N or Al_3N, hydrides such as BH or GaH, and so forth. Instead we find that boron, aluminum, and gallium generally behave as if they had three valence electrons, and form compounds such as BCl_3, Al_2O_3, AlN, and B_2H_6. Furthermore, the three valence electrons seem to be equivalent; there seems to be no way, for example, to associate two of the valence electrons with *s* orbitals and one with a *p* orbital. The bonds formed by the three electrons make angles of 120° with one another.

It is the effect of *sp hybridization* that is responsible for the valence of three (rather than one) in boron and four (rather than two) in carbon. The 4 bonds in CH_4 are *equivalent* and *identical*, which would *not* be expected if we had two *ss* bonds and two *sp* bonds; similarly in BF_3 or BCl_3, the three bonds are identical and are clearly *not* identified with two *sp* bonds and one *pp* bond.

The normal meaning of *hybrid* is an offspring resulting from the union of parents of different types, in which the offspring is not exactly like either parent, but retains some of the attributes of each. In the case of *molecular orbitals*, hybridization refers to a process by which the orbitals can no longer be identified as either *s* or *p* orbitals, but rather are mixtures of *s* and *p* orbitals. The formation of *sp* hybrids is normally as follows:

1. In an atom with the configuration $2s^2 2p^n$, one of the $2s$ electrons is excited to the $2p$ shell, giving a configuration $2s\, 2p^{n+1}$.

2. The hybrid orbitals are formed by taking equal mixtures of the wave functions representing the $2s$ orbital and each of the $2p$ orbitals. For example, in the case of boron, the configuration of $2s^2 2p$ is converted to $2s\, 2p^2$. Assuming the $2p$ states to be $2p_x$ and $2p_y$, the resultant *hybrid* wave functions can be represented as different combinations of ψ_{2s}, ψ_{2p_x}, and ψ_{2p_y}, such as

$$\psi = \psi_{2s} + \psi_{2p_x} + \psi_{2p_y}$$

Other combinations can be formed by subtracting, instead of adding, the individual wave functions.

Illustrations of the probability distributions expected for *sp*, sp^2, and sp^3 hybrid orbitals are shown in Figure 13.16. Keep in mind that these do

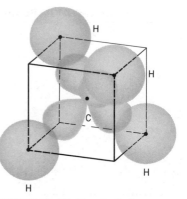

FIGURE 13.15 Tetrahedral arrangement of molecular bonds in CH_4.

sp **Hybridization**

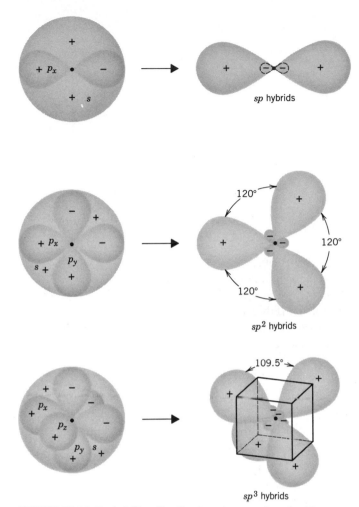

FIGURE 13.16 Probability distributions in sp, sp^2, and sp^3 hybrid orbitals.

not yet represent *molecular* orbitals—they are merely the contribution of one of the atoms to the bonding electronic distributions of the molecule.

The tetrahedral structure of CH_4 is therefore merely the result of the symmetrical spatial arrangement of the four sp^3 hybrid orbitals of C, with each hybrid orbital bonded to one H. The bond angle of such a symmetrical tetrahedron is 109.5°, in good agreement with the measured bond angles of CH_4 and other sp^3 hybrids, shown in Table 13.4.

It is also possible for only two $2p$ electrons in the $2s\,2p^3$ configuration of carbon to become involved in hybrid orbitals; the third $2p$ electron then is available, for example, to form ordinary p-p molecular orbitals. The ethylene molecule C_2H_4 is an example of such a structure. The three sp^2 hybrid orbitals form bond angles of 120°; each carbon atom has two of its three sp hybrid orbitals joined to a H atom, and the third is joined to the other carbon. The unhybridized p orbital also forms a bond between the two carbons, in a manner similar to that shown in Figure 13.11 for the "off-axis" p-p molecular orbitals. Figure 13.17 shows a representation of the bonding in C_2H_4. There are two bonds between the

Table 13.4 Bond Angles of sp^3 Hybrids

Molecule	Bond Angle (°)
CCl_4	109.5
C_2H_6	109.3
C_2Cl_6	109.3
$CClF_3$	108.6
CH_3Cl	110.5
$SiHF_3$	108.2
SiH_3Cl	110.2
$GeHCl_3$	108.3
GeH_3Cl	110.9

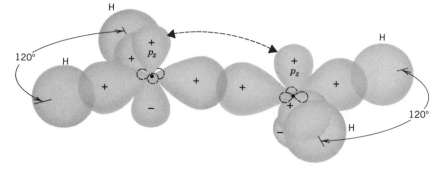

FIGURE 13.17 Molecular bonding in C_2H_4. For clarity the p_z overlap is not shown; the arrow indicates the bond formed by the unhybridized p_z orbital.

pair of carbons, one from the sp^2 hybrid and the other from the unhybridized p_z orbital. Another example of sp^2 hybrids in carbon is benzene (C_6H_6) in which each carbon is joined to one H and two other carbons by the sp^2 hybrids, with again one unhybridized p orbital available to bond the carbons. The basic structure of benzene is a ring of carbon atoms, as shown in Figure 13.18, with the expected angle of 120° between the hybrid orbitals.

Finally, it is also possible for carbon to form sp hybrids, leaving two unhybridized p orbitals. Acetylene (C_2H_2) is an example of such a molecule, with the two carbons now joined by three bonds, one from the sp hybrid and two from the unhybridized p orbitals. Figure 13.19 shows the bonding arrangement in C_2H_2.

This variety of bonds entered into by carbon is the basis for the varied properties of organic molecules, from the simple ones we have studied here, to the complex ones which form the basis of living things. However, it is not only carbon that shows sp hybridization, but other atoms as well. (Indeed, the failure of NH_3 to conform to the expected 90° bond angle could be blamed on sp hybridization rather than on the repulsion of the H atoms.) It is also possible to have $3s$-$3p$ hybrids (silicon) and $4s$-$4p$ hybrids (germanium). It is this hybridization that gives these

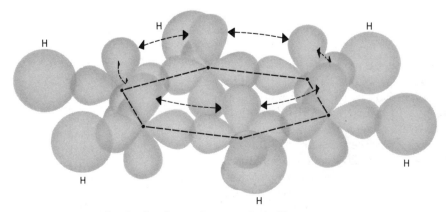

FIGURE 13.18 Molecular bonding in benzene. As in Figure 13.17, the overlap of the unhybridized p_z orbitals is not shown, but the arrows indicate the p_z bonds.

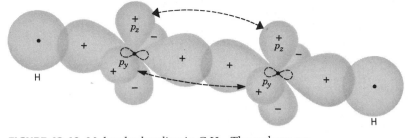

FIGURE 13.19 Molecular bonding in C_2H_2. The carbons are joined by 3 bonds, one from the hybridized p_x orbital and two from the unhybridized p_y and p_z orbitals.

materials, like carbon, a valence of 4 and a symmetrical bonding arrangement, which are partly responsible for the usefulness of Si and Ge as semiconducting materials, as we will learn in the next chapter. It is also interesting to speculate on the possibility of a new type of organic chemistry, including new life forms, based on Si or Ge, rather than C.

In covalent bonding, as we have seen, the bonding electrons do not belong to any particular atom in the molecule, but rather are shared among the atoms. It is also possible to form a molecule that results from the extreme case in which valence electrons are not shared but instead spend all of their time in the neighborhood of only one of the atoms of the molecule.

The most familiar of ionic molecules is NaCl. Suppose we have a neutral sodium atom $(1s^2 2s^2 2p^6 3s)$, with one $3s$ electron outside of a closed shell, and a neutral chlorine atom $(1s^2 2s^2 2p^6 3s^2 3p^5)$, which lacks one electron from a filled $3p$ shell. To remove the outer electron from Na requires 5.14 eV, the *ionization energy* of Na, and we are left with a positively charged Na^+ ion. If we then attach that electron to the Cl atom, creating a negatively charged Cl^- ion, the energy *released* is 3.61 eV, the *electron affinity* of Cl. The energy is released because the filled $3p$ shell **Electron affinity** is an especially stable configuration, which is energetically very favorable. Thus, if we borrow 5.14 eV to ionize the Na, we get back immediately 3.61 eV by attaching the electron to the Cl. We can get back the remaining 1.53 eV ($= 5.14$ eV $- 3.61$ eV) by moving the Na^+ and Cl^- close enough together that their Coulomb attraction energy is 1.53 eV. This distance we can calculate immediately as follows:

$$V = \frac{e^2}{4\pi\varepsilon_0}\frac{1}{R}$$

$$R = \frac{e^2}{4\pi\varepsilon_0}\frac{1}{V}$$

$$= \frac{1.44 \text{ eV·nm}}{1.53 \text{ eV}}$$

$$= 0.941 \text{ nm}$$

That is, as long as the Cl^- and Na^+ are closer together than 0.941 nm, the Coulomb attraction will supply enough energy to overcome the dif-

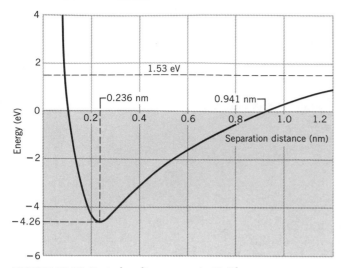

FIGURE 13.20 Ionic bonding energy in NaCl.

ference between the ionization energy of Na and the electron affinity of Cl. Put another way, Na^+ and Cl^- *ions* separated by less than 0.941 nm have less resultant energy, and therefore a more stable configuration, than neutral Na and Cl *atoms*.

It would therefore seem that the closer together we push the Na^+ and Cl^- ions, the more tightly bound they become. However, if they are pushed too close together, the filled 2p shell of Na^+ and the filled 3p shell of Cl^- will begin to overlap. Because electrons are identical particles, when two electron wave functions overlap it is no longer possible to identify which electron came from Na and which from Cl. There is no "Na electron" or "Cl electron"—all electrons are alike. Thus if a 2p electron from Na and a 3p electron from Cl begin to overlap, they will occasionally *both* try to behave as 2p electrons in Na and also occasionally as 3p electrons in Cl. But this is not possible—Na^+ already has a full 2p electron shell, and Cl^- has a full 3p shell; because of the Pauli principle, no additional electrons are permitted in either shell. Forcing the two ions closer together therefore requires that electrons be pushed from the 2p or 3p shells into a higher shell, in order to "make room" for the overlapping electrons. Since this takes energy, we must therefore add energy to the Na^+ + Cl^- system in order to reduce the separation of the ions. We can imagine this energy as a sort of "potential energy" of repulsion, which increases rapidly as we try to force the ions close together.

In summary, when the ions are far apart, they attract one another and when they are too close together, they repel one another; in between there must be an equilibrium position where the attractive and repulsive forces are balanced. It is this equilibrium position that determines the size of an ionic molecule.

Figure 13.20 shows the relative energy of the Na^+ and Cl^- ions as a function of the separation distance. The equilibrium separation is 0.236 nm, and at that distance the energy is −4.26 eV. It takes an energy of 4.26 eV to split the NaCl molecule into neutral Na and Cl atoms; this energy is the *dissociation energy* of NaCl.

Dissociation energy

Table 13.5 Properties of Some Ionic Diatomic Molecules

Molecule	Dissociation Energy (eV)	Equilibrium Separation (nm)
NaCl	4.26	0.236
NaF	4.99	0.193
NaH	2.08	0.189
LiCl	4.86	0.202
LiH	2.47	0.239
KCl	4.43	0.267
KBr	3.97	0.282
RbF	5.12	0.227
RbCl	4.64	0.279

Table 13.5 shows the equilibrium separation and dissociation energy of several ionic molecules.

Remember that we are concerned here with isolated molecules and not with collections of molecules in solids. When we speak of NaCl ions in this section, we mean not the solid salt but rather a gas of NaCl molecules. The spacing between atoms in a solid can be very different from the spacing in a molecule.

13.5 IONIC OR COVALENT?

How can we decide whether two atoms will join by ionic or covalent bonding? The answer to this question must depend on the willingness of the two atoms to share their valence electrons, or equivalently the degree to which one atom dominates the other in its desire to have the valence electrons all to itself.

For homonuclear diatomic molecules, we expect a purely covalent bond; since the two atoms are exactly alike, neither can dominate and the electron is completely shared between them. For heteronuclear molecules, however, the situation is quite different. There can be no purely ionic or purely covalent bond. Since the two atoms have different atomic numbers and different electronic configurations, the wave function of a valence or bonding electron, even if shared, will not be exactly the same near one atom as it is near the other. The electron must therefore spend more time near one atom than near the other; this means that one atom may have a slight excess of negative charge and the bond, even if we think of it as being covalent, will also have a small ionic character. Conversely, a "pure" ionic bond has a small covalent character; we learned in Chapter 7 that electron wave functions do not suddenly drop off to zero amplitude, but instead fall exponentially to zero. Therefore, even in an ionic molecule like NaCl, the wave function of the electron that was transferred to the Cl^- ion is not zero at the location of the Na^+ ion; the wave function may indeed have a very small amplitude at the Na^+ ion, but it is not zero. The electron thus spends some of its time, even if very little, being shared between the atoms, and the bonding, while mostly ionic, has a small covalent character as well.

There are two ways to gauge the relative degrees of ionicity or covalency. In the first, we determine whether one atom has, on the average, more negative charge than the other. We do this by measuring the *electric dipole moment* of the molecule. An electric dipole, as illustrated in

Figure 13.21, consists of equal positive and negative charges q separated by a distance r. The electric dipole moment p is then defined as

$$p = qr \qquad (13.1)$$

In NaCl, for example, we would expect an electric dipole moment of

$$p = e \times 0.236 \text{ nm}$$
$$= 1.60 \times 10^{-19} \text{ C} \times 0.236 \times 10^{-9} \text{ m}$$
$$= 3.78 \times 10^{-29} \text{ C·m}$$

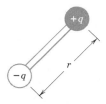

FIGURE 13.21 An electric dipole.

The measured electric dipole moment is

$$p_{\text{meas}} = 3.00 \times 10^{-29} \text{ C·m}$$

It is reasonable to say that the degree or percent ionicity is $p_{\text{meas}}/p = 79$ percent. If NaCl were a pure ionic molecule, the dipole moment would be 3.78×10^{-29} C·m; because NaCl is only partially ionic, the measured electric dipole moment is smaller than the expected value. In a purely covalent molecule, p would be zero; you can either think of this as occurring because the "center of gravity" of all charges is exactly at the midpoint of the molecule, so that $r = 0$, or else because the charges are exactly divided between the two atoms so that the excess charge q is zero.

As another example, consider the molecule HCl. From Table 13.2 we find the separation to be 0.128 nm, and so the expected electric dipole moment is

$$p = 1.60 \times 10^{-19} \text{ C} \times 0.128 \text{ nm}$$
$$= 2.05 \times 10^{-29} \text{ C·m}$$

The measured electric dipole moment is:

$$p_{\text{meas}} = 0.360 \times 10^{-29} \text{ C·m}$$

and so the bond is only $0.360/2.05 = 18$ percent ionic. The bonding in the molecule HCl is therefore mostly covalent.

The second way we can judge whether bonding is ionic or covalent is to determine the relative readiness of an atom to give up or accept an electron. If, as in NaCl, one atom is very willing to give up an electron while the other is eager to accept one, the bond should be mostly ionic. If, on the other hand, the two atoms are about equal in their desire to give up or accept an electron, the bond should be of the covalent type. The quantity that measures the ability of an atom in a molecule to attract or donate an electron is called the *electronegativity* χ and it can be calculated as follows:

$$\chi = \frac{1}{3.17} \cdot \frac{1}{2} \text{ (ionization energy + electron affinity)} \qquad (13.2)$$

The electronegativity can be considered a sort of average between the ionization energy, the energy needed to remove an electron from an atom, and the electron affinity, the energy gained when an electron is attached to a neutral atom. The factor of 3.17 is just a conversion factor to normalize the electronegativities to a convenient scale. One caution here is that the ionization energy and electron affinity must be calculated for the actual orbitals which participate in the molecular bonding; values obtained from free atoms are not suitable, especially when

Linus Pauling (1901–   , United States). Although he is generally considered to be a chemist, his work on molecular bonds transcends the traditional chemistry-physics boundary. He was awarded two Nobel prizes, the chemistry prize in 1954 for his work on bonding and the peace prize in 1963 for his efforts toward a ban on the testing of nuclear weapons.

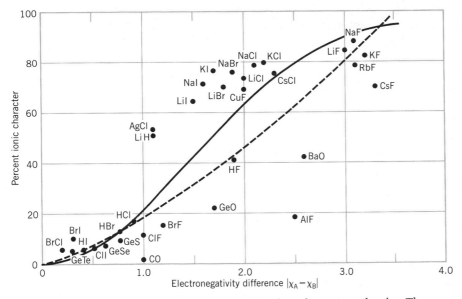

FIGURE 13.22 The fractional ionic character of bonds in diatomic molecules. The solid curve is calculated from Equation (13.3) and the dashed curve from Equation (13.4).

the atomic orbitals which are responsible for the molecular bonding are very different from those of the free atom, such as in the case of sp hybridization.

The degree of ionicity of a molecular bond between atoms A and B depends on the difference between their electronegativities, $|\chi_A - \chi_B|$. If $\chi_A - \chi_B = 0$, neither atom dominates in its desire to have an additional electron, and we expect a covalent bond. If $|\chi_A - \chi_B|$ is large, the bond should be ionic.

There is no exact theory for how the degree of ionicity depends on $|\chi_A - \chi_B|$, but there are empirical relationships. For example, the degree or fraction of ionicity can be approximately

$$1 - e^{-0.25(\chi_A - \chi_B)^2} \tag{13.3}$$

or

$$0.16|\chi_A - \chi_B| + 0.035|\chi_A - \chi_B|^2 \tag{13.4}$$

Both formulas give pure covalent bonds when $\chi_A = \chi_B$. The values of the numerical coefficients are chosen to give reasonable agreement with experiment. Values of the electronegativities of some elements are listed in Table 13.6. In Figure 13.22 are plotted the ionicity fractions determined according to both empirical relationships above. The experimental values of the fractional ionicities determined from the electric dipole moments are also shown. As you can see, both relationships give reasonable estimates of the ionicity.

Degree of ionicity

Table 13.6 Electro-negativities of the Elements

H	2.1		
Li	1.0	F	4.0
Na	0.9	Cl	3.0
K	0.8	Br	2.8
Rb	0.8	I	2.5
Cs	0.7		

Source: L. Pauling, *The Chemical Bond* (Ithaca, Cornell University Press, 1967).

13.6 MOLECULAR VIBRATIONS

A molecule can absorb and emit electromagnetic energy just as an atom does; for example, the molecule H_2 can absorb a photon of the right energy to excite one of the $1s$ electrons to an excited $2p$ state, and it will then return to the ground state by emitting a photon. The energies of such photons are typically of the same order as those from electronic transitions in atoms, 1 to 10 eV. We have discussed such optical transitions (in the visible region of the electromagnetic spectrum) for atomic systems in Chapter 8, and since the optical transitions in molecules are very similar, we will not discuss them any further.

The absorption or emission of an optical photon changes the electronic state of motion in a molecule. There are, however, other ways for molecules, but not atoms, to absorb and emit electromagnetic radiation. It is possible to change the state of motion of the individual atoms themselves, by making them vibrate relative to one another or rotate about the center of mass of the atom. Just like the electronic motion, the vibrational and rotational motions are *quantized*—vibrational energy and rotational energy can only be emitted or absorbed in discrete bundles of a certain size. In the remainder of this chapter, we will study the alternative ways a molecule can absorb or emit electromagnetic energy by changing its rotational or vibrational state of motion. In these discussions, the electrons are completely ignored—both vibration and rotation depend on the mass of the vibrating or rotating object, and the electrons have too little mass to be significant.

We begin with the vibrational motion of a molecule. (We consider only diatomic molecules, in order to simplify the discussions, but the same general principles hold for molecules with more than two atoms.) In Chapter 5 we studied the wave mechanics of the simple harmonic oscillator, and we found that, when the potential is given by $V = \frac{1}{2}kx^2$, the energy levels are

$$E = h\nu(N + \tfrac{1}{2}) \qquad N = 0, 1, 2, \ldots \qquad (13.5)$$ **Energies of vibration**

where ν is the same as the classical frequency of the oscillator:

$$\nu = \frac{1}{2\pi} \sqrt{k/m} \qquad (13.6)$$

Since we think of both atoms of the molecule as participating in the vibration, it is not clear what value to use for the mass in calculating the frequency ν. Suppose the two atoms have masses m_1 and m_2. Then as they pass through their equilibrium positions, the total energy of the molecule is:

$$E_T = \tfrac{1}{2}m_1 v_1^2 + \tfrac{1}{2}m_2 v_2^2$$

$$= \frac{p_1^2}{2m_1} + \frac{p_2^2}{2m_2} \qquad (13.7)$$

If the center of mass of the molecule is fixed (since there is no translational motion of the whole molecule), $p_1 = p_2$, and we can rewrite the energy (with $p = p_1 = p_2$) as

$$E_T = \frac{1}{2} p^2 \left(\frac{1}{m_1} + \frac{1}{m_2} \right) \qquad (13.8)$$

$$= \frac{1}{2} p^2 \left(\frac{m_1 + m_2}{m_1 m_2} \right) \qquad (13.9)$$

$$= \frac{p^2}{2m} \qquad (13.10)$$

where

$$m = \frac{m_1 m_2}{m_1 + m_2} \qquad (13.11)$$

That is, the energy of the system is the same as if it were a single mass of value m, moving with momentum p. This m is a sort of effective mass of the whole molecule and is known as the *reduced mass*. It is the mass we should use in calculating the vibrational frequency.

Notice that $m = \frac{1}{2}m_1$ when $m_1 = m_2$, as in a homonuclear molecule—the effective mass is just half the mass of an individual atom. Whenever one mass is much greater than the other, the reduced mass has a value nearly equal to that of the lighter mass. This is consistent with our expectation, since the inertia of the heavier mass reduces its tendency to move, and most of the vibrational motion is done by the lighter mass.

We now need to evaluate the constant k that appears in the expression for the frequency. There is no precise way of doing this, because a molecule does not behave like a simple harmonic oscillator! An oscillator that has a potential energy of $\frac{1}{2}kx^2$ has a force $F = -kx$ which attracts the mass toward the center. This force increases linearly with x, but we do not expect this to occur for a molecule—the molecular bonding forces grow considerably weaker as the atoms are separated. Nevertheless, the potential $\frac{1}{2}kx^2$ can be a rough approximation to the energy of the molecule, in the region close to the equilibrium position. Figure 13.23 shows the molecular energy of H_2, and in the region of the minimum a parabola $E - E_0 = \frac{1}{2}k(R - R_0)^2$ has been drawn. You can see that the parabola roughly approximates the actual molecular energy in the neighborhood of the minimum. The constant k can be estimated from the graphs by seeing what value of $R - R_0$ is necessary to give a certain value of $E - E_0$. As shown in the figure, when $E - E_0$ is 0.50 eV, the value of $R - R_0$ is about 0.017 nm, and so we estimate that

$$k = \frac{2\,\Delta E}{(\Delta R)^2} = \frac{2 \times 0.50 \text{ eV}}{(0.017 \text{ nm})^2} \qquad (13.12)$$

$$= 36 \times 10^{20} \text{ eV/m}^2 \qquad (13.13)$$

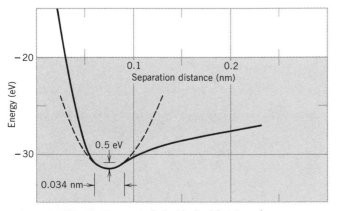

FIGURE 13.23 Fitting a parabola (dashed line) to the energy minimum of H_2.

Since the reduced mass m is just half the hydrogen atom mass, we can estimate the vibrational frequency:

$$\nu = \frac{1}{2\pi}\sqrt{\frac{k}{m}} = \frac{1}{2\pi}\sqrt{\frac{kc^2}{mc^2}} \tag{13.14}$$

$$= \frac{1}{2\pi}\sqrt{\frac{(36 \times 10^{20}\ \text{eV/m}^2)(9.0 \times 10^{16}\ \text{m}^2/\text{s}^2)}{(0.5)(1.008\ \text{u})(931.5\ \text{MeV/u})}} \tag{13.15}$$

$$= 1.3 \times 10^{14}\ \text{Hz} \tag{13.16}$$

This frequency corresponds to a vibrational energy quantum $h\nu$ of 0.54 eV and to a wavelength of 2.3 μm, which is in the infrared region of the electromagnetic spectrum.

A similar calculation in the case of NaCl (see Figure 13.24) gives $\nu = 1.5 \times 10^{13}$ Hz, $h\nu = 0.063$ eV, and $\lambda = 20$ μm, also in the infrared region.

The range of wavelengths in the infrared region, from about 1 to 100 μm, is typical for molecular vibrations.

The excited states of the harmonic oscillator are given by $E = h\nu(N + \frac{1}{2})$. If we were to excite a molecule from its vibrational ground state to an excited state, it would then return to the ground state by emitting one or more photons corresponding to transitions between the different vibrational levels. These transitions are subject to the *selection rule*

$$|\Delta N| = 1 \tag{13.17}$$

Vibrational selection rule

Only transitions that change N by one unit can occur. The vibrational energy levels and the permitted photons are shown in Figure 13.25; notice that all of the transitions have exactly the same energy $h\nu$.

Of course, if we reach a relatively high excited state, we expect that

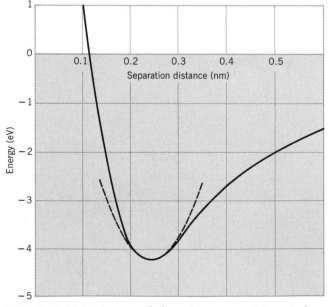

FIGURE 13.24 Fitting a parabola to the energy minimum of NaCl.

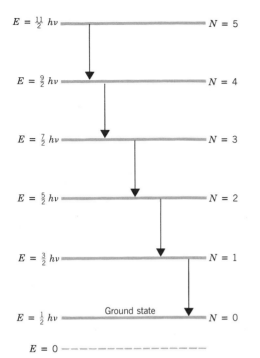

$E = \frac{11}{2} h\nu$ ——————— $N = 5$

$E = \frac{9}{2} h\nu$ ——————— $N = 4$

$E = \frac{7}{2} h\nu$ ——————— $N = 3$

$E = \frac{5}{2} h\nu$ ——————— $N = 2$

$E = \frac{3}{2} h\nu$ ——————— $N = 1$

$E = \frac{1}{2} h\nu$ ——— Ground state ——— $N = 0$

$E = 0$ ------------------

FIGURE 13.25 Energy levels and transitions in a vibrating molecule.

the parabola no longer gives a good approximation to the real energy of the molecule, and we should see differences between the expected and observed transitions. (For NaCl, Figure 13.24 shows that the parabola is a good approximation to the true energy curve for about 0.3 eV; with our estimate of 0.06 eV for the vibrational energy, this corresponds to about $N = 5$.) Above this point, the simple harmonic oscillator no longer accurately describes the vibrational motion; in particular, all the transitions will no longer have the same energy, and transitions with ΔN other than 1 may occur. Nevertheless, our simple model of the oscillating molecule gives results in reasonably good agreement with experiment. We look at some real molecular spectra in the last section of this chapter.

Summarizing our results for the vibrational spectra, we expect many individual transitions to be emitted, all of the same energy, each one corresponding to a change of one unit in the vibrational quantum number N, with wavelengths in the infrared region of the spectrum.

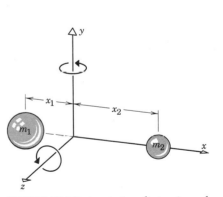

FIGURE 13.26 A rotating diatomic molecule.

A second way that a molecule can change its state of motion when it absorbs or emits radiation is by rotating about the center of mass. Consider the diatomic molecule shown in Figure 13.26. The coordinate system has as its origin the center of mass of the molecule, so that $x_1 m_1 = x_2 m_2$. The rotational kinetic energy is

$$K = \tfrac{1}{2}I\omega^2 \qquad (13.18)$$

where I is the moment of inertia of the molecule and ω is its angular velocity. We can also write the kinetic energy in terms of the angular

13.7 MOLECULAR ROTATIONS

momentum $|\mathbf{L}| = I\omega$

$$K = \frac{|\mathbf{L}|^2}{2I} \qquad (13.19)$$

The moment of inertia of the molecule is

$$I = m_1 x_1^2 + m_2 x_2^2 \qquad (13.20)$$

We can write this in terms of the reduced mass $m = m_1 m_2/(m_1 + m_2)$ and the equilibrium separation $R = x_1 + x_2$ as

$$I = mR^2 \qquad (13.21)$$

The kinetic energy is taken as the total energy of the system which must be quantized, since there is no potential energy. Without going through the solution to the Schrödinger equation for this kind of rotational motion, we can guess at the value of the quantized energies. For the hydrogen atom, the angular momentum $\mathbf{l}$ had a magnitude given by $|\mathbf{l}| = \sqrt{l(l + 1)}\,\hbar$ where $l = 0, 1, 2, 3, \ldots$, and we expect therefore that the energies of the rotating diatomic molecule will be

$$E = \frac{L(L + 1)\hbar^2}{2mR^2} \qquad (13.22) \qquad \textbf{Energies of rotation}$$

where L is a quantum number that takes the values $0, 1, 2, \ldots$, just like the orbital angular momentum of the H atom.

As we excite the molecule to a high rotational state, it drops back toward the ground state by emitting photons corresponding to rotational transitions. The selection rule for these photons is

$$|\Delta L| = 1 \qquad (13.23) \qquad \textbf{Rotational selection rule}$$

The energy of a photon will then just be, as usual, the energy difference between two adjacent levels:

$$\Delta E = E_{L+1} - E_L \qquad (13.24)$$

$$= \frac{(L + 1)(L + 2)\hbar^2}{2mR^2} - \frac{L(L + 1)\hbar^2}{2mR^2} \qquad (13.25)$$

$$= (L + 1)\frac{\hbar^2}{mR^2} \qquad (13.26)$$

You will recall that the transitions among the *vibrational* excitations all had the same energy; however, the energies of the *rotational* transitions depend on L. Figure 13.27 shows a typical sequence of rotational levels and the transitions between them.

The photon energy depends on the quantity $\hbar^2/mR^2$, and we can compute the energies of the photons by estimating this quantity. For H_2, we previously calculated $m \cong 0.5 \times 1.008$ u and $R = 0.074$ nm. Thus:

$$\frac{\hbar^2}{mR^2} = \frac{\hbar^2 c^2}{(mc^2)R^2} = \frac{(\hbar c)^2}{4\pi^2 mc^2 R^2}$$

$$\qquad (13.27)$$

$$= \frac{(1240\text{ eV·nm})^2}{4\pi^2(0.5 \times 1.008\text{ u} \times 931.5\text{ MeV/u})(0.074\text{ nm})^2}$$

$$= 0.0152\text{ eV} \qquad (13.28)$$

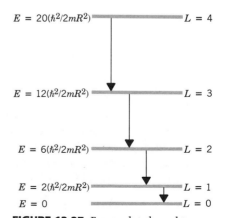

FIGURE 13.27 Energy levels and transitions in a rotating diatomic molecule.

The rotational transitions in hydrogen therefore have the following energies and wavelengths:

$$L = 1 \text{ to } L = 0 \qquad \Delta E = 0.0152 \text{ eV} \qquad \lambda = 81.6 \text{ }\mu\text{m}$$
$$L = 2 \text{ to } L = 1 \qquad \Delta E = 0.0304 \text{ eV} \qquad \lambda = 40.8 \text{ }\mu\text{m}$$
$$L = 3 \text{ to } L = 2 \qquad \Delta E = 0.0456 \text{ eV} \qquad \lambda = 27.2 \text{ }\mu\text{m}$$

and so forth. The emitted energies form a sequence ΔE, $2 \Delta E$, $3 \Delta E$, . . . , and equivalently the emitted wavelengths are λ, $\lambda/2$, $\lambda/3$, ...

The wavelengths are once again in the infrared region, but with values about an order of magnitude larger than those of the vibrational transitions. This region is usually called the "far infrared."

For NaCl, a similar calculation gives $\hbar^2/mR^2 = 53.6 \times 10^{-6}$ eV, and the first few transitions are

$$L = 1 \text{ to } L = 0 \qquad \Delta E = 53.6 \times 10^{-6} \text{ eV} \qquad \lambda = 23.1 \text{ mm}$$
$$L = 2 \text{ to } L = 1 \qquad \Delta E = 107.2 \times 10^{-6} \text{ eV} \qquad \lambda = 11.6 \text{ mm}$$
$$L = 3 \text{ to } L = 2 \qquad \Delta E = 160.8 \times 10^{-6} \text{ eV} \qquad \lambda = 7.71 \text{ mm}$$

In this case the transitions are in the microwave region.

Summarizing, the rotational energy levels are 1 to 2 orders of magnitude closer together than the vibrational energy levels, with transitions in the far infrared or microwave region. The emitted transitions do not all have the same energy, but instead form an ascending sequence of energies (or a descending sequence of wavelengths). The emitted transitions are restricted by the selection rule that permits the rotational quantum number L to change by only one unit.

When we add energy to a molecule, we cannot directly control how much of that energy goes into rotations and how much into vibrations. In fact, both kinds of motion will occur. Moreover, we can't observe a pure rotational or a pure vibrational sequence of emissions. A glance at Figure 13.28 shows why this is true.

Since the rotational energies are much smaller than the vibrational energies, we can represent the molecular excited states as a sequence of equally spaced vibrational levels, with each vibrational level serving as the basis for a sequence of rotational states. The states must be labeled both with the vibrational quantum number N and the rotational quantum number L.

Consider the state with quantum numbers N and L. It cannot make a transition to the next lowest rotational state with quantum numbers N and $L - 1$, because that would vilate the selection rule $|\Delta N| = 1$ for the vibrational part of the motion of the molecule. All transitions must satisfy both selection rules:

$$|\Delta N| = 1 \quad and \quad |\Delta L| = 1 \tag{13.29}$$

The state with quantum numbers N and L can make a transition to the state $N - 1$ and $L - 1$ or $N - 1$ and $L + 1$, as shown in Figure 13.29. This explains why the sequence of rotational transitions $L \rightarrow L - 1 \rightarrow L - 2$ is *not* seen for a given N quantum number.

If we consider the state with quantum numbers N and L, we can

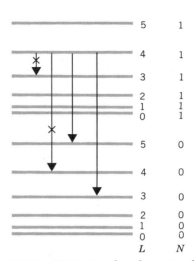

FIGURE 13.28 Combined rotational and vibrational energy levels. Transitions marked with an X violate the selection rules and are not allowed. Note that the rotational spacing is much smaller than the vibrational spacing, so a sequence of rotational states is built on each vibrational state.

13.8 MOLECULAR SPECTRA

Combined vibrational and rotational selection rule

write its energy as the sum of the rotational and vibrational terms:

$$E_{NL} = \left(N + \frac{1}{2}\right) h\nu + \frac{L(L+1)\hbar^2}{2mR^2} \tag{13.30}$$

Since the vibrational term is so much larger than the rotational term, the *emission* wavelengths in the spectrum will usually correspond to $N \to N - 1$, while $L \to L \pm 1$; the *absorption* wavelengths will be those for transitions in which N increases by one unit.

An emission photon will therefore have an energy

$$\Delta E = E_{NL} - E_{N-1,L\pm1} \tag{13.31}$$

$$= \left[\left(N + \frac{1}{2}\right) h\nu + \frac{L(L+1)\hbar^2}{2mR^2}\right]$$

$$- \left[\left(N - \frac{1}{2}\right) h\nu + \frac{(L \pm 1)(L \pm 1 + 1)\hbar^2}{2mR^2}\right] \tag{13.32}$$

$$= h\nu + \frac{\hbar^2}{mR^2}(-L - 1) \qquad \text{for } L \to L + 1 \tag{13.33}$$

Energies of emitted photons

$$= h\nu + \frac{\hbar^2}{mR^2}(L) \qquad \text{for } L \to L - 1 \tag{13.34}$$

We expect to see two sequences of transitions. One sequence, corresponding to $L \to L - 1$, has energies $h\nu + \hbar^2/mR^2, h\nu + 2\hbar^2/mR^2, h\nu + 3\hbar^2/mR^2, \ldots$. Notice that the photon of energy $h\nu$ does not appear in the sequence. This is because we can't have an $L \to L - 1$ transition for $L = 0$; we also can think of $h\nu$ as representing a "pure" vibrational transition, which could only be emitted when $\Delta L = 0$, which is not permitted. The second sequence has energies $h\nu - \hbar^2/mR^2, h\nu - 2\hbar^2/mR^2, h\nu - 3\hbar^2/mR^2, \ldots$. Figure 13.29 shows how these two sequences of photons might look if we measured the energies of the radiation emitted by a molecule.

The same kind of spectrum would be observed if we instead measured the radiation absorbed by the molecule.

Let us compare this expected spectrum with a real spectrum. Figure 13.30 shows the absorption spectrum of the molecule HCl. Although many of the essential features of the expected spectrum are present, there are a number of important differences. We can try, based on our understanding of molecular physics, to understand these discrepancies.

1. *The transitions are not equally spaced.* We expect all of the transitions of the spectrum to be separated by a constant energy $\hbar^2/mR^2$

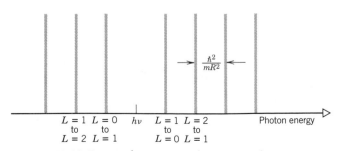

FIGURE 13.29 Expected sequences of transitions between combined rotational and vibrational states.

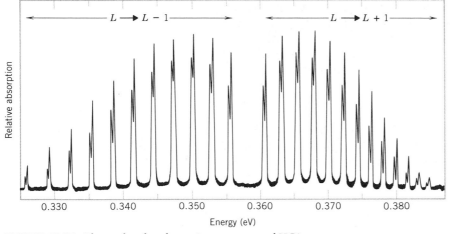

FIGURE 13.30 The molecular absorption spectrum of HCl.

but, as you can see, this is not the case. The explanation for this effect lies with out assumption that the moment of inertia of the molecule is constant. As the molecule rotates, there is an apparent "centrifugal force," which tends to increase the separation of the atoms, increasing the moment of inertia and decreasing the rotational energy. As Figure 13.30 shows, this effect becomes more severe as L increases.

We can calculate the expected energy of these transitions to see how it agrees with experiment. The "missing" central frequency $h\nu$ must be computed from the vibrational properties of the molecule. The force constant k is estimated to be about 11×10^{20} eV/m², and the reduced mass is 0.98 u. The classical oscillation frequency is therefore expected to be:

$$\nu = \frac{1}{2\pi} \sqrt{\frac{k}{m}}$$

$$= \frac{1}{2\pi} \sqrt{\frac{(11 \times 10^{20} \text{ eV/m}^2)(9 \times 10^{16} \text{ m}^2/\text{s}^2)}{(0.98 \text{ u})(931.5 \text{ MeV/u})}} \tag{13.35}$$

$$= 5.2 \times 10^{13} \text{ Hz}$$

and the vibrational energy quantum is:

$$h\nu = 0.22 \text{ eV} \tag{13.36}$$

This is to be compared with the value deduced from the location of the "missing" transition in Figure 13.30:

$$h\nu = 0.358 \text{ eV} \tag{13.37}$$

The agreement is not too bad, considering how difficult it is to estimate k precisely. The rotational spacing should be easier to compute. For HCl, the equilibrium separation is 0.127 nm, so

$$\frac{\hbar^2}{mR^2} = \frac{(hc)^2}{4\pi^2(mc^2)R^2} = \frac{(1240 \text{ eV·nm})^2}{4\pi^2(0.98 \times 931.5 \text{ MeV})(0.127 \text{ nm})^2} \tag{13.38}$$

$$= 0.00265 \text{ eV} \tag{13.39}$$

This is in excellent agreement with the spacing of the transitions near the center of the spectrum, estimated from the figure to be 0.0026 eV.

2. *The heights of the peaks are quite different.* The heights of the peaks give the intensity of the transitions, and the intensity of any transition is proportional to the population of the particular level from which that transition comes. In the case of noninteracting molecules we can assume that the populations are governed by the Boltzmann distribution

$$p(E) = g(E)e^{-E_{NL}/kT} \tag{13.40}$$

In this case the density of states is just the degeneracy of the levels. The vibrational levels are nondegenerate, but the rotational levels have a degeneracy of $2L + 1$, as is usual for angular momentum states. Thus

$$p(E) = (2L + 1)e^{-[(N+1/2)h\nu + L(L+1)\hbar^2/2mR^2]/kT} \tag{13.41}$$

As L becomes large, the populations decrease quite rapidly, as the measured spectrum shows. In the region of intermediate L, the exponential factor causes the populations to decrease while the degeneracy factor $2L + 1$ causes the populations to increase. We can find the location of the maximum of population, and therefore the most intense transitions, by setting dp/dL equal to zero. The result is

$$2L + 1 = \sqrt{\frac{4kT}{\hbar^2/mR^2}} \tag{13.42}$$

For a room temperature measurement, $kT \cong 0.025$ eV, and thus $L \cong 3$ for the maxima, consistent with what is observed in the spectrum.

3. *Each peak appears to be two very closely spaced peaks.* This effect arises because there are two isotopes of Cl, with mass numbers 35 and 37. In our estimates of the rotational and vibrational energies of HCl, we used as the mass of Cl its atomic mass, 35.5. What we should have done was use the mass of one or the other of the two isotopes in the calculation. If we had done so, we would find that the rotational and vibrational frequencies for ^{35}Cl and ^{37}Cl are slightly different, and it is this difference that causes the pairs of peaks that appear in Figure 13.30.

Since ^{37}Cl has a greater mass than ^{35}Cl, the rotational and vibrational energies of H^{37}Cl will be *smaller* than the rotational and vibrational energies of H^{35}Cl. Thus all of the H^{37}Cl transitions appear at energies slightly below the corresponding H^{35}Cl transitions. The isotopic abundance of ^{35}Cl is about three times that of ^{37}Cl, and so the H^{35}Cl transitions have a greater intensity than the corresponding H^{37}Cl transitions.

Just as atomic spectroscopy gives us a way to identify atoms from their characteristic emission or absorption spectrum, *molecular spectroscopy* enables us to identify molecules by the radiation they absorb or emit. Each molecule has its own characteristic "fingerprint" that can be easily recognized. It is important that this technique tells us exactly the composition of the molecule—the number of atoms of each type, the isotopic ratios, even the state of ionization of the molecule. Thus we could easily distinguish CO from CO_2, H^{35}Cl from H^{37}Cl, H_2^+ from H_2.

Molecular spectroscopy

As you might imagine, such a precise identification technique has many applications to areas where it is necessary to identify precisely trace amounts of molecules. Two applications in particular are of interest. The absorption spectra of our atmosphere can be used to identify trace amounts of various pollutants, and thus molecular spectroscopy helps to measure the purity of our air. Similarly, the absorption spectra of interstellar dust are used to identify the molecules which are present in interstellar space. This is the only technique we have (so far) for learning about the formation of complex molecules in our galaxy, because stars are too hot to permit molecules to exist. (Starlight tells us only about the *atoms* that are present in the stellar material.) Unfortunately, our atmosphere absorbs much of the infrared and microwave radiation that characterizes the spectra of these molecules, but spectrometers carried on satellites beyond the atmosphere are permitting the observation of those radiations and the identification of many varieties of molecules, including some relatively complex organic molecules. As an added benefit, when those spectrometers are aimed toward the Earth, they can measure how the infrared radiation emitted by the Earth is absorbed in its atmosphere, and thus detect the presence of various atmospheric pollutants.

SUGGESTIONS FOR FURTHER READING

Some compresensive books on molecular spectroscopy that include introductory as well as advanced material:

G. M. Barrow, *Introduction to Molecular Spectroscopy* (New York, McGraw-Hill, 1962).

M. Karplus and R. N. Porter, *Atoms and Molecules* (Menlo Park, W. A. Benjamin, 1970).

L. Pauling, *The Chemical Bond* (Ithaca, Cornell University Press, 1967).

An advanced but very detailed work:

G. Herzberg, *Molecular Spectra and Molecular Structure* (New York, Van Nostrand, 1950). Volume I covers diatomic molecules.

Two popular-level articles on the search for molecules in space:

B. E. Turner, "Interstellar Molecules," *Scientific American* **228**, 50 (March 1973).

B. Zuckerman, "Interstellar Molecules," *Nature* **268**, 491 (August 1977).

Information on the properties of molecules is tabulated in many locations; see, for example, the *Handbook of Chemistry and Physics* (Chemical Rubber Publishing Co.) or the *Journal of Physical and Chemical Reference Data*.

QUESTIONS

1. Why does H_2 have a smaller radius and a greater binding energy than H_2^+?

2. The molecule LiH has a simple electronic structure. The H atom would like to gain an electron to fill its $1s$ subshell, but the Li atom would similarly like to fill its $2s$ subshell. Based on the atomic structure of H and Li, which would you expect to dominate in the desire for an additional electron? Are the electronegativity values of Table 13.6 consistent with this?

3. In general, would you expect s-s bonds or p-p bonds to be stronger? Why?

4. Explain why the bond angles of the s-p directed bonds (Table 13.3) approach 90° as the atomic number of the central atom increases.

5. How do the molecular force constants k compare with those of ordinary springs? What do you conclude from this comparison?

6. Why is it unnecessary to consider rotations of the diatomic molecule of Figure 13.26 about the x axis? Estimate the moment of inertia for rotations

about the x axis; compare the typical rotational energies with corresponding values for rotations about the y or z axis.

7. Explain how the equilibrium separation in a molecule can be determined by measuring the absorption or emission lines for rotational states.

8. How would the rotational energy spacing of D_2 compare with the rotational spacing of H_2? How would their vibrational energy spacings compare? Their equilibrium separation distances?

9. For a molecule like HCl, estimate the number of rotational levels between the first two vibrational levels.

10. Why does an atom generally only absorb radiation from the ground state, while a molecule can absorb from many excited rotational or vibrational states?

11. If a collection of molecules were all in the $N = 0$, $L = 0$ ground state, how many lines would there be in the absorption spectrum?

PROBLEMS

1. Calculate the ionization energy of H_2.

2. Calculate the Coulomb energy of KBr at the equilibrium separation distance.

3. The ionization energy of potassium is 4.34 eV; the electron affinity of iodine is 3.06 eV. At what separation distance will the KI molecule gain enough Coulomb energy to overcome the energy needed to form the K^+ and I^- ions?

4. Bond strengths are frequently given in units of kilojoules per mole. Find the molecular dissociation energy (in eV) from the following bond strengths: (a) NaCl, 410 kJ/mole; (b) Li_2, 106 kJ/mole; (c) N_2, 945 kJ/mole.

5. (a) Assuming a separation of 0.193 nm, compute the expected electric dipole moment of NaF. (b) The measured dipole moment is 27.2×10^{-30} C·m. What is the fractional ionic character of NaF? (c) Estimate the fractional ionic character using Equations (13.3) and (13.4), and compare with the value found in (b).

6. The equilibrium separation of HI is 0.160 nm and its measured electric dipole moment is 1.47×10^{-30} C·m. Find the fractional ionic character of HI, and compare with estimated values from Equations (13.3) and (13.4).

7. The equilibrium separation of BaO is 0.194 nm and the measured electric dipole moment is 26.5×10^{-30} C·m. Calculate the fractional ionic character, assuming two valence electrons.

8. Show that the angle between bonds in the tetrahedral carbon structure (Figure 13.15) is 109.5°.

9. Derive Equation (13.21) from Equation (13.20).

10. Find the difference in the rotational energy $\hbar^2/mR^2$ for $H^{35}Cl$ and $H^{37}Cl$.

11. Compute the effective force constant k of the HCl molecule from the "missing" transition energy of the absorption spectrum.

12. (a) In a collection of H_2 molecules at room temperature, what is the ratio of the number of molecules in the $N = 1$ vibrational state to the number in the $N = 0$ vibrational state? (Ignore the rotational structure for this problem.) (b) What is the ratio of the number in the $N = 2$ state to the number in the ground state?

13. Repeat Problem 12 for NaCl molecules.

14. (a) In a collection of H_2 molecules at room temperature, find the relative numbers of molecules in the first four rotational states of the $N = 0$ vibrational state. (b) Repeat part (a) if the temperature is 30 K. (*Hint:* Don't forget the degeneracy of the levels.)

15. Repeat Problem 14 for NaCl.

16. At what temperature would 25 percent of a collection of HCl molecules be in the first excited vibrational state? (Ignore the rotational structure.)

17. Figure 13.28 illustrates the combined rotational-vibrational structure when the vibrational energy is much larger than the rotational energy. Make a sketch showing the reverse situation, in which the rotational energy is much larger than the vibrational energy. Use a scale in which $\hbar^2/mR^2 = 20$ units and $h\nu = 2$ units; show rotational levels up to $L = 3$ and vibrational levels to $N = 3$.

18. (a) Sketch a diagram, similar to Figure 13.28, showing all possible absorption transitions from the $N = 0$ to the $N = 1$ states. Include rotational states up to $L = 5$. (b) Use the values $h\nu = 10$ units and $\hbar^2/2mR^2 = \frac{1}{4}$ unit and show the energy spectrum of the absorption, including all transitions from part (a). Label each transition with the initial and final quantum numbers.

19. The most intense absorption line in the rotational-vibrational spectrum of CO at room temperature occurs for $L = 7$. Justify this value with a calculation. (The equilibrium separation of CO is 0.113 nm.)

20. (a) What is the reduced mass of the KCl molecule? (b) With a separation of 0.267 nm, find the spacing of the transitions in the combined rotational-vibrational spectrum.

21. The figure below shows the absorption spectrum of the molecule HBr. Following the basic procedures of Section 13.8, find: (a) the energy of the "missing" transition; (b) the effective force constant k; (c) the rotational spacing $\hbar^2/mR^2$. Estimate the value of the rotational spacing, as in Equation (13.38), and compare with the value deduced from the spectrum. Does the most intense transition occur in the expected location? Why are there only single lines and not double lines as in the case of HCl?

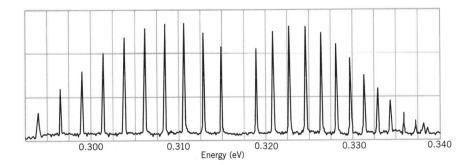

This eerie, futuristic
scene is, in fact, a photograph made with an
electron microscope of
some crystals of the
semiconductor lead tin
telluride.

14

SOLID-STATE PHYSICS

In our study of the structure of matter, we have so far seen how protons and neutrons combine to form atomic nuclei, how nuclei join with electrons to make atoms, and how atoms combine to form molecules, all subject to the laws of quantum physics. The next logical step is to study how the molecules can be combined to form solids, which is the subject of this chapter.

At first thought, there seem to be so many different solids that to classify them and form some general rules for their properties would seem to be a hopeless task. The book you are reading is made of paper and cloth, held together by a glue made from resins, once liquid and now solid. Your desk might be made of wood, of metal, or of plastic; your chair might be made of similar materials, and might perhaps be covered with cloth, leather, or plastic fabric and contain fiber or synthetic foam padding. Around you, there might be many books and papers, pencils made of wood and metal and graphite, rubber erasers, metal and plastic pens, a calculator with a plastic body containing some semiconductors, lights, and switches, and all manner of other objects made of various materials. Looking through the glass window, you see trees, bricks, concrete, metal, selected by humans or by nature for strength, utility, or attractiveness. Each of these solids has a characteristic color, texture, strength, hardness, or ductility; it has a certain measurable electrical conductivity, thermal conductivity, magnetic susceptibility, and melting point; it has certain characteristic emission or absorption spectra in the visible, infrared, ultraviolet, or other regions of the electromagnetic spectrum.

It is a fair generalization to say that all of these properties depend on two features of the structure of the material: the type of atoms or molecules of which the substance is made, and the way those atoms or molecules are joined or stacked together to make the solid. It is the formidable task of the *solid-state* (or *condensed-matter*) *physicist* or *physical chemist* to try to relate the structure of materials with their observed physical or chemical properties.

Many materials have a regular, orderly, and periodic arrangement of atoms or molecules that not only characterizes a substance but also gives it its general properties. Like a brick wall composed of a regular and periodic arrangement of bricks, such substances have a *long-range* order—the placement of an atom or a brick on one side of the material bears a strong relationship to the placement of an atom or a brick on the other side, quite far away compared with the size of an individual atom or brick. This regular arrangement of atoms is called a *lattice,* and materials with such structures are known as *crystals.* Crystals represent only one subset of the entire class of solid substances, but their varieties and properties are relatively easy to list, and so we discuss them briefly in this chapter.

Another class of substances, the *amorphous* solids (amorphous comes from the Greek *without form*) have no such long-range order and resemble more a pile of bricks than a brick wall. Even a random pile of bricks can have a set of well-defined properties—it is relatively strong (although not as strong as a brick wall) and has a certain average spacing between the bricks, determined by the size and shape of individual bricks. It may even have some *short-range order*—a clump of a few bricks attached together in a regular way, like a small piece of brick wall; however, there is no long-range order—the placement of a brick on one side of the pile doesn't depend at all on the placement of bricks on the other side. Because of the lack of form, the properties of amorphous substances (glass and paper are common examples) usually depend more on the properties of individual atoms or molecules and on details of the short-range bonding than on the type of construction of the solid. Since we are interested in the general properties of solids in this chapter, we shall discuss only crystalline materials.

One of the most important effects of quantum physics on the properties of solids is in the area of electrical conductivity. Here the properties of the atomic or molecular wave functions and the quantum statistical distributions (Fermi-Dirac, for electrons) combine to give various materials the properties of conductors, insulators, or the important third category, semiconductors. The quantum mechanics of these systems takes the form of the band theory of solids; we will discuss the band theory and show how it can be applied to understanding the properties of simple semiconducting devices.

Solids are held together by electrostatic forces, and we will characterize those forces by the *cohesive energy* of the solid. The cohesive energy, which is of the order of electron-volts, is the "binding energy" of an atom or a molecule in a solid; it is also the energy per atom we must supply to take the solid apart into individual atoms or molecules.

14.1 IONIC SOLIDS

In our study of ionic bonding in molecules, we learned that the cohesive forces in ionic molecules originated from the electrostatic attraction between a closed-shell ion, such as Na^+, and another closed-shell ion, such as Cl^-. Such materials are expected to form solids readily, because a Na^+ ion could simultaneously attract many Cl^- ions to itself, thereby building up a solid structure. Ionic bonding is a particularly strong type of bonding, and since the forces are electrostatic in origin, we might suppose that the more negative ions we surround a positive ion with, the more stable and strong will be the solid. (Covalent bonds, on the other hand, involve specific electron wave functions and so are limited in the number of near neighbors that can participate in the bonding.)

Ionic solids are crystalline, rather than amorphous, because we can pack ions together more efficiently in a regular array than in a random arrangement. (We can also stack bricks more efficiently in a regular array than in a random pile. There are more bricks per unit volume, and they are closer together, in the regular array.) Also, since the filled shells of the ions are spherically symmetric, the ionic bonds in a solid are not directional—all directions in space are the same for an ion, and we therefore expect ionic solids similarly not to distinguish one direction in space from another.

The type of crystal lattice that is equivalent in all directions in space is the *cubic* lattice, in which we imagine the atoms to be placed at **Cubic lattice** the corners of a succession of cubes which cover the volume of the crystal. Figure 14.1 shows how this would look if we imagine the atoms to stack like hard spheres. This type of stacking is not the most efficient, because there are large gaps at the center of each face of the cube, and also in the middle of the cube itself. We get a better stacking arrangement, which has more atoms per unit volume, if we place another sphere in the gap at the center of each face of the cube, or else at the center of the body of the cube (even though in both cases the spheres at the corners are no longer in direct contact with one another). These two lattices are known as *face-centered cubic* (fcc) and *body-centered cubic* (bcc) and are illustrated in Figures 14.2 and 14.3.

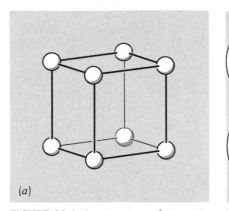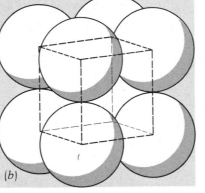

FIGURE 14.1 Arrangement of atoms in a simple cubic crystal. Part (a) shows the relationship of the bonds and part (b) shows the packing of atoms.

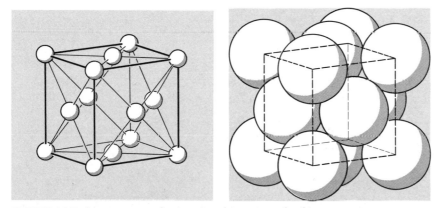

FIGURE 14.2 Arrangement of atoms in a face-centered cubic crystal.

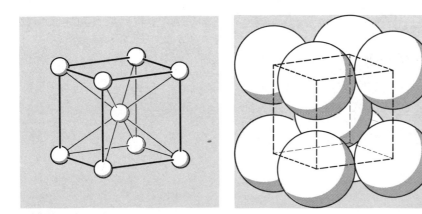

FIGURE 14.3 Arrangement of atoms in a body-centered cubic crystal.

The fcc lattice gives a slightly more efficient packing (more atoms per unit volume) and so it is usually the most stable structure. However, atoms do not stack like hard spheres, and often the bcc structure is preferred. These two crystal types, fcc and bcc, also occur for materials other than ionic solids, and in fact some materials will change their structure from fcc to bcc as the temperature is changed.

The most familiar fcc lattice is NaCl, and for that reason the fcc lattice is often called *NaCl structure.* In order to have the atoms attract one another, we must alternate Na^+ and Cl^- ions, as is shown in Figure 14.4. Notice that a given Na^+ ion is attracted by 6 close Cl^- neighbors, and does not "belong to" any single Cl^- ion. *It is therefore wrong to consider ionic solids as being composed of molecules.*

A typical bcc structure is CsCl, as shown in Figure 14.5, and so the bcc lattice is often known as the *CsCl structure.* In this case each ion is surrounded by 8 neighbors of the opposite charge.

The forces that hold the ionic solid together, like those that hold ionic molecules together, are of simple electrostatic nature—each Na^+ in the NaCl structure feels an attractive potential from 6 nearby Cl^- ions. We therefore expect that the attractive potential energy will be:

$$V = -6\frac{e^2}{4\pi\varepsilon_0}\frac{1}{r} \qquad (14.1)$$

where r is the separation distance of the Na^+ and Cl^- ions. However, just beyond those 6 Cl^- ions, at a distance $r\sqrt{2}$, are 12 Na^+ ions that exert a somewhat weaker *repulsive* electrostatic force on the Na^+ ion, and therefore tend to make the equilibrium position less stable. Beyond those 12 Na^+ are still more Cl^-, with an attractive potential, and so forth. The net effect of all of these attractive and repulsive forces is a resultant attractive electrostatic potential

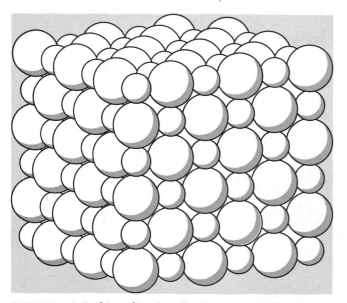

FIGURE 14.4 Packing of Na (small spheres) and Cl (large spheres) in a crystal of NaCl.

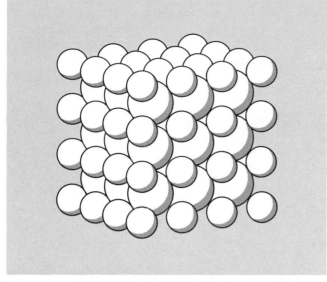

FIGURE 14.5 Packing of Cs (large spheres) and Cl (small spheres) in a crystal of CsCl.

$$V_{\text{attractive}} = -\alpha \frac{e^2}{4\pi\varepsilon_0} \frac{1}{r} \qquad (14.2)$$

where α, called the *Madelung constant*, gives the strength of the net electrostatic attraction and is calculated by adding up the sequence of various attractive and repulsive contributions. (The convergence of the series is, however, extremely slow, so some special tricks must be used in carrying out the sum.) The Madelung constant is merely characteristic of the basic structure of the lattice and does not depend on the type of atoms of which the lattice is composed. For ionic solids it has the following values:

fcc (NaCl structure): $\alpha = 1.7476$

bcc (CsCl structure): $\alpha = 1.7627$

Just as in the case of ionic molecules, there is an additional repulsive force that keeps the atoms from getting too close together. This force arises because of the filled electronic subshells of the ions; the Pauli principle tends to keep those filled subshells from overlapping. This force is quite complicated to calculate precisely, but the repulsive potential can be rather well approximated as

$$V_{\text{repulsive}} = Ar^{-n} \qquad (14.3)$$

where A is the strength of the interaction and n determines how rapidly the potential increases as r decreases. The total interaction of the ions is then the sum of the attractive and repulsive contributions:

$$V_{\text{total}} = -\alpha \frac{e^2}{4\pi\varepsilon_0} \frac{1}{r} + \frac{A}{r^n} \qquad (14.4)$$

The equilibrium separation of the ions comes about when V has its minimum value V_0 (Figure 14.6) and we can impose this condition by setting

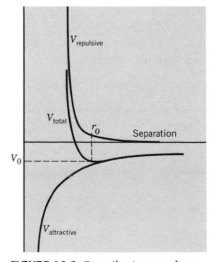

FIGURE 14.6 Contributions to the energy of an ionic crystal.

Table 14.1 Properties of Ionic Crystals

Crystal	Nearest-Neighbor Separation (nm)	Atomic Cohesive Energy (eV)	n	Structure
LiF	0.201	8.52	6	fcc
LiCl	0.257	6.85	7	fcc
NaCl	0.281	6.39	8	fcc
NaI	0.324	5.00	9.5	fcc
KCl	0.315	6.46	9	fcc
KBr	0.330	5.89	9.5	fcc
RbF	0.282	7.09	8.5	fcc
RbCl	0.329	6.34	9.5	fcc
CsCl	0.356	6.46	10.5	bcc
CsI	0.395	5.35	12.0	bcc
MgO	0.210	9.34	7.0	fcc
BaO	0.275	8.90	9.5	fcc

$dV/dr = 0$ at $r = r_0$. This gives a relationship between A, r_0, and n:

$$A = \frac{\alpha e^2 \, r_0^{n-1}}{4\pi\varepsilon_0 \, n} \tag{14.5}$$

and the resulting potential energy at $r = r_0$ is

$$V_0 \equiv V_{\text{total}}(r_0) = -\frac{\alpha e^2}{4\pi\varepsilon_0 r_0}\left(1 - \frac{1}{n}\right) \tag{14.6}$$ **Ionic cohesive energy**

The energy $-V_0$ is the *ionic cohesive energy* of the solid; it is the energy needed to take the solid apart into positive and negative ions, *not* into neutral atoms.

The value of n for most ionic solids is in the range 8 to 10, and so the actual value of n does not greatly affect the value of V_0. The difference between $(1 - \frac{1}{8})$ and $(1 - \frac{1}{10})$ is only 2.5 percent.

Table 14.1 gives some properties of representative ionic solids. The *atomic cohesive energy* given in Table 14.1 is the "binding energy" of the solid relative to *neutral atoms*, not relative to positive and negative ions. The following examples illustrate the relationship between the ionic and atomic cohesive energies.

Calculate the atomic cohesive energy of NaCl. **EXAMPLE 14.1**

From Table 14.1, we find $r_0 = 0.281$ nm and $n = 8$ for NaCl. From Equation (14.6), **Solution**

$$V_0 = -(1.7476)\frac{1.44 \text{ eV·nm}}{0.281 \text{ nm}}\left(1 - \frac{1}{8}\right)$$

$$= -7.84 \text{ eV}$$

This represents the energy needed to remove Na^+ and Cl^- ions from the solid. To form Na^+ from neutral Na atoms requires 5.14 eV (the ionization energy of Na), and to form Cl^- from neutral Cl atoms releases 3.61 eV. Given Na^+ and Cl^-, we supply 3.61 eV to form Cl, and then get back 5.14 eV when we form Na. The energy needed to form the neutral atoms from the solid is

$$-7.84 \text{ eV} - 3.61 \text{ eV} + 5.14 \text{ eV} = -6.31 \text{ eV}$$

and so the computed atomic cohesive energy is 6.31 eV. This is in good agreement with the measured value 6.39 eV given in Table 14.1.

What is the experimental value of the ionic cohesive energy of NaCl in kilojoules per mole? **EXAMPLE 14.2**

The experimental atomic cohesive energy is 6.39 eV from Table 14.1. To convert *Solution*
this to *ionic* cohesive energy we reverse the procedure of the previous example, and find:

$$-6.39 \text{ eV} + 3.61 \text{ eV} - 5.14 \text{ eV} = -7.92 \text{ eV}$$

The ionic cohesive energy is 7.92 eV per Na^+ and Cl^- ion. One mole of NaCl contains 6.02×10^{23} ion pairs, and so the cohesive energy can be expressed as:

$$7.92 \text{ eV} \times 1.60 \times 10^{-19} \text{ J/eV} \times 6.02 \times 10^{23} = 763 \text{ kJ/mole}$$

Based on the cohesive energies of ionic solids given in Table 14.1, we might expect the following properties of the ionic solids:

1. They should form relatively stable and hard cubic crystals.

2. They should be poor electrical conductors, since there are no free electrons available.

3. They should have high vaporization temperatures. Since the cohesive *Properties of ionic solids*
energies are of the order of several electron-volts, in order to vaporize the solid we must heat it to a temperature high enough so that $kT \sim$ 1 eV, or $T \sim$ 12,000 K. This is quite a high estimate; actual values are more like 1000 to 2000 K, still a relatively high temperature.

4. They should be transparent to visible radiation. A photon of visible light will only interact with a material if it can excite an electron from its ground state to an excited state. Since the ionic solids all have filled shells, it is necessary to excite an electron from one shell to the next. The energy required for this is typically much larger than the energies of photons of visible light, and so the photons pass right through ionic solids.

5. Ionic solids absorb strongly in the infrared. We can show this by considering the *force* on an ion in a solid. We can estimate this force from Equation (14.4):

$$F = -\frac{dV}{dr} = -\frac{\alpha e^2}{4\pi\varepsilon_0 r^2}\left(1 - \frac{r_0^{n-1}}{r^{n-1}}\right) \qquad (14.7)$$

Suppose we displace the atom by a small distance x from r_0:

$$F = -\frac{\alpha e^2}{4\pi\varepsilon_0(r_0+x)^2}\left(1 - \frac{r_0^{n-1}}{(r_0+x)^{n-1}}\right) \qquad (14.8)$$

$$= -\frac{\alpha e^2}{4\pi\varepsilon_0 r_0^2}\frac{1}{(1+x/r_0)^2}\left(1 - \frac{1}{(1+x/r_0)^{n-1}}\right) \qquad (14.9)$$

$$\cong -\frac{\alpha e^2}{4\pi\varepsilon_0 r_0^2}\left(1 - \frac{2x}{r_0}\right)(n-1)\frac{x}{r_0} \qquad (14.10)$$

The ion thus experiences a restoring force $F = -kx$ with:

$$k \cong \frac{\alpha e^2}{4\pi\varepsilon_0 r_0^3}(n-1) \qquad (14.11)$$

and oscillates with a frequency $\nu = (1/2\pi)\sqrt{k/m}$. For NaCl, we estimate the force constant to be:

$$k = \frac{7(1.75)(1.44 \text{ eV·nm})}{(0.281 \text{ nm})^3} = 795 \text{ eV/nm}^2$$

and the corresponding frequency is:

$$\nu = \frac{1}{2\pi}\sqrt{\frac{795 \text{ eV/nm}^2 \ (9 \times 10^{16} \text{ m}^2/\text{s}^2)(10^{18} \text{ nm}^2/\text{m}^2)}{23 \text{ u} \times 931.5 \text{ MeV/u} \times 10^6 \text{ eV/MeV}}}$$

$$= 9.20 \times 10^{12} \text{ Hz}$$

(We have used the mass of a Na atom for this estimate.) Photons of this frequency have a wavelength of 32.6 μm, in the infrared region. A beam of infrared radiation of wavelength 32.6 μm will set the Na ions oscillating, and in the process the radiation is absorbed.

6. Ionic solids are usually soluble in polar liquids such as water. The water molecule has an electric dipole moment, which can exert an attractive force on the charged ions, breaking the ionic bonds and dissolving the ionic solid.

14.2 COVALENT SOLIDS

As we have seen in the previous chapter, carbon forms molecules by co-valent bonding of its four outer electrons in sp^3 hybrid orbits. Such bonds are highly *directional* and we have seen how it is possible to calculate the angle between the bonds based on the symmetry of the bonding configuration. Solid carbon, in the form of diamond, is an example of a solid in which the interatomic forces are also of a covalent nature. As in a molecule, the four equivalent sp^3 hybrid orbitals participate in covalent bonds, and since they are equivalent they must make equal angles with one another. The manner in which this is done is shown in Figure 14.7. A central carbon atom is covalently bound to four other carbons that occupy four corners of a cube as shown. The angle between the bonds is 109.5°, as it was in the covalently bonded molecules.

Figure 14.8 illustrates how the solid structure characteristic of diamond is constructed of such bonds. Each carbon has four close neighbors with which it shares electrons in covalent bonds. The basic structure is known as *tetrahedral*, and many compounds have a similar structure as a result of covalent bonding. Table 14.2 shows some of these compounds. The structure is also known as the *zinc sulfide* or *zinc blende* structure.

Some of the covalent solids listed in Table 14.2 have bond energies larger than those of ionic solids. Such substances as diamond and silicon carbide are particularly hard. Other covalent solids with structures similar to carbon are silicon and germanium; the structure of these solids is responsible for their behavior as semiconductors.

The covalent solids do not have the same similarity of characteristics that ionic solids do, and so we cannot make the same generalizations. Carbon, in the diamond structure, has a large bond energy and

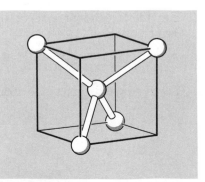

FIGURE 14.7 The tetrahedral structure of carbon.

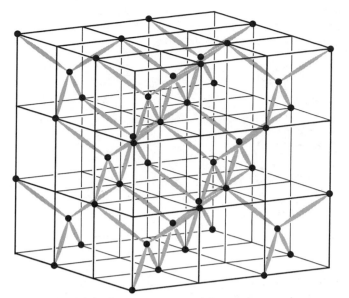

FIGURE 14.8 The lattice structure of diamond.

is therefore very hard and transparent to visible light; germanium and tin have similar structures, but are metallic in appearance and highly reflective. Similarly, carbon (as diamond) has an extremely high melting point (4000 K); germanium and tin melt at much lower temperatures more characteristic of ordinary metals. Of course, these differences depend on the actual bond energy in the solid, which in turn depends on the type of atoms of which the solid is made. Those solids with large bond energies are hard, have high melting points, are poor electrical and thermal conductors, and are transparent to visible light. Those solids with small bond energies may have very different properties.

Metallic Bonds The valence electrons in a metal are usually rather loosely bound, and frequently the electronic shells are only partially filled, so that metals tend not to form covalent bonds. The basic structure of metals is a "sea" or "gas" of approximately free electrons surrounding a lattice of positive ions. The metal is held together by the attractive force between each individual metal ion and the electron gas.

14.3 OTHER SOLID BONDS

Table 14.2 Some Covalent Solids

Crystal	Nearest-Neighbor Distance (nm)	Cohesive Energy (eV)
ZnS	0.235	6.32
C (diamond)	0.154	7.37
Si	0.234	4.63
Ge	0.244	3.85
Sn	0.280	3.14
CuCl	0.236	9.24
GaSb	0.265	6.02
InAs	0.262	5.70
SiC	0.189	12.3

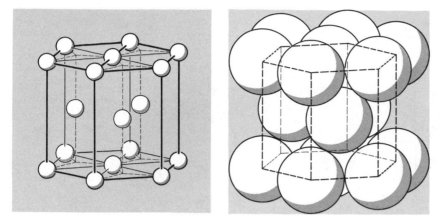

FIGURE 14.9 Arrangement of atoms in a hexagonal close-packed crystal.

The most common crystal structures of metallic solids are fcc, bcc, or a third type known as *hexagonal close packed* (*hcp*). The hcp structure is shown in Figure 14.9; like the fcc structure, it is a particularly efficient way of packing atoms together. Some metals and their characteristics are shown in Table 14.3. The cohesive energy of metal bonds tends to fall in the range 1 to 3 eV, making the metals less strongly bound than ionic or covalent solids. As a result, metals interact strongly with photons of visible light, and so metals are not transparent. In fact, since the electrons behave cooperatively, metals are highly reflective. The free electrons are responsible for the excellent electrical and thermal conductivity of metals. Since metallic bonds don't depend on any particular sharing or exchange of electrons between specific atoms, the exact nature of the atoms of the metal is not as important as it is in the case of ionic or covalent solids; as a result we can make many kinds of metallic alloys by mixing together different metals in varying proportions.

Molecular Solids None of the solids we have discussed so far can be considered as composed of individual molecules. It is, however, possible for molecules to exert forces on one another and to bind together in solids. Since molecules have electrons that are already shared in *molecular* bonds, there are no available electrons to participate in ionic, cova-

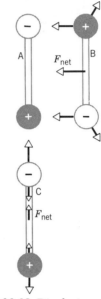

FIGURE 14.10 Dipole A exerts attractive forces on dipoles B and C. Each charge is both attracted and repelled, and the net effect is attraction.

Table 14.3 Structure of Metallic Crystals

Metal	Crystal Type	Nearest-Neighbor Distance (nm)	Cohesive Energy (eV)
Fe	bcc	0.248	4.32
Li	bcc	0.304	1.66
Na	bcc	0.372	1.13
Cu	fcc	0.256	3.52
Ag	fcc	0.289	2.97
Pb	fcc	0.350	2.04
Co	hcp	0.251	4.43
Zn	hcp	0.266	1.35
Cd	hcp	0.298	1.17

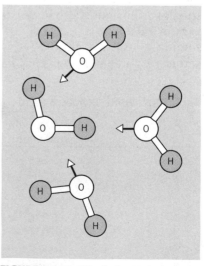

FIGURE 14.11 Dipole forces between water molecules.

lent, or metallic bonds with other molecules. Molecular solids are held together by much weaker forces, which generally depend on the *electric dipole moments* of the molecules. These forces between one molecule and another are much weaker than the internal forces that hold a molecule together; thus a molecule *can* retain its identity in a molecular solid.

The ionic, covalent, and metallic solids are all held together by ordinary $1/r^2$ Coulomb forces; in ionic solids, the force is between positive and negative ions, while in covalent and metallic solids, it is between the positive ions and the shared electrons. In a molecular solid, the molecules are electrically neutral, and so no Coulomb forces are possible. It is, however, possible for the dipole moment of one molecule to exert an attractive force on the dipole moment of another. This cohesive force is proportional to $1/r^3$, so it is in general weaker than the Coulomb force. Molecular solids are therefore more weakly bound and have lower melting points than ionic, covalent, or metallic solids, because it takes less thermal energy to break the bonds of a molecular solid.

Electric dipole forces will be important in molecular bonding for all molecules that have permanent electric dipole moments (*polar molecules*). We can imagine such forces as occurring when the positive end of one dipole exerts an attractive force on the negative end of the next, as shown in Figure 14.10.

The water molecule is an example of a case in which such forces occur. The oxygen atom in water tends to attract all of the electrons of the molecule and so looks like the negative end of the dipole; the two "bare" protons are the positive ends of the dipole, and each can attract the negative oxygen of adjacent water molecules (Figure 14.11). Although it is not obvious from this oversimplified explanation, it is this sort of bonding that causes the characteristic hexagonal crystal structure of ice, and is therefore responsible for the beautiful hexagonal patterns of snowflakes. When bonding of this sort involves hydrogen atoms, as it does in water, it is known as *hydrogen bonding*.

It is also possible to have dipole forces exerted between molecules that have no permanent dipole moments. Consider an atom of helium. The two electrons of helium are in s states, and so helium is spherically symmetric—on the average, it looks like a nucleus of charge $+2e$ surrounded by a sphere of charge $-2e$ of radius 0.1 nm (Figure 14.12). A "snapshot" of a helium atom would in fact look like Figure 14.12. Let us suppose we used instead a camera that could violate the uncertainty principle and show the exact positions of the electrons; a photograph of the helium atom might then look like Figure 14.13. Such an atom may have an *instantaneous* electric dipole moment, although its *average* electric dipole moment is zero. Any measurement we could do in the laboratory would reveal this average value (zero) and not the instantaneous value. A neighboring atom, however, is not subject to such restrictions and it can feel the electric field produced by the instantaneous electric dipole moment. The effect on the neighboring atom might be to attract its negative charge and repel its positive charge, as shown in Figure 14.14. The neighboring atom therefore acquires an *induced* dipole moment, and two atoms can now attract one another through dipole-dipole forces. (A magnet attracts an unmagnetized piece of iron through a similar process.) A snapshot taken an instant later

FIGURE 14.12 The spherically symmetric electronic charge distribution of helium.

FIGURE 14.13 An imaginary "snapshot" of a helium atom, showing the instantaneous positions of the two electrons.

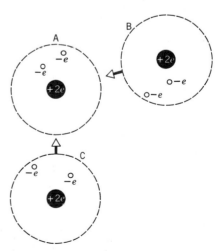

FIGURE 14.14 Electric dipole forces between helium atoms. The instanteous dipole moment of atom A induces dipole moments in atoms B and C. Compare the dipoles with Figure 14.10.

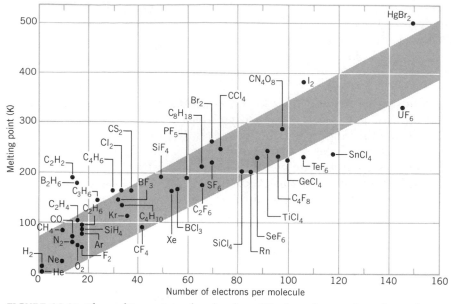

FIGURE 14.15 The melting points of molecular solids depend approximately on the number of electrons per molecule.

might show different positions of the electrons, and a different induced dipole moment of the neighboring atom, but still an attractive force between the neighbors. Such attractive forces, among molecules which have no permanent electric dipole moments, are known as *van der Waals forces* and are responsible for the cohesion of solids containing nonpolar molecules. (They are also responsible for such physical phenomena as surface tension and friction.) Examples of solids that are bound by these forces are those whose atoms are the inert gases, halogens, other gases such as H_2, N_2, or O_2, and also symmetric molecules such as CH_4 or $GeCl_4$.

Van der Waals force

The van der Waals forces are extremely weak, and depend on the separation distance like $1/r^7$. Solids bound by such forces therefore have low melting points (lower even than hydrogen-bonded solids), again because very little thermal energy is required to break the bonds. In fact, since the induced dipole moment of an atom should be approximately proportional to the *total number* of electrons in the atom, we might expect that the melting points of nonpolar molecular solids should be roughly proportional to the number of electrons in each molecule. Figure 14.15 shows this relationship; although the properties of the individual solids cause considerable scatter of the points, there is a rough relationship of the sort expected.

When two identical atoms, such as sodium, are very far apart, the electronic levels in one are not affected by the presence of the other, and we may think of the atoms as being isolated. The $3s$ electron of each atom will have a single energy with respect to its nucleus. As we bring the atoms closer together, the electron wave functions start to overlap, and the interaction between the atoms causes two different $3s$ levels to form, depending on whether the two wave functions add or subtract.

14.4 BAND THEORY OF SOLIDS

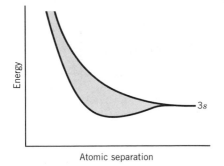

FIGURE 14.16 Splitting of 3s levels when two atoms are brought together.

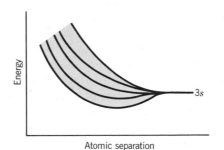

FIGURE 14.17 Splitting of 3s levels when five atoms are brought together.

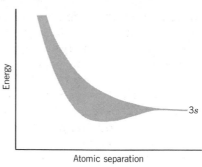

FIGURE 14.18 Formation of 3s band by a large number of atoms.

This is just the effect that is responsible for molecular binding, which we discussed in Section 13.2. A representation of the energy levels is shown in Figure 14.16.

As we bring together large numbers of atoms, in order to form a solid, the same sort of effect occurs. When the sodium atoms are far apart, all 3s electrons have the same energy, and as we begin to move them together, the energy levels begin to "split." The situation for five atoms is shown in Figure 14.17. There are now five energy levels that the five overlapping electron wave functions produce for the combined system. As the number of atoms is increased to the very large numbers which would be characteristic of an ordinary piece of metal (perhaps 10^{22} atoms) the levels become so numerous and so close together that we can no longer distinguish the individual levels, as shown in Figure 14.18. We can regard the N atoms as forming an almost continuous *band* of energy levels. Since those levels were identified with the 3s atomic levels of sodium, we refer to the 3s *band.*

Each energy band has a total of N individual levels. Each level can hold $2 \times (2l + 1)$ electrons (corresponding to the two different orientations of the electron spin and the $2l + 1$ orientations of the electron orbital angular momentum) so that the capacity of each band is $2(2l + 1)N$ electrons.

Figure 14.19 shows a more complete representation of the energy bands in sodium metal. The 1s, 2s, and 2p bands are each full; the 1s and 2s bands each contain $2N$ electrons and the 2p band contains $6N$ electrons. The 3s band *could* accommodate $2N$ electrons as well; however, the N atoms each contribute only one 3s electron to the solid, and so there is a total of only N 3s electrons available. The 3s band is therefore only half full. Above the 3s band is a 3p band, which could hold $6N$ electrons, but which is completely empty.

The situation we have described is, of course, the ground state of sodium metal. When we add energy to the system (thermal or electrical energy, for example) the electrons can move from the filled states to any of the empty states. In this case, electrons from the partially full 3s band could absorb a small amount of energy and move to higher 3s states, still within the 3s band, or could absorb a larger amount of energy and move to the 3p band.

We can describe this situation in a more correct way if we recall our discussions of quantum statistics from Chapter 12. Electrons are

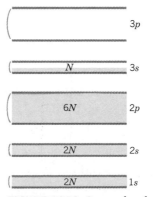

FIGURE 14.19 Energy bands in sodium metal.

described by the Fermi-Dirac distribution. At a temperature of $T = 0$ K, all electron levels below the Fermi energy E_F are filled and all levels above the Fermi energy are empty, as shown in Figure 14.20. In the case of sodium, the Fermi energy would be just in the middle of the $3s$ band, since below that energy all electron levels are occupied. At higher temperatures the Fermi energy gives the level at which the occupation probability is 50 percent; the Fermi energy does not change significantly as we increase the temperature, but the occupation probability of the levels above E_F is no longer zero. Figure 14.21 shows a situation in which the thermal excitation of electrons leads to a small population of the $3p$ band.

Sodium is an example of a substance that is a good electrical conductor. When we apply a very modest potential difference, of the order of 1 V, electrons can easily absorb energy because there are N unoccupied states within the $3s$ band, all within an energy of 1 eV. Electrons absorb this energy as they are accelerated by the applied voltage, and they are therefore free to move as long as there are many unoccupied states within the accessible energy range. In sodium there are N relatively free electrons that can easily move to N unoccupied energy states, and sodium is therefore a good conductor.

A material in which one band is completely full and the next highest completely empty would, on the other hand, be a poor conductor. We refer in this case to the *gap* between the energy bands, and the Fermi energy lies somewhere in the gap as shown in Figure 14.22. Once again, at $T = 0$, all states below E_F are filled. As the temperature is raised, the shape of the Fermi-Dirac distribution function changes; if the energy gap between the bands is large compared with kT, there will be few electrons thermally excited from the lower band, called the *valence band*, to the upper band, called the *conduction band*. This situation is shown in Figure 14.23. There are many electrons in the valence band available for electrical conduction, but there are very few empty states for them to move through, so they do not contribute to the electrical conductivity. There are many empty states in the conduction band, but at ordinary temperatures there are so few electrons in that band that their contribution to the electrical conductivity is also very small. These substances are classified as *insulators* and in general they have two properties: a large energy gap (a few electron-volts) between the va-

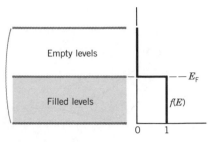

FIGURE 14.20 Detail of a half-filled band, at $T = 0$, showing the Fermi-Dirac distribution on the right.

Energy gap, valence and conduction bands

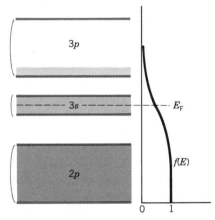

FIGURE 14.21 Population of energy bands in sodium at $T > 0$. The $2p$ band is no longer completely full and the $3p$ band is no longer completely empty.

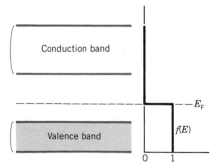

FIGURE 14.22 Band structure in which E_F lies in the gap between bands.

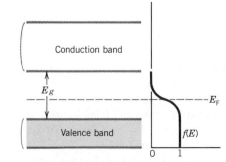

FIGURE 14.23 When $E_g \gg kT$, the conduction band is still unpopulated. This situation is characteristic of an insulator.

lence and conduction bands, and a Fermi level that is in the gap between the bands (i.e., a filled valence band and an empty conduction band).

When a material has the basic structure of an insulator, but with a much smaller energy gap (1 eV or less) the behavior becomes quite different, and these materials are known as *semiconductors*. Figure 14.24 shows a representation of such a substance at ordinary temperatures. There is now a reasonable number of electrons in the conduction band, and of course many empty states accessible to them, so that they can conduct relatively easily. There is also a reasonable number of empty states in the valence band, so that some of the many electrons in the valence band can also contribute to the electrical conductivity by moving about through those states. We consider these two mechanisms of electrical conduction in detail in Section 14.7. For now we note two characteristic properties of semiconductors that relate directly to the band structure as shown in Figure 14.24. (1) Because thermal excitation across the gap is relatively probable, the electrical conductivity of semiconductors depends more strongly on temperature than the electrical conductivity of insulators or conductors. (2) It is possible to alter the structure of these materials, by adding impurities in very low concentration, in such a way that the Fermi energy changes and may move up toward the conduction band or down toward the valence band. This process, known as *doping*, can obviously have a great effect on the conductivity of the semiconductor.

In the examples we have discussed so far, it is not apparent why the band theory is so useful in understanding the properties of a solid. Sodium, for example, is expected to be a good conductor based on its atomic properties alone (a relatively loosely bound $3s$ electron); solid xenon has only filled atomic shells and should be a poor conductor. These conclusions follow either from simple atomic theory or from band theory. However, there are many cases in which atomic theory leads to wrong predictions while band theory gives correct results. We consider two examples. (1) Magnesium has a filled $3s$ shell, and on the basis of atomic theory alone we expect it to be a poor electrical conductor. It is, however, a very good electrical conductor. (2) The $2p$ shell of carbon has only two electrons of the maximum number of six. Carbon should therefore be a relatively good conductor; instead it is an extremely poor conductor. We can understand both of these materials based on the unusual way the bands of these solids behave when the atoms are close enough so that the band gap disappears and the bands overlap. In magnesium (Figure 14.25), for example, the (filled) $3s$ and (empty) $3p$ bands overlap, and the result is a single band with a capacity of $2N + 6N = 8N$ levels. Only $2N$ of those are filled, and so magnesium behaves like a material with a single band filled only to one-fourth its capacity. Magnesium is therefore a very good conductor. In carbon, the extreme overlap of the electronic wave functions at close range first causes mixing of the $2s$ and $2p$ bands, in a way similar to magnesium; a single band is created with a capacity of $8N$ electrons (Figure 14.26). As the atoms approach still closer, the band divides into two separate bands, each with a capacity of $4N$ electrons. Since carbon has four valence electrons (two $2s$ and two $2p$), the lower $4N$ states are completely filled and the upper $4N$ states of the conduction band are completely empty. Carbon is therefore an insulator. Germanium and silicon have

Semiconductors

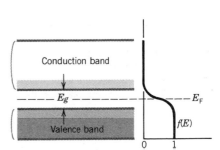

FIGURE 14.24 Band structure of a semiconductor. The gap is much smaller than in an insulator, so there is now a small population of the conduction band.

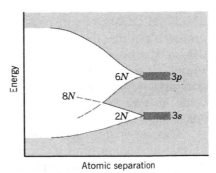

FIGURE 14.25 Band structure in magnesium. The $3s$ and $3p$ bands overlap, forming a single band.

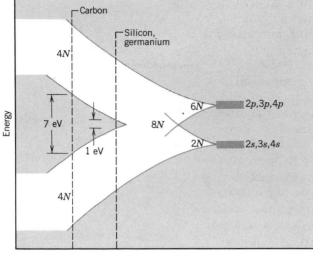

FIGURE 14.26 Band structure of carbon, silicon, and germanium. The combined $ns + np$ band splits into two bands. Carbon is an insulator because the gap is large; Ge and Si are semiconductors because their energy gaps are smaller.

the same type of structure as carbon, but their equilibrium separation is greater, and the gap between the valence and conduction bands is smaller, about 1 eV; it is this feature that causes Ge and Si to be semiconductors.

In this section we have discussed a very schematic approach to the band theory of solids. Let us summarize the main features of this theory:

1. When many atoms are brought together to form a solid, the interactions between the atoms cause the discrete energy levels of the isolated atoms to "smear out" into bands.

2. The properties of the bands are determined by the atomic properties of the isolated atoms and by the equilibrium separations of the atoms in the solid.

3. The properties of the solid are determined by the occupation of the bands, by the spacing between the bands, and by the relative location of the Fermi energy.

Summary of band theory

The band theory of solids has had great success in accounting for the properties of metals, insulators, and semiconductors. Our explanation of the origin of these energy bands was, however, only schematic—somehow the act of bringing the atoms closer together is responsible for the existence of bands, caused by the overlap of the electron wave functions. In this section we consider quite a different approach that also suggests the existence of energy bands.

Our approach here is the rigorous one—we will solve the Schrödinger equation for the electrons that move in a lattice of atoms. When we followed such a procedure in the case of the square well or the hydrogen atom, we found that only discrete values of the energy were permitted. In this case we will find that only certain energy bands are permitted.

*14.5 JUSTIFICATION OF BAND THEORY

To solve the Schrödinger equation in three dimensions for a lattice composed of a large number of atoms or ions all exerting Coulomb forces on the electron is, as you might imagine, quite a complicated process. (Figure 14.27 shows how the potential experienced by the electrons might look in one dimension.) Since we are only interested in showing how energy bands arise in such a situation, we make a number of simplifications. First of all, we consider only the one-dimensional problem; this is just as we did in Chapter 5, where we studied the discrete energies that resulted from solving the one-dimensional Schrödinger equation for various potentials. Second, we replace the Coulomb potential with a square-well potential. This also helps to simplify the calculation without sacrificing physical applicability. We can make a better approximation to the Coulomb potential, in which $V \to -\infty$ as $r \to 0$, by choosing a potential well of width b and depth V_0 and letting $b \to 0$ and $V_0 \to \infty$ in such a way that the product bV_0 remains constant. That is, the well gets simultaneously deeper and narrower but its "area" bV_0 stays constant.

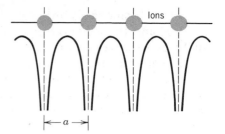

FIGURE 14.27 Potential energy established by ions in a one-dimensional lattice.

Figure 14.28 shows the potential for which we wish to solve the Schrödinger equation: an infinite row of potential wells, each of width b and depth V_0, with center-to-center spacing a. Each well represents one of the ions of the lattice.

The solutions to such a problem of a periodically repeating potential can be found with the aid of *Bloch's theorem*, according to which the wave function of the free electron ($\psi(x) \sim \cos kx$, for example) is modified by the periodic potential to be of the form

$$\psi(x) \sim u(x) \cos kx$$

where the modulating function $u(x)$ is periodic, repeating with the same period as the potential:

$$u(x + a) = u(x)$$

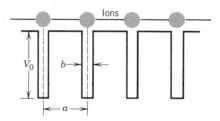

FIGURE 14.28 A simplified version of the potential of Figure 14.27.

Halfway between the ions, $u(x) \cong 1$ and the electron behaves like it is practically free. It is only when we get near an ion that $u(x)$ differs significantly from 1.

We will not go through the mathematics of the solution to this problem, but before discussing the solutions, we must recall two results from Chapter 5. (1) For the free particle, $E = \hbar^2 k^2/2m$; all values of k and therefore of E are permitted and the energy is continuous. (2) When we solved the problem of the square well in Chapter 5, the discrete values of the energy resulted from the application of the restriction that $\psi(x)$ must be continuous at the boundaries of the well. This permitted only certain values of k and therefore of E.

The solution to the problem of the "one-dimensional lattice" is similar. When we apply the boundary conditions, we obtain the following equation:

$$\cos ka = \frac{maV_0b}{\hbar^2}\frac{\sin \alpha a}{\alpha a} + \cos \alpha a \qquad (14.12)$$

where

$$\alpha = \sqrt{\frac{2mE}{\hbar^2}}$$

It is this equation that gives the energy bands.

The left side of Equation (14.12) must take values only between $+1$ and -1, and therefore the right side must do likewise. However, there might be values of α (and therefore of E) for which the right side would become greater than $+1$ or less than -1. Since Equation (14.12) would become meaningless in such cases (assuming all quantities in the equation are real), they must not occur. Those values of the energy are forbidden and give rise to the forbidden bands or energy gaps we discussed previously.

Figure 14.29 is a graph of the right side of Equation (14.12). The function oscillates back and forth, and, for certain values of αa, its amplitude is greater than $+1$ or less than -1, and these are the forbidden regions. In the allowed regions, for which the value is between $+1$ and -1, there are no restrictions on k or on E. Since the allowed regions extend from $+1$ to -1, $\cos ka$ takes all possible values between $+1$ and -1; thus k is not only unrestricted, it takes on a full range of values.

There is another way of interpreting these solutions, which involves the wave nature of the electrons more directly. In this interpretation we consider the relationship between the energy E and the wave number k. For a free particle, $E = \hbar^2 k^2 / 2m$, and the relationship is a parabola, as is shown in Figure 14.30. The relationship between E and k obtained in the solution to the problem of the periodic potential is shown as the S-shaped curve segments in Figure 14.30. The allowed energy bands and forbidden regions are also apparent from this illustration. Notice that the electron often behaves like a free particle; that is, the S-shaped segments frequently lie nearly on the parabola. It is only when the wave number k approaches the values π/a, $2\pi/a$, $3\pi/a$, . . . , that the deviations from the free particle become most noticeable.

To understand why this occurs, we must return to the wave nature of the electron. The movement of electrons within the solid can also be interpreted as the movement of electron deBroglie waves through the

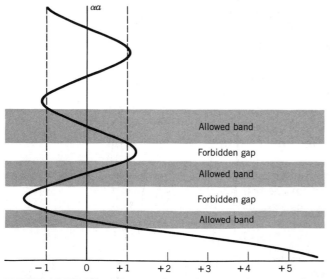

FIGURE 14.29 Allowed energy bands from one-dimensional model of lattice. The solid curve represents the right side of Equation (14.12). The allowed bands correspond to those regions where the curve lies between $+1$ and -1.

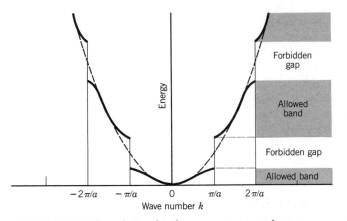

FIGURE 14.30 The relationship between energy and wave number for a one-dimensional lattice. The dashed curve is the free-particle parabola. The solid curves are the solutions to the one-dimensional problem.

solid. Figure 14.31 shows such a wave as it passes through the lattice; in our simple calculation we considered only one dimension, so we will be interested only in what happens along the direction of travel of the wave. Figure 14.31 reminds us of the Bragg reflection of X-ray waves we discussed in Chapter 3; the Bragg condition for the reflection of the waves was

$$2d \sin \theta = n\lambda \qquad (14.13)$$

where d is the atomic spacing and θ is the angle of incidence measured from the plane of atoms (*not* from the normal). In a two-dimensional lattice, the incident wave can be scattered in many different directions, depending on the plane of atoms where we imagine the reflection to occur (recall Figure 3.6); in one dimension, however, only one possible reflection can occur—the incident wave can be reflected back in the opposite direction. We can use the Bragg condition for this case, with $d = a$ (the spacing between the ions or atoms of the lattice) and $\theta = 90°$ (the angle between the "reflecting plane" and the incident wave), and so

$$2a \sin 90° = n\lambda \qquad (14.14)$$

and using $k = 2\pi/\lambda$, we find

$$k = n\frac{\pi}{a} \qquad (14.15)$$

These are just the wave numbers for which the "breaks" in the E versus k curve of Figure 14.30 occurred! For wave numbers that do *not* satisfy this condition, the wave does not get reflected—it does not even *see* the lattice and it propagates freely, like the traveling wave representing a free particle. When the wave satisfies the Bragg condition, it gets reflected back on itself and a *standing wave* results. (Remember that we get a standing wave whenever we superpose two waves of equal wavelengths traveling in opposite directions.) As we have seen repeatedly, the wave functions representing these two electron waves can combine in two ways—their amplitudes can add or subtract—and this gives two different possible standing waves. If we take the origin of the coordinate

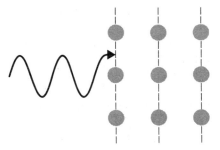

FIGURE 14.31 One-dimensional Bragg scattering. The only possible scattering is a reflection back along the original direction.

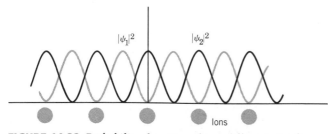

FIGURE 14.32 Probability densities of two different standing waves.

system $(x = 0)$ at one of the ions, then one standing wave has the form $\psi_1 \sim \sin(n\pi x/a)$ and the other has the form $\psi_2 \sim \cos(n\pi x/a)$. The probability densities for these two waves are just the squares of the wave functions and are shown in Figure 14.32. Notice that the solution ψ_1 gives low probability of finding the electron near the ions and a high probability of finding the electron midway between the ions. The solution ψ_2 on the other hand gives a high probability of finding the electron near the ions. Since an electron represented by the wave function ψ_2 spends more time near the positive ions, it is more tightly bound (it has more negative potential energy) and therefore its energy is a bit *below* the energy of the free electron; this corresponds to the branches of the E versus k curves of Figure 14.30, which bend *below* the parabola at the energy gaps. The solution ψ_1 is less tightly bound than the free electron (for which $|\psi|^2$ is a constant) and so its energy lies a bit *above* the free-electron parabola. The difference in energy between ψ_1 and ψ_2 is the energy gap.

Of course, this simple one-dimensional calculation does not give an accurate representation of the behavior of real electrons moving in real three-dimensional solids, but it does give us an idea about the origin of allowed energy bands and forbidden gaps, and it shows us once again that the wave behavior of electrons has important consequences, in this case for the electrical conduction in solids.

14.6 PROPERTIES OF METALS

In Chapter 12 we have already discussed the heat capacity of metals in terms of the free electron theory. In this section we examine in greater detail some of the other properties of metallic substances.

Electrical Conductivity When an electric field is applied to a metal, a current flows. In an ordinary metal the current density j (current per unit cross-sectional area) is proportional to the applied field—more field, more current. The proportionality constant is σ, the electrical conductivity, and the expression that relates j with the electric field E is simply another form of Ohm's law:

$$j = \sigma E \tag{14.16}$$

When an electric field is applied to a collection of electrons, such as exists in a metal, the force $F = eE$ accelerates the electrons. If we measure the current passing any given point of the metal, that current ought to increase with time after the electric field is applied, since the current depends on the speed of the electrons and the speed is increased uniformly by the acceleration. However, what we actually observe is a steady-state current constant in time; the electrons move with a nearly constant velocity $\bar{v}$, known as the *drift velocity*. The explanation for this effect is that, although the electrons are indeed accelerated by the electric field, they only feel that acceleration for a short time τ, after which they are slowed down by collisions with the atoms of the metal. The flow of electric current can therefore be visualized as a sequence of accelerations by the applied electric field and decelerations by collision, the net result of which is the drift velocity $\bar{v}$. The average speed acquired by the electrons is just the acceleration multiplied by the time during which the acceleration occurs, so

$$\bar{v} = \frac{eE}{m}\tau \qquad (14.17)$$

If n is the density of electrons in the conduction band of the metal, then the current density is

$$j = ne\bar{v} \qquad (14.18)$$

$$= \frac{ne^2\tau}{m}E \qquad (14.19)$$

and the conductivity is therefore

$$\sigma = \frac{ne^2\tau}{m} \qquad (14.20)$$

If we knew the time τ between collisions, we could calculate the electrical conductivity. We can express τ in terms of the speed of motion of the electrons and the distance they travel between collisions

$$\tau = \frac{l}{v} \qquad (14.21)$$

where the distance l is known as the *mean free path*. Since only electrons near the Fermi level contribute to the conductivity, we may take for v the speed of electrons whose kinetic energy is equal to the Fermi energy.

Mean free path

Let us try to estimate the electrical conductivity of copper. The Fermi energy is 7.03 eV, and the speed v is therefore 1.57×10^6 m/s. If we estimate the mean free path (the distance between collisions) as the distance between atoms, 0.256 nm, then $\tau = 1.63 \times 10^{-16}$ s. The number of electrons per unit volume is found from the density of copper and the molecular weight, assuming that each atom contributes one electron to the conductivity:

$$n = \frac{(6.02 \times 10^{23} \text{ atoms/mole})(8.96 \text{ g/cm}^3)}{63.5 \text{ g/mole}}$$

$$= 8.49 \times 10^{22} \text{ atoms/cm}^3$$

$$= 8.49 \times 10^{28} \text{ atoms/m}^3$$

and so

$$\sigma = \frac{(8.49 \times 10^{28} \text{ atoms/m}^3)(1.60 \times 10^{-19} \text{ C})^2(1.63 \times 10^{-16} \text{ s})}{9.11 \times 10^{-31} \text{ kg}}$$

$$= 2.43 \times 10^5 \ \Omega^{-1}\text{m}^{-1}$$

This value agrees very poorly with the measured value at room temperature of $5.88 \times 10^7 \ \Omega^{-1}\text{m}^{-1}$. At a low temperature of 4 K, the measured value is about 10^5 times greater than the value at room temperature. We have apparently done something very wrong in this calculation! Not only was our estimate off by two orders of magnitude, but none of the factors in the estimate depend very strongly on the temperature, and so we don't expect such a great difference between the conductivities at room temperature and low temperature.

Our error was in estimating the mean free path l to be the distance between atoms. As we discussed in connection with the band theory, only those electrons that have wave numbers near $k = \pi/a$, $2\pi/a$, $3\pi/a$, . . . are scattered by the lattice. Electrons with wave numbers near the middle of an allowed band do not "see" the lattice at all and behave nearly like free electrons. This applies to metals because in a metal the Fermi energy lies close to the middle of an allowed band (as was the case for sodium in Figure 14.21.) A perfect metallic lattice should have an *infinite* conductivity; the electrons would not be scattered at all!

In a real metallic lattice, there are two contributions to the scattering of electrons, and therefore to the conductivity: the atoms are in thermal motion and therefore do not occupy exactly the positions of a perfectly arranged lattice, and lattice imperfections and impurities cause deviations from the ideal lattice. The first effect is, of course, temperature dependent and dominates at high temperatures; the second effect is independent of temperature and dominates at low temperatures. Figure 14.33 represents the resistivity (the inverse of conductivity) of sodium metal as a function of the temperature.

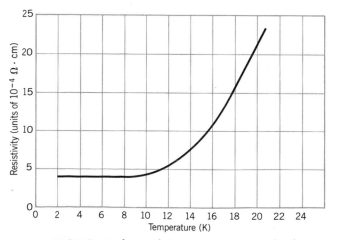

FIGURE 14.33 Dependence of electrical resistivity of sodium metal on temperature.

Thermal Conductivity The electrons in a metal also contribute to its thermal conductivity. The thermal conductivity K of a gas is found from classical kinetic theory to be

$$K = \tfrac{1}{3}Cvl \qquad (14.22)$$

where C is the heat capacity per unit volume, v is the molecular speed, and l is the mean free path. If we, as usual, think of the metal as a gas of electrons, we can use the estimate of Section 12.7 for the *molar* heat capacity of the electron gas, $C \cong \tfrac{27}{32}RkT/E_F$. A more exact calculation gives the factor $\pi^2/2$ rather than $\tfrac{27}{32}$, and therefore

$$K = \frac{\pi^2}{6} \frac{RkTvln}{N_A E_F} \qquad (14.23)$$

where the factor n/N_A converts from molar heat capacity to heat capacity per unit volume. With $R = N_A k$ and $E_F = \tfrac{1}{2}mv^2$, we have

$$K = \frac{\pi^2}{3} \frac{k^2 nlT}{mv} \qquad (14.24)$$

If we try to calculate K using the same estimates as we did for the electrical conductivity, we will face the same problems in estimating the mean free path l. We can eliminate this problem by finding the ratio of the thermal and electrical conductivities:

$$\frac{K}{\sigma} = \frac{\pi^2 k^2}{3e^2} T \qquad (14.25)$$

This remarkable result says that the ratio of the thermal and electrical conductivities depends only on the temperature, and should be the same for all metals at a given temperature, since all of the properties relating to a specific metal (l and v) cancel from the ratio. This result is known as the *Wiedemann-Franz* law; the constant $\pi^2 k^2/3e^2$ is known as the Lorenz number and has the theoretical value 2.44×10^{-8} W·Ω/K². Table 14.4 gives some experimental values of the Lorenz number, determined from the ratio K/σ, for various metals at different temperatures, and it is indeed quite nearly constant, both for different metals and for different temperatures.

Table 14.4 Lorenz Number of Metals (Units of 10^{-8} W · Ω/K²)

Metal	0°C	100°C
Ag	2.31	2.37
Au	2.35	2.40
Cd	2.42	2.43
Cu	2.23	2.33
Pb	2.47	2.56
Sn	2.52	2.49
Zn	2.31	2.33

Wiedemann-Franz law

Contact Potentials Consider what happens when two different metals, with different Fermi energies, are brought into contact, as shown in Figure 14.34. Electrons near the boundary of the metal with the higher Fermi level can easily move to occupy the empty states just above the Fermi level of the other metal. This will occur, like water seeking its own level, until the Fermi levels in the two metals become identical, as shown in Figure 14.35. (This can only happen because the two materials are metals in which the electrons move nearly freely; as we will learn in the next section, in semiconductors the electrons cannot move very far and are confined to move in a small region near the boundary between the materials.) One metal acquires electrons and becomes negatively charged, while the other becomes positively charged, and a potential difference is developed merely by placing the metals in contact. This potential difference is the *contact potential*.

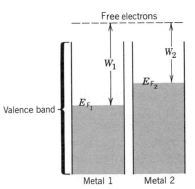

FIGURE 14.34 Two different metals with different Fermi energies and work functions.

Let W be the energy necessary to remove an electron from the Fermi level to outside the metal; this is just the *work function* of the metal. From Figure 14.35 you can see that the contact potential V is determined by the difference between the work functions of the two metals:

$$V = \frac{W_1 - W_2}{e} \qquad (14.26)$$

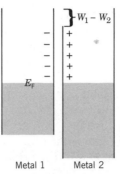

FIGURE 14.35 When the two metals are placed in contact, electrons flow, equalizing the Fermi levels; a potential difference is established at the junction.

Superconductivity At low temperatures, the resistivity of a metal (the inverse of its conductivity) is nearly constant. As the temperature of a material is lowered, the lattice contribution to the resistivity decreases while the impurity contribution remains approximately constant, and as we approach $T = 0$ K the resistivity should approach a constant value. Indeed, many metals, known as *normal* metals, behave in such a way, as illustrated in Figure 14.36.

The behavior of another class of metals is quite different from Figure 14.36. These metals behave normally as the temperature is decreased, but at some critical temperature T_c (which depends on the properties of the metal), the resistivity drops suddenly to zero, as shown in Figure 14.37. These materials are known as *superconductors*. The resistivity of a superconductor is not merely very small at temperatures below T_c; it is vanishing! Such materials can conduct electric currents even in the absence of an applied voltage, and the conduction occurs with no I^2R (joule heating) losses.

The quantum theory of superconductivity is quite involved and beyond the level of this text; we instead indicate schematically how this unusual phenomenon originates. A moving electron interacts with the lattice, possibly by causing lattice vibrations (phonons). This interaction energy can then be transferred to a second electron. We can think of this situation as one in which the two electrons are coupled together through the intermediary of the lattice; no energy is retained by the lattice in this transfer, and so it is as if the electrons can move interacting only with each other and not with the lattice.

In our previous discussions of conductivity, we noted that an electron near the middle of the conduction band ought not to interact with a perfect lattice at all and that such electrons should produce an *infinite* electrical conductivity. In a superconductor, the conductivity is indeed infinite, also because the electrons don't interact with the lattice, but not because the lattice is perfect; superconductivity occurs because the electrons move in highly correlated pairs that do not lose energy by interacting with the lattice.

Ordinary metals can be good conductors when the electrons have a very *weak* interaction with the lattice. Superconductors, on the other hand, have a relatively *strong* electron-lattice-electron coupling. Thus materials that are ordinarily good conductors ought *not* to make superconductors. Superconductivity therefore results from a curious paradox—the electrons interact so strongly with the lattice that they don't interact with it at all!

The critical superconducting temperature T_c is determined by the "binding energy" of the electron pair. (Although the two electrons have correlated wave functions, they are not necessarily bound together physically; they may in fact be separated by large distances and be traveling

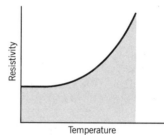

FIGURE 14.36 Resistivity of an ordinary conductor.

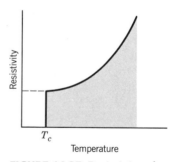

FIGURE 14.37 Resistivity of a superconductor.

in opposite directions, so it is *not correct* to picture a pair of electrons, side by side, moving together.) Typical values of T_c are of order 10 K, which corresponds to a small binding energy of 10^{-3} eV.

Table 14.5 lists some superconducting elements and compounds and their transition temperatures. The normally good conductors Cu, Ag, Au are conspicuously absent from this list; even at temperatures of a small fraction of 1 K, no superconductivity has yet been discovered for these materials. The naturally occurring element with the highest T_c is niobium (the element technetium has a higher T_c of 11.2 K, but this element is naturally radioactive and is not found in nature), and many compounds have even higher values of T_c.

Superconducting materials have many applications that take advantage of their abilities to carry electrical currents without resistive losses. In one application, electromagnets can be constructed that carry large currents and therefore produce large magnetic fields (of order 5 to 10 T). If we use superconducting wires, currents as large as 100 A can be carried by very fine wires, of order 0.1 mm diameter, and thus such magnets can be constructed in a smaller space, using less material, than would be possible with ordinary conductors. Figure 14.38 illustrates such a piece of apparatus. Starting from zero, the current is increased until the magnetic field **B** reaches some desired value. At that point the switch S is closed and the return current, following the path of least resistance, flows through the switch rather than through the power supply. The power supply can then be turned off, and in principle the current will continue to circulate through the coil and switch forever, in practice observations over the course of one year have shown no decrease of the current.

As a second application, superconducting materials should enable the transmission of electric power over long distances without resistive losses. Such an arrangement would be economically reasonable if the cost of keeping the superconducting wire below its T_c were less then the value of the power that would be lost using ordinary conductors. Keeping the costs low requires using a material with the highest possible T_c, and the search for such materials is an active field of research in solid-state physics and materials science.

Ferromagnetism Like many other solid properties, ferromagnetism must be interpreted based first on the properties of individual atoms and second on the interactions of those atoms when they are brought together to form a solid. In Chapter 8 we discussed the magnetic properties of individual atoms, and we learned that both the spin angular momentum S and the orbital angular momentum L are responsible for the magnetic properties of the isolated atom. In solids, however, this is frequently not true, and we find often that it is only the spin S of the atom that brings about its magnetic behavior.

When we apply a magnetic field to a *paramagnetic* substance, the atomic magnetic moments line up with the field; when we remove the field the random thermal motion soon destroys the orderly arrangement of the spins. In some materials, known as *ferromagnets*, there is an additional force that keeps the atoms lined up, even when the applied field is removed. We can think of this force as arising from the overlap of the electron wave functions of neighboring atoms that try to keep their

Table 14.5 Some Superconducting Metals

Metal	$T_c(K)$
Al	1.2
Zn	0.9
Nb	9.1
In	3.4
Sn	3.7
Hg	4.2
Pb	7.2
Nb_3Al	17.5
Nb_3Sn_2	16.6
Nb_3Sn	18.1

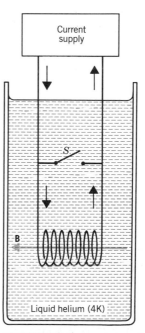

FIGURE 14.38 A superconducting coil immersed in liquid helium can carry a current indefinitely without resistive losses.

spins aligned, and we can express the energy of this interaction as $U = -A\mathbf{S}_i \cdot \mathbf{S}_j$, where $\mathbf{S}_i$ and $\mathbf{S}_j$ are the electronic spins of the ith and jth atoms; A is a parameter that gives the strength of the interaction. Whether A is positive or negative depends on the spacing between neighboring atoms of the solid. When A is positive, the interaction energy is lower when the spins line up, and the ferromagnetic state results; when A is negative, no ferromagnetism occurs. The typical dependence of A on the separation of the atoms is indicated in Figure 14.39.

The value of A is positive for solid iron, cobalt, and nickel, and so these materials are ferromagnetic at ordinary temperatures. Chromium and manganese, which have, like iron, a $3d$ electronic structure, have negative A and so are not ferromagnetic. However, it is possible to change the atomic spacing of Cr and Mn in compounds, which can make them ferromagnetic; materials such as CrO_2 (used in magnetic recording tape) and MnAs are ferromagnetic.

When a ferromagnetic substance is raised to a temperature high enough that $kT > A$, the thermal energy is sufficient to break the coupling between neighboring atoms and the ferromagnetism is destroyed. This temperature is known as the *Curie temperature*, and above its Curie temperature a ferromagnetic material becomes paramagnetic. Iron, for example, has a Curie temperature of 1043 K; above 1043 K, iron is no longer ferromagnetic. Some materials have Curie temperatures below room temperature, so that they are paramagnetic at room temperature and must be cooled in order to become ferromagnetic. The rare earth metals are examples of such materials.

We conclude this discussion with a brief look at how the band theory is related to ferromagnetism. Let us consider a material for which the $3d$ band is only partially full. The $3d$ band has a capacity of 10 electrons; according to the Pauli principle, five electrons can have spin in one direction and five in the opposite direction. We therefore divide the $3d$ band into two parts, a "spin up" part and a "spin down" part, each of which can hold five electrons. Figure 14.40 shows such a case for a nonferromagnet, where the bands are at the same energy and the Fermi level is of course common to the two bands. Since there are as many electrons with "spin up" as with "spin down," there is no net magnetism of the material.

In a ferromagnet the interaction between neighboring atoms causes the energies of the two bands to be slightly different, as shown in Figure 14.41; the energy difference is of order A. Now there are more electrons with "spin up" than with "spin down" and the material is ferromagnetic. Iron has six $3d$ electrons; the band structure and Fermi level in iron are such that there are on the average about four electrons in the "spin up" band and about two electrons in the "spin down" band. Each iron atom therefore contributes about two electron spins to the magnetization, and hence the bulk magnetic moment of iron is about two Bohr magnetons per atom.

A semiconductor is a material with an energy gap E_g of order 1 eV between the valence band and the conduction band. At $T = 0$, *all* states in the valence band are full and all states in the conduction band are empty; recall that the Fermi-Dirac distribution is a step function at $T =$

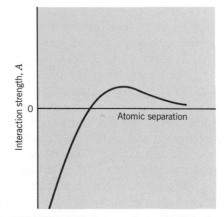

FIGURE 14.39 Interaction energy A of neighboring atomic spins.

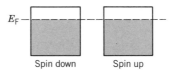

FIGURE 14.40 In a nonferromagnet, the "spin up" and "spin down" parts of the $3d$ band are equally populated.

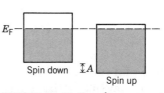

FIGURE 14.41 In a ferromagnet, the "spin up" and "spin down" parts of the $3d$ band are displaced by an energy A and are not equally populated.

14.7 INTRINSIC AND IMPURITY SEMICONDUCTORS

0 and gives an occupation probability of exactly 1 for all states below E_F and exactly 0 for all states above E_F. As the temperature is raised, however, some states above E_F are occupied and some states below E_F are empty. At room temperature, the relationship between the Fermi energy, the valence and conduction bands, and the electron energy distribution might be as we pictured in Figure 14.24.

Although the value of the room-temperature Fermi-Dirac distribution function $f(E)$ is nearly zero in the conduction band (in fact, it is so close to zero that it cannot be seen on the scale of Figure 14.24) it is not *exactly* zero; Figure 14.42 shows a greatly magnified view of $f(E)$ near the bottom of the conduction band. The value of $E - E_F$ is about 0.5 eV if E_F lies near the middle of the 1 eV energy gap, and therefore $E - E_F \gg kT$, since at room temperature $kT \sim 0.025$ eV. The "1" in the denominator of the Fermi-Dirac distribution is therefore negligible, and $f(E)$ is approximately exponential, as shown in Figure 14.42. Assuming the Fermi energy to lie near the middle of the gap, the occupation probability near the bottom of the conduction band is of order $e^{-E_g/2kT} \cong 10^{-9}$. Thus one atom in 10^9 contributes an electron to the electrical conductivity; compare this with a metal in which essentially *every* atom contributes an electron to the conductivity. (On the other hand, consider an insulator, which has a band structure very similar to that of a semiconductor, except the energy gap is perhaps 5 eV instead of 1 eV. This small difference in the size of the energy gap has an enormous effect on the occupation probability of the conduction band at room temperature: $e^{-E_g/2kT} \cong 10^{-44}$. Thus in an ordinary sample containing of order 10^{20} atoms, there may be 10^{11} conduction electrons in a semiconductor, 10^{20} in a conductor, and none in an insulator.)

Figure 14.43 shows the corresponding region near the top of the valence band. If there are a few filled states in the conduction band, there must be a few *empty* states in the valence band, and the Fermi-Dirac distribution is just a tiny bit smaller than 1; in fact it is approximately $1 - e^{(E-E_F)/kT}$. This number is about $1 - 10^{-9}$, based on our discussion for the electrons in the conduction band. (Since all the electrons in the conduction band came originally from the valence band, the number of electrons in the conduction band is *exactly* equal to the number of vacancies in the valence band. Thus the value of $(E - E_F)$ for the valence band must be exactly equal to the value of $-(E - E_F)$ for the conduction band, and so the Fermi energy must lie exactly at the center of the gap.)

When we apply an electric field to a semiconductor, the electrons in the conduction band try to follow the field and therefore give an electric current. The electrons in the valence band also try to follow the field, but in order to do so each electron must move from a filled state to one of the very few vacancies. As the electric field pulls the electrons to one side of the material, the vacancies will end up concentrated on the other side. If we follow the motion of the electrons in the valence band as they hop from one vacancy to the next, we would see an apparent motion of the vacancies in the opposite direction. (Imagine a large parking lot, with cars parked bumper to bumper, and with a single empty spot as in Figure 14.44. As we watch first one car moving to the empty spot, then the next car filling the spot vacated by the first, and so on, there is an apparent movement of the vacancy opposite to the motion of the cars.) These vacancies are known as *"holes"* and it is convenient for us

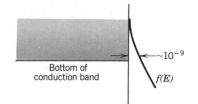

FIGURE 14.42 The tail of the Fermi-Dirac distribution function near the bottom of the conduction band. On the scale of this diagram, the "1" of $f(E)$ is 1000 km off to the right and E_F is about 1 m below the bottom of the page.

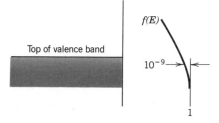

FIGURE 14.43 The Fermi-Dirac distribution function near the top of the valence band, showing the small fraction of empty states.

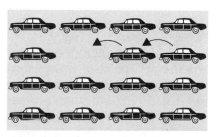

FIGURE 14.44 The motion of cars to the left, each filling the vacant space, is equivalent to the motion of the vacant space to the right.

Holes

to imagine the motion of electrons in the valence band from one atom to the next to be the motion of holes in the opposite direction. (Of course, the atoms themselves don't really move!) The current in a semiconductor therefore consists of two parts: the negatively charged electrons in the conduction band and the positively charged holes in the valence band. Although the number of electrons in the conduction band is equal to the number of holes in the valence band, the two contributions to the current will in general not be equal, since the electrons in the conduction band move more easily than the electrons in the valence band which produce the motion of the holes. Typically, the contribution of the electrons to the current at room temperature is about two to four times the contribution of the holes.

The material we have been describing thus far is an *intrinsic* semiconductor and is characterized by several features: (1) the number of electrons in the conduction band is equal to the number of holes in the valence band; (2) the Fermi energy lies at the middle of the gap; (3) the electrons contribute most to the current, but the holes are important also; (4) about 1 atom in 10^9 contributes to the conduction.

An intrinsic semiconductor is not of great practical value. Because only 1 atom in 10^9 contributes to the conduction, the presence of impurities at the level of 1 part in 10^9 can drastically alter the properties of the semiconductor, and it is simply not practical to manufacture materials at that level of purity. It is a much better procedure to deliberately introduce impurities into the semiconductor. These impurities have well-known properties and can be introduced in such carefully controlled quantities that their contribution to the conductivity of the semiconductor can be precisely determined. If the impurity level is only 1 part in 10^6 or 10^7, the impurity contribution to the conductivity will dominate the intrinsic contribution.

Such materials are known as *impurity* semiconductors, and the process of introducing the impurity is known as *doping*. Impurity semiconductors can be of two varieties: those in which the impurity contributes additional electrons to the conduction band and those in which the impurity contributes additional holes to the valence band.

Let us consider a material such as silicon or germanium, in which there are four valence electrons in hybrid orbitals. In the band theory view, these fill the $4N$ states of the valence band; in the atomic view the lattice is constructed so that each Ge or Si atom has four neighbors with which it shares an electron, and so all electrons participate in covalent bonding (Figure 14.45). Now suppose we replace one of the Si or Ge atoms with an atom that has five valence electrons, such as phosphorous, arsenic, or antimony. Four of the five electrons will form covalent bonds with the neighboring Si or Ge atoms, but the fifth electron is relatively weakly bound to the impurity atom and can be easily detached to contribute to the conductivity (Figure 14.46).

On an energy level diagram, the energy of these loosely bound electrons would appear as discrete levels in the energy gap just below the conduction band, as in Figure 14.47. The energy needed for such electrons to enter the conduction band is relatively small, about 0.01 eV in Ge and 0.05 eV in Si, and so excitations can occur easily at room temperature ($kT \sim 0.025$ eV). These energy levels are known as *donor states* and the impurity is known as a *donor*, because electrons are "donated"

Impurity semiconductors

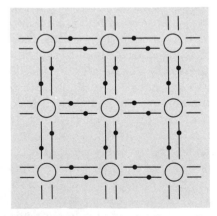

FIGURE 14.45 Covalent bonding in Ge or Si. Each atom shares four electrons in covalent bonds.

Donor states

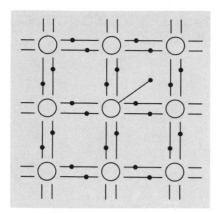

FIGURE 14.46 When a Ge or Si atom is replaced with a valance-5 atom, there is an extra electron that does not participate in covalent bonds.

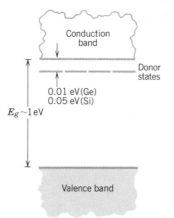

FIGURE 14.47 Energy levels of donor atoms.

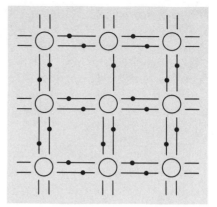

FIGURE 14.48 If a Ge or Si atom is replaced by a valence-3 atom, one covalent bond is lacking.

to the conduction band. A semiconductor that has been doped with donor impurities is known as *n-type* semiconductor, because the conductivity is due mostly to the negative electrons.

It is also possible to use as impurity a valence 3 atom such as boron, aluminum, gallium or indium. When such an atom replaces a Si or Ge atom in the lattice, all of its three electrons form covalent bonds with neighboring Si or Ge atoms. But since the impurity atom is surrounded by four atoms in the lattice, one of the surrounding Si or Ge atoms has an unbound electron (Figure 14.48). Since the completing of the four pairs of covalent bonds is energetically very favorable, an electron is easily captured to complete the symmetry of the lattice. This creates a hole in the valence band and therefore contributes to the conductivity. The energy level diagram is shown in Figure 14.49. These impurity atoms form discrete levels, just above the valence band, known as *acceptor states*. At room temperature, electrons are readily excited from the valence band to the acceptor states.

A material that has been doped with acceptor impurities is known as *p-type* semiconductor, because the conductivity is due mostly to the positively charged holes. (Remember that *n*-type and *p*-type materials are both *electrically neutral* since they are made from neutral atoms. The designations *n* and *p* refer only to the charge carriers, not to the material itself. If, for example, we remove some electrons from *n*-type material, it becomes positively charged.)

At $T = 0$ the Fermi level in *n*-type semiconductors lies between the donor states and the conduction band (remember, all states below E_F are full and all above E_F are empty; at $T = 0$ the donor states are all occupied). In *p*-type semiconductors, the Fermi level at $T = 0$ lies between the valence band and the acceptor states. As the temperature is raised, the thermal excitation of electrons from the valence band to the conduction band (as in an intrinsic semiconductor) causes the Fermi level to move toward the center of the energy gap, as shown in Figure 14.50. For low doping levels and at a high enough temperature, the material may behave like an intrinsic semiconductor.

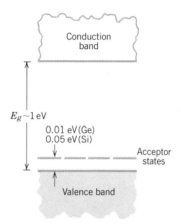

FIGURE 14.49 Energy levels of acceptor atoms.

Acceptor states

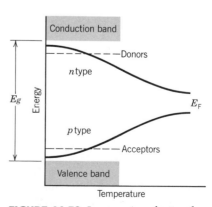

FIGURE 14.50 In a semiconductor the Fermi level changes with temperature.

The* p-n *Junction   When a *p*-type semiconductor is placed in contact with an *n*-type semiconductor (Figure 14.51) electrons will flow from the *n*-type material into the *p*-type material, until equilibrium is established. Just like the case of the two metals in contact, this equilibrium occurs when the Fermi levels in the two substances become identical. Unlike the situation with the two metals, the electrons cannot travel very far from the junction region, because the semiconductor does not conduct particularly well.

The resulting energy level diagram is shown in Figure 14.52. The region between the two materials is known as the *depletion* region, because it has been somewhat depleted of charge carriers. Electrons from the donor states of the *n*-type material fill up the holes of the acceptor states of the *p*-type material. In this region the donor states *do not* provide electrons for the conduction band and the acceptor states *do not* provide holes in the valence band.

Actually, these devices are not made by bringing two different materials into contact, but rather by doping one side of a material so that it becomes *n* type and the other side so that it becomes *p* type. The doping is carefully controlled, and typically depletion layers have a thickness of the order of 1 μm.

The excess electrons that have entered the *p*-type material give that side of the depletion region a negative charge, which tends to repel additional electrons from the *n* region. In equilibrium, enough negative charge builds up to stop the flow of electrons completely, and there is a net electric field in the junction region. (The electric field is generally defined so that its direction is that of the force on a *positive* test charge; electrons therefore feel a force in the opposite direction.) It is useful to picture this situation as the exact cancellation of two currents in opposite directions. (This is true for electrons and holes separately—two opposite electron currents cancel and also two opposite hole currents cancel. We will discuss only the electron currents, but everything in our discussion is true for the hole currents as well.)

14.8 SEMICONDUCTOR DEVICES

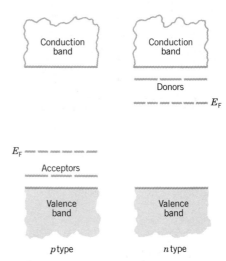

FIGURE 14.51 *n*-type and *p*-type semiconductors before being placed in contact.

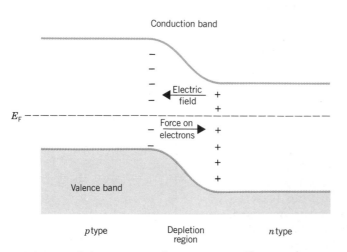

FIGURE 14.52 A *p-n* semiconductor junction. Electrons from the *n*-type semiconductor have traveled a short distance into the *p*-type material, filling the holes and giving it a net negative charge.

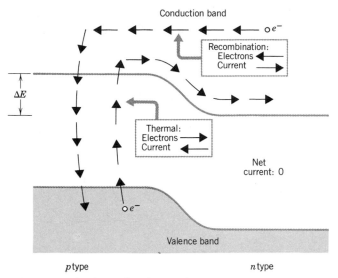

FIGURE 14.53 Thermal and recombination currents in balance in an unbiased junction.

In the tail of the Fermi distribution of electrons in the conduction band of the n region is a small number with energy above the bottom of the conduction band in the p region. These electrons have enough energy to "climb the potential hill" into the p region, where they recombine with holes (i.e., they fall into the valence band); this is the *recombination* electron current. The recombination current is balanced by the *p-to-n thermal* electron current; electrons are thermally excited from the valence to the conduction band of the p-type material, where they are then accelerated by the electric field into the n-type material.

These currents are indicated schematically in Figure 14.53.

Let us now apply an external voltage V_{ext} across the junction. We will first do this in such a way that the p-type material is made more positive than the n-type material; that is, the + terminal of a battery is connected to the p side and the − terminal to the n side. The effect of the external battery is to *lower* the potential hill by an amount V_{ext}.* This situation is called a *forward voltage* or *forward biasing* and is illustrated in Figure 14.54. The thermal current of electrons from the p valence band to the n conduction band is not at all affected by this voltage; this current depends only on the number of electrons on the p side that can travel *down* the potential hill, and changing the size of the hill doesn't change the rate at which electrons flow down. However, the ability of electrons to flow back from the n-type conduction band to the p type is considerably enhanced. The equilibrium of currents flowing in opposite directions is therefore upset, and this latter contribution to the net current will dominate.

Forward and reverse biasing

---

* This confusing statement, that a positive voltage lowers the relative potential, comes about because we are plotting *electron* energy in the vertical scale of our energy level diagrams. In the conventional definition, currents flow from + to −, but electrons flow from − to +; that is, the conventional definition of direction of current is opposite to the direction of motion of the electrons that actually produce the current! This unfortunate convention has been with us for too long to try to change it and set things right.

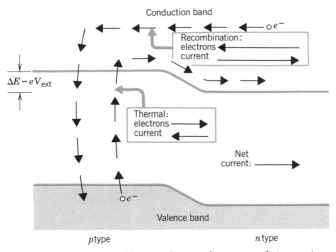

FIGURE 14.54 Forward biasing lowers the potential, increasing the recombination current; the thermal current is not affected by the biasing.

When the p-type region is biased more negative than the n type, the opposite situation occurs; this situation is known as *reverse biasing* (Figure 14.55). Once again, the thermal flow of electrons down the potential hill is not much affected, but now raising the size of the hill makes it more difficult for the recombination electrons to flow back to the p region. The equilibrium of current flows is again upset, but now the small "downhill" part dominates.

Figure 14.56 shows the upper tail of the Fermi-Dirac distribution of the electrons extending into the conduction band of the n-type region. Only those electrons in the portion of the tail above the energy E_0 of the bottom of the conduction band of the p-type region can flow back across into the p-type region and it is these electrons that produce the recombination current. The number of electrons in that tail above the energy

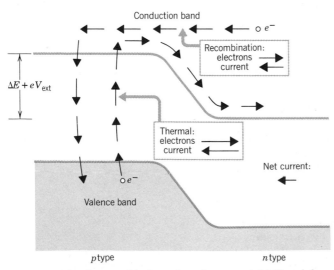

FIGURE 14.55 Reverse biasing raises the potential hill and decreases the recombination current.

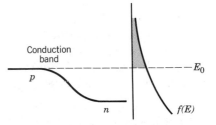

FIGURE 14.56 The recombination current depends on the density of electrons in the tail of $f(E)$ above the energy E_0 of the bottom of the conduction band in the p-type material.

E_0 is approximately

$$N_1 = ne^{-(E_0 - E_F)/kT} \qquad (14.27)$$

where n is some proportionality factor, and where we have approximated the Fermi-Dirac function as an exponential by neglecting the 1 in the denominator. (Since $E_0 - E_F \gtrsim 1$ eV and $kT \cong 0.025$ eV, this is an excellent approximation.) The recombination current is proportional to N_1, and since the thermal and recombination currents are equal, the thermal current is also proportional to N_1. Applying V_{ext} changes the level E_0 to $E_0 - eV_{ext}$, and the number of electrons in the tail above $E_0 - eV_{ext}$ is

$$N_2 = ne^{-(E_0 - eV_{ext} - E_F)/kT} \qquad (14.28)$$

The recombination current is now proportional to N_2; applying the bias did not change the thermal current, so it is still proportional to N_1. The net current is given by the difference:

$$I \propto N_2 - N_1 = ne^{-(E_0 - E_F)/kT}(e^{eV_{ext}/kT} - 1) \qquad (14.29)$$

We can rewrite this expression as

$$I = I_0(e^{eV_{ext}/kT} - 1) \qquad (14.30)$$

This function is plotted in Figure 14.57, and it is immediately obvious why such p-n junctions, also known as *diodes,* have the property of *rectifying* varying currents. When the applied voltage is such that the junction is forward biased, a large forward current can flow. (When $V_{ext} = 1$ V, $I \cong 2 \times 10^{17}I_0$.) When the applied voltage is such that the junction is reverse biased, only a very small current can flow. (When $V_{ext} = -1$ V, $I \cong -I_0$.) Even very small forward voltages can produce large forward currents; even very large reverse voltages can produce only small reverse currents. This ability to rectify a varying voltage is illustrated in Figure 14.58.

The Tunnel Diode

When the p and n regions are very heavily doped, the depletion layer becomes much narrower, perhaps 10 nm, and the energy diagram might look like Figure 14.59. When a small forward bias is applied, there is now a third contribution to the current—an electron from the conduction band of the n region can "tunnel" through the forbidden region directly into the valence band of the p region. This process of course depends on the wave nature of the electron and is an example of the type of barrier penetration we have discussed previously, in Section 5.7. The narrow depletion layer makes the process possible. The wavelength of an electron near the Fermi surface is about 1 nm, and if the thickness of the depletion layer were many orders of magnitude larger than this, tunneling would be unlikely to occur.

As the forward voltage is increased, the potential hill is lowered, and soon it no longer becomes possible for an electron to "tunnel" directly through the forbidden region. For a voltage of a few tenths of a volt, the "tunneling" current becomes zero. At this point the tunnel diode behaves like an ordinary diode.

Figure 14.60 illustrates the characteristic current-voltage relationship for a tunnel diode.

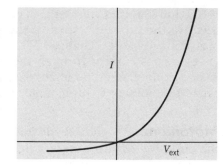

FIGURE 14.57 Current-voltage characteristics of an ideal p-n junction.

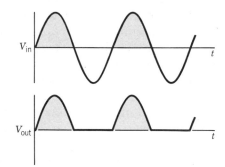

FIGURE 14.58 A p-n junction can rectify an ac voltage.

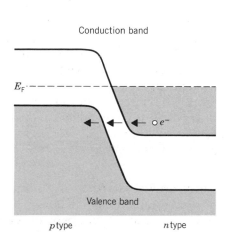

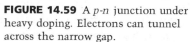

FIGURE 14.59 A p-n junction under heavy doping. Electrons can tunnel across the narrow gap.

Tunnel diodes are useful in electric circuits as high-speed elements, because the characteristics of the device can change as rapidly as the bias voltage can be changed. They can also be used as switches. If we were to pass current through the tunnel diode so that we were on the peak of the characteristic curve, a small increase in the current would cause the voltage to jump suddenly to a much larger value.

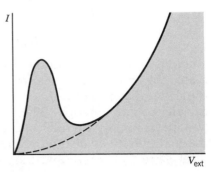

FIGURE 14.60 Current-voltage characteristics of a tunnel diode. The dashed curve shows the characteristics of an ordinary *p-n* junction diode.

Photodiodes A photodiode is a *p-n* junction whose operation involves the emission or absorption of light. These devices operate on principles similar to ordinary atoms. An electron in the valence band may absorb a photon and make a transition to the conduction band. Since photons of visible light have energies of order 2 to 3 eV, a semiconductor with its gap of order 1 eV is just right for such a transition. Conversely, an excited electron from the conduction band can drop back down to the valence band, emitting a photon in the process.

A common device that emits visible light is the LED, or light-emitting diode. An external current supplies the energy necessary to excite electrons to the conduction band, and when the electrons fall back down to recombine with holes, a photon is emitted. The energy is of course equal to the difference in energy of the electronic states. A common material for such devices, which are used for indicator lights, calculator displays, and so forth, is gallium arsenide phosphide, GaAsP, which emits photons in the red region of the spectrum.

Another example of such a device is the semiconductor laser, in which a high enough current produces a population inversion; the ends of the semiconductor can be highly polished to form mirrors, and the device operates in a way very similar to ordinary atomic lasers. (See Section 8.8.)

The reverse case occurs when a beam of photons of sufficient energy is incident on the depletion region of a *p-n* junction. When this occurs, the absorption of the photon gives the electron enough energy to move to the conduction band, and in the process a hole is created in the valence band. The large electric field in the depletion region (such devices are usually operated under reverse bias) immediately forces the hole toward the *p* side and the electron toward the *n* side. A current is generated when the electron travels through an external circuit to recombine with the hole in the *p*-type material. Devices such as this have two common applications: In a silicon solar cell, the purpose is to create electric current from sunlight, so the current in the external circuit is used to supply power to an electric device. In various kinds of photon detecting devices, the purpose is to detect and measure the intensity of a beam of photons incident on the material. Such devices are used as light meters for cameras, as well as for detectors of high-energy photons such as X rays and gamma rays.

Junction Transistors A transistor consists of a thin region of one type of semiconductor sandwiched between two thicker regions of the other type. According to the type of semiconductor used, such transistors are known as *npn* or *pnp* transistors. There is in principle no difference between the two types, other than the reversing of all bias voltages.

An unbiased *npn* transistor is illustrated in Figure 14.61. As with

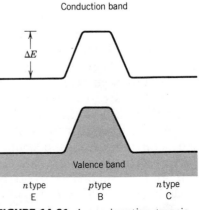

FIGURE 14.61 A *npn* junction transistor.

the *p-n* junction, electrons and holes travel across the two regions of contact until equilibrium is established. Electrical contact is made with each of the three regions. The central region is known as the base (B), and the other regions are known as the emitter (E) and collector (C). It is sometimes helpful to think of a transistor as two *p-n* junctions back to back, although the narrowness of the base has important consequences. In typical operation, the EB junction is forward biased and the BC junction is reverse biased, as shown in Figure 14.62. As with the *p-n* junction, the flow of electrons from E to B is enhanced by the forward biasing. Once into the base, electrons diffuse through its narrow region and are accelerated to the collector by the voltage across the BC junction. Not all of the electrons from the emitter leave the collector; a small fraction may leave through the base into the external circuit.

Transistors are found in so many different applications that to list them would fill many volumes. We mention briefly two applications that can be understood with reference to Figure 14.62. The continuous flow of current from C to E can be interrupted if the EB junction is not forward biased. The transistor can thus serve as a switch, in which the application of a dc voltage level to the base can turn the current on or off. This on-off switching has found application in logic circuits and especially in digital computers. A second application takes advantage of the small current leaving the base. Suppose the transistor is operating in such a way that a constant 5 percent of the current leaves the base while the other 95 percent goes through the emitter. If we are pulling 10 mA of current through C, then 0.5 mA emerges from B while 9.5 mA goes to E.

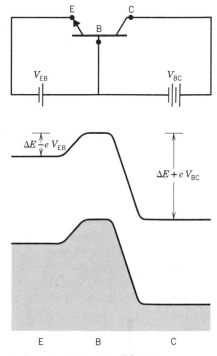

FIGURE 14.62 Typical biasing arrangement of *npn* transistor.

FIGURE 14.63 A modern integrated circuit. This dime-sized, 32-bit microprocessor incorporates 150,000 transistors on a single chip of silicon and is able to perform complex operations similar to today's minicomputers. (Photograph courtesy Bell Laboratories.)

Let us now attach to the base a device that draws an additional 0.1 mA through the base; that is, the base current rises to 0.6 mA. To keep the base current at a constant 5 percent of the collector current, the collector current must rise to 12 mA, and so 11.4 mA must emerge from the emitter. Thus a 0.1-mA variation in the current at the base caused a 1.9-mA variation at the emitter, and the transistor acts like an amplifier with a current gain of 19.

Not only has the transistor revolutionized many aspects of modern life, it is one of the few modern inventions for which technological advances have resulted in lower prices. Almost every aspect of daily life is affected in some way by the transistor, by the modern high-speed digital computer made possible by the transistor, and by the integrated circuit, a collection of transistors reduced to such a small size that thousands of transistors fit into the space formerly occupied by one (Figure 14.63). From pocket calculators to microwave ovens to the 747 aircraft, the integrated circuit has become an essential component of our society. This is only one example of the unexpected directions in which basic research may lead—all of these practical applications followed from the experiments of a few physicists studying the properties of solids.

SUGGESTIONS FOR FURTHER READING

More detailed and comprehensive books on solid-state physics:

A. J. Dekker, *Solid State Physics* (Englewood Cliffs, Prentice-Hall, 1957).

A. Hart-Davis, *Solids: An Introduction* (London, McGraw-Hill, 1975).

C. Kittel, *Introduction to Solid State Physics,* 5th edition (New York, Wiley, 1976).

T. L. Martin and W. F. Leonard, *Electrons and Crystals* (Belmont Cal., Wadsworth, 1970).

R. T. Sanderson, *Chemical Bonds and Bond Energy* (New York, Academic Press, 1976).

The use of superconducting wires to transmit electric power is discussed in D. P. Snowden, "Superconductors for Power Transmission," *Scientific American 226,* 84 (April 1972).

The entire September 1977 issue of *Scientific American* is devoted to integrated circuits and includes articles on their manufacture and applications.

Bulk properties of solids are tabulated in many references, including the *Handbook of Chemistry and Physics* (Chemical Rubber Publishing Co.) and the *American Institute of Physics Handbook* (New York, McGraw-Hill, 1963).

QUESTIONS

1. Compare the equilibrium separations and binding energies of ionic *solids* (Table 14.1) with those of the corresponding ionic *molecules* (Table 13.5). Account for any systematic differences.

2. How should Equation (14.6) be modified to be valid for MgO and BaO?

3. Assuming that its other properties don't also change with temperature, at what temperature would you expect carbon to begin to behave like a semiconductor?

4. Would you expect the Wiedemann-Franz law to apply to semiconductors? To insulators?

5. (a) Why does the electrical conductivity of a metal decrease as the tempera-

ture is increased? (b) How would you expect the conductivity of a semiconductor to change with temperature?

6. Why is it that only the electrons near E_F contribute to the electrical conductivity?

7. In Section 14.6, the origin of ferromagnetism is given as the interaction between neighboring atoms that aligns their spins and therefore their spin magnetic moments. Is it possible for neighboring orbital magnetic moments to interact directly and to align? Consider two neighboring atoms with electrons orbiting in the same plane. What is the effect of one orbital magnetic field on the other atom?

8. How would the ferromagnetic properties of cobalt and nickel compare with iron if all three had the same interaction energy A? The effective magnetic moments of nickel (0.6 Bohr magnetons) and cobalt (1.7 Bohr magnetons) are smaller than for iron. How many electrons are in the spin-up and spin-down parts of the d band for Co and Ni? How do their interaction energies A compare with iron?

9. Do the superconducting elements have any particular electronic structure or configuration in common?

10. Three different materials have filled valence bands and empty conduction bands, and the Fermi energy lies in the middle of the gap. The gap energies are 10 eV, 1 eV, and 0.01 eV. Classify the electrical properties of these materials at room temperature and at 3 K.

11. In what way does a p-n junction behave as a capacitor?

12. What limits the response time of a p-n junction when the external voltage is varied? Why does a tunnel diode not have the same limits?

13. The energy gap E_g is 0.72 eV for Ge and 1.10 eV for Si. At what wavelengths will Ge and Si be transparent to radiation? At what wavelengths will they begin to absorb significantly?

14. Why is a semiconductor better than a conductor for applications as a solar cell or photon detector? Would an insulator be even better?

PROBLEMS

1. Consider the packing of hard spheres, as represented in Figure 14.1. The corners of the basic cube are the centers of the eight spheres. (a) What fraction of the volume of each sphere is inside the volume of the basic cube? (b) Let r be the radius of each sphere and let a be the length of a side of the cube. Express a in terms of r. (c) What fraction of the volume of the cube is taken up by the portions of the spheres? This fraction is called the *packing fraction*.

2. Compute the packing fractions (see Problem 1) of the fcc and bcc structures (Figures 14.2 and 14.3). Which structure fills the space most efficiently?

3. Derive Equations (14.5) and (14.6).

4. By summing the contributions for the attractive and repulsive Coulomb potential energies, show that the Madelung constant has the value 2 ln 2 for a one-dimensional "lattice" of alternating positive and negative ions.

5. Calculate the first 5 contributions to the electrostatic potential energy of an ion in the NaCl lattice, and compare with Equation (14.2).

6. Use the values of the cohesive energies of the ionic solids from Table 14.1, along with other values from the listed references, and see how the boiling points of the ionic solids vary with the cohesive energy. Boiling points may be found in the *Handbook of Chemistry and Physics*.

7. (a) Use Equation (14.6) to find the ionic cohesive energy (in electron-volts)

for CsCl. (b) Find the atomic cohesive energy and compare with the experimental value given in Table 14.1. The ionization energy of Cs is 3.89 eV.

8. (a) Find the ionic cohesive energy from Equation (14.6) for LiF. (b) Find the experimental *ionic* cohesive energy from the value given in Table 14.1, using 5.39 eV for the ionization energy of Li and 3.45 eV for the electron affinity of F.

9. What is the measured value of the atomic cohesive energy of BaO in kJ/mole?

10. Calculate the wavelength at which CsI will absorb energy by means of the vibrational motion of the ions.

11. The electric field of a dipole is proportional to $1/r^3$. Assuming that the induced dipole moment of molecule B is proportional to the electric dipole field of molecule A, show that the van der Waals force is proportional to r^{-7}. (*Hint.* Calculate the potential energy of dipole B.)

12. The density of sodium is 0.971 g/cm³ and its atomic weight is 23.0. In the bcc structure, what is the distance between sodium atoms?

13. Copper has a density of 8.96 g/cm³ and an atomic weight of 63.5. Calculate the center-to-center distance between copper atoms in the fcc structure.

14. Calculate the deBroglie wavelength of an electron with energy E_F in copper, and compare the value with the atomic separation in copper.

15. Use the experimental value of the electrical conductivity of copper to estimate the mean free path of electrons. Compare this value with the lattice spacing of copper (Table 14.3). How many atoms does a typical electron encounter before it is scattered?

16. Use the Wiedemann-Franz ratio to calculate the thermal conductivity of copper at room temperature. The electrical conductivity is 5.88×10^7 $\Omega^{-1}\,m^{-1}$.

17. Calculate the expected value of the Lorenz number.

18. Use the photoelectric work functions of Table 3.1 to find the contact potential of (a) Cu and Al; (b) Zn and Pt; (c) Cu and Ag.

19. We can understand the behavior of donor impurities with the following rough calculation. When we replace an atom of silicon with an atom of phosphorous, the outer electron of phosphorous is screened and sees $Z_{eff} \cong 1$. (a) Compute the binding energy of the electron, assuming that silicon has a dielectric constant of 12 that effectively reduces the electric field experienced by the electron. Compare this value with the value shown in Figure 14.47. (b) Compute the radius of the electron's orbit. How many atoms does it encounter in a sphere of that radius? The lattice spacing of Si is 0.234 nm.

20. When a material such as germanium is used as a photon detector, an incoming photon makes many interactions and excites many electrons across the gap between the valence and the conduction band. (a) ^{137}Cs emits a 662-keV gamma ray. How many electrons are excited across the 0.72 eV gap of germanium by the absorption of this gamma ray? (b) The number N calculated in part (a) is subject to statistical fluctuations of $\sqrt{N}$. Compute the variation in N and the fractional variation in N. (c) What is the corresponding variation in the measured energy of the gamma ray? This result is the experimental resolution of the detector.

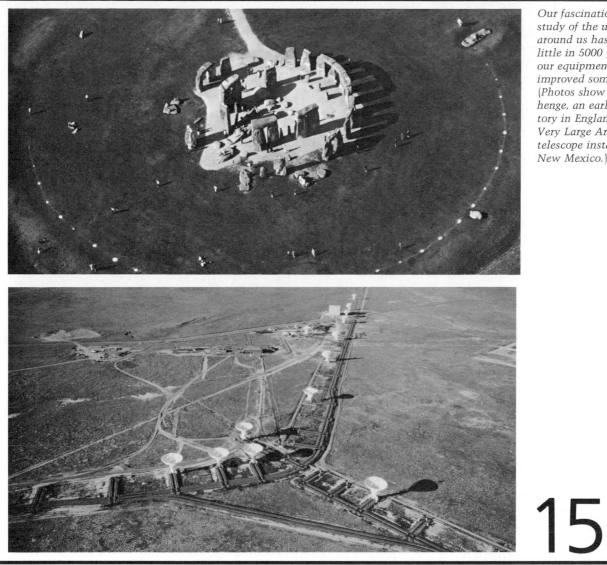

Our fascination with the study of the universe around us has changed little in 5000 years, but our equipment has been improved somewhat. (Photos show Stonehenge, an early observatory in England, and the Very Large Array radiotelescope installation in New Mexico.)

15

ASTROPHYSICS AND GENERAL RELATIVITY

We now discuss what is perhaps the grandest of all human endeavors: the study of our universe on the large scale, and of the stars and galaxies of which it is made. Our study of the properties of matter has taken us from the very small (atoms and their nuclei) through molecules and solids, and now finally to the very large, the stars.

The motions of the stars, moons, and planets can be well understood on the basis of Newton's theory of gravity, and so it may seem surprising that such studies belong in a course of modern physics. However, understanding the life cycle of stars and the origin of the universe involves many of the concepts of nuclear and particle physics that have been previously discussed in this text, and much of the evidence for understanding our universe depends on experiments similar in principle to those we have previously discussed.

It is somewhat unusual that our approach has taken us from the small to the large, because the historical development of classical physical science was just the reverse. Studies of the motion of the planets, by Galileo, Copernicus, Kepler, and Tycho, led to the Newtonian dynamics that subsequently proved so successful in interpreting the physical world. We have seen how the Newtonian world view failed when we tried to apply it to subatomic processes, and we have seen how a new world view, based on special relativity and quantum physics, replaced the Newtonian one. The properties of elementary particles, nuclei, atoms, molecules, and solids are all based on this new world view, and in the two remaining chapters we explore the extension of these new theories to the cosmic scale.

In the present chapter we first discuss one missing ingredient of this new view of our universe, Einstein's *general* theory of relativity, which is basically a new theory of gravitation. The theory, far too mathematical for the present level, will be briefly summarized, and we will see how it is successful in correcting the failures of the Newtonian theory. We then turn to a study of some specific observed and unobserved (but expected) properties of our universe and the objects of which it is made. The more general problems of the origin, age, and composition of the universe as a whole are left to the final chapter.

15.1 SPACE AND TIME

Before we try to understand Einstein's approach to gravity, we need to introduce some new ways of looking at space and time. In Chapter 2 we learned from the *special* theory of relativity that both time and space are relative to the motion of an observer—lengths and time intervals measured by one observer will in general not be the same as lengths and time intervals measured by a different observer. Since *both* length *and* time are relative, and since they depend on relative motion in rather similar ways [recall Equation (2.8) for the Lorentz transformation], it is necessary and convenient to consider not the three dimensions of space and the one dimension of time separately, but rather as more-or-less equivalent components of a single *four*-dimensional *spacetime*. Unfortunately we cannot graph such a system in our three-dimensional world (it is even difficult to graph three dimensions on a two-dimensional page) and so we must be content with graphing the time dimension and one or two spatial dimensions.

With the addition of *general* relativity, the situation becomes even worse. The principal difference between special relativity and general relativity is that special relativity deals only with "flat" spacetimes, while general relativity permits "curved" spacetimes (hence the names—*special relativity* is a special case of the more general *general relativity*). As we illustrate as follows, in order to draw a curved coordinate system, we need yet one higher dimension, and thus to "draw a picture" of a real curved spacetime we would need five dimensions! Since this is an exercise far beyond our graphical abilities, we must be satisfied with some examples of lower dimensionality that can be illustrated more easily.

If, as the science fiction writers would often have us believe, we could "travel" to a higher dimension, we could then "look back" on the curvature of our spacetime. (Such imaginings are not totally fanciful. It was once "accepted scientific fact" that the Earth was flat, and in fact there is hardly any phenomenon we encounter in our daily lives that is inconsistent with that belief. Yet once the first astronauts left the surface of the Earth and turned their cameras back on us, the curvature of the Earth's surface was immediately apparent.) These considerations are *not* a part of general relativity, and we need not perform the mental gymnastics of imagining or illustrating spaces of higher dimensionalities, as long as we can represent them mathematically by the appropriate set of equations. (As an analogy, consider the humble three-dimensional vector. We can easily picture its geometrical properties, and we are familiar with its mathematical properties such as addition or multiplication. It is relatively straightforward to invent a set of similar mathematical properties for N-dimensional vectors, even though we can't also represent easily their geometric properties when $N > 3$.)

Let us begin with some examples of curved space. Figure 15.1 illustrates a "flat" one-dimensional space, which is just a straight line. To curve that space we must bend it through another direction, as illustrated. The curved one-dimensional space requires two dimensions for its illustration. Figure 15.2 shows a "flat" two-dimensional space and illustrates how it would look if it were curved. (Imagine Figure 15.2 to represent a flat sheet of graph paper. In order to curve it, we must lift or bend part of it into a third dimension.) From our perspective, like that of the astronauts, we can directly observe the curvature of these spaces. *Yet the inhabitants of these hypothetical one- and two-dimensional worlds need not "travel" to a higher dimension to observe or measure the curvature of their systems. It is entirely possible to do experiments totally within the coordinate system to measure the curvature.* (In the context of the analogy with the Earth's surface, relatively simple experiments, like the lengths of shadows of measuring rods at different points, permit measurement of the curvature.)

The geometry of flat coordinate systems is Euclidian geometry. It is a well-behaved and familiar system that serves us quite well for nearly all applications. Its rules, illustrated in Figure 15.3, are well known: a straight line is the shortest distance between two points, the sum of the angles of a triangle is 180°, parallel lines never intersect, and so forth. One familiar example of a non-Euclidian geometry is that of the surface of a sphere. Here, as shown in Figure 15.4, the Euclidian rules do not

Spacetime

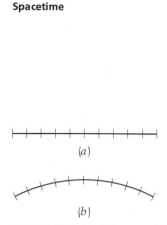

(a)

(b)

FIGURE 15.1 (a) A flat one-dimensional spacetime. (b) A curved, one-dimensional spacetime, which requires two dimensions to illustrate.

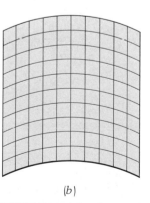

(a)

(b)

FIGURE 15.2 (a) A flat two-dimensional spacetime. (b) A curved two-dimensional spacetime.

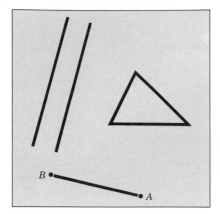

FIGURE 15.3 A flat space and its Euclidian geometrical properties.

FIGURE 15.4 A curved space and its non-Euclidian geometrical properties.

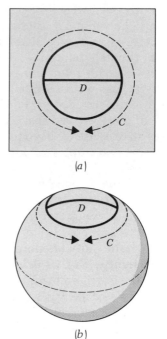

(a)

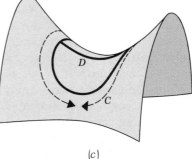

(b)

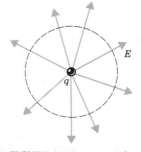

(c)

FIGURE 15.5 (a) In a flat space, $C/D = \pi$. (b) and (c) In a curved space, $C/D < \pi$ or $> \pi$.

apply. The shortest distance between two points is an arc of a great circle, the sum of the angles of a triangle exceeds 180°, and parallel lines can intersect.

It is possible to do experiments to determine whether our spacetime is curved or flat, and in fact to measure its degree of curvature. One way would be to measure the ratio of the circumference to the diameter of a circle. In a flat space, this ratio is π (Figure 15.5); in a curved space that ratio can be less than π or greater than π. From such measurements in our world, we know that our space is flat (or so slightly curved that its curvature is within the limit of error of such experiments). However, as we will learn in the next section, the curvature of space is determined by nearby masses, and in the vicinity of a large mass, the curvature can be large enough to have observable effects.

This connection between gravity and geometry can be viewed in another way. In our three-dimensional world, Coulomb's law for electric forces and Newton's law for gravitational forces have similar forms: the force depends on the product of the charges or masses and varies inversely with the square of their separation. Let us now imagine a two-dimensional world, and try to construct electric and gravitational force laws for such a world. The usual procedure for calculating electric fields is to use Gauss's law, according to which we surround an electric charge by a closed surface and compute the electric flux through the surface. In two dimensions, the "surface" is just a circle (Figure 15.6), and Gauss's law becomes

$$E(2\pi r) = \frac{q}{\varepsilon_0} \qquad (15.1)$$

or

$$E \propto \frac{1}{r} \qquad (15.2)$$

(This compares with $E \propto 1/r^2$ in three dimensions.) Since gravity behaves similarly, we might expect that the gravitational force in a two-dimensional world will also vary as $1/r$. However, this is not the case! General relativity, which for weak fields in three dimensions reduces to the Newtonian $1/r^2$ gravitational force, does not similarly reduce to $1/r$ forces in two dimensions. This suggests perhaps that mass is not simply

FIGURE 15.6 Gauss's law in two dimensions.

a strength parameter, like electric charge, but rather has a more special role in determining the structure of our world.

Special relativity arose from a "thought experiment" in which Einstein imagined trying to "catch up" to a beam of light. The principle of the constancy of the speed of light for all observers, one of the cornerstones of special relativity, follows from Einstein's conclusion that such an event would not be possible. General relativity is based on a principle that follows from another thought experiment. Let us suppose that we are in a rocket at its launching pad on the surface of the Earth (Figure 15.7). We drop an object from rest and observe it to move toward the floor with an acceleration of $g = 9.8$ m/s². Now suppose the rocket is in a region of space in which the gravitational field is negligibly small. We switch on the engines and adjust the thrust so that the ship is given an acceleration of precisely 9.8 m/s². We again release an object, and it again is observed from within the ship to move toward the floor (does the floor move toward the object? does it matter?) with an acceleration of 9.8 m/s². The two experiments yield exactly the same results. Einstein generalized this result into the *principle of equivalence*, which forms the basis of general relativity:

15.2 GRAVITY AND CURVED SPACETIME

Principle of equivalence

> There is no experiment that can be performed
> in a relatively small region of space that
> can distinguish between a gravitational field
> and an equivalent uniform acceleration.

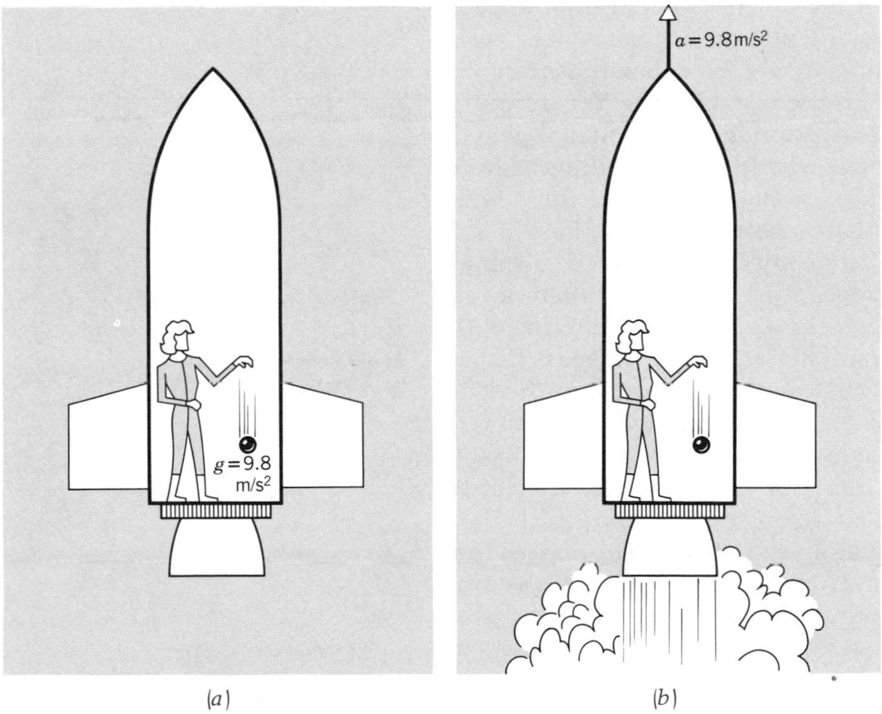

FIGURE 15.7 (*a*) At rest on the Earth, a released object accelerates toward the floor of the rocket ship. (*b*) With the ship accelerating in gravity-free space, the same object is released and approaches the floor.

(The proviso about the "small region of space" is necessary for the following reason: Suppose we release from rest two small objects separated by a small distance D. Near the Earth each object moves along a radius toward the center of the Earth, and so the objects move ever so slightly closer together as they fall. This does not happen if the objects are in the accelerating rocket. However, if the ship is small enough the gravitational field is very nearly uniform and the difference will not be noticeable.)

As a consequence of the principle of equivalence, a freely falling reference frame in the Earth's gravity, such as an elevator in which the cable has broken, is a *local inertial frame.* An observer in the falling elevator could release an object "from rest" (in the observer's frame of reference) and would find it to remain "at rest." (A space capsule orbiting the Earth is in a somewhat similar circumstance; even though the capsule is in a stable orbit, it is still falling freely toward the center of the Earth with a centripetal acceleration which precisely matches the local gravitational acceleration. To refer to the astronauts in such an orbiting capsule as being in a state of "weightlessness" is thus not strictly correct and conveys the false impression that they are somehow beyond the pull of the Earth's gravity. They are, in fact, in a state of free fall and therefore in another sort of local inertial frame. However, this case differs from the freely falling elevator, because there are experiments the astronauts could do to distinguish their circumstance from an inertial state of coasting in gravity-free space. For example, a gyroscope will experience a precession in orbit but not in uniform motion in the absence of external fields.)

This concept of a *local inertial frame* is very different from what we encountered in our previous discussions of inertial frames. The freely falling elevator bears little resemblance to the Newtonian concept of a coordinate frame fixed with respect to the distant stars. Yet, according to the principle of equivalence, there is (from the experimental point of view) no difference. In fact, associated with the freely falling elevator there are infinitely many inertial frames, and we can use the Lorentz transformation to relate the coordinates observed from the different frames, exactly as we did with the "constant-velocity" inertial frames.

The principle of equivalence appears in a slightly different (and weaker) form in introductory physics, where it is stated in terms of the equivalence of *inertial* and *gravitational* mass. That is, the mass that appears in the expression $F = ma$ (inertial mass) is identical to the mass that appears in the expression $F = GMm/r^2$ (gravitational mass). It follows from this form of the principle of equivalence that all objects, regardless of their masses, fall with the same acceleration in the Earth's gravity. This was first tested by Galileo in the famous (and perhaps apocryphal) experiment in which he dropped two different masses from the top of the leaning tower of Pisa and observed them to fall at the same rate. In more recent years other more precise experiments have established the equivalence of gravitational and inertial mass to about 1 part in 10^{11}.

Equivalence of inertial and gravitational mass

Just as the constancy of the speed of light led us to the predictions of time dilation and length contraction, the principle of equivalence leads to some predictions that alter our thinking about space and time. Let us imagine that a light beam is fired across our rocket ship (Figure 15.8) by

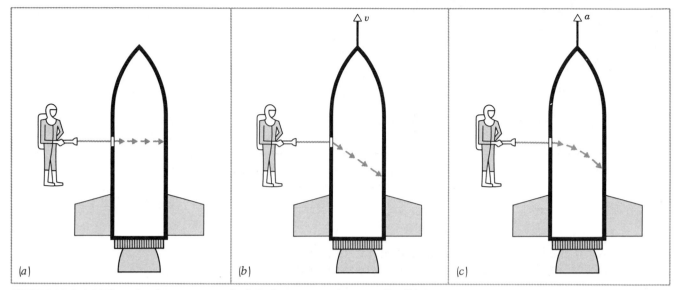

FIGURE 15.8 (a) When the ship is at rest, the light beam moves horizontally. (b) As observed from the ship, motion at constant speed causes the beam to deviate from the horizontal. (c) The beam appears to follow a curved parabolic path when the ship accelerates.

a source that we define to be "at rest" in a region of space with negligible gravity. If the ship were at rest relative to the source, the beam would go directly across; if the ship moves at constant speed, perpendicular to the beam according to an observer at rest with the source, an observer inside the ship sees the beam follow an oblique straight-line path (at an angle v/c with the "horizontal" when v is small). In an *accelerated* ship, v changes and so the angle v/c changes; the beam then follows a *curved* (parabolic) path, according to the observer in the rocket.

If the principle of equivalence is correct, the behavior of the light beam in the accelerated rocket must be the same as in the Earth's gravity. *The light beam must therefore also follow a curved path* in the gravitational field of the Earth.

A light beam has a special place in our notions of space and time, because it must travel the shortest and most direct path between two points. (If it did not, it would be possible for some other projectile to travel between the two points in a shorter time, thereby traveling *faster* than light; since this violates *special* relativity, it is not possible.) If the light beam follows a curved path as the shortest distance between two points, then space itself must be curved, and *it must be the gravitational field* that causes the curvature.

The shortest distance between two points in an arbitrary curved geometry is known as a *geodesic*. In flat (Euclidian) geometry, the geodesics are straight lines; in spherical geometries, the geodesics are arcs of great circles.

Let's consider a simple example that illustrates the difference in language between Newton's gravitation and theories based on curved spacetime. A mass m is constrained to move freely along the x axis (imagine a bead sliding without friction along a wire). Suppose the mass moves (at constant speed) from A to B as shown in Figure 15.9. The

FIGURE 15.9 A bead slides at constant speed on a wire from A to B.

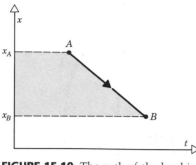

FIGURE 15.10 The path of the bead in spacetime.

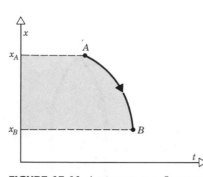

FIGURE 15.11 A mass at $x = 0$ causes the bead to accelerate as it moves from A to B. Its path in spacetime is now curved.

"spacetime" of the bead can be represented by the two-dimensional xt plane; since nothing of interest happens in the y or z directions, it is not necessary to show them. In spacetime, the "path" of m is the straight line shown in Figure 15.10. The mass begins at $x = x_A$ and moves toward the origin.

In the language of classical physics, the mass moves according to the expression $x = x_A - vt$. In the language of spacetime, we define the *interval*

$$(ds)^2 = v^2(dt)^2 - (dx)^2 \qquad (15.3)$$

We assert that, according to our theory, the particle is allowed to move only through those points of spacetime that are connected by intervals of zero length; the locus of all such points defines the particle's "geodesic."

Now suppose a large mass M is placed at the origin; M attracts m with the usual Newtonian gravitational force. As the particle moves from A to B it accelerates, and so the spacetime path is now curved, as shown in Figure 15.11. (In this figure, the particle is obviously *not* traveling along the "shortest path" from A to B.)

According to classical physics, the motion of the particle can be described by

$$x(t) = (x_A^{3/2} - \tfrac{3}{2}\sqrt{2GM}\; t)^{2/3} \qquad (15.4)$$

which follows from solving Newton's second law $F = m\, d^2x/dt^2$ for the force $F = GmM/x^2$. Figure 15.11 is a plot of Equation (15.4). The presence of the mass M curves the xt spacetime, so that it might look like Figure 15.12. The particle is then simply following the shortest path from A to B as it moves in its curved spacetime. The interval now becomes

$$(ds)^2 = \frac{2GM}{x}(dt)^2 - (dx)^2 \qquad (15.5)$$

Once again, the "geodesic" path connects those points with zero relative interval, and you should be able to show that Equations (15.5) and (15.4) are consistent. The "strength" of the "gravitational force" is no longer meaningful from this viewpoint; all that matters is the curvature of the spacetime, which is determined by the mass M.

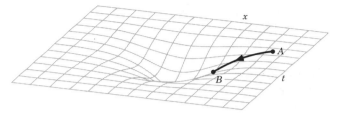

FIGURE 15.12 The curved, two-dimensional spacetime of the bead.

This simple example, while it does illustrate the difference in approach between classical Newtonian gravitation and the physics of curved spacetime, is *not* correct from the standpoint of general relativity. Even in the case of a single point mass M, the geodesics of massive $(m \neq 0)$ particles in curved spacetime are mathematically complicated and beyond the scope of this text. In the spirit of the two alternative descriptions of this simple case, we can however begin to see the distinction between classical Newtonian gravitation and general relativity, each of which would describe a given situation using its own unique terminology:

Newton	Einstein
"The particle is accelerated, as it moves from A to B, by a gravitational force that varies as $1/x^2$, and that is caused by and proportional to the mass M located at $x = 0$."	"The particle moves from A to B along a geodesic path of a curved spacetime; the curvature is caused by and is proportional to the quantity of mass M at $x = 0$."

The basic problem of motion in a gravitational field is, within the framework of general relativity, the problem of finding the coefficients of $(dt)^2$, $(dx)^2$, $(dy)^2$, and $(dz)^2$ in the interval $(ds)^2$. These coefficients, taken as a group, form the *metric* of the spacetime; the metric is in part determined by the gravitating masses which are present and in turn determines the motion of masses and light beams.

Have we really gained anything by all of this? The justification for any physical theory is not its intrinsic appeal or its aesthetic beauty (general relativity scores well on both points), but rather how well it succeeds in its comparisons with experiment. As we learn in the next sections, general relativity has withstood all experimental tests. (To be completely fair, there are competing theories of gravity that may also be consistent with these experiments. The most stringent tests of general relativity are those done in extremely strong gravitational fields, and these have not yet been done.)

15.3 THE GENERAL THEORY OF RELATIVITY

Einstein's equations of general relativity are tensor equations, which require mathematical techniques beyond the level of this text. The basic equation of general relativity can be written symbolically as

$$\left(\begin{matrix} \text{curvature} \\ \text{of space} \end{matrix} \right) = \frac{8\pi G}{c^4} \left(\begin{matrix} \text{mass-energy} \\ \text{density} \end{matrix} \right) \qquad (15.6)$$

Note that this expression incorporates gravitation (with Newton's gravitational constant G) and special relativity (with the speed of light c). In the limit of classical kinematics ($c \to \infty$), and in the case of no matter or energy, the right-hand side is zero, and space is flat.

There are two applications of general relativity that we will discuss. In the first case, the solutions to Einstein's equation are derived in the region near a single nonrotating mass M. The metric of the curved four-dimensional spacetime near such a mass was first worked out by Karl Schwarzschild in 1916; for this case the interval can be written in the following form (using spherical coordinates r, θ, ϕ):

Schwarzschild solution

$$(ds)^2 = c^2 \left(1 - \frac{2GM}{c^2 r} \right) (dt)^2 - \frac{(dr)^2}{\left(1 - \frac{2GM}{c^2 r} \right)}$$

$$- r^2 (d\theta)^2 - r^2 \sin^2 \theta \, (d\phi)^2 \qquad (15.7)$$

The existence of curved spacetime near massive objects can be tested in a variety of experiments described in the next section.

The second application concerns the universe itself—is the entire spacetime of the universe (not just in the vicinity of massive objects) curved or flat? (In this sense general relativity follows the philosophy of Ernst Mach, who claimed that *all* inertial forces, including the mass inertia of objects as well as centripetal forces, are caused by the rest of the matter in the universe.) In this case what we are after is the "size" or "radius" R of the universe, and how it changes with time. Experiments that we will discuss in Chapter 16 show that our universe is expanding; we would like to solve for $R(t)$ to project that expansion into the indefinite past and the infinite future in order to study the origin and fate of the universe.

The solutions to Einstein's equations for this case can be written

$$\left(\frac{dR}{dt} \right)^2 = \frac{8\pi}{3} G\rho R^2 - kc^2 \qquad (15.8)$$

Here $R(t)$ represents the size or distance scale factor of the universe at time t and ρ represents the mass-energy density of the universe *at the same time*. (It is important to remember that in an evolving universe the density changes with time.) The density should be interpreted as a large-scale average density, measured over a region of space large compared with the spacing between galaxies (just as, in the case of a solid, we wouldn't be successful if we tried to measure the density by choosing a small representative volume between the atoms or inside a nucleus).

The constant k that appears in Equation (15.8) specifies the overall geometrical structure of the universe: $k = 0$ if the universe is flat, like Figure 15.5a; $k = +1$ if the universe is curved and closed, like Figure 15.5b; $k = -1$ if the universe is curved and open, like Figure 15.5c. When $k = +1$, the distance factor $R(t)$ is directly related to the size or "radius" of the universe, but its meaning is not so apparent when $k = 0$ or $k = -1$, since in both of the latter cases the universe is infinite in extent. In these cases $R(t)$ should be regarded as a scale factor that represents the expansion of the space; the absolute magnitude of R in this case is not significant, and only its variation with time is of interest, since the distances between galaxies will vary as R varies. (Imagine an infinite three-dimensional xyz lattice in which the entire coordinate system is being stretched.)

To solve Equation (15.8) for $R(t)$ we must therefore specify the constant k and we must also make some assumptions about the density ρ. Let us assume either that $k = 0$ or else that the size and density of our universe are such that the kc^2 term of Equation (15.8) is negligible compared with the other terms. We can't integrate Equation (15.8) directly to find $R(t)$, since we don't yet know about the time dependence of ρ. To simplify still further, let us assume that R is expanding according to a simple power law: $R(t) = At^n$, where A and n are constants to be determined. Inserting this form of $R(t)$ into Equation (15.8), we find

$$(nAt^{n-1})^2 = \frac{8\pi}{3} G\rho A^2 t^{2n} \tag{15.9}$$

and thus

$$\rho = \frac{3n^2}{8\pi G} t^{-2} \tag{15.10}$$

The mass-energy density ρ can be of two forms: ordinary matter ρ_m or radiation ρ_r (or, of course, a mixture of both). In the present universe, the contribution from matter may be dominant. Since matter is presumably conserved (no new matter is being created), as the universe expands the density must decrease so as to keep $\rho_m R^3$ constant. Thus $\rho_m \propto R^{-3}$, and since Equation (15.10) demands that $\rho \propto t^{-2}$, it follows that $R \propto t^{2/3}$. Thus $n = \frac{2}{3}$, and Equation (15.10) gives

$$t = \frac{1}{\sqrt{6\pi G\rho_m}} \tag{15.11}$$

In the early history of the universe, the mass-energy was mainly in the form of radiation (i.e., $\rho_r \gg \rho_m$). The energy density of thermal radiation at a temperature T is given by the blackbody formula, for example Equation (3.20); we therefore expect $\rho_r \propto T^4$. According to Wien's displacement law, $\lambda_{max}T = $ constant, so that $\rho_r \propto (\lambda_{max})^{-4}$. As the universe expands, all distances vary with $R(t)$, including wavelengths; thus $\rho_r \propto R^{-4}$. Combining this result with Equation (15.10), $\rho \propto t^{-2}$, we conclude $R \propto t^{1/2}$. In this case therefore $n = \frac{1}{2}$, and Equation (15.10) gives

$$t = \sqrt{\frac{3}{32\pi G\rho_r}} \tag{15.12}$$

Equations (15.10) to (15.12) show how the density of the universe varies with time.

A useful way to characterize the expansion of the universe is by means of *Hubble's parameter* H (usually known as *Hubble's constant*, even though it is not constant according to most cosmological models): **Hubble's constant**

$$H = \frac{1}{R}\frac{dR}{dt} \tag{15.13}$$

H has units of (time)$^{-1}$.

If the universe were expanding at a constant rate (and has been doing so since $t = 0$), then $H = t^{-1}$, and so H^{-1} would be the age of the universe. In the slightly more general case we have been using, in which $R(t) = At^n$, we obtain $H = nt^{-1}$ and the age of the universe would be nH^{-1}; with $n = \frac{2}{3}$ or $\frac{1}{2}$, the age of the universe is less than, but still of the order of H^{-1}. For this reason, H^{-1} is often taken as a rough measure of the age of the universe. If the rate of expansion has been slowing (as we would expect, under the influence of the mutual gravitational attraction of the galaxies), then the universe is somewhat younger than H^{-1}.

The *rate of change* of the expansion is proportional to d^2R/dt^2, which is a sort of acceleration. It is convenient to define a dimensionless *deceleration parameter q* as

Deceleration parameter

$$q = \frac{-R(d^2R/dt^2)}{(dR/dt)^2} \tag{15.14}$$

which, with $R = At^n$, gives

$$q = \frac{1}{n} - 1 \tag{15.15}$$

Differentiating Equation (15.8), keeping in mind that ρ is a function of t, and assuming that we are at present in a matter-dominated universe, we can find a relationship between the deceleration parameter and the density:

$$q = \frac{4\pi G\rho_m}{3H^2} \tag{15.16}$$

Notice that, since the term proportional to kc^2 vanishes when Equation (15.8) is differentiated, Equation (15.16) applies more generally than Equation (15.15), which was derived for the $k = 0$ model.

This special $k = 0$ model of the universe, called the *Einstein-deSitter* model, is only one of many possible models, each of which gives very different results for $R(t)$ and the other derived parameters. The Einstein-deSitter model is useful, not because it necessarily represents the actual state of affairs in our universe, but because it permits us to use general relativity to derive expressions for some of the parameters of our universe in relatively simple forms. In Section 16.6, we return to a discussion of other possible forms of $R(t)$ and how they can be distinguished. In the next section we look at direct local experimental tests of the predictions of general relativity; further discussions of the applications to the origin, evolution, and future of the universe are given in the next chapter.

Newtonian gravitation and Einstein's general relativity each give predictions that can be tested against experiment, but under most circumstances the differences between the two predictions are extremely small. In weak gravitational fields, general relativity reduces to Newton's gravity, just as special relativity reduces to Newtonian kinematics ($K = \frac{1}{2}mv^2$, and so forth) at low speeds. At the surface of the Earth, space is curved by only about 1 part in 10^8; that is, the ratio of the curvature of space caused by the Earth's mass to the curvature of the Earth's surface itself is about 10^{-8}, and deviations from Euclidian geometry would occur only at about that level. Even at the surface of the sun, the curvature is only about 1 part in 10^6.

15.4 TESTS OF GENERAL RELATIVITY

Nevertheless, there are experiments we can do that are precise enough to detect the difference between flat spacetime and curved spacetime. In this section we discuss several of these experiments.

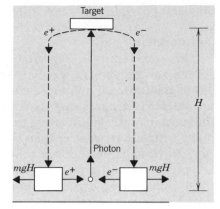

FIGURE 15.13 A source of free energy.

Gravitational Spectral Shifts Figure 15.13 illustrates a thought experiment that could be a source of free energy—a sort of perpetual motion machine. A photon of energy 1.022 MeV $(2m_ec^2)$ is fired upward from the surface of the Earth a distance of $H = 1000$ m, where it collides with a target and produces a positron and an electron by *pair production* as we discussed in Section 3.5. The positron and electron each then fall back to the surface of the Earth, each gaining a kinetic energy of m_egH in the process. The energy of $2m_egH$ is extracted, and the positron and electron (now at rest) are recombined to give a photon of energy 1.022 MeV, which then repeats the cycle. Each time through the cycle we therefore "gain" an energy $2m_egH$, in apparent violation of the principle of conservation of energy; the energy gain is not large, about 2×10^{-7} eV, but if we believe in conservation of energy as a fundamental law, even this small energy production must not be possible.

We can restore conservation of energy if we assume that the photon, traveling upward against the Earth's gravity, loses an amount of energy equal to $2m_egH$. The energy E of the photon is equal to $2m_ec^2$, and thus its energy loss is

$$\Delta E = \frac{EgH}{c^2} \qquad (15.17)$$

According to *special* relativity, we assign a mass $m = E/c^2$ to the photon, and ΔE is just the kinetic energy lost as m rises against the Earth's gravitational field.

Based on this analysis, it might seem that the energy loss is an effect of *special* relativity, rather than *general* relativity. Let us instead consider the experiment shown in Figure 15.14. A light source at point A of a rocket in gravity-free interstellar space emits a light beam of frequency ν and simultaneously begins to accelerate uniformly. By the time an observer at B sees the light, a time $\Delta t \cong H/c$ later, the ship is moving with speed $v = a\Delta t = aH/c$. (We assume $v \ll c$.) The observer detects the Doppler-shifted frequency $\nu' \cong \nu(1 + v/c)$, and thus the change in frequency of the light is

$$\Delta\nu = \nu\frac{v}{c} = \nu\frac{aH}{c^2} \qquad (15.18)$$

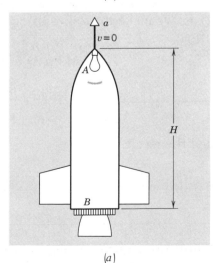

FIGURE 15.14 (*a*) Light leaves the source at A as the ship begins to accelerate. (*b*) When the light is received at B, the ship is moving at speed v.

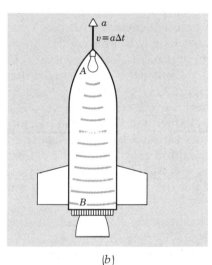

Multiplying on both sides by h to convert frequency to photon energy, this is identical to Equation (15.17) with $g = a$. As the principle of equivalence demands, experiments in uniform gravitational fields and in uniformly accelerated systems give identical results. The observation of such a change in frequency is therefore a test of the principle of equivalence.

In 1959, R. V. Pound and G. A. Rebka allowed 14.4-keV photons from the radioactive decay of ^{57}Co to fall down the Harvard tower, a distance of 22.6 m. The expected fractional change in frequency, $\Delta\nu/\nu = gH/c^2$, was 2.46×10^{-15}; that is, to detect the effect, they had to measure the frequency or energy of the photon at the bottom of the tower to a precision of about 1 part in 10^{15}! The Mössbauer effect (Section 9.12) makes it possible to achieve such a level of precision, and the measured result was $\Delta \nu / \nu = (2.57 \pm 0.26) \times 10^{-15}$, consistent with the equivalence principle.

Photons falling in a gravitational field gain energy and are therefore shifted toward shorter wavelengths ("blue shifted"). Photons climbing out of gravitational fields lose energy and are "red shifted." Light leaving the surface of a star is slightly red shifted by this gravitational effect, but the motion of the star relative to us causes a Doppler shift in its emitted light, which is far greater than the expected gravitational shift (see Section 16.1), and the thermal motion of the atoms of the star causes a Doppler broadening of the spectral lines (see Section 12.3). Together, these effects make it practically impossible to observe the small gravitational shift in spectral lines.

Deflection of Starlight   The last section gave an example of a case in which we combined special relativity and Newtonian gravitation to give a result identical with general relativity. Here we examine a different case in which the results are not identical and in which the experimental results favor general relativity.

Consider a beam of light that passes close to the sun, as indicated in Figure 15.15. According to *special* relativity, the photons in the beam have an effective mass $m = E/c^2$, and should be deflected by a Newtonian gravitational field like any ordinary projectile. Because Newton's gravitational force law has the same form as Coulomb's law for electrical forces, gravitational deflection will occur exactly in the same way as electric deflection, so we can calculate the deflection angle using Equation (6.17) for Coulomb (Rutherford) scattering

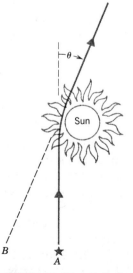

$$b = \frac{zZ}{2K} \frac{e^2}{4\pi\varepsilon_0} \cot \frac{\theta}{2} \qquad (15.19)$$

If a photon of effective mass $m = E/c^2$ just grazes the surface of the sun, $b = R$ (the radius of the sun). From the relationship of the two force laws, we make the following substitutions:

$$ze \rightarrow m = E/c^2$$
$$Ze \rightarrow M \text{ (mass of sun)}$$

$$\frac{1}{4\pi\varepsilon_0} \rightarrow G \text{ (gravitational constant)}$$

$$2K = mv_0^2 \rightarrow mc^2 = E \text{ (energy of photon)}$$

and the result is

$$\cot \frac{\theta}{2} = \frac{.ER}{GmM}$$
$$= \frac{c^2 R}{GM} \qquad (15.20)$$

Substituting the numerical values, we find $\theta = 0.87''$; this is the prediction of special relativity and Newtonian gravitation for the deflection.

General relativity gives a very different view of this deflection. Spacetime in the vicinity of a mass like the sun is curved; a curved two-dimensional spacetime can be represented as shown in Figure

FIGURE 15.15 A light beam passing near the sun is deflected. The star at A appears to be at B.

15.12. A light beam follows a geodesic in this spacetime, which is *not* the hyperbola of a mass subject to Newtonian gravitation (or the charged particle in Rutherford scattering). According to general relativity, the deflection is 1.74″, exactly twice the value computed from Newtonian gravity.

Measuring this effect requires the observation of a beam of light, such as from a star, which passes near the edge of the sun. Starlight near the sun can only be observed during a total solar eclipse. In 1919, just a few years after Einstein completed his general theory, two expeditions of British astronomers traveled to Africa and to South America to observe the solar eclipse and to measure the apparent changes in positions of stars whose light grazed the sun. Their results for the deflection angles, 1.98″ ± 0.18″ and 1.69″ ± 0.45″, gave strong support for the new general theory. In the years since those early results, this experiment has been repeated at nearly each total solar eclipse, and the overall agreement with general relativity is within 10 percent. Radio emission from quasars has also been used to confirm this effect, and here the agreement with general relativity is within 3 percent.

These experimental results give a clear distinction between Newtonian gravity (even with special relativity included) and general relativity.

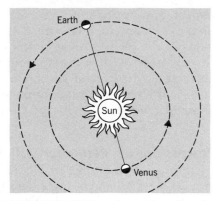

FIGURE 15.16 Superior conjunction of Earth and Venus.

Delay of Radar Echoes

When a line joining Earth and another planet (Venus, for example) passes through the sun, the situation is known as "superior conjunction" and is illustrated in Figure 15.16. We can use Newtonian mechanics to determine the orbits of the two planets, and so we can calculate the distance between Earth and Venus. If we send a radar signal from the Earth to Venus, we therefore know (based on Newtonian gravitation) exactly how long it will take that signal to be reflected from Venus and return to Earth (about 20 minutes). Near superior conjunction, the signal passes close to the sun, and therefore, according to general relativity, it does not travel in an Euclidian straight line, but instead follows the geodesic path of the curved space of Figure 15.12. It therefore takes the signal a bit longer than the expected 20 min to make the round-trip journey, since the curved geodesic is somewhat longer than the Euclidian straight line (Figure 15.17). This time delay is

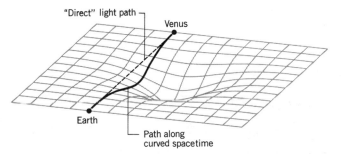

FIGURE 15.17 Path of signal between Earth and Venus in curved spacetime. The actual path (solid line) is slightly longer than the Euclidian direct path. (Diagrams like this, called *embedding diagrams,* represent one way of picturing a two-dimensional slice of four-dimensional spacetime.)

expected to be about 10^{-4} s and can be measured with sufficient precision to test general relativity. The experiment is a difficult one (the positions of the planets must be known to a precision of a few kilometers; even worse, on Earth the time difference between a signal reflected from the top of Mt. Everest and one reflected from sea level is about 10^{-4} s —how high are the mountains of Venus?), but once again, the predictions of general relativity are confirmed to within about 10 percent.

Precession of Perihelion of Mercury

Consider a simple solar system, such as shown in Figure 15.18, consisting of a single massive star (of mass M) and a single light planet in orbit around it. According to Newtonian gravitation, the orbit is a perfect ellipse with the star at one focus. The equation of the ellipse is

$$r = r_{min} \frac{1 + e}{1 + e \cos \phi} \qquad (15.21)$$

where r_{min} in the minimum distance between planet and star and e is the *eccentricity* of the orbit (the degree to which the ellipse is noncircular; $e = 0$ for a circle). When $r = r_{min}$, the planet is said to be at *perihelion*; this occurs regularly, at exactly the same point in space, whenever $\phi = 0, 2\pi, 4\pi, \ldots$. According to general relativity, the orbit is not quite a closed ellipse; the effect of the curved space near the star is to cause the perihelion direction to *precess* somewhat, as shown in Figure 15.19. After completing one orbit, the planet returns to r_{min}, but at a slightly different ϕ; the difference $\Delta\phi$ can be computed from general relativity, according to which the orbit is

$$r = r_{min} \frac{1 + e}{1 + e \cos(\phi - \Delta\phi)} \qquad (15.22)$$

where

$$\Delta\phi = \frac{6\pi GM}{c^2 r_{min}(1 + e)} \qquad (15.23)$$

For the sun, $6\pi GM/c^2 = 27.80$ km, and thus even for the smallest value of r_{min} (for Mercury, 46×10^6 km) $\Delta\phi$ is of order 10^{-6} rad, an extremely small quantity. However, this effect is *cumulative*; that is, it builds up orbit after orbit, and after N orbits, the perihelion has advanced by $N\Delta\phi$. We usually express this precession in terms of the total precession per century (per 100 Earth years), and some representative values are shown in Table 15.1.

The expected precessions are very small, of the order of seconds of arc per century, but nevertheless have been measured with great accuracy; for the three planets closest to the sun, and for the asteroid Icarus, the measured values are in agreement with the predictions of general relativity. In the best case, the agreement is within about 1 percent.

These experiments are very difficult to do because (except for Mercury and Icarus) the eccentricities are small and locating the perihelion is difficult. A more serious problem is that other effects, not associated with general relativity, also cause an apparent precession of the perihelion. In the case of Mercury, the observed precession is actually about 5601″ per century; of that, 5026″ are due to the precession of the Earth's

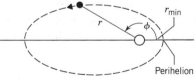

FIGURE 15.18 The elliptical orbit of a planet about a star.

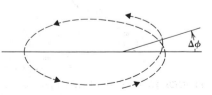

FIGURE 15.19 Precession of the perihelion (greatly exaggerated). After each orbit, the perihelion advances by an angle $\Delta\phi$.

Table 15.1 Precession of Perihelia

Planet	N (Orbits per Century)	e	r_{min} (10^6 km)	$N\Delta\phi$ (arc seconds per century) General Relativity	Observed
Mercury	415.2	0.206	46.0	43.0	43.1 ± 0.5
Venus	162.5	0.0068	107.5	8.6	8.4 ± 4.8
Earth	100.0	0.017	147.1	3.8	5.0 ± 1.2
Mars	53.2	0.093	206.7	1.4	
Jupiter	8.43	0.048	740.9	0.06	
Icarus	89.3	0.827	27.9	10.0	9.8 ± 0.8

equinox (a classical Newtonian effect of the spinning Earth) and 532″ are due to the gravitational pull of the other planets on Mercury (also a classical Newtonian effect). Only the difference of 43″ is due to general relativity.

The Schwarzschild Singularity The Schwarzschild solution for the interval between points of curved spacetime near a mass M was given by Equation (15.7) and includes an apparently annoying problem: a term of $(1 - 2GM/c^2r)$ in the denominator causes the separation to appear to "blow up" at the *Schwarzschild radius*

$$r_s = 2GM/c^2 \qquad (15.24)$$

where r_s is the distance measured from the center of M, as represented in Figure 15.20. This situation is partly a problem of the coordinates we used in Equation (15.7); none of the physical coordinates actually goes to infinity at $r = r_s$, and an object falling toward a gravitating mass would feel nothing unusual as it passed through $r = r_s$. For external observers of the falling object, however, the situation can be quite different. For example, general relativity predicts that clocks in differing gravitational potentials run at different rates; to an external observer at rest very far from M, clocks carried by the falling object would appear to run ever slower as the object falls toward r_s, and when the object reaches r_s the clocks (and the object itself) would appear to stop and be frozen forever in time at that point! At the same time, the light emitted from the falling object becomes ever more red shifted, and the red shift becomes infinite, according to the distant observer, when the object reaches r_s; thus the object disappears from view! The object itself notices nothing unusual as it falls, with one exception: At any time before r_s is reached, the falling object could escape the gravitational pull of M, for example by firing the engines of its rockets. Once r_s is passed, escape is not possible. In the language with which we are familiar, the escape velocity would be greater than c, and nothing (*not even light*) can escape. No travel or communication is possible from inside r_s to the outside world. Anything within r_s, even the brightest star, has in effect disappeared from the rest of the universe! (It continues, however, to exert gravitational forces on objects outside r_s or, in the language of general relativity, to curve spacetime outside r_s.) An object whose mass M lies totally within the corresponding radius r_s, and which could therefore

Schwarzschild radius

FIGURE 15.20 A mass M and its event horizon (Schwarzschild radius) r_s. This is a black hole only if all of M is concentrated within r_s.

Table 15.2 Black Hole Event Horizons

Object	Mass (kg)	Ordinary Radius (m)	r_s (m)
^{238}U nucleus	4.0×10^{-25}	7×10^{-15}	6×10^{-52}
Physics book	1	0.1	1.5×10^{-27}
Earth	6×10^{24}	6×10^6	8.9×10^{-3}
Sun	2×10^{30}	7×10^8	3×10^3
Galaxy	$\sim 2 \times 10^{41}$	$\sim 10^{20}$	3×10^{14}
Universe	$\sim 10^{51}$	$10^{26}(?)$	$\sim 10^{24}$

Black hole

exhibit these properties, has become known as a *black hole*. The radius r_s is known as the *event horizon*, since events that occur inside r_s are hidden from the view of external observers.

Whether black holes exist in nature, or are merely curiosities arising from the mathematics of general relativity, is presently the subject of active investigation by astrophysicists. To form such an object may require that ordinary matter be compressed to very high density. Table 15.2 shows the values of r_s for some representative objects. For the Earth to become a black hole, its mass must be compressed to a radius of less than 1 cm, and the sun would become a black hole only if compressed to a 3-km radius! It has been conjectured that the extreme densities and pressures needed to create black holes may occur as a natural end-point in the evolution of sufficiently massive stars (see Section 15.9) or else may have occurred in the early stages of the evolution of the universe (Section 16.4).

Far from a black hole, the gravitational field, like that of any ordinary object, is Newtonian in character and the effects of curved spacetime would be small. For example, a planet or another star could orbit a black hole at $r \gg r_s$ subject, to a very high degree of approximation, to the usual $1/r^2$ gravitational force. Close to a black hole (or, in fact, close to any massive, compact object), the effects of curved spacetime can become substantial, and the differences between general relativity and Newtonian gravitation would be large and perhaps easy to observe. The discovery of a massive black hole could therefore provide an ideal "laboratory" to test general relativity where the curvature of spacetime may be large.

15.5 STELLAR EVOLUTION

The problem of understanding the structure and evolution of stars is exceedingly complex. We have identified nuclear fusion as the basic source of energy production in stars, but our knowledge of the cross sections for many of the nuclear reactions of the fusion cycle is not sufficiently complete for us to be confident of calculations of the rates at which these reactions occur. (We usually think of the center of the sun as a very hot place, full of energetic particles, but its temperature of about 10^7 K corresponds to particle kinetic energies of only about 1 keV. We routinely study nuclear reactions in our Earthbound laboratories at MeV energies. It is ironic that our understanding of the energy production in stars is limited, not because we can't reach *high* enough energies in our laboratories, but rather because it is very difficult to study these reactions at *low* energies!)

Even if our knowledge of nuclear physics were more complete, we would still have the difficulties of understanding the physical structure of a star. For example, we know the masses and radii of many stars, so we can find their *average* density, but this does not tell us how this density varies inside the star. The rate at which nuclear reactions occur will depend on the higher density at the center of the star, not on its average density. The temperature of the surface of a star can be determined from the radiation emitted by the star, but the surface temperature (10^3 to 10^4 K) is much lower than the interior temperature (10^7 K), and we can't determine how the temperature varies over the interior of the star. The brightness of a star depends on how the nuclear energy (mostly gamma rays) produced in nuclear reactions near the center gets absorbed and reradiated as visible light, a process that will also vary over the interior of a star.

In spite of these difficulties, we can construct a reasonably good theoretical model of the evolution and structure of stars. We explore some of the details of our model in this section and will look at the fate of stars in the following sections.

We begin with a large, cold, diffuse cloud of hydrogen gas, the end product of the early history of the universe. (As we learn in the next chapter, the cloud contains about 25 percent helium, but that is not important for this calculation.) The atoms are electrically neutral, and their nuclei are far apart, so the only force that can operate is the gravitational force between the atoms. Under its influence the cloud begins to contract. As the average distance between the atoms decreases, the gravitational potential energy $V = -Gm_1m_2/r$ decreases, and to balance the total energy, the kinetic energy must increase, with a corresponding increase in temperature. As the cloud contracts, more atoms will be attracted toward the center, and so the density and temperature near the center increase more rapidly than near the outer regions.

As the temperature slowly rises, the gas begins to radiate energy, much like a blackbody: the higher the temperature, the more radiation is emitted. Each change in gravitational potential energy is accompanied by an increase in the kinetic energy K and in the radiant energy U:

$$|\Delta V| = \Delta K + \Delta U \qquad (15.25)$$

Just like an ordinary object when we heat it, the cloud begins to glow with a dull red color. Its size is still much larger than the final star, perhaps 10 times as large, and its temperature might be of the order of 1000 K.

As the cloud continues to contract, the temperature rises especially rapidly near the center. Eventually, temperatures in the range of 10^7 K are reached, nuclear fusion reactions begin, and a star is born. Perhaps 10^6 y have passed since the contraction began.

The star now enters a stable period, in which further contraction is halted by the outward pressure due to the radiation (photons) traveling from the core to the surface. It continues to generate energy in fusion reactions at a rate determined by its mass; for a star such as the sun, this period may last for 10^{10} y, while a much heavier star (10 to 100 solar masses) may burn in this phase for only 10^7 y.

The basic fusion reactions discussed in Section 10.6 provide the energy radiated by the star, as four protons combine to make one helium nucleus, with the liberation of 26.7 MeV. Stars like our sun produce most of their energy from the proton-proton cycle; such stars are either not hot enough for the carbon cycle, or else have no carbon present. (Recall from Section 10.6 that the carbon acts only as a catalyst.)

Depending on the availability of ^{4}He, other proton-proton cycles may occur:

$$^3\text{He} + {}^3\text{He} \longrightarrow {}^4\text{He} + 2\ {}^1\text{H}$$

$$^3\text{He} + {}^4\text{He} \longrightarrow {}^7\text{Be} + \gamma$$
$$^7\text{Be} + e^- \longrightarrow {}^7\text{Li} + \nu$$
$$^7\text{Li} + {}^1\text{H} \longrightarrow 2\ {}^4\text{He}$$

$$^1\text{H} + {}^1\text{H} \longrightarrow {}^2\text{H} + e^+ + \nu$$
$$^2\text{H} + {}^1\text{H} \longrightarrow {}^3\text{He} + \gamma$$

$$^3\text{He} + {}^4\text{He} \longrightarrow {}^7\text{Be} + \gamma$$
$$^7\text{Be} + {}^1\text{H} \longrightarrow {}^8\text{B} + \gamma$$
$$^8\text{B} \longrightarrow {}^8\text{Be} + e^+ + \nu$$
$$^8\text{Be} \longrightarrow 2\ {}^4\text{He}$$

Proton-proton cycles

Our sun probably produces most of its energy by the second of these possible sets of reactions.

The helium which was present in the initial gas cloud, along with that produced in fusion, forms a core at the center of the star, since helium is heavier than hydrogen (see Figure 15.21). Surrounding the helium is a layer in which the hydrogen fusion reactions take place. The outer radius of this shell is less than 10 percent of the radius of the star. The fusion energy is liberated mostly in the form of gamma rays (95 percent) and neutrinos (5 percent), which must make their way to the surface to be radiated into space. The gamma rays undergo repeated Compton scatterings in the dense solar material, and may take 10^5 to 10^6 y to arrive at the surface—the light that we see today from the sun comes from energy generated in nuclear reactions perhaps a million years ago! Moreover, the gamma rays lose energy until they become photons of visible light so what we see is more characteristic of the solar surface than of its interior. The neutrinos, on the other hand, have a very small cross section for interaction, and they go directly through the solar material at the speed of light. Studying the solar neutrinos is therefore a way of studying the sun's core and its nuclear processes.

The electron capture decay of ^{7}Be (in the second set of proton-proton reactions) has a Q value of 0.862 MeV; the recoil energy of the ^{7}Li is negligible, and so a monoenergetic neutrino of energy 0.862 MeV should be observed. For the last several years, R. Davis of the Brookhaven National Laboratory has been searching for these solar neutrinos in a laboratory one mile underground (to shield against cosmic rays) at the Homestake Gold Mine in South Dakota. His apparatus is shown in Figure 15.22. The neutrinos are incident on a 100,000-gal tank of tetrachloroethylene (C_2Cl_4) and give the following reaction:

$$\nu + {}^{37}_{17}\text{Cl}_{20} \longrightarrow {}^{37}_{18}\text{Ar}_{19} + e^-$$

The Q value for this reaction is -0.814 MeV, so the 0.862-MeV solar neutrinos have enough energy to produce the reaction. The Ar gas is

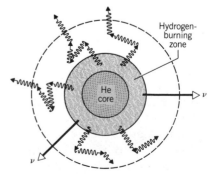

FIGURE 15.21 A cross section of the sun. The fusion energy is generated in a region near the center. Emitted gamma rays are scattered many times in the outer zones, but neutrinos penetrate easily.

FIGURE 15.22 The solar neutrino experiment. The large tank, one mile underground, contains 100,000 gallons of C_2Cl_4. (Photograph courtesy Brookhaven National Laboratory.)

inert, and it can be removed from the fluid by bubbling. It is also radioactive with a half-life of 35 days. The experiment is done by periodically purging the tank and determining the ^{37}Ar content in the removed gas. The neutrinos penetrate the solar material because they have an extremely small cross section for interacting with matter; for the same reason, their chance of being captured by a Cl nucleus is very low, and even though Davis's tank contains about 10^{30} atoms of Cl, the expected production rate of ^{37}Ar is about *one atom per day!*

From the amount of ^{37}Ar, Davis can deduce the flux of neutrinos from the sun. His result, after nearly 10 y of experiments, is only about one-third of the value expected based on our current understanding of fusion processes and of the properties of the neutrinos. This puzzling result has not yet been explained, and may indicate either that our knowledge of the properties of the sun's core and the processes that occur there is not yet complete, or else that we are not correctly accounting for the properties of the neutrinos themselves.

15.6 NUCLEOSYNTHESIS

The fate of a star that has used up its hydrogen fuel depends on the mass of the star. What is true in general is that once the hydrogen fusion stops, gravitational contraction sets in again, raising the temperature and making possible fusion reactions with helium and heavier elements. The star may in the process settle into periods of stability while these other processes occur, but one thing is clear: the star is entering its old age and will soon use up all its available fusion energy and die, either by slowly dissipating its remaining energy as a white dwarf star, by contracting to an unimaginably large density as a neutron star or black hole, or perhaps by throwing out all its remaining energy (and a good deal of its matter) in one explosion as a supernova, and for a brief few days outshining an entire galaxy.

The physical processes that occur in the collapse of a star are discussed in detail in the next few sections of this chapter. In this section we look at the nuclear processes that accompany the collapse.

The primordial matter resulting from the birth of the universe (see Section 16.3) consisted of about 75 percent hydrogen and 25 percent helium; all of the other chemical elements were formed by nuclear reactions in the interiors of stars. The original hydrogen and helium condensed into stars, where nuclear fusion converts most of the hydrogen into helium by reactions we considered in the previous section. The ignition temperature for these reactions is about 10^7 K, where the thermal kinetic energy is large enough to overcome the Coulomb repulsion of the protons; this thermal kinetic energy resulted from the loss in gravitational potential energy of the cloud of gas from which the star formed.

When the hydrogen is converted to helium, gravitational collapse begins again, and the core of the star heats up from about 10^7 K to about 10^8 K. At this point there is enough thermal kinetic energy to overcome the Coulomb repulsion of the helium nuclei, and helium fusion can begin. In this process three ^{4}He are converted into ^{12}C by the two-step process

$$^4\text{He} + {}^4\text{He} \longrightarrow {}^8\text{Be}$$

$$^8\text{Be} + {}^4\text{He} \longrightarrow {}^{12}\text{C}$$

The first reaction is endothermic, with a Q value of 92 keV. The nucleus ^{8}Be is unstable and decays back into two alpha particles in a time of the order of 10^{-16} s. Even so, the Boltzmann factor $e^{-\Delta E/kT}$ suggests that at 10^8 K there will be a small concentration of ^{8}Be. The second reaction proceeds through a resonance and therefore has a particularly large cross section; in spite of the rapid breakup of ^{8}Be, there is still a good chance to form ^{12}C. The net Q value for the process is 7.3 MeV, or about 0.6 MeV per nucleon, much less than the 6.7 MeV per nucleon produced by hydrogen burning.

Once enough ^{12}C has formed in the core, other alpha particle reactions become possible, such as

$$^{12}\text{C} + {}^4\text{He} \longrightarrow {}^{16}\text{O}$$

$$^{16}\text{O} + {}^4\text{He} \longrightarrow {}^{20}\text{Ne}$$

$$^{20}\text{Ne} + {}^4\text{He} \longrightarrow {}^{24}\text{Mg}$$

Each of these reactions is exothermic, releasing a few MeV of energy and contributing to the star's energy production. At still higher temperatures (10^9 K) carbon burning and oxygen burning begin:

$$^{12}\text{C} + {}^{12}\text{C} \longrightarrow {}^{20}\text{Ne} + {}^4\text{He}$$

$$^{16}\text{O} + {}^{16}\text{O} \longrightarrow {}^{28}\text{Si} + {}^4\text{He}$$

Eventually ^{56}Fe is reached, at which point no further energy is gained by fusion (Figure 9.4).

If this explanation of the formation of elements is correct, we expect the abundances of the elements to have the following properties:

1. Large relative abundances of the light, even-Z elements; small relative abundances of odd-Z elements.

2. Little or none of the elements between He and C (Li, Be, B), which are not produced in these reactions.

3. Large relative abundance of Fe, the end product of the fusion cycle.

Figure 15.23 shows the relative abundances of the light elements in the solar system, and they are in agreement with all of the three above expectations. Each even-Z element is 10 to 100 times more abundant than its odd-Z neighbors, there is a prominent peak at Fe, the heavy elements with $Z > 30$ *combined* are less abundant than every element but one in the range C to Zn, and the three elements Li, Be, B are far less abundant than the elements in the range C to Zn.

The light odd-Z elements can be produced by alternative reactions among the fusion products, for example:

$$^{12}C + {}^{12}C \longrightarrow {}^{23}Na + {}^{1}H$$

$$^{16}O + {}^{16}O \longrightarrow {}^{31}P + {}^{1}H$$

The abundance of nitrogen is nearly equal to that of its neighbors C and O, which are the most abundant of elements beyond H and He; nitrogen has a greater abundance than any other odd-Z element shown, and greater than all even-Z elements with $Z > 8$. The formation of nitrogen must therefore be a relatively common process in stars. Since the element B is so rare, alpha particle reactions are of no help in forming nitrogen. The most likely sources of N are

$$^{12}C + {}^{1}H \longrightarrow {}^{13}N + \gamma$$

$$^{13}N \longrightarrow {}^{13}C + e^{+} + \nu$$

$$^{13}C + {}^{1}H \longrightarrow {}^{14}N + \gamma$$

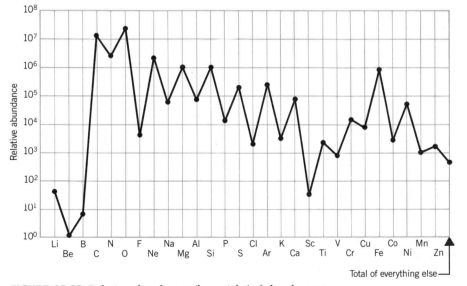

FIGURE 15.23 Relative abundances (by weight) of the elements beyond helium in the solar system.

and

$$^{16}O + {}^{1}H \longrightarrow {}^{17}F + \gamma$$

$$^{17}F \longrightarrow {}^{17}O + e^{+} + \nu$$

$$^{17}O + {}^{1}H \longrightarrow {}^{14}N + {}^{4}He$$

The stable isotopes ^{13}C and ^{17}O are found in natural carbon and oxygen with abundances of 1.1 percent and 0.04 percent, which suggests that reactions of this sort can indeed take place.

The production of the elements beyond iron requires the presence of neutrons, which are not produced in the reactions we have studied so far, because neutrons can only be emitted in reactions with nuclei that have an excess of neutrons. If enough of the heavier isotopes, such as ^{13}C, ^{17}O, or ^{21}Ne, are formed, the following reactions can produce neutrons:

$$^{13}C + {}^{4}He \longrightarrow {}^{16}O + n$$

$$^{17}O + {}^{4}He \longrightarrow {}^{20}Ne + n$$

$$^{21}Ne + {}^{4}He \longrightarrow {}^{24}Mg + n$$

How are the heavy elements built up by neutron capture? Let us consider the effect of neutron capture on ^{56}Fe, as we did in Chapter 10:

$$^{56}Fe + n \longrightarrow {}^{57}Fe \quad \text{(stable)}$$

$$^{57}Fe + n \longrightarrow {}^{58}Fe \quad \text{(stable)}$$

$$^{58}Fe + n \longrightarrow {}^{59}Fe \quad (t_{1/2} = 45 \text{ d})$$

What happens next depends on the number of available neutrons. If that number is small, the chances of ^{59}Fe encountering a neutron *before* it decays to ^{59}Co are small, and the process might continue as follows:

$$^{59}Fe \longrightarrow {}^{59}Co + e^{-} + \bar{\nu}$$

$$^{59}Co + n \longrightarrow {}^{60}Co \quad (t_{1/2} = 5 \text{ y})$$

$$^{60}Co \longrightarrow {}^{60}Ni + e^{-} + \bar{\nu}$$

On the other hand, if the number of neutrons is very large, a different sequence might result:

$$^{59}Fe + n \longrightarrow {}^{60}Fe \quad\quad (t_{1/2} = 3 \times 10^{5} \text{ y})$$

$$^{60}Fe + n \longrightarrow {}^{61}Fe \quad\quad (t_{1/2} = 6 \text{ m})$$

$$^{61}Fe \longrightarrow {}^{61}Co + e^{-} + \bar{\nu}$$

$$^{61}Co \longrightarrow {}^{61}Ni + e^{-} + \bar{\nu}$$

If the density of neutrons is so low that the chance of encountering a neutron is, on the average, less than once every 45 days, the first process ought to dominate, with the production of ^{60}Ni. If the chance of encountering a neutron is more like once every few minutes, the second process should dominate, and no ^{60}Ni is produced.

The first type of process, which occurs *slowly* and allows the nuclei time to beta decay, is known as the *s process* (s for slow); the second process occurs very rapidly and is known as the *r process* (r for rapid).

s-process and r-process

The neutron densities inside ordinary stars are sufficiently low that the average time necessary to capture a neutron is perhaps of the order of years, so the s process dominates. Suppose we have a nucleus of mass number A with abundance N_A. It is produced by neutron capture from mass number $A - 1$. The abundance N_A is *decreased* by neutron capture, which produces mass number $A + 1$. The rate of change in N_A is therefore

$$\frac{dN_A}{dt} \propto \sigma_{A-1}N_{A-1} - \sigma_A N_A \tag{15.26}$$

where σ_{A-1} and σ_A are the neutron capture cross sections of the nuclei with mass numbers $A - 1$ and A. The *larger* the cross section σ_A, the *smaller* is the abundance N_A; a large cross section means that a large fraction of N_A will react to produce nuclei with $A + 1$, thus decreasing N_A. Since the s process occurs over a very long period of time, an equilibrium situation can be reached in which A is produced at the same rate as it disappears. For $dN_A/dt = 0$,

$$\sigma_{A-1}N_{A-1} = \sigma_A N_A = \text{constant} \tag{15.27}$$

For the nuclei just beyond the iron peak, the product σN should decrease, because the nuclei are produced quickly (*not* by the s process) and then react slowly, so equilibrium is not reached. Only for the heavier nuclei is equilibrium reached. Figure 15.24 shows the product σN for some elements with $A > 56$, and the expected dependence is apparent from the variation of the experimental points. Just above $A = 56$, the product σN decreases, but above $A = 100$, σN levels off as equilibrium is reached.

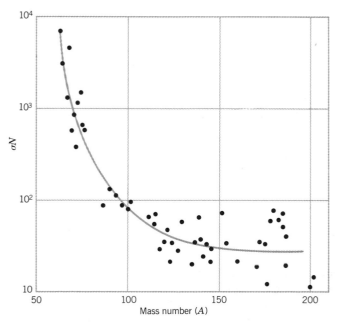

FIGURE 15.24 Product of abundance and cross section for elements beyond Fe. As expected for elements produced in the s process, the product becomes approximately constant for heavy nuclei.

The heaviest element that can be built up out of s-process neutron captures is ^{209}Bi; the half-lives of the isotopes beyond ^{209}Bi are too short to allow the s process to continue. The presence in nature of heavier elements such as thorium or uranium suggests that the r process must operate as well. The r process is also important for lighter elements. For example, ^{70}Zn cannot be formed by s-process neutron capture in Zn isotopes, because the half-life of ^{69}Zn is too short (56 min). Other examples of nuclei for which the r process is important are ^{76}Ge, ^{82}Se, ^{86}Kr, ^{96}Zr.

The r process most likely occurs during supernova explosions, following the breakdown and implosion of a star that has used up its fusion reserves. In a very short time, lasting of the order of seconds, the star implodes, produces an enormous flux of neutrons (perhaps 10^{32} n/cm^2/s), and builds up all elements to about $A = 260$. When the final explosion occurs, these elements are hurled out into space, to become part of new star systems. The heavy atoms of which the Earth is made may have been produced in such an explosion.

15.7 WHITE DWARF STARS

Once a star uses up its hydrogen fuel, and is composed mostly of helium, hydrogen fusion can no longer occur, and the radiation pressure is no longer able to prevent further gravitational collapse. The helium core therefore begins to collapse, and once again it heats up as gravitational energy is converted into thermal kinetic energy. The density of the core increases from 100 g/cm^3 to about 10^5 g/cm^3 and the temperature rises to about 10^8 K. At this point helium burning begins.

There is one other effect of the increased temperature associated with this collapse of the core—the increased radiation pressure pushes the outer layers of a star outward until its radius becomes hundreds or thousands of times as large. The total energy production in the star must be spread over a much larger surface area and so the energy per unit area decreases—the surface of the star seems to reach a lower temperature or to become redder. This is the "red giant" stage; when it becomes a red giant, our sun will become large enough to swallow up the orbits of Mercury and Venus.

The star now continues to burn helium, and perhaps heavier elements, in the red giant stage until the fusion cycle ends with the formation of ^{56}Fe.

At this point there is no further fusion energy available to prevent the gravitational collapse of the entire star, and so gravity takes over again. After perhaps going through a brief stage as a nova, the star eventually contracts to a *white dwarf*, with a density of perhaps 10^6 g/cm^3 and a surface temperature of the order of 10^4 K. A white dwarf of the sun's mass would be about the size of the Earth. Eventually it cools down as its energy is radiated away and becomes a black dwarf, a dark cold chunk of the ash left over from the original star.

What keeps further gravitational collapse from occurring? At such enormous densities, the atoms are crowded so close together that their electron wave functions begin to overlap. Just as in the case of the formation of molecules, there is a repulsive force which tends to oppose that overlap, because the Pauli exclusion principle prevents the electrons from having identical quantum numbers. The best way to picture this situation is as a solid, in which the energies of the electrons are governed by the Fermi-Dirac distribution; the electrons occupy very closely spaced, nearly continuous, energy levels from 0 up to the Fermi energy E_F, where

$$E_F = \frac{\hbar^2}{2m_e}\left(\frac{3\pi^2 N_e}{V}\right)^{2/3} \qquad (15.28)$$

The electron density is N_e/V in this expression. The average electron energy is $\frac{3}{5}E_F$, and therefore the total electron energy is $\frac{3}{5}N_e E_F$. The total gravitational potential energy can be shown (see Problem 9 at the end of the chapter) to be $-\frac{3}{5}GM^2/R$, where M is the total mass and R is the radius, and thus the total energy of the white dwarf, which we assume to be spherical, of uniform density and at a constant temperature is

$$E = \frac{3}{5}N_e\frac{\hbar^2}{2m_e}\left(\frac{3\pi^2 N_e}{V}\right)^{2/3} - \frac{3}{5}\frac{GM^2}{R} + \frac{3}{2}N_a kT + E_{rad} \quad (15.29)$$

The first term represents the motion of the electrons, the second term is the gravitational potential energy, the third term takes into account the thermal motions of the atoms (N_a = number of atoms), and the fourth term gives the energy radiated by the star. For the moment we neglect the last two terms (later we will show them to be small), and we will assume the star to be composed of N nucleons (in the form of fusion products such as iron and lighter elements) and approximately $\frac{1}{2}N$ electrons. To find the equilibrium radius we set $dE/dR = 0$ and find

$$R = \frac{3^{4/3}\pi^{2/3}}{8}\frac{\hbar^2}{Gm_e m_n^2}N^{-1/3} \qquad (15.30)$$ **Radius of white dwarf star**

For a star of the mass of the sun ($M = 2.0 \times 10^{30}$ kg, so $N = 1.2 \times 10^{57}$) we find $R = 7.1 \times 10^3$ km, about the same as the Earth's radius, and this corresponds to a density of 1.1×10^6 g/cm³. Going back to Equation (15.28) we find $E_F = 0.194$ MeV, so the total contribution of the electrons is

$$E_{electrons} = \frac{3}{5}N_e E_F = 7 \times 10^{55} \text{ MeV} \qquad (15.31)$$

We conclude this section by justifying our assumptions in neglecting the last two terms of Equation (15.29). The thermal energy can be easily computed at $T \cong 10^8$ K (the central temperature is $\sim 10^9$ K, while the surface temperature is only 10^4 K):

$$E_{thermal} = \frac{3}{2}N_a kT$$

$$= \frac{3}{2}\left(\frac{1.2 \times 10^{57}}{30}\right)\left(8.6 \times 10^{-5}\frac{eV}{K}\right)10^8 \text{ K}$$

$$= 5 \times 10^{53} \text{ MeV} \qquad (15.32)$$

(where we take the average atom to contain 30 nucleons) and so $E_{thermal} \ll E_{electrons}$, as we assumed. The radiant energy can be estimated from Stefan's law as σT^4, which gives the *power per unit area*; the power is just radiant energy per unit time, so

$$E_{rad} = (\sigma T^4)(4\pi R^2)t \qquad (15.33)$$

where t is the time over which the star has been radiating. The observed surface temperatures of white dwarfs are of the order of 10^4 K, and so we

estimate

$$E_{rad} = (2.3 \times 10^{36} \text{ MeV/s})t \qquad (15.34)$$

Even if the star had been radiating uniformly for the entire age of the universe $(15 \times 10^9 \text{ y} \cong 5 \times 10^{17} \text{ s})$ the total power radiated would be only about 10^{54} MeV, and so $E_{rad} \ll E_{electrons}$.

The white dwarf star represents an extraordinary state of matter; its average density is about 10^6 g/cm^3, and its central density may approach 10^8 g/cm^3. A cubic centimeter of such matter would, on the Earth, weigh 100 tons! We have based our analysis of this unusual condition on expressions that we derived for ordinary matter, and it is surprising, perhaps comforting, to learn that we can understand this state of matter from these expressions. Quantum physics works well even in the interiors of white dwarf stars! A white dwarf is prevented from collapse by the Pauli principle, which prevents the electron wave functions from being squeezed too close together. Nature is, however, not through surprising us, and in the next section we examine a still denser state of matter.

15.8 NEUTRON STARS

In the last section, we made a slight oversight in calculating the properties of white dwarf stars. The expression for E_F was derived in Chapter 12 using the classical expression $p^2/2m_e$ for the electron's kinetic energy. When we evaluated the Fermi energy we found $E_F = 0.194$ MeV, so it is not true that $E_F \ll m_e c^2$. This is not a serious problem for the estimates we made in the previous section, but as the star gets more massive, E_F increases and relativistic effects become more important. Nevertheless, it is still true that as M (and therefore N) increases, E_F increases. Will the Pauli principle be able to prevent collapse of any star, no matter how massive?

There is a mass limit of about 1.4 solar masses, called the *Chandrasekhar mass*, beyond which the Pauli principle applied to electrons cannot prevent further gravitational collapse. Let us look at how this might occur. Our classical calculation gives $E_F = 0.304$ MeV for a white dwarf having the Chandrasekhar mass, but remember that this is only the point at which the Fermi-Dirac distribution has the value 0.5. There is a high-energy tail of the Fermi-Dirac distribution, and some of the electrons in that tail have enough energy to produce the inverse beta decay reaction

$$e^- + p \longrightarrow n + \nu$$

for which the Q value is 0.782 MeV, not too far above E_F. This reaction will in effect remove some electrons from the star, reducing the effects of the Pauli principle, and allowing the star to collapse a bit (R depends on $N_e^{5/3}$). The Fermi energy *increases*, pushing more electrons above the 0.782-MeV Q value, resulting in more electrons being lost, and so on, until all (or very nearly all) of the electrons vanish. The star is now composed of neutrons, instead of protons and electrons. The "pressure" of the electrons no longer can oppose gravitational collapse, and so the star contracts until the Pauli principle applied to the *neutrons* (which also obey Fermi-Dirac statistics) prevents further collapse. The situation is now described by expressions identical with (15.28) to (15.30) but ap-

J. Robert Oppenheimer (1904–1967, United States). A theoretical physicist with a philosophical nature, it is ironic that he was the director of the laboratory at Los Alamos that developed the first nuclear weapons. His early theoretical work speculated on the existence of black holes.

plied to neutrons; in particular

$$E_F = \frac{\hbar^2}{2m_n}\left(\frac{3\pi^2 N}{V}\right)^{2/3} \tag{15.35}$$

$$R = \frac{3^{4/3}\pi^{2/3}}{2^{4/3}}\frac{\hbar^2}{Gm_n^3}N^{-1/3} \tag{15.36}$$ **Radius of neutron star**

For a star of 1.5 solar masses, for example, $R = 11$ km and its density is about 4×10^{14} g/cm³. This is about the same density as in the interior of atomic nuclei, so in this sense the star has become a giant nucleus, 20 km across, with a mass number of about 10^{57}! Such objects have become known as *neutron stars*.

Are these neutron stars merely figments of the physicist's imagination or do they really exist? In 1967, radio astronomers at Cambridge University discovered a unique signal among their observations—an extremely sharp and regular pulsation, such as is shown in Figure 15.25, with a period of 1.34 s. No previously known astronomical object could produce such sharp and regular pulses, and at first the Cambridge group suspected that they might have discovered signals from an extraterrestrial intelligent civilization. (The object emitting the pulses was at first called LGM-1; LGM stands for "Little Green Men.") This notion was later discarded (unfortunately) and the object became known as a *pulsar*. **Pulsars** Since 1967, over 100 other pulsars have been discovered; all have extremely regular periods of the order of one second.

The connection between pulsars and neutron stars was made by astronomer Thomas Gold soon after their discovery. Gold proposed that the collapse of a rotating star to a neutron star ought to cause the neutron star to rotate much more rapidly. (Since angular momentum is conserved in the collapse, a decrease in the moment of inertia, as the star contracts, must be balanced by an increase in the angular speed.) The intense magnetic fields of such a rapidly rotating object will trap any emitted charged particles and accelerate them to high speeds, where

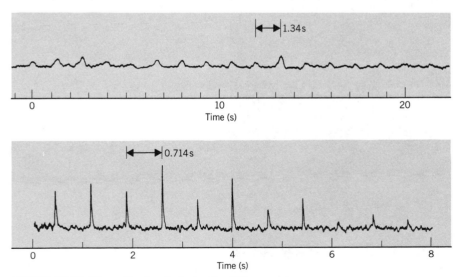

FIGURE 15.25 The radio signals from two different pulsars. The top signal is the record of the first pulsar discovered.

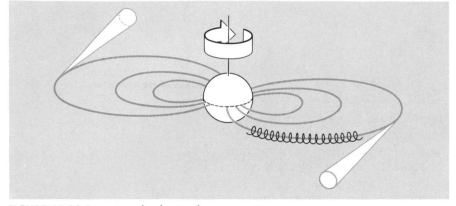

FIGURE 15.26 Emission of radiation from a rotating neutron star. Charged particles are trapped as they spiral around the magnetic field lines. Since the magnetic field rotates with the same rapid angular velocity as the star, the centripetal acceleration far from the star can become quite large, and the accelerated charged particles emit radiation.

they will give off radiation (Figure 15.26). As the star rotates, this beam of emitted radiation sweeps around like a searchlight or a lighthouse, and we see it whenever the beam sweeps through the Earth. The observed interval between the pulses is, according to this interpretation, just half the rotational period of the neutron star.

If this explanation of a pulsar as a rotating neutron star is correct, we ought to see the pulsars slowing down somewhat, as the radiated en-

FIGURE 15.27 The Crab Nebula, remnant of a supernova observed in the year 1054. (Photograph courtesy Kitt Peak National Observatory).

ergy is compensated by a decrease in the neutron star's rotational kinetic energy. This effect has been seen for nearly all pulsars, and amounts to about 1 part in 10^9 per day.

Although the exact mechanism of the collapse of a star to a neutron star is not yet understood, we suspect that the violent explosions known as *supernovas* leave a neutron star as a remnant. In 1054 Chinese astronomers observed a supernova explosion (which they called a "guest star") that was visible in the daytime over many days. Today we see the expanding shell of that explosion as the Crab Nebula (Figure 15.27). At the center of the Crab Nebula is a pulsar, rotating with a period of 30 Hz. It is remarkable that none of the many photographs of the Crab that were taken before 1967 have revealed this pulsar blinking on and off every 0.033 s; all of these photographs were taken over long exposure times, and so the pulsations were not observable. When careful measures are taken, however, the blinking effect can be seen quite clearly (Figure 15.28). This suggests that, at least in this instance, pulsars may be identified as supernova remnants.

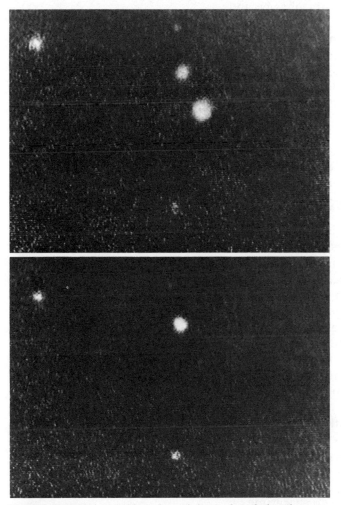

FIGURE 15.28 The visible pulsar of the Crab Nebula. The two exposures show the pulsar blinking on and off relative to the other stars in the photograph. (Courtesy Hale Observatories)

Pulsars have now been observed at many different wavelengths (optical, X ray, γ ray, radio) and with such great precision of timing that the slowing down of 10^{-9} per day is easily observable. (Even "starquakes" have been observed, in which the hard outer crust of the neutron star suddenly shifts in response to the slowing; this sudden decrease in the radius causes an immediate increase in the rotational speed, which astronomers call a "glitch.")

A neutron star is still not the ultimate fate of the collapse of massive stars. If the star is so massive that the gravitational force is strong enough to overcome even the neutron Pauli principle, there is now nothing to prevent the material in the star from suffering *complete* collapse, down to a single point in space. If we could observe such an event, we would see matter in its most elemental forms, quarks and leptons, approaching a single, dimensionless, mathematical point, a singularity. (This process is, in a way, and on a very small scale, the reverse of the Big Bang to be described in Chapter 16.) We are not sure what happens during that final collapse, because the laws of quantum physics and gravitation have not been combined in a way that permits us to study that final moment; nor are we sure exactly what the mass limits are on the final collapse—for a star of less than two to three solar masses, a white dwarf or neutron star probably is the ultimate fate, but stars of greater mass may suffer this complete collapse. Whether or not a singularity is reached can never be observed, because once the collapse passes the event horizon (Schwarzschild radius), Equation (15.24), no further light, radiation, particles, or information of any sort can reach the rest of the universe from the star. It has become a *black hole.*

How do we go about finding a black hole? To find an isolated black hole is virtually impossible (a black object against a black sky). The only chance to find a black hole would be as part of a binary system—two objects revolving about their common center of mass. From the orbital period, the relative masses of the two objects may be determined, and if one of the objects is a visible star, its mass can be estimated from its color and luminosity. The mass of the unseen companion of the star can thus be deduced. Many such binary systems have been observed, and they are frequently intense X-ray sources; the unseen companion pulls material from the large star, and this material heats up and emits X rays as it falls in toward the companion. Of several binary X-ray sources which have been discovered, one has an invisible member heavy enough to be a black hole. Located in the constellation Cygnus (the Swan), this source is called Cygnus X-1; it is at present the leading candidate for a black hole, although the evidence in its favor is not direct and many astrophysicists doubt its interpretation as a black hole.

The existence of black holes has not yet been verified, and many are skeptical that black holes exist at all; they are perhaps the most curious and enigmatic objects in our universe and the subject of much speculation by scientists. Does a black hole lose energy when its intense gravitational field creates particle-antiparticle pairs? Does the matter that falls through a singularity reappear at some other point in spacetime, or in some "other" universe? Can a black hole be harnessed as an energy source? If the calculations of astrophysicsts are correct, there are in our

15.9 BLACK HOLES

galaxy perhaps 10^9 black holes that have evolved from massive stars, and as we refine our abilities to survey the skies from X-ray and γ-ray satellites beyond the atmosphere, perhaps further evidence of their existence will soon emerge.

There are many introductory, advanced, and popular-level books and articles that cover in more detail the subjects touched only briefly in this chapter. To pursue any of these subjects, you may first need some background material on astronomy and astrophysics:

SUGGESTIONS FOR FURTHER READING

F. Golden, *Quasars, Pulsars, and Black Holes* (New York, Scribner's, 1976). A popular-level, nonmathematical introduction.

M. Harwit, *Astrophysical Concepts* (New York, Wiley, 1973). Nearly everything you would want to know about astrophysics, using undergraduate math.

L. Motz, *Astrophysics and Stellar Structure* (Waltham, Ginn & Co., 1970). An advanced presentation of astrophysics, complete but mathematically sophisticated.

L. Motz and A. Duveen, *Essentials of Astronomy* (New York, Columbia University Press, 1977). A complete introductory undergraduate astronomy text.

H. L. Shipman, *Black Holes, Quasars, and the Universe* (Boston, Houghton Mifflin, 1976). An excellent popular-level survey; see especially Chapter 4 on clock rates and red shifts during the descent into a black hole.

Some additional detail on general relativity, without the advanced math:

P. G. Bergmann, *The Riddle of Gravitation* (New York, Scribner's, 1968). Very descriptive and successful presentation of the philosophy of general relativity by a student of Einstein's.

M. Berry, *Principles of Cosmology and Gravitation* (Cambridge, Cambridge University Press, 1976). Reduces the mathematics to undergraduate level; see especially Chapter 5 on the tests of general relativity.

M. G. Bowler, *Gravitation and Relativity* (Oxford, Pergamon Press, 1976). An intermediate-level text, with some tensor math.

P. C. W. Davies, *Space and Time in the Modern Universe* (Cambridge, Cambridge University Press, 1977). Excellent nonmathematical introduction to spacetime and general relativity.

R. P. Feynman, R. B. Leighton, and M. Sands, *The Feynman Lectures on Physics* (Reading, Addison-Wesley, 1964). Chapter 42 contains an interesting elementary discussion on the properties of curved spacetimes.

R. Gerach, *General Relativity from A to B* (Chicago, University of Chicago Press, 1978). Emphasizes use of spacetime diagrams; no math.

W. J. Kaufmann, III, *The Cosmic Frontiers of General Relativity* (Boston, Little, Brown and Co., 1977); also by Kaufmann, *Black Holes and Warped Spacetime* (San Francisco, W. H. Freeman and Co., 1979). Both are excellent as elementary introductions to the properties of black holes.

W. Rindler, *Essential Relativity* (New York, Van Nostrand, 1969). Uses a bit of tensor math; a good beginning if you really want to get into the complete theory.

R. v. B. Rucker, *Geometry, Relativity, and the Fourth Dimension* (New York, Dover, 1977). A good nonmathematical introduction to higher dimensions; includes a particularly comprehensive list of references, including Abbott's classic work *Flatland.*

R. M. Wald, *Space, Time, and Gravity* (Chicago, University of Chicago Press, 1977). Brief but thorough introduction using a bit of calculus but no tensors; particularly good on black holes.

If you want to see the complete theory, in all its mathematical glory:

S. W. Hawking and G. F. R. Ellis, *The Large-Scale Structure of Space-Time* (Cambridge, Cambridge University Press, 1973).

C. W. Misner, K. S. Thorne, and J. A. Wheeler, *Gravitation* (San Francisco, Freeman, 1973). Figure 33.2 shows how to use a black hole as a combined energy source and garbage dump.

S. Weinberg, *Gravitation and Cosmology* (New York, Wiley, 1972).

Each of these is a difficult but standard work, and each contains a little that you should understand.

As with many of the subjects covered in this book, the best source for current, popular-level articles is the journal *Scientific American*. From time to time, articles appear covering techniques and results in current astronomy and astrophysics. Two articles that relate directly to this chapter are:

S. W. Hawking, "The Quantum Mechanics of Black Holes," *Scientific American* 236, 34 (January 1977).

K. S. Thorne, "The Search for Black Holes," *Scientific American* 231, 32 (December 1974).

QUESTIONS

1. If we were to measure the equivalence of gravitational and inertial mass, would we show that $m_{\text{inertial}} = m_{\text{gravitational}}$ or merely that $m_{\text{inertial}} \propto m_{\text{gravitational}}$?

2. Do tidal effects distinguish between Newtonian gravity and curved space-time? What would be the shape of a drop of liquid following a geodesic path in a curved spacetime? Can such a drop distinguish between a uniform gravitational field and a uniform acceleration?

3. Is Hubble's constant really a constant? Does it vary over large distances in space? Over long intervals in time?

4. Suppose that the first measurement of deflection of starlight during a solar eclipse had been done after 1905, when the special theory of relativity was introduced, but before 1916, when the general theory was introduced. What would have been the effect of this measurement on the special theory?

5. If we could make a precise comparison of light from the sun with light from the moon, would the moonlight be red shifted, blue shifted, or unshifted relative to sunlight?

6. What difficulties might arise in the Pound and Rebka experiment on the gravitational red shift if the temperature of the source or the absorber varied?

7. How "sharp" is the Fermi distribution for the electrons in a white dwarf star? That is, how does E_F compare with kT? Are we justified in using the $T = 0$ expressions for E_F and E_{av}?

8. Why does the radius of a white dwarf or neutron star depend inversely on the number of nucleons N? Shouldn't a star with more matter have a larger radius?

9. Do the neutrons in a typical neutron star have enough energy to create pi mesons in collisions such as $n + n \rightarrow n + n + \pi$? As a neutron star collapses toward a black hole, the neutrons become increasingly energetic, and greater numbers of pi mesons are produced. Why doesn't the Pauli principle applied to the pi mesons prevent collapse?

10. Explain why ^{76}Ge and ^{82}Se cannot be produced by means of the s process.

1. Derive Equation (15.4) by solving Newton's second law for the force $F = GmM/x^2$.

2. A satellite is in orbit at an altitude of 150 km. We wish to communicate with it using a radio signal of frequency 10^9 Hz. What is the gravitational change in frequency between a ground station and the satellite? (Assume g doesn't change appreciably.)

3. According to the uncertainty principle, what is the minimum time interval necessary to measure a change in frequency of the magnitude in the Pound and Rebka experiment?

4. A certain star is 80.0 light-years from Earth. Directly along the line of sight from Earth to the star, and at a distance of 20.0 light-years from the star, is a white dwarf star, with a mass equal to the sun's mass and a radius of 7.0×10^3 km. The deflection of light from the star by the white dwarf causes us to observe two images of the star:

Find the angle α between the two images. (Such an effect, called a gravitational lens, has been observed for galaxies; see *Scientific American*, November 1980.)

5. Derive Equation (15.16) by differentiating Equation (15.8).

6. (a) Show that the reaction $^7Be + e^- \rightarrow {}^7Li + \nu$, which occurs in the proton-proton cycle, results in a neutrino of energy 0.862 MeV. (b) Compute the Q value of the reaction $\nu + {}^{37}Cl \rightarrow {}^{37}Ar + e^-$, which is used to detect the neutrinos, and show that the neutrino energy found in part (a) is sufficient to produce the reaction.

7. Trace the path of the s process from the stable isotope ^{63}Cu to the stable isotope ^{75}As, showing the neutron capture and beta decay processes.

8. Show how the s process proceeds from stable ^{81}Br to stable ^{95}Mo.

9. Consider a spherical distribution of mass of uniform density ρ and radius R. (a) An element of mass dm is at a radius $r < R$. Assuming there is a "Gauss's law" for gravitation, what is the quantity of mass that attracts dm? (b) What is the gravitational potential dV of the mass element dm? (c) Express dm in terms of the volume element in spherical coordinates. (d) Integrate over the volume of the sphere to obtain the total gravitational potential energy that was used in Equation (15.29).

10. Neglecting the last two terms of Equation (15.29), show that $dE/dR = 0$ at the value given by Equation (15.30).

11. Show that Equation (15.30) can be written

$$R = (7145 \text{ km}) \left(\frac{M}{M_\odot}\right)^{-1/3}$$

where $M_\odot$ is the mass of the sun.

12. (a) For a white dwarf star of the mass of the sun, compute the deBroglie wavelength of electrons at the Fermi energy. (Use nonrelativistic kinematics.) (b) Assuming the star to be made of iron atoms and to be of uniform density, compute the distance between atoms and compare with the electron deBroglie wavelength. What do you conclude from this comparison? Do the electrons "see" the lattice of iron atoms? Are they easily scattered by the atoms?

13. In deriving Equation (15.30), it was assumed that $N_e = \frac{1}{2}N$, as would be true for a star composed of lightweight elements with equal numbers of protons and neutrons. When electrons begin to be consumed in the development of a neutron star, this is not true. Repeat the derivation of Equation (15.30) for the case $N_e = fN$, where f is the fraction of electrons. Show that $R \propto N_e^{5/3}$, as claimed in Section 15.8.

14. Estimate the angular speed of a neutron star in the following way. Consider a star 1.5 times as massive as the sun, rotating on its axis about once per year. (This is quite a slow rate of rotation—our sun rotates about once per month.) Assume the star to have about the same radius as the sun $(7 \times 10^5$ km) and to be relatively uniform in density. If angular momentum is conserved in the collapse, what will be the final angular velocity? Is this of the right order of magnitude for a neutron star? What errors are made in this estimate, and how do they affect the final result?

15. (a) Compute the value of Gm_n^2 in units of J·m. (b) Compute the value of $\hbar^2/m_n$ in MeV·m². (Multiply top and bottom by c^2.) (c) Combine the results of (a) and (b) to find $\hbar^2/Gm_n^3$ in meters. (d) Combine the results of (a) and (b) to find $G^2m_n^5/\hbar^2$ in MeV.

16. Use the results of the previous problem to rewrite Equations (15.35) and (15.36) as:

$$E_F = (56.27 \text{ MeV}) \left(\frac{M}{M_\odot}\right)^{4/3}$$

and

$$R = (12.34 \text{ km}) \left(\frac{M}{M_\odot}\right)^{-1/3}$$

where $M_\odot$ is the mass of the sun.

17. (a) Compute the neutron Fermi energy in a neutron star of 2.0 solar masses. (b) What is the average neutron energy? [See Equation (12.36).] (c) What is the total energy of the neutrons? (d) Calculate the total gravitational potential energy of the neutron star. (e) The gravitational potential energy resulted from the collapse of the original star (or of the gas cloud that formed the star.) Only part of that energy went into the neutron kinetic energies. What fraction of the gravitational energy ended with the neutrons? Where did the rest go?

18. Find the Fermi energy of the neutrons in a neutron star of 1.5 solar masses. Are we justified in using nonrelativistic kinematics to describe their motion?

19. Find the deBroglie wavelength of the neutrons at the Fermi energy in a neutron star of mass $1.80M_\odot$. Compare the deBroglie wavelength with the distance between neutrons. Do the neutrons "see" one another's deBroglie waves? Are they scattered by the lattice of neutrons?

20. A neutron star of 2.00 solar masses is rotating at a rate of 1.00 revolutions per second. (a) What is the radius of the neutron star? (b) Assuming its density is uniform, find its moment of inertia and its angular momentum. (c) Find its rotational kinetic energy. (d) If its rotational speed slows by 1 part in 10^9 per day, find the loss in rotational kinetic energy per day. (e) Assuming that the entire energy loss goes into radiation, find the radiative power.

21. Assume that the neutron star of Problem 20 is 10^4 light-years from Earth. (a) If its radiative power were distributed uniformly in space, what power would be delivered to 1.0 m² of antenna? (b) A typical pulsar, 10^4 light-years from Earth, has an observed average power level of 5.0×10^{-20} W/m². What does the comparison of this power with the value deduced in part (a) tell us about the radiative power?

Today we scan the skies at all wavelengths, from the very long (radioastronomy) to the very short (X-ray and gamma-ray astronomy); many new and unexpected phenomena have been discovered at these wavelengths. Quasars, pulsars, novas, supernovas, black holes—all these suggest that the universe is not at all static and eternal but instead is active and evolving and teeming with radiation. Van Gogh's painting The Starry Night suggests exactly that view, even though it was painted in 1889, long before any of these discoveries were made.

16

COSMOLOGY: THE ORIGIN AND FATE OF THE UNIVERSE

Each major advance in the understanding of our universe has placed us further from its center and reduced our importance in it, and yet each advance simultaneously brings us a new feeling of awe when we try to comprehend the natural phenomena which are responsible for the present character of our universe. Seventeenth-century astronomy revealed that the Earth was not the center of the solar system, but rather one of many planets circling the sun. In the nineteenth century astronomers turned their telescopes to the stars and used their newly developed spectroscopes to measure the wavelengths of starlight. These observations and others suggested that the sun is merely a rather ordinary star, quite unremarkable on the galactic scale, one of perhaps 10^{11} stars in our little island of space known as the Milky Way galaxy.

These same astronomers also turned up some mysterious objects from their telescopes—faint, wispy "nebulae," broad patches of light much too large to be stars. Some of these nebulae were later deduced to be clouds of gas in our own galaxy, representing either new material out of which young stars are being formed, or the remnants of stars which underwent fiery, explosive deaths. Other nebulae, much fainter, did not yield easily to understanding. As larger and larger telescopes were put into operation, culminating with the 100-in. telescope on Mt. Wilson and the 200-in. on Mt. Palomar, both in California, more and more of these mysterious nebulae were discovered. One opinion was that these objects were inside our own galaxy; another opinion was that they were island galaxies like our own, containing at least as many stars and separated from us by the incomprehensible distance of millions of light-years ($\sim 10^{20}$ km). To settle this question would require the resolution of the light from these faint objects into individual stars, an experimental problem of practically insurmountable difficulty, requiring the exposure of a photographic plate for a full night while the astronomers struggled in the cold of a California mountaintop to keep the focus of the telescope fixed on the nebula, as the Earth turned and as the temperature variations caused changes in the dimensions of the telescope. In the 1920s Edwin Hubble succeeded in his efforts to resolve individual stars in our nearest neighboring galaxies, and in the process deduced their size, brightness, and distance.

As more regions of the sky were studied with these large telescopes, more and more of these galaxies were discovered and our place in the universe shrunk still further—not only are we one star of 10^{11} in our galaxy, but our galaxy is one of perhaps 10^{11} in the universe. Beyond the stars of our own galaxy are yet again as many entire galaxies!

Hubble's observations held still another even more startling result—each of these galaxies is hurtling away from us (and from one another) at unimaginable speeds, some approaching the speed of light. Indeed, the further a galaxy is from us, the faster it appears to be moving away. This remarkable deduction, perhaps one of the most significant in the history of physical science, leads to what has become the accepted current model of the universe and its origin. If the galaxies are now growing further apart, they must previously have been closer together. If we go back far enough in time, all matter may have originated from a single point in a violent explosion that began the expansion of the universe, dragging the galaxies along their present course.

Still more startling news was to follow. In 1965, two radioastronomers, Arno Penzias and Robert Wilson, at the Bell Laboratories in New Jersey, finished some rather ordinary tasks associated with developing radio links with Bell's new Telstar communications satellite. Turning their radio telescope on the sky itself, they discovered (in an event as unexpected as Hubble's was carefully planned and preconceived), that the radiation from that primordial explosion fills the universe and bathes the Earth in its glow, even though it has cooled considerably in the intervening 15 billion years. (This work earned Penzias and Wilson the 1979 Nobel prize in physics.)

The experimental work begun by Hubble and by Penzias and Wilson form the two cornerstones on which our present speculations about the origin, evolution, and fate of the universe are based. Such theories are within the realm of *cosmology,* in which experimental results from such diverse fields as astronomy and particle physics are combined with mathematical results such as from thermal physics, statistical physics, electrodynamics, and general relativity.

In this chapter, after we examine the two principal experiments in greater detail, we will discuss the present leading theory of cosmology, now known (at first perhaps pejoratively but now affectionately) as the "Big Bang" theory. Independent measurements that support this theory will be discussed; after examining the present composition of the universe, we conclude with some cosmological speculations on the future of our universe.

One warning before we begin: we often poke fun at our scientific ancestors who believed that the Earth was the center of the universe and that everything revolved about us. Our present knowledge of the universe shows how far we have come from this geocentric view! And yet through it all, one form of geocentrism still persists—we must accept that the laws of physics here on Earth are the same as those elsewhere in the universe. In particular, quantum physics, special and general relativity, nuclear physics, and particle physics must follow identical rules everywhere and at all times. Is it chauvinistic of us to believe that the laws we have discovered in our tiny and insignificant corner of the universe are those that apply everywhere? Certainly! Is it possible that there are other physical laws, not yet even imagined by us, which apply elsewhere in the universe? Of course! Do we have any choice but to *assume* our laws to be universal? Unfortunately, no. There is, happily, no evidence to suggest that this assumption is false: Other stars *seem* to shine by the same mechanism that makes our sun shine, and other galaxies *seem* to be bound together by the same gravitational force which holds our galaxy together. Although it is not often explicitly stated, this new Geocentric Principle is the ultimate cornerstone of our cosmology.

16.1 THE EXPANSION OF THE UNIVERSE

The evidence for the expansion of the universe comes from the Doppler shift of light from distant galaxies. Of the stars in our own galaxy, some may happen to be moving toward us, with their light shifted toward the shorter wavelengths (blue), and some may happen to be moving away from us, with their light shifted toward the longer wavelengths (red). The average speed of these stars relative to us is about 30 kilometers per second or 0.0001 of the speed of light. In Chapter 2 we learned that the

relativistic Doppler shift of light is

$$\nu' = \nu \sqrt{\frac{1 - v/c}{1 + v/c}} \qquad (16.1)$$

where now v represents the relative speed and we have assumed that the source and observer are moving away from one another. We can rewrite Equation (16.1) in terms of wavelength

$$\lambda' = \lambda \sqrt{\frac{1 + v/c}{1 - v/c}}$$
$$= \frac{\lambda(1 + v/c)}{\sqrt{1 - v^2/c^2}} \qquad (16.2)$$

Here λ' is the wavelength we measure on Earth and λ is the wavelength emitted by the moving star or galaxy in its own rest frame. For $v/c \cong 0.0001$, the change in wavelength is very small, and we do not expect to see such effects. (This is consistent with observations; if v/c were large, we should observe many more red and blue stars than we do.)

If we turn next to the light from nearby galaxies, those of the "local group" for example, we again find some with rather small blue shifts and others with small red shifts, consistent with more-or-less random motions.

It is only when we turn to the very distant galaxies that the spectral shifts become both large and systematically red.

Before we turn to the experimental evidence for these large red shifts, let us review how they are measured. The sun emits radiation with a continuous spectrum. As that light travels through the sun's atmosphere, some of the light is absorbed by the gases which surround the sun, and so the continuous *emission* spectrum has a few dark *absorption* lines superimposed. *It is these absorption lines that show the evidence of the red shifts.* Were it not for the presence of the impurity atoms, even in very small quantities, in the distant galaxies, it would be very difficult for us to be certain of the red shifts.

Figure 16.1 shows an example of the experimental determination of the red shifts of several galaxies.

How do we know that it is the expansion of the universe which causes these red shifts? The following three arguments support this belief.

1. There are *not* equivalent numbers of red and blue shifts—all distant galaxies are moving away from us. We therefore cannot explain these red shifts as the Doppler shifts of many galaxies subject to a random, statistical distribution of motions.

2. The red shifts of the *galaxies* are probably not the *gravitational* red shifts of general relativity. The matter in galaxies is not sufficiently compact to produce such large red shifts, which would require extremely large gravitational fields. Although there *are* gravitational red shifts in the light from *any* massive glowing object (including our own sun), such red shifts are extremely small (recall Section 15.4) and could not account for the size of the observed shifts. (There is at present a serious disagreement over whether this same reasoning ap-

Edwin Hubble (1889–1953. United States). His observational work with large telescopes revealed the existence of galaxies, and he was the first to measure their size and distance. His discovery of the recessional motion of the galaxies was one of the most exciting and important in the history of astronomy.

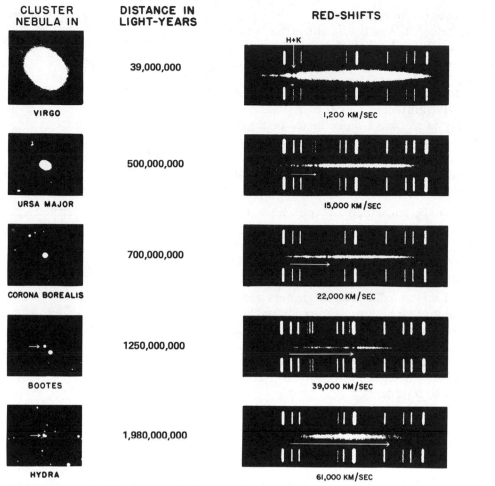

CLUSTER NEBULA IN	DISTANCE IN LIGHT-YEARS	RED-SHIFTS
VIRGO	39,000,000	1,200 KM/SEC
URSA MAJOR	500,000,000	15,000 KM/SEC
CORONA BOREALIS	700,000,000	22,000 KM/SEC
BOOTES	1250,000,000	39,000 KM/SEC
HYDRA	1,980,000,000	61,000 KM/SEC

FIGURE 16.1 Red shifts for several galaxies. The spectra from the galaxies are absorption line spectra, continuous emission spectra containing dark absorption lines. For these galaxies, the two dark lines indicated by the arrows are from calcium. The emission lines above and below each absorption spectrum are for calibration. Below each spectrum is the velocity determined from the red shift. (Courtesy Hale Observatories)

plies to the *quasars*, which might be compact enough to have large gravitational red shifts; see Section 16.5.)

3. The observed red shifts are proportional to the distance of the objects from us. This is probably the most important step in the chain of reasoning which supports the idea of the expanding universe, and is usually expressed in the form known as *Hubble's law*

Hubble's law

$$v = Hd \qquad (16.3)$$

where v is the speed of the galaxy (deduced from its red shift) and d is its distance from us. The proportionality constant H is *Hubble's constant* of Equation (15.13).

The most difficult (and controversial) aspect of verifying Hubble's law is the measurement of the distances to the remote galaxies. Our distance scale is built on a complex sequence of experimental information, each one depending on all of those before it. As a result, the last 50 years have seen many changes in the "accepted" value of the Hubble constant, some by as much as a factor of 5; all these changes were caused by adjustments in the distance scale.

The difficulty in measuring distance comes about because, for example, when we view an apparently faint object in the sky we don't know whether it is an intrinsically faint object close to us, or a very bright object very far away. If we have a way to determine the *intrinsic brightness*, or *luminosity*, of distant stars and galaxies, we could then deduce their distance from their apparent brightness. The chain of reasoning that gives the distance scale proceeds as follows:

1. The distance to nearby stars can be measured by the *method of parallax*—as the Earth moves in its orbit, the nearby stars will appear to move against the background of "fixed" distant stars. From measuring the apparent brightness of these nearby stars, we can deduce their intrinsic brightness or luminosity.

2. Some of the nearby stars are *variable stars*, whose brightness varies regularly with periods of the order of days. From the study of these variable stars, whose distance and therefore luminosity have been determined by the parallax method, we learn that there is a regular relationship between period and luminosity. By measuring the period of a variable star, we can deduce its luminosity, and from its luminosity and apparent brightness, we deduce its distance.

3. Powerful telescopes enable us to observe variable stars in nearby galaxies, and using the period-luminosity relationship, we can deduce the distance to the nearby galaxies. This does not work for very distant galaxies, for we can't resolve individual stars even with the most powerful telescopes.

4. The final steps in the reasoning rely on statistical arguments. Although the luminosity of the individual stars in a galaxy may vary considerably, there are so many different varieties of stars in *every* galaxy that the brightest stars in one galaxy should have about the same luminosity as the brightest stars in any other galaxy. Similarly, novas in one galaxy should be about as bright as novas in other galaxies. We can use this assumption to estimate the distances to nearby galaxies.

5. Finally, galaxies are observed to occur in clusters, and applying the same statistical argument suggests that the brightest galaxy in any one cluster should have about the same luminosity as the brightest galaxy in any other cluster.

As you might guess, with such a connected chain of reasoning, there are many opportunites for error, and indeed such errors have in the past 50 years resulted in substantial changes in the Hubble constant. (At one point the value of the Hubble constant led to a deduced age of the universe that was *smaller* than the age of the Earth, as deduced from radioactive dating!) There is now a general agreement on the distance scale and therefore on the Hubble constant, and it is probably correct to within 10 to 20 percent.

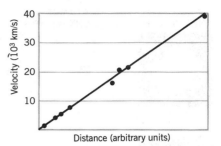

FIGURE 16.2 Velocity-distance relationship for galaxies, illustrating the Hubble law. These results are based on Hubble's early data for galaxies. The distance scale has been corrected several times, so no distance units are shown, but the linear relationship is apparent.

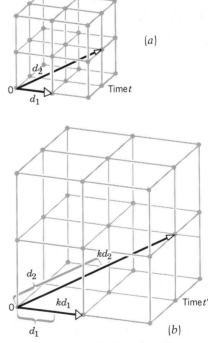

FIGURE 16.3 The expansion of a coordinate space, showing that the apparent speed of recession depends on the distance; d_2 is greater than d_1, and d_2 increases faster than d_1.

Figure 16.2 shows the relationship between the speed of recession v, deduced from red shifts, and the galactic distance, deduced as described above. As you can see, these results are quite consistent with the linear relationship suggested by Hubble's law. With the present distance scale, the value of the Hubble constant is about

$$H = 75 \frac{\text{km/s}}{\text{Mpc}}$$

to within an uncertainty of order ± 20 percent. (A parsec, pc, is a measure of distance on the cosmic scale that is used more frequently than the light-year. One parsec is the distance that corresponds to one angular second of parallax. Since the parallax method involves the Earth's motion about the sun, the parallax angle, defined as 2α, is the diameter $2R$ of the Earth's orbit divided by the distance d to the star or galaxy, so α (in radians) is just R/d. One radian is 57.3° or 206, 265″, and it follows that one parsec is $206,265R$ or 3.26 light-years. One *mega*-parsec, Mpc, is 10^6 parsec.)

How does the Hubble law show that the universe is expanding? Consider the unusual universe represented by the three-dimensional coordinate system shown in Figure 16.3a, where each point represents a galaxy. With the Earth at the origin, we can determine the distance d to each galaxy. If this universe were to expand, with all the points becoming further apart, as in Figure 16.3b, the distance to each galaxy would be increased to d'. Suppose the expansion were such that every dimension increased by a constant ratio k in a time t; that is, $x' = kx$, and so forth. Then $d' = kd$, and a given galaxy moves away from us by a distance $d' - d$ in a time t, so its apparent recessional speed is

$$v = \frac{d' - d}{t}$$

$$= d\frac{k - 1}{t} \tag{16.4}$$

If we compare two galaxies 1 and 2,

$$\frac{v_1}{v_2} = \frac{d_1}{d_2} \tag{16.5}$$

a relationship identical with Hubble's law, Equation (16.3). Thus, in an expanding universe, it is perfectly natural that, the further away from us a galaxy might be, the faster we observe it to be receding.

Notice also from Figure 16.3 that this is true no matter which point we happen to choose as our origin. From *any* point in the "universe" of Figure 16.3, the other points would be observed to satisfy Equation (16.5) and thus also Hubble's law. We can further demonstrate this with two analogies. If we paint some spots on a balloon (Figure 16.4) and then inflate it, *every* spot observes every other spot to be moving away from it, and the farther away a spot is, the faster its separation grows. For a three-dimensional analogy, consider a loaf of raisin bread shown in Figure 16.5 rising in an oven. As the bread rises, every raisin observes all the others to be moving away from it, and the speed of recession varies with the separation.

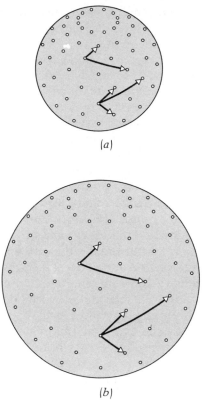

(a)

(b)

FIGURE 16.4 As a balloon is inflated, every observer on the surface observes a velocity-distance relationship of the form of the Hubble law.

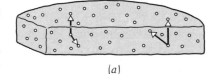

(a)

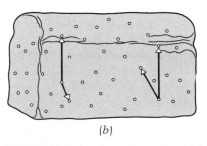

(b)

FIGURE 16.5 Another system in which the Hubble law is valid.

The information presented here for the expansion of the universe, most of which was gathered in the 1920s and 1930s, shows that the galaxies are moving away from one another as the universe expands. We must therefore regard this expansion as experimentally established, and indeed it has been accepted by the scientific community since its introduction. There are, however, two interpretations of this expansion. (1) If the galaxies are moving apart, long ago they must have been closer together. The universe was much denser in its past history, and if we look back far enough we find a single point of infinite density. This is the "Big Bang" hypothesis, first developed by George Gamow and his colleagues. (2) The universe has always had about the same density it does now. As the galaxies move apart, additional matter is continuously created in the empty space between the galaxies, to keep the density more or less constant. This is the "Steady State" hypothesis, of astronomer Fred Hoyle and others. New galaxies created from this new matter would make the universe look the same not only from all vantage points, but also *at all times* in the present and future. (To keep the density constant, the rate of creation need be only about one hydrogen atom per cubic meter every billion years.)

Big Bang versus Steady State cosmologies

Both hypotheses had their supporters, and during the 1940s and 1950s the experimental evidence did not seem to favor either one over the other. (Note that both hypotheses violate conservation of energy, the first at a single time by an infinite amount, the other at all times by a large number of very small amounts.) In the 1960s, the new field of radio astronomy revealed the presence of a universal background radiation in the microwave region, which is believed to be the remnant radiation from the Big Bang. This single observation has propelled the Big Bang theory to the forefront of cosmological models.

16.2 THE COSMIC MICROWAVE BACKGROUND RADIATION

In the early times immediately following the Big Bang, the universe was filled with elementary particles—quarks, antiquarks, leptons, antileptons, and photons. The matter and radiation were in equilibrium—for example, electrons and positrons were being created by photons as rapidly as existing electrons and positrons were annihilating to create photons. As the universe expanded, it cooled; that is, the energies of the photons decreased. First the energies of the photons dropped below the threshold for production of particles and antiparticles, then below the threshold for splitting primitive nuclei (deuterium and helium) into their constituent nucleons (an energy of several mega electron-volts), and finally below the hydrogen atom ionization energy (13.6 eV). At this point (about 10^6 y after the Big Bang) the radiation and the matter stopped interacting (decoupled), and the radiation temperature continued to drop unaffected by the presence of matter. During the ensuing 15×10^9 y, the expansion of the universe has "stretched out" or red shifted the wavelengths of those photons, but none have been created or destroyed since the decoupling time. When we observe these photons we are therefore looking back to a time just 10^6 y after the Big Bang.

Before the decoupling time, the universe was uniformly filled with photons and particles traveling more or less randomly, perhaps similar to a container of gas molecules. The slow expansion of the universe, or of the gas container, changes the temperature, but not the distribution of the particles. They are not "coming from over there" or "going over here," but still today, as they did originally, fill the universe uniformly. Even though the matter has clumped together into galaxies, the photons are still distributed uniformly but are today much "cooler" than they were at the decoupling time.

The wavelength spectrum of those photons must have been that of a blackbody at whatever temperature T characterized the universe at that time. Although the wavelengths changed as the universe expanded, the spectrum should still be that of a blackbody, but of course at a much lower temperature. In the 1940s, the Big Bang cosmologists (Gamow and others) predicted that this "fireball" would today be at a temperature of the order of 5 to 10 K; such photons would have a typical energy kT of the order of 10^{-3} eV or a wavelength of order 1 mm, in the microwave region of the spectrum.

Let us first review some of the properties of blackbody radiation we discussed in Chapter 3. The wavelength spectrum is given by the Planck distribution, Equation (3.26) or (12.24):

George Gamow (1904–1969, United States). His significant contributions to nuclear physics (theories of alpha decay, beta decay, and nuclear structure), astrophysics (nucleosynthesis, stellar structure), and cosmology (the Big Bang theory) would alone be enough to put him among the first rank of scientists, but he is also one of the most successful writers of popular science, to which he brings unusual and amusing perspectives.

$$u(\lambda)\,d\lambda = \frac{8\pi hc}{\lambda^5}\frac{1}{e^{hc/\lambda kT}-1}\,d\lambda \tag{16.6}$$

where $u(\lambda)\,d\lambda$ is the energy density (energy per unit volume) of the radiation emitted between the wavelengths λ and $\lambda + d\lambda$. (As we have discussed previously, this differs from the spectral radiancy $R(\lambda)$ by the factor $c/4$.) The wavelength distribution for a specific temperature has a peak at λ_{max}, determined by Wien's displacement law, Equation (3.21):

$$\lambda_{max}T = 2.898 \times 10^{-3}\ \text{m·K} \tag{16.7}$$

The total radiant energy density U at all wavelengths is found from Stefan's law, Equation (3.20):

$$U = \int u(\lambda)\,d\lambda$$

$$= \frac{\sigma}{c/4}\,T^4 \tag{16.8}$$

where σ is the Stefan-Boltzmann constant, with a value of 5.67×10^{-8} W/m²·K⁴.

As our universe expands, λ increases. If the present wavelength λ' is greater than the wavelength λ at decoupling by the factor f ($\lambda' = f\lambda$) then you can see from Equation (16.6) that the present energy spectrum will still be a blackbody spectrum, but characterized by a lower temperature $T' = T/f$. The energy density is decreased by the factor f^4.

Before discussing the experimental results, let us use Equations (16.6) to (16.8) to derive some other basic results. First we calculate the *energy* spectrum of the photons from Equation (16.6) by replacing λ with hc/E

$$u(E)\,dE = \frac{8\pi E^3}{(hc)^3}\frac{1}{e^{E/kT}-1}\,dE \tag{16.9}$$

Since this gives the *energy density*, we can divide by the energy E to find the *number* of photons of energy E per unit volume, $n(E)$

$$n(E)\, dE = \frac{u(E)\, dE}{E}$$

$$= \frac{8\pi E^2}{(hc)^3} \frac{1}{e^{E/kT} - 1}\, dE \qquad (16.10)$$

To find the *total* number of photons of all energies per unit volume, N, we integrate (16.10) over energy:

$$N = \int_0^\infty n(E)\, dE$$

Total number density of blackbody photons

$$= \frac{8\pi}{(hc)^3} \int_0^\infty \frac{E^2 dE}{e^{E/kT} - 1}$$

$$= \frac{8\pi}{(hc)^3} (kT)^3 \int_0^\infty \frac{x^2 dx}{e^x - 1} \qquad (16.11)$$

where we have substituted $x = E/kT$. The definite integral is a standard form and is approximately equal to 2.404. Equation (16.11) shows that the total number of photons per unit volume is proportional to the cube of the temperature, and evaluating the constants we find

$$N = 2.03 \times 10^7\ T^3\ \text{photons/m}^3 \qquad (16.12)$$

We can write Equation (16.8) in the same form by evaluating the constants:

$$U = 4.73 \times 10^3\ T^4\ \text{eV/m}^3 \qquad (16.13)$$

and the *average* energy per photon is just

$$E_{\text{av}} = \frac{U}{N} = 2.33 \times 10^{-4}\ T\ \text{eV} \qquad (16.14)$$

We turn now to the experimental evidence for the existence of this microwave radiation and the determination of its temperature. From Equation (16.6) we see that the measurement of the blackbody radiant energy density at *any* wavelength is enough for a determination of the temperature T, although to demonstrate that the radiation actually has a blackbody spectrum requires measurement over a range of wavelengths.

In their 1965 experiment, Penzias and Wilson used a radiotelescope tuned to a wavelength of 7.35 cm. At this wavelength they recorded an annoying "hiss" from their telescope that could not be eliminated, no matter how much care they took in refining the measurement. After painstaking efforts to eliminate the "noise," they concluded that it was coming from no identifiable source and was striking their telescope from all directions, day and night, summer and winter. (The story of this discovery and of other recent discoveries in this field can be found in the references listed at the end of this chapter.) From the radiant energy at that wavelength they deduced a temperature of 3.1 ± 1.0 K, and it was later concluded that the radiation was the present remnant of the Big Bang "fireball."

Since that original experiment there have been many additional studies, at various wavelengths in the range 0.1 to 100 cm, all giving

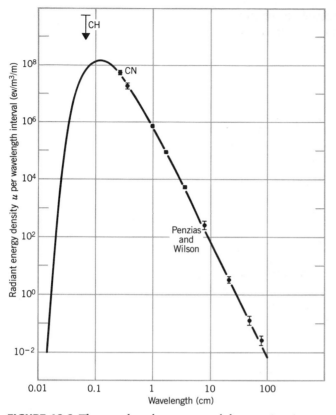

FIGURE 16.6 The wavelength spectrum of the cosmic microwave background radiation. The solid line is the Planck blackbody spectrum for $T = 2.7$ K.

about the same temperature. The present best value for the temperature of the cosmic background radiation is 2.7 ± 0.1 K. These experiments are summarized in Figure 16.6. (Note the logarithmic scale, which enables the low-intensity regions to be seen more clearly.) The agreement with the expected blackbody spectrum is striking, although it would be more satisfying if measurements could be made at wavelengths below 0.1 cm. Unfortunately, the Earth's atmosphere absorbs strongly at these wavelengths, and so ground-based radiotelescopes are not useful for such experiments. However, balloon flights to the upper atmosphere have confirmed that the spectrum does decrease in intensity below 0.1 cm, as expected for a 2.7 K blackbody.

There is an independent set of measurements that confirms the temperature deduced from radiotelescopes. One of the diatomic molecules identified from its absorption spectrum in interstellar space is cyanogen, CN. As we learned in Chapter 13, the energy levels of such a molecule are combinations of electronic, vibrational, and rotational excited states. The molecule in its ground state absorbs at a certain characteristic wavelength, $\lambda = 387.46$ nm, in the blue end of the visible spectrum (Figure 16.7). The first rotational state is at an energy of 4.70×10^{-4} eV above the ground state. From this state, the absorption line has a wavelength of 387.40 nm. If we measure the absorption spectrum, the ratio of the intensities of these two lines is a measure of the ratio of the number of molecules in the ground state and first rotational state.

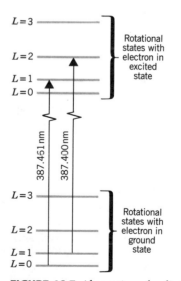

FIGURE 16.7 Absorption of radiation by cyanogen (CN). The molecule can absorb in the ground state or in the first excited state, and the relative intensity of the two absorption lines gives the population of the first rotational level.

If the CN were at $T = 0$, we expect all of the molecules to be in the ground state, but at a temperature T, the population of the excited states is determined by the Boltzmann factor $e^{-E/kT}$ and by the statistical weights of the levels, as in Equation (13.41):

$$\frac{N_1}{N_2} = \frac{2L_1 + 1}{2L_2 + 1} e^{-(E_1 - E_2)/kT} \qquad (16.15)$$

The relative number of molecules in the two levels is therefore a way of determining the temperature. The observed relative strengths of the two absorption lines suggests that about 25 percent of the molecules are in the excited state, so

$$\frac{0.25}{0.75} = \frac{2 \times 1 + 1}{2 \times 0 + 1} e^{-(4.70 \times 10^{-4}\text{eV})/kT}$$

from which it follows that $kT = 2.13 \times 10^{-4}$ eV, or $T = 2.5$ K. That is, in the extreme cold of interstellar space, something is "warming" these molecules to this temperature! As in the case of the Penzias and Wilson experiment, it is necessary to eliminate all other possibilities which might cause the excitation of these molecules, but after doing so the conclusion is inescapable—the molecules are experiencing the "warm" glow of the fires of the primordial universe.

Many experiments of this sort have been done, with the result $T = 2.7 \pm 0.2$ K, in exact agreement with the result of radiotelescope experiments. Other experiments suggest that the radiation is *isotropic* to about 1 part in 10^3; that is, it comes equally from all directions, as we would expect for radiation from the Big Bang that filled the universe then as it does now.

Using the deduced temperature of 2.7 K, we can calculate from Equations (16.12) to (16.14) that there are about 4.0×10^8 of these photons in every cubic meter of space (there are about 10^8 of them in your body right now!), that they contribute to the universe an energy density of about 2.5×10^5 eV/m³ (about half the mass energy of an electron) and that each photon has an average energy of about 0.00063 eV. The number of photons is particularly important, since for nearly all of the last 15×10^9 y the *ratio* of the number of nucleons (protons and neutrons) to photons has been constant. This has important consequences for the Big Bang cosmology.

16.3 THE BIG BANG COSMOLOGY

As we run the cosmic clock backward, we try to examine at what point the universe would be *significantly different* from what it is now. One hundred thousand years ago humans appeared on the scene, but surely we can't imagine that insignificant event to have had a major influence on the universe! Primitive life first developed on Earth 10^9 y ago, and our solar system was born 5×10^9 y ago. The oldest stars formed perhaps 10×10^9 y ago; many have since died fiery deaths as supernovas or evolved into dwarf stars or black holes. Galaxy formation occurred $13-14 \times 10^9$ y ago; before that time the universe consisted entirely of hydrogen gas, helium gas, and radiation, but *gravity* was still the major influence on the structure of the universe. Nearly 15×10^9 y ago, a mere 700,000 y following the Big Bang, the decoupling of radiation and matter occurred. Before primitive protons and electrons joined to form neutral hydrogen, the electromagnetic fields of the radiation pushed the great masses of positive protons and negative electrons about the universe, and in their turn the moving charged particles contributed to the electromagnetic waves. Hydrogen atoms may have formed, with photons emitted in the process, but they were soon burst apart by the surrounding radiation. *Electromagnetic forces at that time played a major role in shaping the evolution of the universe,* until the expansion had lowered the temperature to a point at which the photons had not sufficient energy to dismantle the primordial hydrogen. The protons and electrons then quickly joined, forming neutral atoms, which were not affected by the electromagnetic fields of the radiation.

As we go back still further, to times when the universe was hotter and denser, the strong and weak nuclear interactions in their turns became the dominant forces in the primitive universe. At still earlier times, we reach two barriers beyond which we cannot look. The first barrier occurs at some time *before* the radiation field cooled to the point at which nucleon-antinucleon creation was common. As we go back beyond this time, at some point the radiation will be energetic enough to create quark-antiquark pairs. No one has yet seen a free quark, so we don't yet know what their masses are. Therefore we don't know at what time or at what temperature this point is reached. In fact, there is a school of thought that holds that free quarks cannot exist at all! This barrier is one which results from our limited knowledge of fundamental processes involving quarks; someday we may learn enough about these processes to be able to see beyond this barrier.

As we go back still further toward $t = 0$, we reach another, more fundamental barrier. This is the *Planck time,* about 10^{-43} s, beyond which the universe was so dense and so hot that we cannot even speak of distinct particles. Quantum theory and gravity are so hopelessly intertwined that none of our present physical theories gives us any clue about the structure of such a universe. The quantum wavelength of the total mass of the universe at that point was the same size as the radius of the universe, and the random, statistical processes of quantum physics controlled the properties of the universe at that time.

In this same context, it makes no sense to ask what came "before" the Big Bang, in the same way that we cannot speak of what is "beyond" the edge of the universe. By definition, we measure time *and* space only from the Big Bang; we *define* that instant to be $t = 0$, and times "before" that have no more meaning than do temperatures "below" the absolute zero of the temperature scale.

Let us now try to recreate the history of the early universe, based on what we know about fundamental particles, nuclear physics, and thermal physics. The very early universe was radiation dominated. Even though particles and matter were in equilibrium, with one creating the other, the matter was so energetic as to behave ultrarelativistically. As we learned in Chapter 2, such ultrarelativistic particles have $E \cong pc$, just as is true for radiation. Since the matter behaves much like radiation, we analyze the early universe as if it were pure radiation. For such an energy-filled universe, the density is just the energy density of radiant energy, Equation (16.13).

From our calculations of the radius of the universe according to general relativity (Section 15.3), we learned that during the radiation-dominated era, the radius of the universe depended on the time as $t^{1/2}$. We also found that as the universe evolved, there was a relationship between its density ρ_r and the time:

$$t = \sqrt{\frac{3}{32\pi G \rho_r}} \qquad (16.16)$$

In the radiation-dominated era, we should use the radiation density U instead of the density ρ_r. (To make the units come out right, we substitute $\rho_r = U/c^2$, which is analogous to $m = E/c^2$.) Using Equation (16.13) for $U(T)$ and putting in all of the numerical constants, we find

$$T = \frac{1.5 \times 10^{10}}{t^{1/2}} \qquad (16.17)$$

where T is in Kelvins and t is in seconds. This expression allows us to relate the age of the early universe to its temperature.

(There is one feature of the universe that has been neglected in this formula—any physical system can "cool," that is lose energy, by undergoing a *phase transition*. Water at 0°C, for example, can "cool" to ice at 0°C without its temperature changing. The same is true for the matter of the early universe. Two "hot" protons and two "hot" neutrons will "cool" when they form a helium nucleus; the energy from this transition goes into the radiation field, so the radiation field cools more slowly while matter is condensing.)

The early radiation field was coupled to the matter by two processes:

$$2 \text{ photons} \rightarrow \text{particle} + \text{antiparticle}$$

$$\text{particle} + \text{antiparticle} \rightarrow 2 \text{ photons}$$

In each case the total mass energy of the two particles, $2mc^2$, is equal to the total photon energy, $2E$. Thus to produce a certain species of particle requires a certain energy E and a corresponding temperature $T = E/k$. When the temperature falls below that value, the first process is no longer possible, although the second one may still occur. No new particles and antiparticles of that species are created; after all possible annihilation reactions have occurred, the leftover particles (or perhaps antiparticles) become, if they are stable against decay, the constituents of the material universe.

Table 16.1 shows some of the more common particle species that

Table 16.1 Particles of the Early Universe

Particle	Rest Energy (MeV)	Threshold Temperature (10^9 K)
Photons	0	0
$\nu, \bar{\nu}$	$\cong 0$	$\cong 0$
e^-, e^+	0.511	5.9
μ^-, μ^+	106	1230
π^-, π^+	140	1620
$p, \bar{p}$	938	10,880
$n, \bar{n}$	940	10,910

may have been present in the early universe, along with the threshold temperatures for their production.

Let us begin our story at a time of about 10^{-6} s, when the universe was hot enough ($T = 1.5 \times 10^{13}$ K) to permit formation of all of the particles listed in Table 16.1 (and of course some others which are not listed). If the universe is closed and finite its radius was then smaller than its present radius by the red shift (1.5×10^{13} K/2.7 K $= 5.5 \times 10^{12}$). The present radius is probably about 10^{26} m, so its radius then was about 2×10^{13} m, about the size of the solar system. The numbers of protons and neutrons, an important quantity for the later development of the universe, were about equal, since the radiation could produce $n + \bar{n}$ about as easily as it could produce $p + \bar{p}$. There was perhaps a slight imbalance of matter over antimatter; that imbalance has persisted to this time in our universe, which appears to be dominated by matter (see Section 16.4).

The ratio of neutrons to protons is determined by three factors.

1. *The Boltzmann factor* $e^{-\Delta E/kT}$; since protons have less mass energy than neutrons, there will be more of them at any given temperature. The energy difference ΔE is just $(m_n - m_p)c^2 = 1.3$ MeV, so the neutron-to-proton ratio can be expressed as $e^{-1.5 \times 10^{10}/T}$ for T in Kelvins. For $T \sim 10^{13}$ K, this ratio is very nearly 1, but it becomes different from 1 as T approaches 10^{10} K.

2. *Nuclear reactions*, such as $n + \nu \rightleftarrows p + e^-$ and $n + e^+ \rightleftarrows p + \bar{\nu}$. These reactions can go in either direction and tend to make it easy for protons to turn into neutrons and visa versa, as long as there are plenty of e^-, e^+, ν, and $\bar{\nu}$ around.

3. *Neutron decay*. The neutron half-life is about 15 min, which is going to be important only at later times. For $t = 1$ s, there has not yet been enough time for an appreciable number of neutrons to decay.

At $t = 0.01$ s, for which $T = 1.5 \times 10^{11}$ K, nucleons and pi mesons are no longer being manufactured. The antinucleons have largely disappeared; $p + \bar{p} \rightarrow 2\gamma$ and $n + \bar{n} \rightarrow 2\gamma$ are still possible, but the radiation no longer has enough energy for the reverse reactions. The energy released in the annihilation of *all* of the antimatter and *most* (99.999999 percent) of the matter has slowed the expansion somewhat, and the temperature might therefore be somewhat larger than 1.5×10^{11} K.

At $t = 1$ s, the temperature is about 1.5×10^{10} K and the neutron-to-proton ratio is about e^{-1}—there are about 73 percent protons and 27 percent neutrons. In the interval from 0.01 to 1 s, the neutrinos and antineutrinos begin to go out of thermal equilibrium with the matter. The increasing size of the universe causes the spacing between the particles to increase, and the simultaneous cooling decreases the neutrino energy; these two changes have the effect of greatly reducing the probability of the neutrinos to interact with the matter. This is the era of "neutrino decoupling," following which the neutrinos and anti-neutrinos will continue to expand and cool but will not interact with matter. (The electromagnetic radiation will similarly decouple from the matter at about $t = 700,000$ y.) After $t = 1$ s, the weak interaction does not play a significant role in determining the structure of the universe.

Over the next 5 s, the temperature drops to 6×10^9 K, and electrons

Neutrino decoupling

and positrons stop forming. All of the positrons and 99.999999 percent of the electrons begin to annihilate. Except for the very slow decay of the neutron, weak interactions are now totally unimportant. With no neutrinos available (they are decoupled) and electrons much less available, neutrons and protons can no longer be converted from one to the other; the relative number of neutrons and protons can no longer vary. This number is determined approximately by the Boltzmann factor at this temperature ($e^{-2.5} = 0.08$), so there are about 92 percent protons and 8 percent neutrons. This estimate is not quite correct, because ever since $t = 1$ s, the weak interaction rate has been falling and it has been becoming increasingly difficult for neutrons to turn into protons to satisfy the Boltzmann distribution. A better estimate is about 83 percent protons, 17 percent neutrons.

Let us review progress up to this point. (1) A hot, dense universe, full of elementary particles of all varieties, has cooled to below 10^{10} K. (2) Most of the unstable elementary particles decayed away. (3) All of the original antimatter and nearly all of the original matter annihilated one another, leaving a small number of protons, an equal number of electrons (to make the net charge zero) and about one-fifth as many neutrons. (4) A neutrino density about the same as the photon density decoupled at about 1 s and will continue cooling and expanding. This cosmic neutrino background is slightly cooler today than the microwave photon background, because the photons were "heated up" after $t = 1$ s by annihilation processes.

The universe has (at $t = 6$ s) reached the point at which nuclear processes (strong interactions) become important. With nothing but neutrons and protons present, the only nuclear reaction that will occur is deuterium formation

$$n + p \longrightarrow d + \gamma$$

We recall from Chapter 9 that the deuterium binding energy is 2.22 MeV. In order to have any appreciable buildup of deuterium, the photons present must first cool until their energies are below 2.22 MeV; otherwise the deuterium will be broken up as quickly as it can be formed. The energy 2.22 MeV corresponds to a temperature $T = 2.5 \times 10^{10}$ K, and we therefore might expect deuterium to be formed as soon as the temperature drops below 2.5×10^{10} K. However, this does not happen. The radiation does not have a single energy, but rather has a blackbody spectrum. A small fraction of the photons will have energies *above* 2.22 MeV, and these photons will continue to break apart the deuterium nuclei (Figure 16.8).

Before matter-antimatter annihilation occurred, there were about as many photons as nucleons and antinucleons, but after $t = 0.01$ s, the ratio of nucleons to photons is about 10^{-8}; about $\frac{1}{6}$ of the nucleons are neutrons. If the fraction of photons above 2.22 MeV is greater than $\frac{1}{6} \times 10^{-8}$, there will be enough energetic photons to prevent deuterium formation. Our next job is to calculate to what temperature the photons must cool before fewer than $\frac{1}{6} \times 10^{-8}$ of them are above 2.22 MeV.

The Planck spectrum in terms of energy was given by Equation (16.9). We expect that the temperature must be much less than 2.5×10^{10} K, and so we are interested in the Planck distribution where $E \gg$

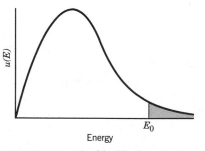

FIGURE 16.8 The blackbody radiation spectrum. The photons above $E_0 = 2.22$ MeV are energetic enough to break apart deuterium nuclei.

kT, for which it is approximately

$$u(E)\, dE = \frac{8\pi E^3}{(hc)^3}\, e^{-E/kT}\, dE \qquad (16.18)$$

That is, the distribution falls exponentially. Since each photon has energy E, the number density of photons at each energy is $u(E)dE/E$, or

$$n(E)\, dE = \frac{8\pi E^2}{(hc)^3}\, e^{-E/kT}\, dE \qquad (16.19)$$

and the total number density above some energy E_0 is

$$N_{E>E_0} = \int_{E_0}^{\infty} n(E)\, dE \qquad (16.20)$$

which can be shown to be

$$N_{E>E_0} = \frac{8\pi}{(hc)^3}\, (kT)^3\, e^{-E_0/kT} \left[\left(\frac{E_0}{kT}\right)^2 + 2\left(\frac{E_0}{kT}\right) + 2 \right] \qquad (16.21)$$

From Equation (16.11) we find the *total* number density of photons, and thus the fraction above E_0 is $N_{E>E_0}/N$

$$f = 0.42 e^{-E_0/kT} \left[\left(\frac{E_0}{kT}\right)^2 + 2\left(\frac{E_0}{kT}\right) + 2 \right] \qquad (16.22)$$

which has the solution, when $f = \frac{1}{6} \times 10^{-8}$,

$$\frac{E_0}{kT} \cong 25 \qquad (16.23)$$

With $E_0 = 2.22$ MeV, the required temperature is about 10^9 K; when $T > 10^9$ K, the number of photons with $E > 2.22$ MeV is greater than the number of neutrons, and deuterium formation is prevented. When T drops below 10^9 K (which occurs at about $t = 225$ s), deuterium begins to form. From 6 s to 225 s, very little (except expansion and the corresponding temperature decrease) happens in the universe, but after $t = 225$ s things happen very quickly. Deuterium begins to form; with many neutrons and protons available, the deuterium then reacts to give

Deuterium formation

$$d + p \longrightarrow {}^3\text{He}$$

and

$$d + n \longrightarrow {}^3\text{H}$$

The energies of formation of these nuclei are 5.49 MeV and 6.26 MeV, well above the 2.22 MeV threshold of the deuterium formation. If the photons are not energetic enough to break apart the deuterium, they are certainly not energetic enough to break apart ${}^3\text{He}$ and ${}^3\text{H}$. The final steps in the formation of the heavier nuclei are

$$^3\text{He} + n \longrightarrow {}^4\text{He}$$

and

$$^3\text{H} + p \longrightarrow {}^4\text{He}$$

There are no stable nuclei with $A = 5$, so no further reactions of this sort are possible. Nor is it possible to have ${}^4\text{He} + {}^4\text{He}$ reactions since ${}^8\text{Be}$

is highly unstable. (It would be possible to form stable ^{6}Li and ^{7}Li, but these are made in very small quantities relative to H and He; from Li further reactions are possible, such as ^{7}Li + ^{4}He → ^{11}B, and so forth, but these occur in still smaller quantities. The end products d and He, along with the leftover original p, make up about 99.9999 percent of the nuclei after the era of nuclear reactions.)

By $t = 225$ s, the original 17 percent neutrons present at $t = 6$ s had beta-decayed to about 12 percent, leaving 88 percent protons. Since most of the deuterium was "cooked" into heavier nuclei, its relative abundance is quite low, and we can assume the universe to be composed only of hydrogen and helium nuclei. Of the N nucleons originally present at $t = 225$ s, $0.12N$ were neutrons and $0.88N$ were protons. The $0.12N$ neutrons combined with $0.12N$ protons, forming $0.06N$ ^{4}He, and leaving $0.76N$ protons. The universe then consisted of $0.82N$ nuclei, of which 7.3 percent were ^{4}He and 92.7 percent were protons. Since helium is about four times as massive as hydrogen, by *mass* the universe is about 24 percent helium.

Helium abundance

At this point the universe begins a long and uneventful period of cooling, during which the *strong* interactions cease to be of importance.

The final step in the evolution of the primitive universe is the formation of neutral hydrogen and helium atoms from the p, d, ^{3}He, and ^{4}He nuclei and the free electrons. In the case of hydrogen, this takes place when the photon energy drops below 13.6 eV; otherwise any atoms that might happen to form will be immediately ionized by the radiation. There are still about 10^8 photons for every proton, and so we must wait for the radiation to cool until the number of photons above 13.6 eV is less than about 10^{-8}. We can use the previous result (16.22) to find the value of E_0/kT which corresponds to $f = 10^{-8}$, and the result is

$$\frac{E_0}{kT} = 24$$

With $E_0 = 13.6$ eV, the corresponding temperature is $T = 6600$ K, which occurs at time $t = 5.2 \times 10^{12}$ s $= 160,000$ y. These final estimates are actually not quite correct. We have been considering only the energy density of radiation present in the universe. As the universe cools, the contribution of the matter to the total energy density becomes more significant, and so the temperature drops more slowly than we would estimate. This contribution may increase this time by about a factor of 4 to about 700,000 y, and the radiation temperature is decreased by about a factor of 2, to $T \cong 3000$ K.

Now that neutral atoms have formed, there are virtually no charged particles left in the universe, and the radiation field is not energetic enough to ionize the atoms. This is the time of the decoupling of the radiation field from the matter, and now electromagnetism, the third of the four basic forces, is no longer important in shaping the evolution of the universe. The large-scale development of the universe is from this point governed only by gravity.

Decoupling of electromagnetic radiation

The intervening 14.9993×10^9 y have been comparatively uneventful, at least from the point of view of cosmology. Fluctuations in the density of the hydrogen and helium trigger the condensation of galaxies, and then first-generation stars are born. Supernova explosions of the

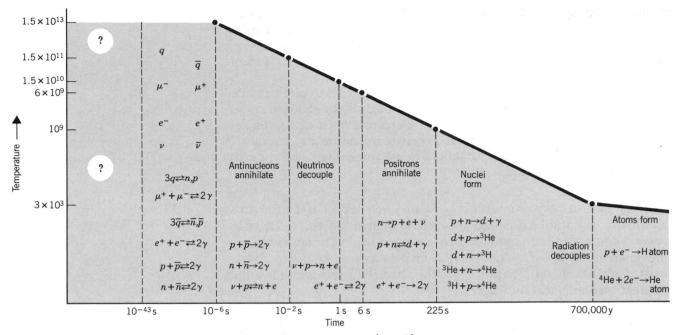

FIGURE 16.9 Evolution of the universe according to the Big Bang cosmology. The heavy solid line shows the temperature and time in the radiation dominated era before decoupling. The most important reactions in each era are shown.

material from these stars permits the formation of second generation systems, among which planets will form from the rocky debris.

Meanwhile, the decoupled radiation field, unaffected by the gravitational coming and going of matter, begins the long journey that will eventually take it, cooled again by a factor of 1000, to the radio telescopes of twentieth-century Earth.

The details of the Big Bang cosmology are summarized in Figure 16.9. It is a remarkable story, all the more so because we can understand most of its details, with the possible exception of the first instant, with nothing more than some basic theories of modern physics, most of which we can study (on a much smaller scale!) in our laboratories on Earth. The details of the Big Bang are consistent with all of these experiments and with nearly all astronomical results as well, and it is fair to say that the Big Bang is today the cosmology accepted by most of the scientific community, as the weight of experimental evidence continues to support it. Still, there are those who do not accept all of the premises of the Big Bang, and the puzzling evidence of the quasars, discussed in Section 16.5, shows that we do not yet know all there is to know about our universe.

We have already discussed the 2.7 K background radiation from the Big Bang as the best single piece of experimental evidence for this particular cosmology. Yet the universe carries other memories of its birth, which can perhaps serve as other tests of the theory.

16.4 ECHOES OF THE BIG BANG

Neutrino Background   The neutrinos decoupled from matter much before the radiation, and the radiation was heated somewhat by particle-antiparticle annihilation. The neutrinos should be red shifted just like the photons, but should still be at a somewhat lower temperature than the photons, perhaps 2 K.

Neutrinos are extremely elusive particles, difficult to catch and detect, but the density of these early neutrinos ought to be about the same as the photon density, perhaps $10^8/m^3$. Detecting such neutrinos and measuring their energy spectrum and their temperature are beyond the realm of current technology, but would be a stringent test of the theory if such experiments could be done.

Gravity Waves   When matter is born, or when matter and antimatter annihilate, gravitational radiation is produced, just like the electromagnetic radiation which is produced when positive and negative charges are made or disappear. This gravitational radiation, considerably red shifted, should also still be present in the universe, but this also is a field of great technological difficulty.

Helium Abundance   Much of the matter in the universe has been formed and reformed, and so has lost its "memory" of the Big Bang. There may, however, be "first generation" matter in stars and galaxies, which should show the roughly 24 percent helium abundance that characterized the formation of matter.

A variety of experiments suggests that the abundance of helium in the universe is 25 to 30 percent by mass, quite close to our rough estimate of 24 percent. These experiments include the emission of visible light from gas clouds near stars and the emission of radio waves by interstellar gas, both of which permit us to compare the amounts of hydrogen and helium present. In addition, the dynamics of stellar formation depends on the initial hydrogen and helium concentrations; present theories permit us to estimate their ratio from the observed properties of stars. The 25 to 30 percent abundance seems to be rather constant throughout the universe, as we would expect if it were predetermined by the Big Bang. (Not enough helium has been produced by nuclear fission in stars in the last 15×10^9 y to change this ratio significantly.)

In fact (and here physics comes nearly full circle, from the very old and large to the very new and small), the early helium abundance is very much a function of the conditions before 10^{-6} s, when perhaps quarks and antiquarks were being produced in great numbers. It has been calculated that the ~ 25 percent observed helium abundance is not consistent with the existence of more than eight kinds (up, down, strange, charmed, . . .) of quarks. Particle physicists and astrophysicists have much in common!

Antimatter   The present model of the Big Bang assumes that there was a slight imbalance of antimatter and matter in the early universe, perhaps 1 part in 10^8 or 10^9. Our numerical estimates are based on *all* of the antimatter and 99.999999 percent of the matter disappearing in mutual annihilation, with the leftover 0.000001 percent of matter constituting the present universe. There is no evidence to indicate that there are still large quantities of antimatter present in our universe, but neither is the evidence *against* antimatter particularly strong. Whether there are antistars and antigalaxies we cannot say; we can only study distant objects by observing their light and their gravity, and since antimatter emits exactly the same light and experiences exactly the same gravity as matter, we can't tell the difference from our observations.

Where did the 10^{-8} or 10^{-9} excess of matter in the early universe come from? We really can't answer this question, but evidence gathered in particle physics experiments in 1964 may provide a clue. The decay of the neutral K meson shows interference effects between matter and antimatter, but only at the very low level of 1 part in 10^3. (J. W. Cronin and V. L. Fitch received the 1980 Nobel prize in physics for their work on this experiment.) The weak decay of the K^0 is so far the only case in which this asymmetry between matter and antimatter has been observed; all other experiments would yield identical results when performed with antimatter as they do when performed with matter. This experiment has suggested the possibility of a fifth force in nature, with a strength at least 10^{-3} below that of the weak interaction, sometimes known as the "superweak" force. The era of the superweak force, which would have been before 10^{-6} s, might have been responsible for the slight preference for matter over antimatter.

Mini Black Holes   The enormous energy densities present at the earliest stages of the Big Bang could have compressed matter to a high enough density to have allowed tiny black holes to form, possibly with masses very much less than one gram. This is pure speculation, of course, since no black hole of *any* sort has been conclusively identified, but if such mini black holes exist, they might provide the "missing mass" necessary to "close" the universe (see Section 16.6).

In 1908 in the Tunguska region of Siberia there occurred a violent explosion of sufficient force to level the forest over roughly 1000 km². It was thought to be caused by a meteor impact, yet no crater or meteor fragments were ever found. It has been suggested that this explosion was caused by the impact of one of these primordial mini black holes, which would have generated enormous heat and an explosive shock wave as it passed through the atmosphere. It could then have traveled through the Earth, gravitationally attracting and consuming tiny amounts of matter as it went, and finally emerged out the other side, perhaps in the mid-Atlantic, to continue its journey through space.

This is an interesting possibility, but in the absence of any confirming evidence it must be regarded as no more probable than any other explanation of this unusual explosion (such as the impact of a comet, an antimatter meteor, or even an alien spacecraft!). However, if such mini black holes are ever found, they could provide clues to the origin of the universe.

16.5 THE PROBLEM OF THE QUASARS

There are many sources of radio waves in the sky, but the most powerful seem to be those associated with astronomically "large" objects, such as galaxies or gaseous nebulae in our own galaxy. As the ionized atoms and their electrons swirl about in these clouds, radio waves are emitted that can be detected on Earth, and indeed much of the early work in radio astronomy was devoted to the study and mapping of these nebulae. In the 1960s, it became known that some of the radio sources were smaller and more compact than the nebulae, and when optical astronomers examined them with their telescopes, they discovered small, starlike objects at the location of the radio emissions. Since it was not clear just what exactly was the nature of these objects, they became known as "quasi-stellar radio sources" or *quasars* (sometimes also called QSO's, for quasi-stellar objects).

The optical spectra from the quasars were particularly puzzling, since they did not seem to fit any of the known lines of the optical spectrum. It was soon realized that these spectra could be explained if they were caused by very large red shifts, so large that the ordinary lines seen in optical spectra were shifted out of the visible region into the infrared, while their place in the visible spectrum was taken by new lines which were formerly in the ultraviolet. In fact, red shifts as large as a factor of 3 have been discovered, so that the first transition in the Lyman series of hydrogen ($2p \rightarrow 1s$, or 10.2 eV), which normally appears in the *ultraviolet* at about 120 nm, would appear red shifted to 360 nm, in the violet end of the *visible* spectrum.

As we have seen, there are three possible origins of red shifts. Each of these has been invoked by various theories to "explain" the red shifts of the quasars, but each has serious problems in accounting for the observational evidence.

Cosmological Red Shift

The red shifts of the quasars, according to this explanation, originate from the expansion of the universe, just like the red shifts of the galaxies. We cannot measure the distance to the quasars, but we can use the Hubble law to infer their distance, *if* the red shifts are indeed cosmological in origin. A red shift of a factor of 3 corresponds to a speed $v = 0.8c$; the distance from us is therefore, from the Hubble law with $H = 75$ km/s·Mpc, about 3200 Mpc or 11×10^9 light-years. This places the quasars near the very edge of the observable universe, making them the *most distant,* and therefore the *oldest* objects we can examine. In order for the quasars to be observable at such distances, their energy output must be enormous, perhaps millions of times larger than that of an ordinary galaxy. Yet another puzzling fact emerged from quasar observations in the 1960s: their energy output varies over a period of days. Since it is not possible for one side of the quasar to "know" that the other side is fluctuating and to fluctuate in unison with it, unless a light signal can travel from one side to the other to carry that information, this variation suggests that the diameter of the quasars is of the order of a few "light-days," that is, only a few times the diameter of the solar system. How can an energy output millions of times larger than that of our *galaxy* come from a region the size of our *solar system*?

Doppler Red Shift

Perhaps the quasar red shifts are not cosmological at all, but are just local Doppler shifts, caused by true motion of these objects relative to us. If so, we cannot deduce their distance at all—they could be quite close, or very far away. Evidence has been gathered by astronomers that quasars seem to occur more often "near" galaxies. (That is, they appear close to galaxies on the photographic plates we make with our telescopes, but since the photograph does not measure depth, the quasars may in fact be far "in front of" or "behind" the galaxy which happens to appear nearby on the photograph.) It has been hypothesized that the quasars are hurled from the core of galaxies at great speeds in violent explosions. Quasars shot out from the core of the Milky Way and other nearby galaxies a long time ago would all appear to be moving away from us (those that were originally moving "toward" us have passed by already) and so would all be red shifted, in agreement with observations. Quasars hurled from distant galaxies, however, should still be moving toward us. Why then don't we observe any blue-shifted quasars?

Gravitational Red Shifts   Perhaps the quasars are not moving away from us at all, but are red shifted because they are small, massive objects with intense gravitational fields at their surface. These red shifts are those associated with general relativity and have nothing at all to do with the relative motion. (This would of course explain why no blue shifts are observed—there are none in general relativity.) Why don't these small and massive objects collapse, as they should, into black holes and thus disappear from observation?

At present no one knows the answer to the questions raised by the quasars, but many exotic explanations have been proposed to explain the observed energy production or red shifts—the quasars may be "white holes," in which the matter falling into a black hole at some other place and time in the universe gushes forth at the observed place in our time, or they may be "nurseries" for newly born matter, in which the electrons and nuclei have masses different from ordinary electrons and nuclei, and therefore emit light at different wavelengths. Whatever the correct explanation, it is possible that, in the quasars, we have an example of the operation of some new principle of physics.

16.6 THE FUTURE OF THE UNIVERSE

What of the future? Does the universe go on expanding forever, or will it eventually stop its expansion and begin contraction? Will there be a Big Bang in reverse, a sort of Big Crunch, as all the matter in the universe rushes to a point, while the 2.7 K radiation heats up again? Following the Big Crunch will there be another Big Bang, which will begin the evolution of a new universe? If so, this process may have been going on continuously, and the Big Bang is more like a Big Bounce.

According to Newtonian gravitation, the expansion of the universe will slow under the influence of the gravitational attraction of its components. Whether there is enough deceleration to reverse the expansion is determined by how much mass is present. In a similar way, the curvature of spacetime in general relativity is determined by the density ρ_m of the universe; according to Equation (15.16) the deceleration parameter q, which is proportional to ρ_m, determines the rate of change of dR/dt and thus tells us how much the expansion is slowing.

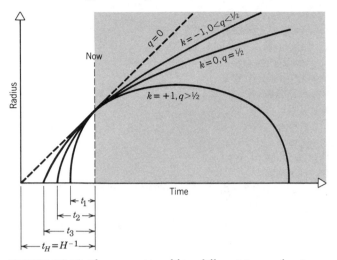

FIGURE 16.10 The expansion of four different types of universe. The dashed line represents a universe that has been expanding at a constant rate since $t = 0$, which occured a time H^{-1} before the present. The solid lines represent three different curved universes, all of which are younger than H^{-1}.

The dependence of $R(t)$ on t for possible types of universe is illustrated in Figure 16.10. A relationship between k and q can be found by combining Equations (15.8) and (15.16):

$$\left(\frac{dR}{dt}\right)^2 (1 - 2q) = -kc^2 \qquad (16.24)$$

A flat universe with $k = 0$ must have $q = \frac{1}{2}$, and will continue to expand forever. The curved, open universe with $k = -1$ must have $0 < q < \frac{1}{2}$ and will likewise continue its expansion. The closed universe with $k = +1$ has $q > \frac{1}{2}$; this universe would reach a maximum radius and then begin to contract.

The critical value of the deceleration parameter is therefore $\frac{1}{2}$, and we can use Equation (15.16) to compute the density that corresponds to $q = \frac{1}{2}$

$$\rho_m = \frac{3qH^2}{4\pi G} \qquad (16.25)$$

Taking $H = 75$ km/s·Mpc, $H^{-1} \cong 4.1 \times 10^{17}$ s and thus the critical density is

$$\rho_{cr} = \frac{3(\frac{1}{2})(4.1 \times 10^{17} \text{ s})^{-2}}{4\pi(6.67 \times 10^{-11}\text{N·m}^2/\text{kg}^2)} \qquad \textbf{Critical density}$$

$$\cong 1 \times 10^{-26} \text{ kg/m}^3$$

$$\cong 1 \times 10^{-29} \text{ g/cm}^3$$

If the density of the universe is less than ρ_{cr}, the expansion cannot be reversed.

From the visible galaxies we estimate the density of matter in the universe to be

$$\rho_{gal} = 3 \times 10^{-31} \text{ g/cm}^3$$

far smaller than the critical value. There may be additional invisible, nonluminous matter present in the universe whose presence we are not aware of; this matter could be in the form of gas, dust, black holes, and so forth. The 2.7 K microwave background radiation contributes an energy density of 0.25 eV/cm³, equivalent to a density of about $\rho_{rad} = 4 \times 10^{-34}$ g/cm³. Neutrinos are another possibility; the sea of neutrinos left over from the Big Bang has a temperature of about 2 K and about the same number density as the microwave photons. Experimental upper limits on the mass of the neutrinos are not particularly good; although we generally assume $m_\nu c^2 = 0$, a value of 10 eV is well within the best experimental limits. In this case ρ_ν would exceed ρ_{cr}, and the neutrinos would provide the gravitational attraction necessary to reverse the expansion of the universe.

It is also possible to measure the deceleration parameter directly by looking at the red shifts of the most distant objects. If the expansion rate is slowing, distant objects should have relatively *larger* red shifts than we would deduce from applying the Hubble law to nearby objects. The amount by which distant objects deviate from the extrapolation of the Hubble relationship is a measure of how much their motion has been slowed and therefore how rapidly the universe is decelerating. Figure 16.11 shows such an analysis, from which it is difficult to draw any definite conclusions.

Based on the best evidence we have available, we still can't be sure whether the universe is open or closed, whether it will expand forever or halt its expansion and begin to contract. The Hubble program of measuring red shifts at large distances is inconclusive, in part because of uncertainty in the distance scale. The search for the mass sufficient to close the universe is similarly inconclusive; the necessary mass has not been found, but there are several ways it could be hidden from our view.

The scientific method, which was discussed in Chapter 1, suggests that progress in science results from the asking of questions and the

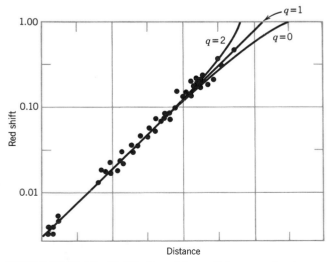

FIGURE 16.11 The Hubble law at large distances and red shifts. The red shift is $\Delta\lambda/\lambda$, which is linear in v only for small v. The distance scale depends on the deduced apparent magnitude of the galaxy.

search for answers through experiments. Some of those questions are *heuristic*—their answers lead to new questions, then to new experiments, and so on. Most of the work discussed in this book is of a heuristic nature. The discovery of a new elementary particle, for example, leads to new questions about its properties and its relationship to other particles.

Other questions posed by science are of a different nature. Their answers are given in terms of only two mutually exclusive possibilities, and the purpose of experiment is to choose one or the other. Often these questions are of such a fundamental nature that contemplating either possibility fills us with awe, fright, wonder, and reverence. The existence of other civilizations in the universe is such a question. Either we are alone in this vast and empty cosmos, surely a frightening and awesome notion, or else the universe is teeming with forms of life beyond our imagination, a similarly awesome notion. The ultimate future of the universe is another such question with two possible answers, each of which inspires fear and wonder. (1) The universe expands forever; all stars and galaxies eventually use up their energy and become black dwarfs or black holes. The universe becomes cold and dark, and all life ends. (2) The expansion slows and eventually stops, and is followed by gravitational contraction, with the entire universe collapsing to a single point and (presumably) a new Big Bang. The cosmic dice are again thrown and a new universe is born, perhaps even with new and different physical laws; perhaps intelligent life arises, and perhaps one or two generations of that intelligence will be privileged (as we are) to live at that special time when the fundamental laws of the universe begin to be explored and understood.

SUGGESTIONS FOR FURTHER READING

Many of the reference books on general relativity listed in Chapter 15 also include material on cosmology. Some nonmathematical introductory books on cosmology are:

T. Ferris, *The Red Limit* (New York, Wm. Morrow & Co., 1977). Certainly among the best-written works on *any* scientific topic; rich in historical and personal detail.

W. J. Kaufmann III, *Relativity and Cosmology* (New York, Harper and Row, 1977). Another well-written general work; includes general relativity and black holes.

S. Weinberg, *The First Three Minutes* (New York, Basic Books, 1977). A modern classic; if you read only one book on cosmology, choose this one.

All of the above are available in paperback. There are several works on cosmology which include only calculus-level math (no tensors):

P. T. Landsberg and D. A. Evans, *Mathematical Cosmology* (Oxford, Clarendon Press, 1977). A superior nontensor mathematical treatment.

P. J. E. Peebles, *Physical Cosmology* (Princeton, Princeton University Press, 1971). Emphasizes observational evidence for the Big Bang.

M. Rowan-Robinson, *Cosmology* (Oxford, Clarendon Press, 1977). Includes summaries of other models and discusses some of the unsolved problems of cosmology.

D. W. Sciama, *Modern Cosmology* (Cambridge, Cambridge University Press, 1971). Compact and highly readable; touches all aspects of cosmology.

A few popular articles:

J. R. Gott III, J. E. Gunn, D. N. Schramm, and B. M. Tinsley, "Will the Universe Expand Forever?," *Scientific American 234*, 62 (March 1976).

R. A. Muller, "The Cosmic Background Radiation and the New Aether Drift," *Scientific American 238*, 64 (May 1978).

J. D. Barrow and J. Silk, "The Structure of the Early Universe," *Scientific American 242*, 118 (April 1980).

QUESTIONS

1. Can we look out into the distant universe without also looking back into time?

2. We have used throughout this chapter an age of the universe of about 15×10^9 y and a Hubble constant of about 75 km/s·Mpc. Are these numbers consistent?

3. All natural processes are governed by the rule that the entropy must increase; the increase of entropy, as the universe "runs-down," defines for us a direction of time. If the universe begins to contract and therefore to heat up, will the entropy of natural processes therefore decrease? Will the inhabitants of that universe observe time to be running backward?

4. The hydrogen in the universe contains a small fraction of deuterium. Assuming the deuterium originated in the Big Bang, what era of the Big Bang would we learn about by measuring the deuterium abundance? Can we accomplish this measurement using terrestrial hydrogen? What properties of deuterium could we use to determine its presence in distant regions of the galaxy?

5. Between $t = 1$ s and $t = 6$ s, the neutron fraction should drop from 27 to 8 percent; instead it only drops to about 17 percent. Why don't more neutrons turn into protons during this era? Is it as difficult for protons to turn into neutrons?

6. If we were able to observe the neutrinos from the early universe, would they have a blackbody spectrum?

PROBLEMS

1. Use Hubble's law to estimate the wavelength of the 590.0 nm sodium line as observed emitted from galaxies whose distance from us is (a) 1.0×10^6 light-years; (b) 1.0×10^8 light-years; (c) 1.0×10^{10} light-years.

2. Find the peak wavelength of the 2.7 K blackbody radiation.

3. (a) Differentiate Equation (16.9) to find the *energy* at which the maximum of the radiation spectrum occurs. (b) Evaluate the peak photon energy of the 2.7-K microwave background.

4. Evaluate the numerical constants in Equations (16.8) and (16.11) in order to obtain Equations (16.12), (16.13), and (16.14).

5. Photons of visible light have energies between about 2 and 3 eV. (a) Compute the number density of photons from the 2.7-K background radiation in that interval. [Use Equation (16.21) to find $N_{>2\,eV}$ and $N_{>3\,eV}$.] Would such photon densities be visible to the eye? (b) Assume the eye could detect about 100 photons/cm³. At what temperature would the background radiation be visible? How long ago would this have occurred?

6. The first rotational state of cyanogen is at an energy of 4.70×10^{-4} eV above the ground state. Compute the relative populations of the ground state and the first three rotational states at $T = 2.7$ K.

7. Derive Equation (16.17).

8. At what time did the universe cool below the threshold temperature for (a) nucleon production; (b) pi meson production?

9. (a) At what temperature was the universe hot enough to permit the photons to produce K mesons $(mc^2 = 500$ MeV)? (b) At what age did the universe have this temperature?

10. Derive Equations (16.21) and (16.22).

11. Suppose the *number density* of neutrinos from the Big Bang were the same as the present number density of photons. Find the rest energy of the neutrinos that would provide the critical density needed to close the universe.

12. Suppose the difference between matter and antimatter in the early universe were 1 part in 10^9 instead of 1 part in 10^8. (a) From Equation (16.22) evaluate the temperature at which deuterium begins to form. (b) At what time does this occur? (c) Evaluate the temperature and the corresponding time of radiation decoupling (when hydrogen atoms form).

13. Because we don't yet have a quantum theory of gravity, we cannot analyze the properties of the universe before the Planck time, about 10^{-43} s. If we assume that the properties of the universe during that era were determined by quantum theory, relativity, and gravity, the Planck time should be characterized by the fundamental constants of those three theories: h, c, and G. We can therefore write $t \propto h^i c^j G^k$, where i, j, and k are exponents to be determined. (a) Do a dimensional analysis to determine i, j, and k. (b) Assuming the proportionality constant is of order unity, evaluate t.

14. The Hubble constant could be as low as 50 km/s·Mpc or as high as 100 km/s·Mpc. Compute the critical density necessary to close the universe for these two extremes.

15. Derive Equation (16.24).

APPENDIX  A

CONSTANTS AND
CONVERSION FACTORS

CONSTANTS*

Speed of light	c	2.99792458×10^8 m/s
Charge of electron	e	1.602189×10^{-19} C
Boltzmann constant	k	1.38066×10^{-23} J/K
		8.6174×10^{-5} eV/K
Planck's constant	h	6.62618×10^{-34} J·s
		4.13570×10^{-15} eV·s
	$\hbar = h/2\pi$	1.054589×10^{-34} J·s
		6.58217×10^{-16} eV·s
Gravitational constant	G	6.6726×10^{-11} N·m²/kg²
Avogadro's number	N_A	6.022045×10^{23} mole⁻¹
Universal gas constant	R	8.3144 J/mole·K
Stefan-Boltzmann constant	σ	5.6703×10^{-8} W/m²·K⁴
Rydberg constant	R_∞	1.0973732×10^7 m⁻¹
Hydrogen ionization energy		13.60580 eV
Bohr radius	a_0	5.291771×10^{-11} m
Bohr magneton	μ_B	9.27408×10^{-24} J/T
		5.78838×10^{-5} eV/T
Nuclear magneton	μ_N	5.05084×10^{-27} J/T
		3.15245×10^{-8} eV/T
Fine structure constant	α	$1/137.0360$
	hc	1239.853 eV·nm (MeV·fm)
	$e^2/4\pi\varepsilon_0$	1.439976 eV·nm (MeV·fm)

SOME PARTICLE REST MASSES

	kg	u	MeV/c^2
Electron	9.10953×10^{-31}	5.485803×10^{-4}	0.511003
Proton	1.672649×10^{-27}	1.00727647	938.280
Neutron	1.674955×10^{-27}	1.00866501	939.573
Deuteron	3.34364×10^{-27}	2.01355321	1875.628
Alpha	6.64477×10^{-27}	4.00150618	3727.409

CONVERSION FACTORS

1 eV = 1.602189×10^{-19} J

1 u = 931.502 MeV/c^2
   = 1.660566×10^{-27} kg

1 y = 3.156×10^7 s $\cong \pi \times 10^7$ s

1 Å = 10^{-10} m

1 b = 10^{-28} m²

1 Ci = 3.7×10^{10} decays/s

1 light-year = 9.46×10^{15} m

1 parsec = 3.26 light-year

---

* The number of significant figures given for the numerical constants indicates the precision to which they have been determined; in each case there is an experimental uncertainty, typ·ally of a few parts in the last digit.

APPENDIX B

TABLE OF ATOMIC MASSES

The table gives the atomic masses of some isotopes of each element. All naturally occurring stable isotopes are included (with their natural abundances shown in italics in the last column). Some of the longer-lived radioactive isotopes of each element are also included, with their half-lives. Each element has many other radioactive isotopes that are not included in this table. More complete listings can be found in the sources from which this table was derived: *Table of Isotopes* (Seventh Edition), edited by C. M. Lederer and V. S. Shirley (New York, Wiley, 1978); A. H. Wapstra and K. Bos, *The 1977 Atomic Mass Evaluation*, in *Atomic Data and Nuclear Data Tables 19*, 177 (1977).

In the half-life column, $My = 10^6$ y.

	Z	A	Atomic mass (u)	Abundance or Half-life		Z	A	Atomic mass (u)	Abundance or Half-life
H	1	1	1.007825	*99.985%*	Ne	10	18	18.005710	1.7 s
		2	2.014102	*0.015%*			19	19.001880	17.3 s
		3	3.016049	12.3 y			20	19.992439	*90.51%*
							21	20.993845	*0.27%*
He	2	3	3.016029	*1.38×10^{-4}%*			22	21.991384	*9.22%*
		4	4.002603	*99.99986%*			23	22.994466	37.6 s
							24	23.993613	3.4 m
Li	3	6	6.015123	*7.5%*	Na	11	21	20.997654	22.5 s
		7	7.016005	*92.5%*			22	21.994435	2.60 y
		8	8.022487	0.84 s			23	22.989770	*100%*
Be	4	7	7.016930	53.3 d			24	23.990964	15.0 h
		8	8.005305	0.07 fs			25	24.989955	60 s
		9	9.012183	*100%*			26	25.992606	1.1 s
		10	10.013535	1.6 My	Mg	12	22	21.999577	3.86 s
		11	11.021660	13.8 s			23	22.994127	11.3 s
B	5	8	8.024608	0.77 s			24	23.985045	*78.99%*
		9	9.013329	0.85 as			25	24.985839	*10.00%*
		10	10.012938	*19.8%*			26	25.982595	*11.01%*
		11	11.009305	*80.2%*			27	26.984343	9.46 m
		12	12.014353	20.4 ms			28	27.983879	21.0 h
C	6	10	10.016858	19.2 s	Al	13	25	24.990432	7.18 s
		11	11.011433	20.4 m			26	25.986895	0.72 My
		12	12.000000	*98.89%*			27	26.981541	*100%*
		13	13.003355	*1.11%*			28	27.981913	2.24 m
		14	14.003242	5730 y			29	28.980448	6.6 m
		15	15.010599	2.45 s	Si	14	26	25.992332	2.21 s
N	7	13	13.005739	9.96 m			27	26.986704	4.13 s
		14	14.003074	*99.63%*			28	27.976928	*92.23%*
		15	15.000109	*0.366%*			29	28.976496	*4.67%*
		16	16.006099	7.1 s			30	29.973772	*3.10%*
		17	17.008449	4.2 s			31	30.975364	2.62 h
O	8	14	14.008597	71 s			32	31.974137	650 y
		15	15.003065	122 s	P	15	29	28.981804	4.1 s
		16	15.994915	*99.76%*			30	29.978310	2.50 m
		17	16.999131	*0.038%*			31	30.973763	*100%*
		18	17.999159	*0.204%*			32	31.973908	14.3 d
		19	19.003576	26.9 s			33	32.971726	25.3 d
		20	20.004078	13.5 s	S	16	30	29.984904	1.2 s
F	9	17	17.002095	64.5 s			31	30.979555	2.6 s
		18	18.000937	110 m			32	31.972072	*95.02%*
		19	18.998403	*100%*			33	32.971459	*0.75%*
		20	19.999982	11 s			34	33.967868	*4.21%*
		21	20.999949	4.3 s			35	34.969033	87.4 d
							36	35.967079	*0.017%*
							37	36.971113	5.0 m

	Z	A	Atomic mass (u)	Abundance or Half-life
Cl	17	33	32.977453	2.51 s
		34	33.973765	1.53 s
		35	34.968853	*75.77%*
		36	35.968307	0.30 My
		37	36.965903	*24.23%*
		38	37.968011	37.3 m
		39	38.968006	56 m
Ar	18	34	33.980269	0.844 s
		35	34.975256	1.78 s
		36	35.967546	*0.337%*
		37	36.966776	35.0 d
		38	37.962732	*0.063%*
		39	38.964315	269 y
		40	39.962383	*99.60%*
		41	40.964501	1.83 h
		42	41.96305	33 y
K	19	37	36.973377	1.23 s
		38	37.969080	7.61 m
		39	38.963708	*93.26%*
		40	39.963999	1.28 Gy
		41	40.961825	*6.73%*
		42	41.962402	12.4 h
		43	42.960721	22.3 h
Ca	20	38	37.976318	0.44 s
		39	38.970711	0.86 s
		40	39.962591	*96.94%*
		41	40.962278	0.10 My
		42	41.958622	*0.647%*
		43	42.958770	*0.135%*
		44	43.955485	*2.09%*
		45	44.956189	165 d
		46	45.953689	*0.0035%*
		47	46.954543	4.54 d
		48	47.952532	*0.187%*
		49	48.955677	8.72 m
Sc	21	43	42.961154	3.89 h
		44	43.959409	3.93 h
		45	44.955914	*100%*
		46	45.955174	83.8 d
		47	46.952410	3.42 d
		48	47.952230	43.7 h
Ti	22	44	43.959693	47 y
		45	44.958128	3.09 h
		46	45.952633	*8.2%*
		47	46.951765	*7.4%*
		48	47.947947	*73.7%*
		49	48.947871	*5.4%*
		50	49.944786	*5.2%*
		51	50.946610	5.80 m
		52	51.946893	1.7 m
V	23	48	47.952257	16.0 d
		49	48.948517	330 d
		50	49.947161	*0.250%*
		51	50.943963	*99.750%*
		52	51.944779	3.76 m
		53	52.944323	1.6 m
Cr	24	48	47.954033	21.6 h
		49	48.951338	41.9 m
		50	49.946046	*4.35%*
		51	50.944769	27.7 d
		52	51.940510	*83.79%*
		53	52.940651	*9.50%*
		54	53.938882	*2.36%*
		55	54.940842	3.55 m
		56	55.940671	5.9 m
Mn	25	52	51.945567	5.59 d
		53	52.941291	3.7 My
		54	53.940361	312 d
		55	54.938046	*100%*
		56	55.938906	2.58 h
		57	56.938285	1.6 m
Fe	26	52	51.948114	8.27 h
		53	52.945310	8.51 m
		54	53.939612	*5.8%*
		55	54.938295	2.7 y
		56	55.934939	*91.8%*
		57	56.935396	*2.15%*
		58	57.933278	*0.29%*
		59	58.934878	44.6 d
		60	59.934045	0.3 My
		61	60.93665	6.0 m
Co	27	57	56.936294	271 d
		58	57.935755	70.8 d
		59	58.933198	*100%*
		60	59.933820	5.27 y
		61	60.932478	1.65 h
Ni	28	56	55.942134	6.10 d
		57	56.939775	36.0 h
		58	57.935347	*68.3%*
		59	58.934350	0.075 My
		60	59.930789	*26.1%*
		61	60.931059	*1.13%*
		62	61.928346	*3.59%*
		63	62.929670	100 y
		64	63.927968	*0.91%*
		65	64.930087	2.52 h
Cu	29	61	60.933462	3.41 h
		62	61.932586	9.73 m
		63	62.929599	*69.2%*
		64	63.292766	12.7 h
		65	64.927792	*30.8%*
		66	65.928871	5.10 m
		67	66.927746	61.9 h
Zn	30	62	61.934333	9.2 h
		63	62.933214	38.1 m
		64	63.929145	*48.6%*
		65	64.929244	244 d
		66	65.926035	*27.9%*
		67	66.927129	*4.10%*
		68	67.924846	*18.8%*
		69	68.926552	56 m
		70	69.925325	*0.62%*
		71	70.927725	2.4 m

	Z	A	Atomic mass (u)	Abundance or Half-life		Z	A	Atomic mass (u)	Abundance or Half-life
Ga	31	67	66.928204	78.3 h	Sr	38	82	81.918413	25.0 d
		68	67.927982	68.1 m			83	82.917620	32.4 d
		69	68.925581	*60.1%*			84	83.913428	*0.56%*
		70	69.926028	21.1 m			85	84.912942	64.8 d
		71	70.924701	*39.9%*			86	85.909273	*9.8%*
		72	71.926365	14.1 h			87	86.908890	*7.0%*
		73	72.92514	4.87 h			88	87.905625	*82.6%*
							89	88.907458	50.5 d
Ge	32	68	67.928104	288 d			90	89.907746	28.8 y
		69	68.927970	39.0 h					
		70	69.924250	*20.5%*	Y	39	87	86.910889	80.3 h
		71	70.924954	11.2 d			88	87.909503	106.6 d
		72	71.922080	*27.4%*			89	88.905856	*100%*
		73	72.923464	*7.8%*			90	89.907160	64.1 h
		74	73.921179	*36.5%*			91	90.907301	58.5 d
		75	74.922860	82.8 m					
		76	75.921403	*7.8%*	Zr	40	88	87.910230	83.4 d
		77	76.923549	11.3 h			89	88.908900	78.4 h
							90	89.904708	*51.5%*
As	33	73	72.923834	80.3 d			91	90.905644	*11.2%*
		74	73.923930	17.8 d			92	91.905039	*17.1%*
		75	74.921596	*100%*			93	92.906477	1.5 My
		76	75.922393	26.3 h			94	93.906319	*17.4%*
		77	76.920649	38.8 h			95	94.908037	64.0 d
							96	95.908272	*2.80%*
Se	34	72	71.927113	8.4 d			97	96.910946	16.9 h
		73	72.926775	7.1 h					
		74	73.922477	*0.87%*	Nb	41	91	90.906992	
		75	74.922524	118.5 d			92	91.907195	32 My
		76	75.919207	*9.0%*			93	92.906378	*100%*
		77	76.919908	*7.6%*			94	93.907282	0.020 My
		78	77.917304	*23.5%*			95	94.906832	35.0 h
		79	78.918497	<0.065 My					
		80	79.916521	*49.8%*	Mo	42	90	89.913938	5.67 h
		81	80.917991	18.5 m			91	90.911757	15.5 m
		82	81.916709	*9.2%*			92	91.906809	*14.8%*
		83	82.919045	22.5 m			93	92.906814	3000 y
							94	93.905086	*9.3%*
Br	35	77	76.921373	57.0 h			95	94.905838	*15.9%*
		78	77.921141	6.46 m			96	95.904676	*16.7%*
		79	78.918336	*50.69%*			97	96.906018	*9.6%*
		80	79.918528	17.6 m			98	97.905405	*24.1%*
		81	80.916290	*49.31%*			99	98.907709	66.0 h
		82	81.916803	35.3 h			100	99.907473	*9.6%*
		83	82.915164	2.39 h			101	100.910342	14.6 m
Kr	36	76	75.92582	14.8 h	Tc	43	95	94.907662	20.0 h
		77	76.924599	75 m			96	95.907868	4.3 d
		78	77.920397	*0.356%*			97	96.906362	2.6 My
		79	78.920087	35.0 h			98	97.907211	4.2 My
		80	79.916375	*2.27%*			99	98.906252	0.214 My
		81	80.916578	0.21 My			100	99.907656	15.8 s
		82	81.913483	*11.6%*					
		83	82.914134	*11.5%*	Ru	44	94	93.911357	52 m
		84	83.911506	*57.0%*			95	94.910411	1.65 h
		85	84.912537	10.7 y			96	95.907596	*5.5%*
		86	85.910614	*17.3%*			97	96.90760	2.88 d
		87	86.913358	76 m			98	97.905287	*1.86%*
							99	98.905937	*12.7%*
Rb	37	83	82.915205	86.2 d			100	99.904218	*12.6%*
		84	83.914384	32.9 d			101	100.905581	*17.0%*
		85	84.911800	*72.17%*			102	101.904348	*31.6%*
		86	85.911178	18.8 d			103	102.906322	39.4 d
		87	86.909184	*27.83%*			104	103.905422	*18.7%*
		88	87.911324	17.8 m			105	104.907743	4.44 h

	Z	A	Atomic mass (u)	Abundance or Half-life
Rh	45	101	100.906162	3.3 y
		102	101.906842	2.9 y
		103	102.905503	*100%*
		104	103.906653	42.3 s
		105	104.905684	35.4 h
Pd	46	100	99.908502	3.6 d
		101	100.908290	8.5 h
		102	101.905609	*1.0%*
		103	102.906089	17.0 d
		104	103.904026	*11.0%*
		105	104.905075	*22.2%*
		106	105.903475	*27.3%*
		107	106.905130	6.5 My
		108	107.903894	*26.7%*
		109	108.905952	13.4 h
		110	109.905169	*11.8%*
		111	110.90765	22 m
Ag	47	105	104.906522	41.3 d
		106	105.906678	24.0 m
		107	106.905095	*51.83%*
		108	107.905956	2.4 m
		109	108.904754	*48.17%*
		110	109.906113	24.4 s
Cd	48	104	103.90998	58 m
		105	104.909462	56.0 m
		106	105.906461	*1.25%*
		107	106.906616	6.50 h
		108	107.904186	*0.89%*
		109	108.904949	453 d
		110	109.903007	*12.5%*
		111	110.904182	*12.8%*
		112	111.902761	*24.1%*
		113	112.904401	*12.2%*
		114	113.903361	*28.7%*
		115	114.905429	53.4 h
		116	115.904758	*7.5%*
		117	116.907230	2.4 h
In	49	111	110.905094	2.83 d
		112	111.905529	14.4 m
		113	112.904056	*4.3%*
		114	113.904911	71.9 s
		115	114.903875	*95.7%*
		116	115.905257	14.1 s
Sn	50	110	109.907854	4.1 h
		111	110.907740	35 m
		112	111.904823	*1.01%*
		113	112.905172	115.1 d
		114	113.902781	*0.67%*
		115	114.903344	*0.38%*
		116	115.901744	*14.6%*
		117	116.902954	*7.75%*
		118	117.901607	*24.3%*
		119	118.903310	*8.6%*
		120	119.902199	*32.4%*
		121	120.904239	27.1 h
		122	121.903440	*4.56%*
		123	122.905721	129 d
		124	123.905271	*5.64%*
		125	124.907782	9.62 d
		126	125.907651	0.1 My

	Z	A	Atomic mass (u)	Abundance or Half-life
Sb	51	119	118.903937	38.0 h
		120	119.905077	15.8 m
		121	120.903824	*57.3%*
		122	121.905182	2.68 d
		123	122.904222	*42.7%*
		124	123.905944	60.2 d
		125	124.905259	2.7 y
Te	52	118	117.905882	6.00 d
		119	118.906400	16.0 h
		120	119.904021	*0.091%*
		121	120.904983	16.8 d
		122	121.903055	*2.5%*
		123	122.904278	*0.89%*
		124	123.902825	*4.6%*
		125	124.904435	*7.0%*
		126	125.903310	*18.7%*
		127	126.905222	9.4 h
		128	127.904464	*31.7%*
		129	128.906595	69 m
		130	129.906229	*34.5%*
		131	130.908533	25.0 m
I	53	125	124.904626	60.2 d
		126	125.905624	13.0 d
		127	126.904477	*100%*
		128	127.905815	25.0 m
		129	128.904986	16 My
		130	129.906713	12.4 h
Xe	54	122	121.90857	20.1 h
		123	122.90844	2.08 h
		124	123.90612	*0.096%*
		125	124.90649	17 h
		126	125.904281	*0.090%*
		127	126.905190	36.4 d
		128	127.903531	*1.92%*
		129	128.904780	*26.4%*
		130	129.903510	*4.1%*
		131	130.905076	*21.2%*
		132	131.904148	*26.9%*
		133	132.905892	5.25 d
		134	133.905395	*10.4%*
		135	134.907132	9.1 h
		136	135.907219	*8.9%*
		137	136.911739	3.82 m
Cs	55	131	130.905457	9.69 d
		132	131.906415	6.47 d
		133	132.905433	*100%*
		134	133.906700	2.06 y
		135	134.905888	3 My
		136	135.907292	13.1 d
Ba	56	128	127.908232	2.43 d
		129	128.908624	2.2 h
		130	129.906277	*0.106%*
		131	130.906897	12.0 d
		132	131.905042	*0.101%*
		133	132.905992	10.7 y
		134	133.904490	*2.42%*
		135	134.905668	*6.59%*
		136	135.904556	*7.85%*
		137	136.905816	*11.2%*
		138	137.905236	*71.7%*
		139	138.908830	82.9 m

	Z	A	Atomic mass (u)	Abundance or Half-life		Z	A	Atomic mass (u)	Abundance or Half-life
La	57	136	135.90764	9.87 m	Gd	64	150	149.918663	1.8 My
		137	136.90646	0.06 My			151	150.920378	120 d
		138	137.907114	*0.089%*			152	151.919803	*0.20%*
		139	138.906355	*99.911%*			153	152.921505	242 d
		140	139.909479	40.3 h			154	153.920876	*2.1%*
		141	140.910888	3.90 h			155	154.922629	*14.8%*
Ce	58	134	133.90900	76 h			156	155.922130	*20.6%*
		135	134.90923	17.8 h			157	156.923967	*15.7%*
		136	135.90714	*0.190%*			158	157.924111	*24.8%*
		137	136.90777	9.0 h			159	158.926397	18.6 h
		138	137.905996	*0.254%*			160	159.927061	*21.8%*
		139	138.906639	137.2 d			161	160.929676	3.7 m
		140	139.905442	*88.5%*	Tb	65	157	156.924029	150 y
		141	140.908279	32.5 d			158	157.925416	150 y
		142	141.909249	*11.1%*			159	158.925350	*100%*
		143	142.912389	33.0 h			160	159.927171	72.1 d
		144	143.913654	284 d			161	160.927573	6.90 d
Pr	59	139	138.908907	4.4 h	Dy	66	154	153.924432	10 My
		140	139.909079	3.39 m			155	154.925758	10.0 h
		141	140.907657	*100%*			156	155.924287	*0.057%*
		142	141.910048	19.2 h			157	156.925470	8.1 h
		143	142.910827	13.6 d			158	157.924412	*0.100%*
Nd	60	140	139.90958	3.37 d			159	158.925743	144.4 d
		141	140.909605	2.5 h			160	159.925203	*2.3%*
		142	141.907731	*27.2%*			161	160.926939	*19.0%*
		143	142.909823	*12.2%*			162	161.926805	*25.5%*
		144	143.910096	*23.8%*			163	162.928737	*24.9%*
		145	144.912582	*8.3%*			164	163.929183	*28.1%*
		146	145.913126	*17.2%*			165	164.931712	2.33 h
		147	146.916110	11.0 d	Ho	67	163	162.928739	33 y
		148	147.916901	*5.7%*			164	163.930287	29.0 m
		149	148.920157	1.73 h			165	164.930332	*100%*
		150	149.920900	*5.6%*			166	165.932296	26.8 h
		151	150.923838	12.4 m			167	166.933102	3.1 h
Pm	61	143	142.910941	265 d	Er	68	160	159.929091	28.6 h
		144	143.912597	349 d			161	160.930009	3.24 h
		145	144.912754	17.7 y			162	161.928787	*0.14%*
		146	145.914717	5.5 y			163	162.930040	75.1 m
		147	146.915148	2.62 y			164	163.929211	*1.56%*
		148	147.917477	5.37 d			165	164.930737	10.4 h
		149	148.918343	53.1 h			166	165.930305	*33.4%*
Sm	62	142	141.915215	72.5 m			167	166.932061	*22.9%*
		143	142.914642	8.83 m			168	167.932383	*27.1%*
		144	143.912009	*3.1%*			169	168.934603	9.40 d
		145	144.913413	340 d			170	169.935476	*14.9%*
		146	145.913061	103 My			171	170.938041	7.52 h
		147	146.914907	*15.1%*	Tm	69	167	166.932865	9.25 d
		148	147.914832	*11.3%*			168	167.934186	93.1 d
		149	148.917193	*13.9%*			169	168.934225	*100%*
		150	149.917285	*7.4%*			170	169.935813	128.6 d
		151	150.919942	90 y			171	170.936442	1.92 y
		152	151.919741	*26.6%*	Yb	70	166	165.933889	56.7 h
		153	152.922107	46.8 h			167	166.934962	17.5 m
		154	153.922218	*22.6%*			168	167.933908	*0.135%*
		155	154.924642	22.4 m			169	168.935201	32.0 d
Eu	63	149	148.917940	93.1 d			170	169.934774	*3.1%*
		150	149.919747	36 y			171	170.936338	*14.4%*
		151	150.919860	*47.9%*			172	171.936393	*21.9%*
		152	151.921756	13 y			173	172.938222	*16.2%*
		153	152.921243	*52.1%*			174	173.938873	*31.6%*
		154	153.922999	8.5 y			175	174.941287	4.19 d
		155	154.922893	4.9 y			176	175.942576	*12.6%*
		156	155.924764	15 d			177	176.945265	1.9 h

	Z	A	Atomic mass (u)	Abundance or Half-life
Lu	71	173	172.938947	1.37 y
		174	173.940353	3.3 y
		175	174.940785	97.39%
		176	175.942694	2.61%
		177	176.943766	6.71 d
Hf	72	172	171.93953	1.87 y
		173	172.94066	24.0 h
		174	173.940065	0.16%
		175	174.941441	70 d
		176	175.941420	5.2%
		177	176.943233	18.6%
		178	177.943710	27.1%
		179	178.945827	13.7%
		180	179.946561	35.2%
		181	180.949111	42.4 d
Ta	73	179	178.945951	1.7 y
		180	179.947489	0.0123%
		181	180.948014	99.9877%
		182	181.950170	115 d
W	74	178	177.94586	21.5 d
		179	178.947093	38 m
		180	179.946727	0.13%
		181	180.948215	121 d
		182	181.948225	26.3%
		183	182.950245	14.3%
		184	183.950953	30.7%
		185	184.953441	75.1 d
		186	185.954377	28.6%
		187	186.957174	23.9 h
Re	75	183	182.950841	71 d
		184	183.952559	38 d
		185	184.952977	37.40%
		186	185.955008	90.6 h
		187	186.955765	62.60%
		188	187.958126	16.9 h
Os	76	182	181.95214	22.0 h
		183	182.95331	13 h
		184	183.952514	0.018%
		185	184.954066	93.6 d
		186	185.953852	1.6%
		187	186.955762	1.6%
		188	187.955850	13.3%
		189	188.958156	16.1%
		190	189.958455	26.4%
		191	190.960936	15.4 d
		192	191.961487	41.0%
		193	192.964158	30.6 h
Ir	77	189	188.95869	13.1 d
		190	189.96060	11.8 d
		191	190.960603	37.3%
		192	191.962613	74.2 d
		193	192.962942	62.7%
		194	193.965095	19.2 h

	Z	A	Atomic mass (u)	Abundance or Half-life
Pt	78	188	187.959434	10.2 d
		189	188.96074	10.9 h
		190	189.959937	0.013%
		191	190.961677	2.9 d
		192	191.961049	0.78%
		193	192.963008	50 y
		194	193.962679	32.9%
		195	194.964785	33.8%
		196	195.964947	25.3%
		197	196.967332	18.3 h
		198	197.967879	7.2%
		199	198.970564	30.8 m
Au	79	195	194.965032	183 d
		196	195.966546	6.18 d
		197	196.966560	100%
		198	197.968233	2.696 d
		199	198.968756	3.14 d
Hg	80	194	193.965426	260 y
		195	194.96666	10 h
		196	195.965812	0.15%
		197	196.967005	64.1 h
		198	197.966760	10.0%
		199	198.968269	16.8%
		200	199.968316	23.1%
		201	200.970293	13.2%
		202	201.970632	29.8%
		203	202.972864	46.8 d
		204	203.973481	6.9%
		205	204.976062	5.2 m
Tl	81	201	200.970816	73 h
		202	201.972101	12.2 d
		203	202.972336	29.5%
		204	203.973856	3.77 y
		205	204.974410	70.5%
		206	205.976094	4.20 m
Pb	82	202	201.972150	0.3 My
		203	202.973382	52.0 h
		204	203.973037	1.42%
		205	204.974475	14 My
		206	205.974455	24.1%
		207	206.975885	22.1%
		208	207.976641	52.3%
		209	208.981080	3.25 h
Bi	83	207	206.978467	38 y
		208	207.979733	0.368 My
		209	208.980388	100%
		210	209.984110	5.01 d
		211	210.987263	2.15 m
Po	84	207	206.981589	5.7 h
		208	207.981240	2.90 y
		209	208.982422	102 y
		210	209.982864	138.4 d
At	85	209	208.986165	5.4 h
		210	209.987143	8.3 h
		211	210.987490	7.21 h

Chapter 5

7. 62 MeV

9. $\dfrac{w}{L} - \dfrac{1}{2\pi n}\left[\sin\dfrac{2\pi n(x+w)}{L} - \sin\dfrac{2\pi nx}{L}\right]$

15. $5E_0$

19. $x_{av} = 0$; $(x^2)_{av} = \dfrac{\hbar}{2m\omega}$;

$\Delta x = \sqrt{\dfrac{\hbar}{2m\omega}}$

21. $\Delta x\, \Delta p = \tfrac{1}{2}\hbar$

27. $D = B = -A\sqrt{\dfrac{E}{V_0 - E}}$

Chapter 6

1. (a) 45.7 nm
   (b) 86.2 nm

5. 33 MeV

7. (a) 3.37×10^{-5}
   (b) 4.41×10^{-3}
   (c) 1.33×10^{-2}
   (d) 0.982

9. 2.84

11. 4.39×10^{-3} MeV

13. 59/s

15. Lyman: $\lambda_{limit} = 91.13$ nm
    Paschen: $\lambda_{limit} = 820.1$ nm

17. $5\rightarrow4$: $\Delta E = 0.306$ eV
    $5\rightarrow3$: $\Delta E = 0.97$ eV
    $5\rightarrow2$: $\Delta E = 2.86$ eV
    $5\rightarrow1$: $\Delta E = 13.1$ eV

21. (a) 1.51 eV
    (b) 13.6 eV
    (c) 7.65 eV

23. 0.178 nm

25. $E_1 = -54.4$ eV
    $E_2 = -13.6$ eV
    $E_3 = -6.04$ eV
    $E_4 = -3.40$ eV

27. $E \sim 7 \times 10^{-8}$ eV

31. $a_0 = 1.19 \times 10^{29}$ m
    $E_2 - E_1 = 2.0 \times 10^{-78}$ eV

33. 0.440 nm

37. $n = 745$

Chapter 7

1. (4, 0, 0), (4, 1, 1), (4, 1, 0), (4, 1, −1), (4, 2, 2), (4, 2, 1), etc.

3. (a) $\sqrt{12}\hbar$
   (b) 7 different z components
      $l_z = +3\hbar, +2\hbar, +\hbar, 0,$
      $-\hbar, -2\hbar, -3\hbar$
   (c) $\theta = 30°, 55°, 73°, 90°,$
      $107°, 125°, 150°$
   (d) no

9. $r = (3 \pm \sqrt{5})a_0$

11. $P_{20} = 0.651$
    $P_{21} = 0.440$

13. $r_{av}(2s) = 6a_0$
    $r_{av}(2p) = 5a_0$

15. For silver atoms, assuming
    $\mu \cong \mu_B$ the separation is 3.3 mm

19. (a) 656.043 nm, 656.113 nm,
    656.183 nm

21. (a) 656.113 nm + 0.0087 nm
    656.113 nm + 0.0066 nm
    656.113 nm − 0.0066 nm
    656.113 nm − 0.0087 nm

Chapter 8

1. (a) N, P, As, Sb, Bi
   (b) Co, Rh, Ir

5. Ca: 3.68 keV
   Zr: 15.5 keV
   Hg: 63.7 keV

9. $L = 0, S = 3$

11. (a) $L = 1, S = \tfrac{1}{2}$
    (b) $L = 0, S = 0$
    (c) $L = 3, S = 1$
    (d) $L = 2, S = 2$

13. 18 T

15. (a) 0.045 eV
    (b) $3s$: 3.374 eV
       $4s$: 4.373 eV
       $5s$: 4.750 eV
    (c) $2p$: 3.54 eV
       $3s$: 2.02 eV

17. (a) $\Delta E_{4p\rightarrow3p} = 0.655$ eV
    (b) $\Delta E_{4d\rightarrow3d} = 0.663$ eV

19. 36

Chapter 9

1. (a) $^{19}_{9}\text{F}_{10}$
   (b) $^{199}_{79}\text{Au}_{120}$
   (c) $^{107}_{47}\text{Ag}_{60}$

3. (a) 15 MeV
   (b) 824 MeV

5. $r_{min} = 12.6$ fm

7. (a) $B = 28.30$ MeV;
   $B/A = 7.074$ MeV

(b) $B = 160.6$ MeV;
   $B/A = 8.032$ MeV

(c) $B = 342.1$ MeV;
   $B/A = 8.551$ MeV

(d) $B = 482.1$ MeV;
   $B/A = 8.765$ MeV

9. (b) ^{17}O: $S_n = 4.144$ MeV
   ^{7}Li: $S_n = 7.250$ MeV
   ^{57}Fe: $S_n = 7.646$ MeV

11. 789 MeV

13. (a) $\frac{1}{4}$
    (b) $\frac{1}{16}$
    (c) $\frac{1}{1024}$

15. 3.9×10^{-5} s^{-1}

17. 1460 Ci

19. (a) 0.85 μCi
    (b) 0.13%

25. 1.93×10^{-3} W

29. 11.509 MeV

31. 0.864 MeV

33. (a) 6
    (b) 4
    (c) 42.658 MeV
    (d) 27.8 μW

35. (a) 4.67×10^{-9} eV
    (b) 1.95×10^{-3} eV
    (c) 0.097 mm/s

Chapter 10

1. (a) $^1_1$H$_0$
   (b) $^1_0$n$_1$
   (c) $^{30}_{15}$P$_{15}$
   (d) $^2_1$H$_1$

5. (a) $\frac{1}{2}$
   (b) $\frac{3}{4}$
   (c) $\frac{15}{16}$

7. 2.59×10^{-3} b

9. 1.58 μCi

11. ^{63}Cu: $\sigma = 4.49$ b
    ^{65}Cu: $\sigma = 2.20$ b

13. (a) $dI = -I\sigma \dfrac{\rho N_A}{M}\, dx$

17. (a) -10.313 McV
    (b) 2.315 MeV

19. $K_p = 12.208$ MeV, $K_\alpha = $
    11.152 MeV or $K_p = $
    22.711 MeV, $K_\alpha = 0.647$ MeV

21. 8.662 MeV

23. (a) 6.546 MeV
    (b) 4.807 MeV

25. 180.76 MeV

27. 5.8×10^9 K

29. 14.1 MeV

31. (a) 1.94 MeV, 7.55 MeV, 7.30 MeV
    (b) 1.199 MeV, 1.732 MeV
    (c) 3.725 MeV

Chapter 11

1. (a) strong
   (c) weak
   (e) strong

3. (a) L_e
   (b) S
   (c) B

5. (a) ν_e
   (c) π^0

9. (a) 2.6 keV
   (c) 11 keV

11. (a) 33.9 MeV
    (c) 218.5 MeV
    (e) 76.9 MeV

13. (a) K^+
    (c) K^0
    (e) μ^+

17. (a) -2201.2 MeV
    (c) 1802.2 MeV

19. $K_n = 310.7$ MeV
    $K_\pi = 57.4$ MeV
    $\theta_n = 9.7°$

21. (a) $s \rightarrow d + u + \bar{u}$
    (c) $u + u \rightarrow$ energy
    (e) energy $\rightarrow d + \bar{d}$

Chapter 12

3. $\Delta v = \left(\dfrac{kT}{m}\right)^{1/2} \sqrt{3 - 8/\pi}$

5. (a) $e^{-E/kT}$
   (b) $E/(1 + e^{E/kT})$
   (c) $NE/(1 + e^{E/kT})$
   (d) $R(E/kT)^2[e^{E/kT}/(1 + e^{E/kT})^2]$

7. (a) $p(+1) = 0.3295$
   $p(0) = 0.3333$
   $p(-1) = 0.3372$

9. $u(E) = \dfrac{\pi E^3}{(\hbar c \pi)^3} \dfrac{1}{e^{E/kT} - 1}$

11. $E_{\max} = (2.4313 \times 10^{-4} \text{ eV·K}^{-1})T$

15. 7.12 eV

17. (a) 4.37×10^{-36}
    (b) 6.00×10^{-6}

19. 244 keV

Chapter 13

1. 15.4eV
3. 1.13 nm
5. (a) 30.9×10^{-30} C·m
   (b) 88.0%
   (c) 91%; 83%
7. 42.7%
11. 2.98×10^{21} eV/m²
13. (a) 8.2×10^{-2}
    (b) 6.8×10^{-3}
15. (a) 1/2.99/4.97/6.91
    (b) 1/2.94/4.70/6.31
21. (b) 2.37×10^{21} eV/m²
    (c) 2.1×10^{-3} eV

Chapter 14

1. (a) $\frac{1}{8}$
   (b) $a = 2r$
   (c) 0.5236
5. $V = -\dfrac{e^2}{4\pi\epsilon_0 r}\left(6 - \dfrac{12}{\sqrt{2}} + \dfrac{8}{\sqrt{3}} - \dfrac{6}{2} + \dfrac{24}{\sqrt{5}} - \cdots\right)$
7. (a) -6.45 eV
   (b) -6.17 eV
9. 857 kJ/mole
13. 0.255 nm
15. $l = 38.7$ nm
19. (a) -0.094 eV
    (b) 7.62 nm

Chapter 15

3. 1.8×10^{-5} s
9. (a) $\frac{4}{3}\pi r^3 \rho$
   (b) $-\dfrac{Gm}{r}\,dm$
   (c) $dm = \rho\, r^2\, dr \sin\theta\, d\theta\, d\phi$
15. (a) 1.8721×10^{-64} J·m
    (b) 4.1443×10^{-29} MeV·m²
    (c) 3.5468×10^{22} m
    (d) 3.2944×10^{-74} MeV
17. (a) 140 MeV
    (b) 85 MeV
    (c) 2.0×10^{59} MeV
    (d) -4.0×10^{59} MeV
19. $\lambda = 2.58$ fm
    $d = 1.27$ fm
21. (a) 6.21×10^{-15} W

Chapter 16

1. (a) 590.0 nm
   (b) 594.5 nm
   (c) 1624 nm
3. (a) $E_{max} = (2.431 \times 10^{-4}$ eV/K$)T$
   (b) 6.56×10^{-4} eV
5. (a) $\sim 10^{-3746}$/nm³
   (b) $T \sim 1000$ K
      $t \sim 7 \times 10^6$ y
9. (a) 5.80×10^{12} K
   (b) 6.7×10^{-6} s
11. 14 eV
13. (a) $i = \frac{1}{2}, j = -\frac{5}{2}, k = \frac{1}{2}$
    (b) 1.3×10^{-43} s

PHOTO CREDITS

CHAPTER OPENING PHOTOS

Chapter 1 Frontispiece from Galileo's *Dialogue Concerning the Two Chief World Systems*, 1632.

Chapter 2 © 1970 by Sidney Harris, *American Scientist* Magazine.

Chapter 3 Groove Tube Photos.

Chapter 4 Courtesy Garth L. Nicolson, The University of Texas, M.D. Anderson Hospital & Tumor Institute, Houston.

Chapter 5 Courtesy Carleen Maley Hutchins from *Scientific American*, October 1982, page 177.

Chapter 6 AIP-Niels Bohr Library.

Chapter 7 Courtesy William P. Spencer, Massachusetts Institute of Technology, from *Scientific American*, May 1981, page 132.

Chapter 8 Photo by T. Kallard, from *Laser Art & Optical Transforms* (Optosonic Press). Live laser image from Laserium, produced by Laser Images, Inc.

Chapter 9 Courtesy Ronald Y. Cusson and Joachim Maruhn, Oak Ridge National Laboratory, from *Scientific American*, December 1978, page 59.

Chapter 10 U.S. Air Force Photo.

Chapter 11 Top, CERN, Switzerland; bottom, "Autumn Rhythm" by Jackson Pollack, The Metropolitan Museum of Art, George A. Hearn Fund, 1957.

Chapter 12 Courtesy Las Vegas News Bureau, Convention Center.

Chapter 13 Courtesy M. H. F. Wilkins.

Chapter 14 Specimen prepared by R. W. Bicknell, photo by N. S. Griffin. Allen Clark Research Center, The Plessey Co., Limited.

Chapter 15 Top, Tony Howarth/Woodfin Camp & Associates; bottom, Georg Gerster/Photo Researchers.

Chapter 16 The Museum of Modern Art.

PORTRAITS

Page 20, AIP-Niels Bohr Library.

Page 22, The New York Times.

Pages 65 & 69, AIP-Niels Bohr Library, W. F. Meggers Collection.

Pages 73, 84, 97, & 115, AIP-Niels Bohr Library.

Page 154, *Nature*, Courtesy AIP-Niels Bohr Library.

INDEX

Units Used in This Book

Unit	Abbreviation	Quantity Measured
gram	g	mass
meter	m	length
second	s	time
newton	N	force
joule	J	energy
watt	W	power
electron-volt	eV	energy
hertz	Hz	frequency
kelvin	K	temperature
coulomb	C	electric charge
ampere	A	electric current
volt	V	electric potential
ohm	Ω	electric resistance
tesla	T	magnetic field
atomic mass unit	u	mass
curie	Ci	activity
barn	b	cross section

Prefixes of Units

Prefix	Abbreviation	Meaning	Prefix	Abbreviation	Meaning
atto	a	10^{-18}	centi	c	10^{-2}
femto	f	10^{-15}	kilo	k	10^{3}
pico	p	10^{-12}	mega	M	10^{6}
nano	n	10^{-9}	giga	G	10^{9}
micro	μ	10^{-6}	tera	T	10^{12}
milli	m	10^{-3}			